Basic Macroscopic Principles of Applied Superconductivity

T0187777

V.R. Romanovskii

National Research Center Kurchatov Institute,
Moscow, Russia

CRC Press

Taylor & Francis Group

Boca Raton   London   New York

CRC Press is an imprint of the
Taylor & Francis Group, an **informa** business

A SCIENCE PUBLISHERS BOOK

Cover illustration provided by the author, V.R. Romanovskii

First edition published 2021
by CRC Press
6000 Broken Sound Parkway NW, Suite 300, Boca Raton, FL 33487-2742

and by CRC Press
2 Park Square, Milton Park, Abingdon, Oxon, OX14 4RN

© 2021 Taylor & Francis Group, LLC

CRC Press is an imprint of Taylor & Francis Group, LLC

Library of Congress Cataloging-in-Publication Data

Names: Romanovskii, V. R., 1952- author.
Title: Basic macroscopic principles of applied superconductivity /
V. R. Romanovskii.
Description: First. | Boca Raton : CRC Press, 2020. | Includes
bibliographical references and index.
Identifiers: LCCN 2020024230 | ISBN 9781138541832 (hardcover)
Subjects: LCSH: Superconductivity. | Electrodynamics. | Superconductors.
Classification: LCC QC611.92 .R66 2020 | DDC 537.6/23--dc23
LC record available at https://lccn.loc.gov/2020024230

ISBN: 978-1-138-54183-2 (hbk)
ISBN: 978-0-367-53812-5 (pbk)

Typeset in Times
by TVH Scan

Preface

Superconductivity of certain materials (pure metals, metal-alloys, compounds, ceramics and others) losing resistance at extremely low temperatures is an important research field, both in experimental and theoretical physics, with many engineering applications. Since its discovery, there are a big number of laboratories and research groups all over the world performing research on superconductivity. Superconductivity permits one to get high magnetic fields, which are of the basic tools that allow one to investigate new properties of materials. Indeed, conventional electromagnets can supply DC magnetic fields up to about 2 T only in small volumes. To produce higher fields, it is possible to use the water-cooled high-power solenoids or the cryoresistive windings. At the same time, superconducting magnets allow higher fields to be produced with lower power utilization. It is because superconducting materials have a very high critical current densities than the operational current densities of the conventional normal metals of the same size. This means that the superconducting magnets consume no power working at high current density without iron core. Because of these features, superconducting magnets offer lower power costs, higher magnetic fields and higher gradients in the large operating volumes.

The macroscopic behavior of superconductors, which plays a key role in the stable performances of superconducting magnets, is studied by means of various theoretical models. In general, this book outlines the basic principles of applied superconductivity underlying the design and operation of the superconducting magnets for large-scale applications. It provides detailed coverage of the major investigation results of the macroscopic states of superconducting materials used in high-field applications. The relevant topics discussed are helpful for all researchers interesting in basic aspects of applied superconductivity. They may also allow one to do new researches of the macroscopic phenomena in superconducting materials to develop their novel potential applications.

The outline of the book is as follows: Preface, Introduction and eight chapters in total. They generalize studies about macroscopic thermo-electrodynamic phenomena in technical superconductors.

Chapter 1 presents a brief overview of the main results of the fundamental and applied superconductivity and highlights those problems, which arise in the development of superconducting magnets. The importance of the stability analysis of the superconducting state is underlined.

Chapter 2 is devoted to the main problems arising in the macroscopic electrodynamic of nonideal type-II superconductors. The laws of the stable penetration of magnetic flux into superconductors, which are in a changing external magnetic field or when a transport current is charged into them, are discussed. Stable modes of the superconducting state formation are considered in the framework of the Bean model, viscous flux-flow model and models based on the real voltage-current characteristic of superconductor. The analysis of the sweep-rate and relaxation modes is performed in the scaling approximation. The corresponding results of the numerical simulations performed in a more general setting prove the scaling solutions. As a result, the characteristic physical features of the electrodynamic phenomena in low-temperature and high-temperature superconductors are formulated.

In Chapter 3, the self-field losses are analyzed taking into account the nonlinear nature of the voltage-current characteristic of the technical superconductors. It is shown that the model of the critical state not only underestimates the heat losses from the transport current but also does not allow one to calculate correctly the energy losses in superconductors with the real V-I characteristics in the currents range from the full penetration current to the critical current. The error depends on the smoothness parameters of the V-I characteristic, the current charging rate and the transverse size of the superconductor. The models, which allow one to estimate the self-field losses with satisfactory accuracy in the closed-form approximation both at the partial and full penetration modes, are proposed.

Chapter 4 is a discussion of the basic physical regularities of the formation of thermal and electrodynamic states of low- and high-temperature superconductors when changing the external magnetic field. Considering them, the conditions of the magnetic instability onset are formulated in the nonisothermal approximation. Thermal features of their development are investigated. It establishes a nontrivial connection between the conditions of the magnetic instability onset, the permissible heat generation and overheating of the superconductor before the instability onset. It leads to the general concept of superconductivity destruction under the action of the disturbances of different nature. The causes of the oscillations of temperature and electromagnetic field in superconductors are formulated. The significant influence of the thermal state of the superconductor on the conditions of their occurrence is shown. As a consequence, there exists the effect of the thermal self-suppression of oscillations, when they are absent if there is an intense stable increase in the temperature of the superconductor. As a result, the oscillations disappear at intensive heat dissipation. Namely, they are absent in the thermally insulated superconductors and at high rising rates of the external magnetic field. They can be also absent at the intensive heat removal of the thermal losses.

In Chapter 5, the basic physical regularities governing the formation of the thermal and electrodynamic states of composite superconductors based on low-temperature superconductors at variation in the external magnetic field are discussed. Considering them, the conditions for the occurrence of the magnetic instabilities are written in a nonisothermal approximation. Thermal peculiarities of their development are discussed.

Chapter 6 investigates the physical features of the formation and destruction of stable superconducting states in composites on the basis of low-temperature and high-temperature superconductors at current charging. The influence of the stable overheating of superconducting composites on their current-carrying capacity is studied. Using zero-dimensional and the one-dimensional models, the stability conditions of the current modes of superconducting composites and tapes, depending on their thermal properties, transverse dimensions, cooling conditions and current charging rate are formulated. They allow one to determine the limiting currents, which can stably flow in composites with voltage-current characteristics described by the exponential or power equations with different temperature dependencies of the critical current density. It follows from them that the current charged into the superconducting magnetic system can be both subcritical and supercritical. The existence of the characteristic values of electric field, which determines the thermal role of the matrix and coolant in the character of the current state formation of superconducting composites, is shown.

Chapter 7 discusses the destruction features of the superconducting states of composite superconductors carrying a constant transport current by external thermal perturbations. In the framework of the model of anisotropic superconducting composites with an ideal voltage-current characteristic, the conditions of the full and partial thermal stabilization of composite superconductors are formulated in one-dimensional and two-dimensional approximation. Their dependence on the length and duration of the thermal disturbance is shown. The characteristic regularities of the transient phenomena occurring both in single composites and in superconducting multicomponent current-carrying elements during irreversible propagation of a region with normal conductivity are discussed. Special attention is paid to the analysis of the thermal mechanisms underlying the spatial formation of the region with normal conductivity in the longitudinal and transverse directions of superconducting current-carrying elements.

In Chapter 8, the influence of the voltage-current characteristic of a superconductor on the conditions of the occurrence and development of thermal instabilities in technical superconductors is studied. The obtained results are compared with the conclusions of the thermal stabilization theory based on the model with the ideal voltage-current characteristic. It is shown that it leads to overestimated values of the Joule heat release in a wide range of the temperature changes of composite. Moreover, it appears that when used in the thermal stabilization theory, the concepts of the critical current, defining the maximum value of a transport current and the temperature of the resistive transition beyond which the transport current starts to share between the superconducting core and matrix have no physical meaning for technical superconductors. As a result, the stable states are possible at currents that are higher than the conditionally set value of the critical current of the composite with the ideal voltage-current characteristic. The analysis of the dissipative phenomena occurring in the technical superconductors with continuously increasing voltage-current characteristics due to the redistribution of current across the cross-section of the composite in response to any external thermal disturbance is performed.

The analysis of dissipative processes is carried out that occurs in superconducting composites as a result of the redistribution of the current across the composite section, which occurs as a response to the action of an external disturbance, taking into account the emergence of a stable value of the electric field due to the existence of voltage-current characteristic. It is shown that regardless of the nature of disturbances initiating current diffusion, the induced additional heat dissipation can significantly exceed the Joule losses determined in the approximation based on the constant current distribution over the entire cross-section of the composite. As a consequence, dissipative phenomena occurring in superconducting composites in response to external disturbances can lead to a significant reduction in the boundary of stable states.

The main part of the results presented is based on the investigations carried out at the Kurchatov Institute in the period from 1980 to the present time. Since 1994, our researches on applied superconductivity were supported by the Russian Foundation for Basic Research of the Russian Academy of Sciences. This support allowed to formulate the most important principles of macroscopic electrodynamics of superconductors in a non-isothermal approximation developing the necessary methods for modeling interrelated thermal and electrodynamic phenomena in superconductors and composites based on them. Besides, it should be pointed out that the results presented in Chapter 6 were obtained in the High Field Laboratory for Superconducting Materials, Institute for Materials Research, Tohoku University, Sendai, Japan with the active participation of professors Kazuo Watanabe and Satoshi Awaji. It was a pleasure working with them. Their various suggestions and important critical comments have been extremely useful. I would like to express my deepest thanks for their contributions and valuable discussions.

In concluding note that it is difficult to avoid some omissions and errors when writing such a manuscript. Therefore, I will be happy to listen to the criticisms made after its publication.

V.R. Romanovskii,
Moscow,
August 2020

Introduction

Superconductivity opens new fields in science and engineering, especially in the material processing technologies responsible for changing the properties of materials in high magnetic fields. One of the first ideas proposed by H.K. Oness, who discovered superconductivity, was to wind the superconducting magnet free from dissipation to generate a magnetic field of about 10 T using lead. During the hundred years after this experiment, thousands of superconducting materials have been found, namely, pure elements, metal-alloys, intermetallic compounds, ceramics, inorganic polymers, pnictides, heavy fermion compounds and many others. The discoveries, crucial to the practical use of superconductivity, were made after 1960s. Superconductivity was found in such materials as metal-alloys Nb-Zr, Nb-Ti and in intermetallic compounds as Nb_3Sn, Nb_3Ge, Nb_3Al, V_3Ga and others. They have high values of the critical currents and magnetic fields. Superconductivity of many oxide cuprates was discovered after an unexpected finding of the superconducting properties in ceramic $La_{1.85}Ba_{0.15}CuO_4$ that was discovered by J.C. Bednorz and K.A. Müller in 1986. The superconducting ceramics have many important characteristics. First of all, their critical temperature and critical magnetic field are very high. Today, metallic elements and alloys, which exhibit superconducting properties below 30 K, are usually named as low-temperature superconductors (LTS) in contrast to cuprates with the critical temperature exceeding 30 K named as high-temperature superconductors (HTS).

Modern superconducting current-carrying elements are the conductors of a composite structure. There is a wide range of superconducting wires from which superconducting devices for various purposes can be made.

Wires based on LTS can contain up to tens of thousands of filaments of micron diameter and consist of superconducting and nonsuperconducting components with different physical properties. Superconductors inside the LTS composites are arranged in a strict order in a matrix. Usually, copper or its alloys are used as a matrix. In addition, diffusion barriers protecting the purity of the copper stabilizing shell or resistive barriers can be introduced into the cross-section of the wires to increase the resistance of the wire, reducing the possibility of current flowing between the filaments. Their industrial production began in the mid-1970s and is now generally developed. Wires based on Nb-Ti are widely used in magnet systems that induce fields up to 10 T at operating temperatures of 1.8-4.2 K. The intermetallic compound Nb_3Sn is used to create magnets with induction up to 20 T.

The first generation (1G) of HTS wires were made of bismuth-based materials, namely, $Bi_2Sr_2CaCu_2O_{8+x}$ (Bi2212) and $Bi_2Sr_2Ca_2Cu_3O_{10+x}$ (Bi2223) by the powder-in-tube method. This superconducting material is embedded in a nonsuperconducting matrix (typically silver). After packing the precursor powders into a tube, the composite is drawn into a wire with a hexagonal cross-sectional shape. Then the wire is cut and plased into another silver tube to be drawn and flat-rolled. The flattened composites are subsequently heat-treated at high temperatures to form Bi2212 or Bi2223 phases. However, there are major drawbacks of 1G HTS which seriously confine its application and development. First, the labor-intensive processing and a large amount of expensive metals used for the matrix make 1G conductors expensive. Second, the irreversibility field of bismuth-based HTS materials (BSCCO), at which the critical current density drops to very low values, is very low at higher temperatures. It makes 1G HTS highly sensitive to magnetic fields. Third, their superconducting properties are very low at 77 K and have very low applied fields, which are problematic to the construction of superconducting magnets and other high field applications. Accordingly, 1G wires are cooled, as a rule, to temperatures in the neighborhood of 4.2-30 K for high-field applications.

To overcome the huge variation of the critical current occurring in the presence of external fields, ReBCO compounds are used instead of bismuth-based materials. Here, 'Re' refers to rare earth elements such as yttrium, lanthanum, dysprosium, samarium and gadolinium. The most common ReBCO compound is made of yttrium, namely, $YBa_2Cu_3O_{7-x}$ (YBCO). It has a better intrinsic flux pinning and irreversibility field than BSCCO. In addition, YBCO is able to carry high currents in strong magnetic fields while being cooled by liquid nitrogen. However, YBCO is brittle and highly anisotropic, which implies that electric current does not flow well from grain to grain if they are not perfectly aligned. Such a compound is produced by a coating of YBCO on a metallic substrate. This has led to the emergence of the second generation (2G) HTS conductors, which are also named as coated conductors.

It is also worth spending some words on the discovered superconducting material MgB_2 at the beginning of 2001 having a critical temperature at zero magnetic fields of about 39 K. Research on MgB_2 has attracted much of the interest of the scientific community. On the one side, this material shows interesting properties from the point of view of basic solid-state physics since it has a completely different structure with respect to the cuprates, but it has a critical temperature above the critical temperature of LTS. On the other side, it seems promising for applications due to its easiness of fabrication and to the satisfactory properties that have already been achieved.

The LTS and HTS have led to many research and development efforts in practical applications of superconducting magnet system (SMS) and devices. It may be expected that new phenomena will be discovered by the extreme conditions combined with high pressure and ultra-low temperature in high fields. However, some problems encountered in applications of HTS remain unsettled. The key issue to overcome their practical use is the broad transition from the superconducting to the normal state, manufacturing problems of HTS wires which significantly delay the development of the HTS industry.

Thereby, a high field technique is progressing rapidly and now static fields over 20 T are available for many types of research using low-temperature and high-temperature superconductors, mentioned above, that may be called as technical superconductors. The investigations can be made by conduction-cooled superconducting magnets, which allow one to make long-term experiments without any labor for charging liquid helium. However, technical superconductors usually exhibit a finite resistance at intensive operating modes. The nature of this resistive state attracts much attention since it allows one to understand the fundamental phenomena occurring in technical superconductors, such as the mechanisms of superconductivity, thermal fluctuations, macroscopic electrodynamics, etc. Therefore, thermal and electrodynamics investigations of technical superconductors, taking into account of the resistive state, is important for the advancement of fundamental science and the development of applications. Such investigations play a key role in applied superconductivity, and they are studied by means of various theoretical models. Their basics are presented below.

For the obvious reasons, the book cannot claim to be a complete description of all the problems arising in applied superconductivity. To fill the inevitable gap in the material presented in this publication, the author recommends readers to use additional literature, among which they should pay attention to the monographs (Brechna 1973, Carr, 1983; Buckel and Kleiner 2004, Narlikar 2014, Poole C.P. et al. 1995, Wesche 1998b). The author tried not to repeat the discussion of the results given in these monographs as much as possible.

Resume

The investigation of macroscopic phenomena in superconductors and composite conductors or tapes based on them is one of the most important topics in applied superconductivity. Its basic principles were discussed in detail in the presented chapters. They are not only important for understanding the essence of electrodynamic and thermal processes in technical superconductors but also allow to determine the limits of the workability of superconducting magnets. In particular, the conditions underlying the preservation of superconductivity under the influence of magnetic, current and thermal (in general, thermo-electromagnetic) disturbances were discussed. In general, the solution methods proposed allow one to study the problems arising in the analysis of nonlinear phenomena in a metastable multiphase dissipative medium with unknown moving boundaries between the phases.

The complete description of the qualitative and quantitative features of the development of nonlinear modes happening in low-temperature and especially in high-temperature superconductors must be based on the analysis of the dynamics of superconducting state, taking into account the collective variation of the temperature and electromagnetic field in the superconductors or superconducting composites/tapes during the diffusion of the magnetic flux induced by any perturbations. The main specificity of these phenomena is that they have the character of a chain reaction of an explosive nature.

The main purpose of the chapters examined is the discussion of the formation mechanisms of stable electrodynamic states in low- and high-temperature superconductors under consideration of their thermal history. The proposed general models and numerical-analytical methods for the analysis of essentially nonlinear thermo-electrodynamic phenomena occurring in superconducting materials with different V-I characteristics allow one to describe the bifurcation formation of stable and unstable states in current-carrying elements of superconducting magnets. They may occur during the change of the external magnetic field, transport current, the action of the heat disturbances of any power, length and duration, as well as at any combination thereof.

To determine the instability conditions of the superconducting state, the unified method based on the finite perturbation of the initial equilibrium state is proposed. It allows one to correctly determine the stability boundary of the collective evolution of thermal and electrodynamic states both in the superconductors and in the superconducting composites or tapes with various types of V-I characteristics under different operating modes and action of arbitrary external perturbations. The characteristic physical features of the change of stable overheating of the superconductor, preceding the destruction of the superconducting state, which affects the conditions of its stability, are formulated. They show the non-trivial relationship between the permissible increase in the superconductor temperature, conditions of the superconducting state stability, and heat losses released stably in the superconductors before the magnetic or current instabilities onset.

The results of such an analysis essentially extend the physical understanding of the development of macroscopic phenomena in low- and high-temperature superconductors. In general, the results obtained allow one to combine the independently developing theories of losses, magnetic and current instabilities and the theory of thermal stabilization. In this case, the stability conditions may be formed on the basis of a single theoretical concept that assumes the existence of any critical disturbance. It is independent neither of the nonlinearity type of V-I characteristics nor of the perturbation nature but abide by the limiting transfers to known stability criteria. Accordingly, as discussed in the corresponding chapters, the phenomena in technical superconductors are characterized by the following basic macroscopic electrodynamic and thermal laws.

In the creep of the magnetic flux, the differential resistivity of the superconductor in the magnetization region decreases monotonously in the direction to its moving boundary and takes zero value on it. Therefore, the electromagnetic field generated by the continuously increasing external magnetic field or by the charged current as well as during the relaxation states penetrates the superconductor at a non-uniform velocity, which determines the formation features of the stable states of superconductors and superconducting composites.

There are the electrodynamic states of superconductors with strong and weak creep, which differ in the effect of the creep on the distribution of the electromagnetic field within the magnetization region. If the V-I characteristics are described by the power or exponential equations, the difference between their corresponding states increases with the decrease of the growth parameters of V-I characteristics. The most noticeable difference between them will be observed in superconductors with strong creep ($n < 10$, $J_\delta/J_c > 0.1$).

The Bean model of the critical state is a zero approximation, which describes the spatial distribution of the electric field in the superconductor even with an arbitrary equation of its V-I characteristic. However, to adequately describe the formation of electrodynamic states, equation of the moving magnetization boundary and the distributions of the current and magnetic field in the superconductor must be written taking into account the corresponding V-I characteristic. This simplified approach allows the losses in the superconductors to be calculated with sufficient accuracy.

The consideration of the permissible change in the temperature of the hard superconductors during stable formation of its critical state shows that:

- the destruction conditions of the critical state of cooled superconductor depend on the sweep rate of the external magnetic field due to the corresponding stable change in the superconductor temperature before the instability onset (in the adiabatic mode this dependence does not exist);
- there are characteristic temperatures that determine the thermal structure of the magnetic instability conditions of the critical state when, in particular, instability may not occur under the action of intensive external thermal perturbations (effect of the thermal stabilization);
- the consequence of the thermal stabilization effect is the stabilization of the critical state when the critical state of the hard superconductor is stable with respect to arbitrary temperature perturbations whose amplitude changes in the range from the coolant temperature to the critical temperature of the superconductor at the conventional condition of the adiabatic stability;
- taking into account the matrix and real shape of the voltage-current characteristic of superconductor lead to a smoother mode of the transition of superconducting composite from the superconducting state to the normal one as well as a 'softer' oscillation evolution in them. The presence of the matrix extends the magnetic stability boundaries.

The explanation for the physical causes of the oscillations onset in superconducting materials may be given only taking into account the interrelated evolution of the thermo-electrodynamic states in superconductors. Accordingly, they are based on the difference in the rates of the conductive-convective removal of heat releases from the bulk of the superconductor to the coolant, compared to the increase rate of the Joule losses induced by the varying external magnetic field. In turn, this leads to the existence of the characteristic stages of the oscillation onset, which determine the interrelated changes in space and time of the electric field, the superconductor temperature, and the density of the screening current.

There is an influence of the temperature of superconductor on the conditions of oscillation onset. As a result, there exists an effect of thermal self-suppression of oscillations. Namely, they will not occur if the permissible overheating of the superconductor is high. Accordingly, the probability of the oscillation occurrence decreases when the cooling conditions deteriorate, in particular, they are completely absent under the adiabatic conditions or at a high sweep rate.

During the current charging, instability (quench) may occur both at the partial and full penetration modes of a transport current. The evolution of their stable and unstable states also has a bifurcation nature. In the full penetration mode, instability conditions do not depend on the current charging rate but depend on the temperature margin and the features of the current sharing between the superconducting core and matrix. During the partially penetrated current instabilities, quenching currents decrease monotonically with increasing current charging rate. If it is high, the instability conditions are close to adiabatic ones leading to a high stable temperature rise. Overall, both during the partial and full penetrated current instabilities, the operating states of superconducting composites or tapes depend on the stable increase in the temperature of them. The critical current density of a superconductor at coolant temperature, the cooling conditions, the current charging rate, the filling coefficient, the composite cross-section or the smoothness character of the V-I characteristics affect the modes when the electrodynamic states of a composite become non-isothermal.

In general, during current charging, the operating states of superconducting composites or tapes are characterized by the existence of the following features:

- there are characteristic values of the electric field, which determine the mechanisms in the formation of stable electrodynamic states during current charging, including the role of the convective-conductive heat transfer conditions;
- in the continuous current charging, the dependence of the heat capacity on the temperatures of the superconductor and matrix appreciably influences on the shape of the V-I characteristic;
- the permissible maximum values of the electric field and charged current can be both subcritical and supercritical in relation to the conventionally set critical parameters, that is a priory set critical quantities have no physical meaning;
- regardless of the shape of the V-I characteristics, the quenching current is determined by the heat balance between the Joule heat release and heat flux to the coolant;
- there is a direct correlation between the current depth penetration, heat losses and the permissible overheating before quench;
- the existence of the permissible overheating leads to the effect of the thermal degradation of current-carrying capacity of a superconducting composite, and as a consequence, its quenching currents do not increase proportionally to the increase in its critical current;
- there is a non-trivial relationship between the properties of superconductor and liquid coolant, resulting in different mechanisms for the onset of the current instability.

The quenching currents of superconducting composites having power and exponential V-I characteristics may not be equivalent. The noticeable difference will be seen in the strong creep due to the corresponding difference in the stable increase in temperature of a superconductor. The quenching currents of HTS composites depend not only on negative value of dJ_c/dT but also on its variation in the intermediate temperature range, which is not close to the critical temperature of a superconductor. By that, the quenching currents may be absent in HTS composite superconductors at intermediate operating temperatures.

In the subcritical states, the stable temperature rise of the superconducting composite is mainly determined by the permissible overheating of the superconductor. In the supercritical current range, the stable overheating is also influenced by the matrix properties. As a result, the stable overheating of HTS composites changes in a wide range. The subcritical electric fields (stable electric fields that do not exceed a priori set critical quantity) are more probable in the high magnetic fields or when a composite with the high-resistivity matrix has a relatively high value of the filling coefficient. In these modes, the stable value of the current flowing in a composite may be lower or higher than those defined by a priori set critical voltage criterion.

The static V-I characteristics of HTS composites may have multistable sections due to small but finite values dJ_c/dT in the range of the intermediate temperatures. They can be observed in the composites with a low filling coefficient as well as at high values of the external magnetic field, at relatively high temperatures of the coolant. As a result, the current instabilities of the HTS composites may be completely absent in the intermediate range of coolant temperatures, when the temperature of the composite may change stably up to its critical temperature in the static modes.

There are characteristic current instability mechanisms of AC regimes. Their stable formation has the stages, which are defined by the characteristic time windows. They exist due to the dynamic thermal equilibrium between the average values of the heat generation and the average values of heat removal to a coolant despite the high stable overheating of a superconductor and the induced electric field. Therefore, the HTS may save the superconducting state during the AC overloaded modes. This feature defines the existence of the maximum value of the peak current of stable AC overloaded regimes at the given frequency and cooling conditions. In all cases, this value is higher not only the critical current of a superconductor but also the corresponding quenching current defining the current stability boundary in the DC regimes. Besides, stable peak values of the electric field and temperature are also higher than the corresponding DC quenching values.

Under the action of thermal perturbations of an arbitrary length and duration, there is a boundary value of the energy of thermal perturbation, below which the occurring normal zone does not destroy the superconducting properties of composites. If it is exceeded, the normal zone spontaneously propagates throughout the composite. This leads to the concept of the critical energy of thermal perturbation, which is more general than the concepts of the minimum quench energy. The critical energy is finite even under the action of infinitely extended perturbations, that is, when the perturbation is distributed over the entire volume of the composite. Physically, this is due to the removal of the Joule heat into the refrigerant. Under adiabatic conditions, overheating to the critical temperature of the superconductor corresponding to a given magnetic field is permissible.

The development of thermal processes in composites at thermal impulses with an energy close to the critical one is characterized by the tendency of the normal zone to the region of an unstable equilibrium state after the termination of the perturbation. It has a characteristic spatial size (the length in the 1D approximation) that determines the specificity of the kinetics of the subcritical and supercritical propagation of normal zones initiated by perturbations of any length. For its correct definition, it is necessary to consider the current sharing mechanism.

Under the action of supercritical perturbations, the irreversible transition of the composite from the superconducting state to the normal state is a self-sustaining process. These modes are characterized by the occurrence of a thermal autowave propagating at a constant velocity in an extended superconducting composite, which does not depend on the nature of the initial thermal disturbances.

In the approximation that assumes a jump-like transition from a superconducting state to a normal one (a composite with an ideal V-I characteristic), the critical energies take on an infinitely large value at the full stabilization current, which follows from the equal area theorem in the quasi-linear approximation, and are zero at the critical current. The critical energies decrease with the increase in the thermal stabilization parameter α (the Stekly parameter). In the limiting case $\alpha = 1$, known as the Stekly cryostabilization condition, the superconducting properties of composites will always be preserved up to the critical current under the action of external thermal perturbations with arbitrary energy.

Critical energies are minimal under the action of instantaneous point-like perturbations, which are, therefore, the most dangerous. They increase with the length and duration of the disturbance. In particular, under the action of extended instantaneous perturbations, the critical energies approach their asymptotic values, which are described by an infinitely long temperature perturbation in the approximation of instantaneous heating. In general, the critical energies depend on the type of disturbance, but they are finite regardless of its nature over the entire range of currents from the full stabilization current to the critical one. At the same time, the widely used MPZ-concept cannot give an answer about the influence of the length and duration of the thermal disturbance on the thermal stabilization conditions. Moreover, for their correct quantitative analysis, the current sharing between the superconducting core and the matrix must be taken into account, considering the kinetics of the occurring normal zones.

For a wide range of practical applications, taking into account the heat capacity of the insulation will have little effect on the final results when calculating thermal stabilization conditions. However, even a small thickness of insulation can significantly modify the conditions for the thermal stabilization of intensely cooled current-carrying elements. In other words, the conditions for the occurrence of thermal instabilities depend primarily on the value of the thermal resistance of the insulation over the entire range of changes in the transport current. Therefore, to determine them for a superconducting multicomponent region with good accuracy, one can use a simplified model, in which the thermal interaction between superconducting turns is described by a model based on the concept of the thermal resistance of the contact layer.

The irreversible propagation of the normal zone inside a multiwire superconducting composition has a number of features that are not described by the expanding ellipsoid model. They are based on the following features:

1. The zone with normal conductivity formed as a result of the irreversible transition to the normal state of superconducting elements of a composition made up of composite superconductors separated by a finite thermal resistance has the shape of a truncated oval.

2. The propagation velocity of the normal zone in the longitudinal direction of a multiwire composition is lower than the corresponding value of the propagation velocity of the normal zone in a single composite. The latter is the asymptotic limit for all values of the velocities at which the normal zones will propagate in each element of the composition. This limiting case is reached only after the transition of all elements of the composition to the normal state when the moving front separating the superconducting region from the non-superconducting one acquires a flat shape.

3. In the range of currents close to the full stabilization current of a single composite (the current following from the equal area theorem), states are possible in which the superconducting properties are not destroyed for all elements of the composition. As a result, a limited number of sub-domains with normal conductivity can stably exist in a superconducting medium with a discrete structure, in which the normal zone propagates at a constant velocity.

4. The propagation velocity of the normal zone in the cross-section of a multicomponent superconducting medium is practically not dependent on the longitudinal thermal conductivity of its elements. First of all, it is determined by the thermal conductivity and thickness of the contact layer.

These regularities lead to physical features of the formation of the boundary of the region with normal conductivity in spatially limited superconducting composition. Their characteristic feature is the obligatory flattening of the boundary of the resistive region after it reaches one of the outer boundaries of the composition and begins to propagate in a direction free from restrictions. In this case, the flattening of the boundary occurs the faster, the better the conditions of heat transfer (less thermal resistance) between the elements of the composition. Obviously, if the increase in the size of the resistive region is limited in all directions, the increase in the size of the region with normal conductivity will be completed after reaching their boundaries, and all superconducting composition will pass in the normal state. The regularities noted should be taken into account when describing transients in superconducting magnets, since the model of an expanding ellipsoid is not correct for spatially limited superconducting composition with a discrete structure.

When developing large SMS with massive current-carrying elements, the corresponding analysis of their thermal stability should be carried out under consideration of the multi-dimensional nature of the heat flux propagation. In one-dimensional approximation, the thermal stability conditions of the superconducting state primarily depend on the Stekly parameter. When analyzing the thermal stability conditions taking into account the inhomogeneous temperature distribution over the cross-section of the composite, the thermal stabilization of the composite superconductor depends on the dimensionless parameter that is equal to the ratio of the conductive heat flux in the transverse direction to the heat flux into the refrigerant.

For massive composite superconductors, one-dimensional theory gives a significantly overestimated estimate of the boundary of their full stabilization (the Maddock condition in 1D approximation). For them, this condition depends on the thermal state of the surface of the current-carrying element. Moreover, the true value of the full stabilization current is always smaller than the corresponding value calculated under the assumption of uniform temperature distribution over the cross-section. The difference between them increases not only with an increase in the transverse dimensions of the current-carrying elements but also with an improvement in the conditions of their heat exchange with the refrigerant. This feature should be taken into account when designing large magnetic systems that use massive current-carrying elements. If, in this case, the conditions for the complete preservation of superconducting properties by the winding are calculated on the basis of one-dimensional theory, then, in reality, the irreversible propagation of thermal instability in a magnetic system with parameters set in this way can lead to irreversible destruction of the entire SMS.

Under the action of point-like thermal perturbations, the stability of the superconducting state of composites with a small transverse size depends on the mechanism of the longitudinal thermal conductivity of the composite. But this mechanism is less significant in the destruction of superconductivity in massive composites. In this case, the transverse thermal conductivity of the composite plays the main stabilizing role.

For cooled massive current-carrying elements, there is a significant discrepancy between one-dimensional and two-dimensional calculations of the thermal autowave velocity. This difference can be not only quantitative but also qualitative due to a significant increase in the calculated values of the normal zone velocity in two-dimensional approximation. Therefore, if transients in a massive composite were studied on the basis of one-dimensional models, in particular, the currents, at which its superconducting states are stable to arbitrary perturbations, were determined, then, for these parameters, it may be in an unstable state in real operating conditions.

The existence of the size effect in the thermal stabilization conditions of the superconducting state of cooled composites is explained by an inevitable increase in the average temperature of massive current-carrying elements.

The analysis of the thermal stabilization conditions of composite superconductors with a real V-I characteristic shows that they not only differ quantitatively from the corresponding conditions obtained in the approximation of an ideal V-I characteristic but are also characterized by qualitative differences. First of all, the critical current of the composite, introduced on the basis of a jump-like approximation of its transition from a superconducting state to a normal one, does not lead to a zero value of the critical energy. The actual value of the instability current can be either greater or less than this fictitious value. It is equal to the quenching current calculated for a superconducting composite with a real V-I characteristic. Along with this, the temperature of the resistive transition, after which the transport current begins to share between the superconductor and the matrix in the model with an ideal V-I characteristic, also has no physical meaning. This is because within the model with an ideal V-I characteristic, the current sharing mechanism between the superconductor and the matrix depends only on the decrease with the temperature of the critical current of the superconductor. In superconductors with a real V-I characteristic, the current sharing mechanism depends on the character of the increase in temperature of the V-I characteristic of the superconductor and the resistance of the matrix. As a result, in the temperature range up to the critical value, the current in its superconducting part not only is not equal to the critical current, as is assumed in the theory based on the ideal V-I characteristic but also exceeds this value. That is why the Joule losses calculated in the framework of the model with a continuously increasing V-I characteristic are always smaller than the corresponding values determined according to the model with an ideal V-I characteristic. As a result, there are more optimistic conditions for the stability of the superconducting state with respect to external thermal disturbances. However, the calculated values of the velocities of irreversible propagation of the thermal instability along the composite, which are determined taking into account the continuous increase in the V-I characteristic, are lower than the corresponding values calculated in the framework of the ideal V-I characteristic. Moreover, the thermal instability propagates along the composite in the form of a thermo-electric autowave at both subcritical and supercritical currents.

The condition of full thermal stabilization of composite superconductors (an analog of the Stekly cryostabilization condition), which takes into account the essentially nonlinear and continuous increase in the character of their V-I characteristic, significantly expands the range of currents that are stable against the external thermal disturbances during intensive cooling. As a result, the Stekly condition, according to which a superconducting composite is completely stable to thermal disturbances with arbitrary energy, does not allow one to correctly determine the parameters of a composite with a real V-I characteristic that ensure its full thermal stabilization.

Intensely cooled current-carrying elements may lack the minimum currents of existence and propagation of the normal zone, which determine the range of currents in which the superconducting state is stable to arbitrary thermal disturbances. This feature of the stability conditions will be observed due to an intense increase of the specific electrical resistance of the stabilizing matrix with temperature when the increase in the Joule heat release exceeds the increase in the heat flux into the refrigerant at temperatures above the critical temperature of the superconductor. As a result, under these modes, the autowave propagation velocities of thermal instabilities do not take negative values over the entire range of current changes.

The analysis of the interrelated increase in the Joule heat release and heat flux into the refrigerant with temperature shows that after the onset of instability, the change in the thermal state of technical superconductors intensively cooled by liquid refrigerants can occur under conditions close to adiabatic in their most heated part. It is this feature that leads to the possible burnout of technical superconductors. The probability of burnout increases with an increase in the current density in the superconductor and with a faster increase in the matrix resistivity with temperature. In such operating modes of technical superconductors, the formation of a thermoelectric autowave may not occur during the development of thermal instability.

In superconducting composites/tapes, in which the additional transport current diffusion is initiated by thermal disturbances during current charging, there may be intense heat losses. They can occur already at the stage of stable states. The heat release generated in these modes may differ by several orders of magnitude from the amount of heat determined in the framework of the theory of thermal stabilization based on the ideal V-I characteristic, leading to a noticeable increase in the temperature of the composite. Therefore, the expression commonly used in the theory of thermal stabilization for determining the Joule heat releases leads to their significantly underestimated values over the entire range of temperature variation of the superconducting composite at high rates of current charging. The consequence of these dissipative phenomena will be a corresponding decrease in the current of the instability onset. In particular, under the action of external thermal perturbations that do not transfer the superconductor to the normal state, a premature violation of the stable current charging is possible, as a result of which the currents of instability onset become the smaller, the higher the subcritical energy of the external thermal perturbation.

Instead of Conclusion

The macroscopic theory of applied superconductivity is one of the most interesting areas of modern applied science. Because of the limited space, only the basic formation laws of the thermo-electrodynamic states in superconducting elements of SMS were discussed. All of them form the foundation of the fundamental and applied superconductivity investigations in macroscopic approximation. Therefore, the models proposed may be successfully applied in the simulation of the operating modes of current-carrying elements of modern superconducting magnets and in the analysis of their stable operability. They will allow one to keep superconductivity in a wide range of operating temperature changes under intensive external disturbances of different nature. In general, when calculating the conditions for thermo-electromagnetic stabilization of a real SMS, it is necessary to take into account the nonlinear nature of the change in the properties of the superconductor, the stabilizing matrix, and the coolant with temperature. However, for their qualitative assessment, the quasi-linear approximation can be used (the exception is the description of burnout phenomena). Both of these methods were widely used in this publication and may be used in the new investigations.

List of Principal Symbols

a – half-thickness of a superconducting slab or radius of a cylindrical conductor

a_s – characteristic thickness of a superconductor

b – half-width of a superconducting slab

B – magnetic field

B_a – external magnetic field

B_c – critical magnetic field of a superconductor of a type-I

B_{c1} and B_{c2} – first (lower) and second (upper) critical fields of a type-II superconductor

B_i – initial distribution of a magnetic field

B_m – magnetic field of a first magnetic-flux jump

B_p – magnetic field of a full penetration

$\dot{B}_a = dB_a/dt$ – sweep rate of an external magnetic field

C_s, C_m and C_k – volumetric heat capacities of a superconductor, matrix and composite, respectively

dI/dt – current charging rate

E – electric field in a superconductor

e – dimensionless value of an electric field

E_a – electric field on the surface of a superconductor

E_c – critical electric field of a superconductor

E_x – characteristic value of an electric field

G – heat generation function

G_p – hysteresis losses simulated in a framework of scaling approximation for a superconductor with an actual voltage-current characteristic

G_c – hysteresis losses calculated in a framework of the critical state model

G_f – hysteresis losses calculated using a full current penetration model

G_m – heat release before flux jump

H – intensity of an external magnet field

H_s, H_k – dimensionless coefficients of heat transfer from a surface of a superconductor and composite, respectively

h – heat transfer coefficient

I – transport current in a superconductor

I_c – critical current of superconductor or composite

I_{c0} – critical current at a temperature of refrigerant

I_f – value of an applied current when it completely penetrates into a superconductor

I_m – limiting value of a current in a composite when it is charged at a finite rate

I_q – limiting value of a current in a composite when it is charged at an infinitesimal rate

i – normalized current in a superconducting composite (transport current divided by I_{c0})

J – current density in a superconductor

J_c – critical current density of a superconductor

J_{c0} – critical density of a superconductor at a refrigerant temperature

J_s – current density in a superconductor

J_m – current density in a matrix

l – half-length of a composite

L_x – characteristic value of a penetration depth of magnetic flux inside a superconductor

L_z – characteristic thermal length of a composite

M – magnetic moment

M_c – magnetic moment of a slab at an incomplete penetration of a magnetic flux calculated in the Bean approximation

n – exponent of the increase in a voltage-current characteristic of a superconductor described by a power equation

p – cooling perimeter

r_p – moving boundary of a superconductor's magnetization region with actual voltage-current characteristic in a cylindrical coordinate system

R_k, Q_k – eigenfunctions

S – cross-sectional area

t – time

T – temperature

T_a – temperature on a surface of superconductor or composite

T_{cB} – critical temperature of a superconductor at a given magnetic field

T_{cs} – current sharing temperature of a superconducting composite

T_i – superconductor temperature after terminating increase of an external magnetic field

T_i' and T_i'' – characteristic superconductor's temperatures that determine the development of magnetic instability

T_k – final temperature of a superconducting composite after development of magnetic instability

T_m – temperature of a superconductor before the flux jump

T_0 – the temperature of the refrigerant

t_f – full penetration time of a magnetic flux into a superconductor

t_h – characteristic cooling time

t_m – characteristic time of a magnetic flux diffusion in a viscous flux-flow model

t_s – time of increasing an external magnetic field

t_x – characteristic time of a magnetic flux penetration into a superconductor

t_λ – characteristic time of heat flow diffusion.

$V(Z)$ – sought function of the self-similar solution

V_c – penetration velocity of a magnetic flux into a superconductor calculated in a framework of the critical state model

V_f – penetration velocity of a magnetic flux into a superconductor calculated in a framework of the viscous flux-flow model

V_n – penetration velocity of a magnetic flux into a superconductor with an actual voltage-current characteristic

x_c – coordinate of a penetration boundary of external magnetic field calculated in a framework of the critical state model in a rectangular coordinate system

X_0 – dimensionless penetration depth of an external magnetic field in a rectangular coordinate system

x_0 – penetration depth of an external magnetic field in a rectangular coordinate system

x_f – moving boundary of a magnetization region in a framework of the viscous flux-flow model in a rectangular coordinate system

x_p – moving boundary of a magnetization region with an actual voltage-current characteristic in a rectangular coordinate system

z – longitudinal variable

Z – self-similar variable

Z_0 – moving boundary of a magnetization region in a self-similar approximation

α – Stekly parameter

α_{eff} – parameter of a current instability

β – magnetic instability parameter of a superconductor

β_c – critical value of a magnetic instability parameter

β_k – magnetic instability parameter of a superconducting composite

δ – dimensionless rise parameter of a voltage-current characteristic of a superconductor described by an exponential equation

Φ_0 – elementary magnetic flux in a vortex

η – filling factor of superconductor in a composite

Λ – dimensionless parameter equal to a ratio of a characteristic diffusion time of a magnetic flux in a superconductor to the characteristic diffusion time of a heat flux

Λ_k – dimensionless parameter equal to a ratio of a characteristic diffusion time of a magnetic flux in a superconducting composite to a characteristic diffusion time of a heat flux

λ_L – London penetration depth

λ_s, λ_m and λ_k – thermal conductivity coefficients of a superconductor, matrix, and composite, respectively

μ – relative magnetic permeability

μ_0 – magnetic permeability of vacuum

μ_k, υ_k – eigenvalues

θ – dimensionless temperature

$\rho_d = \partial E / \partial J$ – differential resistivity of a superconductor

ρ_f – electrical resistivity of a superconductor in the viscous flux-flow mode

ρ_k – electrical resistivity of a composite

ρ_m – electrical resistivity of a matrix

ρ_x – characteristic value of an electrical resistivity of a superconductor

τ – dimensionless time

ξ – coherence length

$\varkappa$ – Ginzburg-Landau parameter

List of Abbreviations

V-I characteristic – voltage-current characteristic

HTS – high-temperature superconductor

CSM – critical state model

LTS – low-temperature superconductor

SMS – superconducting magnet system

Contents

**Chapter 7:   Thermal Phenomena in Composite Superconductors with an
                Ideal Voltage-Current Characteristic and their Thermal
                Stabilization Conditions   255**

**Chapter 8:   Electrophysical Features of Thermal Instabilities Occurrence in
                Technical Superconductors with Various Nonlinearity Types of
                V-I Characteristics   329**

1

Concepts and Principal Provisions of Fundamental and Applied Superconductivity

1.1 SUPERCONDUCTIVITY

Superconductivity was discovered by Heike Kamerlingh Onnes in 1911 (Onnes, 1911a, 1911b, 1911c) at the University of Leiden, The Netherlands. He found that the resistance of mercury sharply disappeared below a certain temperature, which he called the *critical temperature* (T_c). As it was determined later, superconductivity is a new physical phase at $T < T_c$. This state is called the *superconducting state* and such materials are called *superconductors*. Today, the main part of pure elements having superconducting properties is known as *type-I superconductors* or *soft superconductors*. Note that the superconductors have an electrical resistivity ρ above T_c higher than a resistivity of the good-conductivity metal, such as copper or silver in the temperature range above T_c. The superconducting state is a stable thermodynamic phase, and it is a quantum mechanical phenomenon on a macroscopic scale when the motion of all superconducting electrons is correlated with lattice vibrations. As a result, the superconducting electrons bump into nothing and create no friction. Thereby, they can transmit current with no loss of energy since the superconducting electrons pass through the complex lattice despite the fact that there are impurities and lattice inside a superconductor. Several experiments have proved the existence of permanent currents over years so that the resistance of superconducting pure metals seems practically zero, more exactly, $\rho < 10^{-25}$ Ωm.

Soon after the discovery of superconductivity, it was found that superconductivity may be destroyed, that is, superconductors can become normal again not only by its heating but also by the magnetic field, which calls the *critical magnetic field* (B_c). As a result, the field-temperature phase diagram characterizes superconductors. Onnes (Tuyn and Onnes 1926, Sizoo et al. 1926) investigated such diagrams of pure metals. According to his experiments, the following relation $B_c = B_{c0}[1 - (T/T_c)^2]$ was formulated, where B_{c0} is the approximation of the critical magnetic field at zero temperature. Superconductivity of pure metals exists below the dependence of $B_c(T)$.

The superconducting state differs qualitatively from the normal (nonsuperconducting) state. Namely, besides zero electrical resistivity, the magnetic field does not penetrate a superconductor (*perfect diamagnetic properties*). The exclusion of magnetic fields from the bulk of a superconductor is known as the Meissner-Ochsenfeld effect (Meissner and Ochsenfeld 1933), and it is the feature of superconductors that distinguishes superconductivity from ideal conductivity. The nature of this effect cannot be explained using the macroscopic Maxwell equations because it is a quantum phenomenon. The phenomenological theory facilitates understanding of the Meissner-Ochsenfeld effect. This theory was developed by F. London and H. London (London F. and London H. 1935). They proposed the system of equations describing the microscopic electric and magnetic fields in superconductors (the so-called London's electrodynamics). The first equation describes the perfect electrical conductivity. The second equation leads to the solution according to which a magnetic field has an exponentially decaying form, in particular, in the 1 D approximation $B = B_a \exp(-x/\lambda_L)$, that is, the external magnetic field B_a is screened from the interior of a superconductor within a distance λ_L, known as the *London penetration depth*, which has the values $10^{-6} - 10^{-5}$ cm.

In 1950, Ginzburg and Landau (Ginzburg and Landau 1950) proposed the theory using thermodynamic concepts. It is based on Landau's theory of the phase transitions of the second-order to predict the superconducting electron density n_s. The Ginzburg-Landau equations lead to the characteristic parameter $\kappa = \lambda_L/\xi$, which is equal to the ratio of *the penetration depth λ_L and the coherence length ξ*. The latter quantity may be defined as the length scale over which Ginzburg-Landau's order factor varies. The coherence length is also a measure of the length scale over which the gradual change from normal to superconducting state occurs at the external boundary of a superconductor. Therefore, it can be considered as the scale over which the superconducting electron of the density n_s goes from zero at the external boundary to a constant value inside the superconductor. The Ginzburg-Landau theory is an alternative to London's theory. However, the Ginzburg-Landau theory does not explain the microscopic mechanisms of superconductivity. It examines the macroscopic properties of the superconductor with the aid of general thermodynamic equations. This theory is the *phenomenological theory* in the sense of its assumptions for describing the state transition. It is based on quantum mechanics instead of the macroscopic electromagnetic phenomena.

Bardeen, Cooper and Schrieffer advanced the understanding of superconductivity through the microscopic theory of superconductivity (Bardeen et al. 1957). It explains superconductivity at temperatures close to absolute zero. Their theory offered that atomic lattice vibrations affect the entire current in the superconductor. They force the electrons to pair up into teams that could pass all the obstacles, which cause the resistance of a normal conductor.

Abrikosov used the Ginzburg-Landau theory and investigated the properties of superconductors with $\kappa \gg 1$ (Abrikosov 1957). Earlier such superconducting alloys were experimentally observed by De Haas and Voogd (De Haas and Voogd 1929) and Shubnikov and his colleagues (Rjabinin and Schubnikow 1935). Abrikosov theoretically justified the existence of these materials. They were called the *type-II superconductor*. Later, the type-II superconductors, which may be used in practical applications, came to be called *nonideal type-II superconductors* or *hard superconductors*. Type-II superconductors are characterized by two critical fields: the *first or lower* (B_{c1}) and *second or upper* (B_{c2}) critical fields. They depend on temperature. The value of B_{c2} is zero at the critical temperature of the superconductor. Type-II superconductors are in the Meissner state and exclude the external magnetic field from the bulk of the superconductor below the value of B_{c1}. Magnetic fields between B_{c1} and B_{c2} penetrate the superconductor in the form of vortexes containing the single elementary quantum ($\Phi_0 = 2.0678 \cdot 10^{-15}$ Wb). The magnetic field is surrounded by a vortex supercurrent. The Cooper pair density is equal to zero at the center of the vortex meaning that the core of a vortex is a normal state. Abrikosov had shown that the normal regions organize a triangular lattice of vortexes. Such a state is called *a mixed state*. These states were experimentally observed in LTS and HTS (Essmann and Träuble 1966, Hess et al. 1991, Hug et al. 1994, Eskildsen et al. 2002). The triangular lattice is the close-packed configuration. There are large numbers of type-II superconductors. In particular, LTS and HTS are the type-II superconductors.

Since the discovery of superconductivity, many efforts were made to find superconducting materials with critical temperatures as high as possible to solve cryogenic problems. In 1973, the record value of the critical temperature reached for Nb_3Ge film with a value of 23.2 K. Remarkable discovery was achieved by G. Bednorz and A. Müller (Bednorz and Müller 1986). It was shown that a copper oxide compound with the stoichiometry $La_{2-x}Ba_xCuO_4$ had the superconducting properties at about 29 K which is strongly dependent on x. The superconductor with the stoichiometry $YBa_2Cu_3O_{7-x}$ was found later. Its critical temperature is about 90 K that is higher than the boiling temperature of liquid nitrogen. Ceramics $Bi_2Sr_2CaCu_2O_{8+x}$ (Bi2212) and $Bi_2Sr_2Ca_2Cu_3O_{10+x}$ (Bi2223),

$Tl_2Ba_2Ca_2CuO_{8+x}$ and $HgBa_2Ca_2Cu_3O_{8+x}$ raised the upper limit of the critical temperature. The critical current of about 164 K was measured in $HgBa_2Ca_2Cu_3O_{8+x}$ under the pressure of 31 GPa.

The HTS cuprates have a significant interest in applied superconductivity. However, they are highly anisotropic granular materials. Herewith, they are characterized by a weak-link effect between their grain boundaries which decreases allowable current density. They are also brittle. These features cause difficulties for applications. In spite of these handicaps, the technology of HTS fabrication makes progress. As a result, the critical currents of HTS have increased to be useful for applications. Many pilots HTS devices were manufactured: electric power cables, rotating electrical machines, generators and others. Today, HTS is used in large-scale LTS magnets as high-field inserts to produce a magnetic field higher than 22 T and are used more in the future at liquid helium temperature level, like effective current leads, as fault current limiters for electrical engineering.

Thus summarizing, it is important to note that the search for new high-temperature superconductors will undoubtedly lead to new and unexpected discoveries. Today, the scientific and technical programs of developed countries provide a lot of activities, including fundamental and applied researches aimed at solving the problems of practical superconductivity applications in energy and electrical engineering, medicine, electronics and others. Thus, applied superconductivity becomes a large branch of the modern industry which allows for the emerging energy needs of society.

1.2   IDEAL AND NONIDEAL TYPE-II SUPERCONDUCTORS RESISTIVE STATES OF TECHNICAL SUPERCONDUCTORS AND THEIR VOLTAGE-CURRENT CHARACTERISTICS

As discussed above, there exist two types of superconductors: type-I ($\kappa \ll 1$) and type-II ($\kappa \gg 1$). Type-I superconductors have the critical field that is related thermodynamically to the free-energy difference between the normal and superconducting states (Buckel and Kleiner, 2004). The critical magnetic field B_c for these superconductors is very low. Magnetic fields, weaker than the critical field B_c, are completely pushed out from the bulk of the superconductor by screening current flowing in a very thin layer at the surface. The superconductors used for applications are of type-II. For applications, they are operated in a mixed state. Their critical parameters T_c, B_{c1} and B_{c2} are intrinsic material properties. Bulk superconductivity is lost at the upper critical magnetic field B_{c2}. This value is essentially higher than the thermodynamic critical magnetic field B_c of type-I superconductors.

Another important characteristic of a superconductor is the *critical current density* (J_c). For pure metallic elements, it is the maximum current density which can flow in a superconductor without dissipation. Their critical current density depends on the magnetic field and temperature of the element.

The temperature dependence for the critical current density of pure metallic elements (type-I superconductors) was defined by Onnes (Onnes, 1913d) as follows: $J_c = J_0(1 - T/T_c)$. Here, J_0 is the critical current density approximation at zero temperature. He also observed the influence of a magnetic field on J_c. The theoretical formulation of this dependence was made by Silsbee (Silsbee 1916) who analyzed details of Onnes' published results. It was affirmed that the critical current density of pure metallic elements has the value at which the magnetic field stated by current itself equals the critical magnetic field. Therefore, all the critical parameters of the type-I superconductors are interrelated.

As above-mentioned, the linear dependence $J_c(T)$ can be also used to describe the critical current density of type-II superconductors. However, the value of J_c is not correlated with T_c, B_{c1} and B_{c2} in contrast to type-I superconductors. It depends on the material properties.

If a current is passed through the type-II superconductor in the mixed state, then it will interact with the vortices causing their movement in the perpendicular direction to the current and magnetic field due to the Lorentz force. The Lorentz force density between the current J and the flux is given by $F_L = J \times B = J \times n\Phi_0$, where n is the total number of vortices. This is a viscous motion that leads to heat generation. Under features of a flow flux, there exist ideal (without structure defects) and nonideal (with structure defects) type-II superconductors.

A current, which flows through the ideal type-II superconductor, generates constant heat because the vortices continuously move due to the Lorentz force. In other words, the superconductor has a resistance because there exists the moving electric field, which is equal to $E = v \times B$ and parallel to J. Here, v is the velocity of the vortices. As a result, the critical current density of ideal type-II superconductors is equal to zero.

In nonideal type-II superconductors, the vortexes are pinned by the impurities or defects under the so-called pinning force F_p. The sites, in which the vortices stay at a fixed position, are called pinning centers. There are various types of pinning centers (both naturals and artificially created): zero-dimensional pinning centers like a crystal lattice point defects, the one-dimensional pinning centers like displacement lines, the two-dimensional pinning centers like grain and domain boundaries or surface defects and the three-dimensional pinning centers like artificially created precipitations and columnar defect channels. For the specific current, the Lorentz force overcomes the pinning force and the vortexes start the movement. As a result, a dissipation phenomenon occurs. They will continually move through the superconductor until the vortexes are pinned by other pinning centers. The higher the pinning forces F_p, the higher the current density, which can be passed through the superconductor before the transition to a normal state. In other words, the flux pinning phenomenon determines the critical current of nonideal type-II superconductors. Therefore, the critical current density of nonideal type-II superconductors corresponds to the current when the Lorentz force equals the average pinning force, that is, it is equal to $|J_c| = |F_p|/|B|$. Besides, the pinning force, the pinning center density also influences the critical current.

These features show that there exists the characteristic state of nonideal type-II superconductors. It is called the *critical state*. The latter indicates that the distributions of the magnetic field and the current (transport or screening) lead to the balance between the Lorentz and flux pinning forces. That is why ideal type-II superconductors without defects have zero critical currents.

Not only the magnetic mechanism influences the movement of the vortexes, they can hop out of the pinning centers by the thermal activation mechanism due to which the movement of the vortexes occurs even without a magnetic field or a transport current (Kim et al. 1963). For evaluating a similar effect, the depth of the potential well U_0 of the pinning centers has to be compared with the thermal energy kT of the vortices (k is the Boltzmann constant). For LTS, the ratio U_0/kT is quite large, so that the vortices are confined in the pinning centers up to temperatures close to T_c. For HTS, the operating temperature is higher so that the ratio U_0/kT is smaller for the same pinning energy U_0. As a result, if the current becomes locally larger than J_c in this site, then the excessive current will be redistributed into the regions where the current density J is still below J_c. That is why many experiments have shown that there exists the so-called flux-creep state (the thermally activated flux flow) in LTS and HTS. The redistribution is determined by many reasons: a thermal activation of magnetic flux, pinning heterogeneity, vortex structure defects, etc. They lead to a random local drift of the magnetic flux under the action of the Lorentz force which is accompanied by corresponding energy dissipation. As a result, the electric field is continually increased inside the nonideal type-II superconductor due to the random nature of the vortices' movements. Because of this, its resistance has a finite value even at currents smaller than J_c. Consequently, the corresponding dependence of $E(J)$ (commonly known as the voltage-current characteristic or V-I characteristic) is measured in experiments for a wide range of electric fields, even for their very small values. Thereby, the macroscopic electromagnetic properties of LTS and HTS may become sensitive to the subcritical electric field region.

Thus, the theoretical models described the electrodynamics states of nonideal type-II superconductors should be based on the investigation of the microscopic dynamics of the vortexes (Blatter et al. 1994, Yeshurun et al. 1996). However, the proposed approaches lead to significant difficulties. At the same time, if the density of vortices is large on the spatial scales exceeding the London penetration depth, then the evolution of the magnetic flux is a result of its macroscopic diffusion occurring in the superconductors in response to any perturbation. Therefore, the electrodynamic states of these superconductors can be investigated using macroscopic models. Then the Maxwell equations, supplemented by the corresponding V-I characteristic, facilitate analyzing evolution of electrodynamic states using the phenomenological models. In this case, the V-I characteristic, which, first of all, exist due to the fluctuating nature of the vortices penetration in the superconductor that takes into account integrally the features of the vortex lattice formation as the whole. Let us define the superconductors and composites based on them having the real V-I characteristic as the *technical superconductors*.

In this regard, the innovative investigations of Bean should be noted. To explain the hysteresis phenomena observed in the technical superconductors, Bean proposed the so-called *critical state model* (Bean 1962a, 1964b). This concept was the result of many experimental measurements of the voltage-current characteristics of low-T_c superconductors when their change in the vicinity of a certain current has a sharply increasing character. The critical state model made it possible to understand in a simple and clear form many features of the development of electrodynamics phenomena in technical superconductors like to estimate losses (Carr 1983) and formulate the stability criteria of the superconducting state (Wilson 1983, Gurevich et al. 1997c). Note that the critical current density of the superconductor in the Bean model is independent of the magnetic field. This dependence is taken into account in the Anderson-Kim model (Anderson et al. 1964). This model describes more accurately the electrodynamic states of superconductors.

Using the concept of the critical state, the real V-I characteristic is described by an ideal dependence written in the form

$$E = \begin{cases} 0, & |J| \le J_c(T, B) \\ J_c, & |J| > J_c(T, B) \end{cases} \tag{1.1}$$

It describes an abrupt transition from the superconducting state to the normal one with infinitely large differential resistivity (a jump-like transition). In other words, the critical state model indicates that the distributions of the magnetic field and the current density (transport or induced current) are based on the balance of the Lorentz forces (or flux driving force) and their pinning forces. In the framework of this approximation, any change of the magnetic flux in a superconductor leads to the appearance of the screening currents with the current density which is equal to the critical one. These are macroscopic currents flowing in superconductors.

Modification of this model is given by the relations

$$E = \begin{cases} 0, & |J| \le J_c \\ J\left[\rho_f\left(|J| - J_c\right)\right]/|J|, & |J| > J_c \end{cases} \tag{1.2}$$

which is known as the viscous flux-flow model. Under this approximation, the superconductor is in a dissipative regime termed as viscous flux-flow. This model is characterized by a linear rise of V-I characteristic at $|J| > J_c$ because it assumes that the electrical field is proportional to the number of the vortices in the superconductor up to the normal state. As a result, a new physical quantity called the flux-flow resistivity ρ_f,—which is associated with normal resistivity of the superconductor ρ_n by the relation $\rho_f = \rho_n B/B_{c2}$—is introduced.

Idealized models (1.1) and (1.2) are intensively used to get analytical and numerical solutions for many practical cases. They show that the vortices do not move as long as current density is less than the critical current density and thus it completely neglects the energy dissipation in the subcritical region $J < J_c$. To describe the flux-creep electrodynamics of the technical superconductors more exactly, one may investigate the macroscopic evolution of the electromagnetic field in the superconductor taking into account the real form of the V-I characteristics. Various models considering the microscopic quantities of a superconductor are used for explanations of these dependencies caused by specific mechanisms of the flux-creep. Among them, the model of a thermally activated uncorrelated hopping of point-like vortex bundles (Anderson 1962, Anderson and Kim 1964), the model of the creep activation barrier when the vortex motion is controlled by many spatial intrinsic defects (Zeldov et al. 1989, Fisher 1989, Blatter et al. 1994), the vortex glass and the collective creep models (Zeldov et al. 1990, Nattermann 1990, Fisher D. S. et al. 1991, Feigel'men et al. 1991) are most successfully and widely used.

These are simplest models taking into account the resistivity of superconductors, when the drift of vortices determines the existing electric field. Unfortunately, a complete theory describing the nonlinear part of the real V-I characteristics is not developed yet. Therefore, the various phenomenological relations describe the electric field in superconductors measured in an experiment. In particular, the power and exponential equations are often used. In the simplest cases, it can be written as follows

$$E = E_c (J/J_c)^n \tag{1.3}$$

$$E = E_c \exp[(J - J_c)/J_\delta] \tag{1.4}$$

Here, E_c is a priori defined electric field criterion which states the critical current density J_c, namely, $E(J_c) = E_c$, n is the exponent of the power V-I characteristic (n-value) and J_δ is the creep current density (rise parameter) determining the slope of the exponential V-I characteristic.

The power equation (1.3) corresponds to a logarithmic current dependence of the potential barrier $U = nT \ln(J_c/J)$ in the Arrhenius law written as follows: $E = E_c \exp[-U(J, T, B)/T]$. This dependence is measured in many experiments that deal with both LTS and HTS. In LTS, the values of n are typically above 50, and the critical current weakly depends on the definition criteria E_c. The HTS has noticeable smaller values of n. Therefore, their critical current is ill-defined. As a result, the influence of the n-value cannot be ignored, and it is an important factor in addition to the critical parameters of HTS. The n-value is defined by fitting of the voltage-current characteristic, namely, $n = d(\ln E)/d(\ln J) = (J/E)dE/dJ$. The n-value may be determined also by fitting of the V-I characteristic described it in logarithmic coordinates. Then, $n = \ln(E/E_c)/\ln(J/J_c)$.

The Anderson–Kim model with the linear current dependence of the potential barrier ($U = U_0[1 - J/J_c]$) lies at the basis of the exponential relation (1.4). This state is also observed for both LTS and HTS. Usually, the inequality $J_\delta \ll J_c$ is valid for all of practically interest cases. The exponential V-I characteristic will be sufficiently steep if the condition $J_\delta/J_c \ll 0.1$ occurs.

The written relations (1.3) and (1.4) facilitate in describing the experimental data with sufficient accuracy. Quantities n and J_δ define the steepness of the transition from the superconducting to the normal state: the higher the n-value (the lower the J_δ value), the better the superconductor because of a steep transition. It can carry a current very close to J_c generating a very low voltage. In general, J_c and n-value are the functions of the magnetic field and temperature. The dependence of $n(T)$ may be essentially nonlinear near liquid nitrogen operating temperature. The n and J_δ values are isotropic in LTS and anisotropic in HTS. Equations (1.3) and (1.4) are also convenient because they can interpolate the limiting transitions to the critical state model at $n \to \infty$ or $J_\delta \to 0$. Expanding the dependence (1.4) in series by taking into account that $\delta = J_\delta/J_c(T_0, B_a) \ll 1$ and keeping the first two terms, one can find that $n \propto 1/\delta$ with the accuracy of $\delta^2 \ll 1$.

As a rule, the critical current is experimentally measured by the conventional four-probe technique and determined by two equivalent criteria. They are the electric field criterion, which is equal to $E_c = 1 \mu V/cm$, or with the resistivity $\rho_c = 10^{-11} \Omega \cdot cm$. Given the value of ρ_c is much smaller than the resistivity of copper at 4.2 K. That is why this resistivity criterion may be adopted as almost zero resistance for measurements.

Although the critical current and n-value are easily obtained by the four-probe technique for short HTS samples, it is difficult or almost impossible to measure both parameters in each section of long HTS tapes by this method. In this case, to measure the critical current and n-value of HTS tapes, the contact-free methods or the statistical analysis methods taking into account the inhomogeneity of the critical current and n-value used in long HTS tapes.

It should be noted that the relations (1.1)–(1.4) do not exhaust all the possible types of the V-I characteristic. At the same time, they permit the experimental data to be described with sufficient accuracy. Therefore, they are widely used in theoretical analyses.

Thus, the transition of nonideal type-II superconductors (hence, the technical superconductors) from the normal state to the superconducting one depends on three critical parameters: the current density, the external magnetic field and the temperature. They lead to the critical surface of the superconductor. As a rule, the critical surface of technical superconductors is depicted as a dependence of their critical current density $J_c(B, T)$ on the magnetic field induction and the temperature. According to the experimental studies of the critical parameters of LTS and HTS, various analytical expressions are also offered to calculate them. Here are some of them.

In the uncomplicated cases, the critical current density of superconductors may be described by linear approximation as follows

$$J_c(T, B) = J_{cB}(B)[1 - T/T_{cB}(B)] \tag{1.5}$$

offered by Onnes (1913d) or in the form

$$J_c(T, B) = J_{c0}(B)(T_{cB} - T)/(T_{cB} - T_0), \quad J_{c0}(B) = J_{cB}(B)[1 - T_0/T_{cB}(B)] \tag{1.6}$$

Here, the current density J_{cB} and the temperature T_{cB} are the fitting critical parameters at a given value of the applied magnetic field, T_0 is the operating temperature.

Different equations are used to describe J_{cB}. In the framework of the Bean model, it is stated as a constant. At the same time, other critical state models were formulated. First of all, it should mention the Anderson-Kim model. According to this model, the dependence of the critical current density on the magnetic field is described as follows

$$J_{cB} = \alpha/(B + B_0) \tag{1.7}$$

where α and B_0 are the fitting parameters for each superconductor.

According to Onnes (Tuyn and Onnes 1926, Sizoo and Onnes 1926), the magnetic field dependence on the critical temperature may be described as follows

$$T_{cB} = T_{c0}\sqrt{1 - B/B_{c2}(0)}$$

or it may be found by the linear fitting of the corresponding experimental data in which the dependence $J_c(T, B)$ is measured. Here, $B_{c2}(0)$ is the approximation of the upper critical magnetic field at zero temperature and T_{c0} is the critical temperature at zero magnetic field. Then, the temperature dependence of the second critical field is expressed as:

$$B_{c2}(T) = B_{c2}(0)[1 - (T/T_{c0})^2].$$

It should be noted that the expressions (1.5)-(1.7) do not allow approximating the experimental Curves $J_c(B, T)$ in the entire variation range of the values T and B. Therefore, the models that are considered more general are used.

According to Lubell (Lubell 1983), the temperature dependence of the second critical field, the field dependencies of the critical temperature and the critical current density of the Nb-Ti alloy can be described by the expressions

$$B_{c2}(T) = B_{c2}(0)[1 - (T/T_{c0})^n], \quad T_{cB} = T_{c0}(1 - B/B_{c2}(0))^{1/n},$$
$$J_c(T, B) = J_{cB}(B)[1 - (T/T_{cB})]^n$$

at $n = 1.7$, $B_{c2}(0) = 14.5$ T, $T_{c0} = 9.2$ K.

In the Bottura approximation (Bottura 1999a), the critical current density of Nb-Ti can be calculated by the formula

$$J_c(T, B) = \frac{C_0}{B} \left[\frac{B}{B_{c2}(T)} \right]^\alpha \left[1 - \frac{B}{B_{c2}(T)} \right]^\beta \left[1 - \left(\frac{T}{T_{c0}} \right)^n \right]^\gamma.$$

Here, the parameters n, $B_{c2}(0)$, J_{cB}, T_{c0}, C_0, α, β and γ are chosen based on the experimental data as they depend on the structure of the superconductor.

For Nb_3Sn superconductors, it is possible to use the Summers formulae (Summers et al. 1991), which take into account the influence of voltages acting on the brittle Nb_3Sn. According to them, the temperature dependence of the second critical field can be calculated as follows

$$B_{c2}(T, \varepsilon) = B_{c2}(0, \varepsilon) \left[1 - \left(\frac{T}{T_{c0}(\varepsilon)} \right)^2 \right] \left\{ 1 - 0.31 \left(\frac{T}{T_{c0}(\varepsilon)} \right)^2 \left[1 - 1.77 \ln \left(\frac{T}{T_{c0}(\varepsilon)} \right) \right] \right\}.$$

Here, the second critical magnetic field at zero temperature and the critical temperature at zero magnetic fields are described by formulae,

$$B_{c2}(0, \varepsilon) = B_{c2}(0, 0) \left[1 - \alpha |\varepsilon|^{1.7} \right], \quad T_{c0}(\varepsilon) = T_{c0}(0) \left[1 - \alpha |\varepsilon|^{1.7} \right]^{1/3}$$

respectively, in which $\alpha = 900$ at compression deformation and $\alpha = 1.25$ at tensile deformation. For binary compounds $B_{c2}(0, 0) = 24$ T and $T_{c0}(0) = 16$ K and for triple compounds $B_{c2}(0, 0) = 28$ T and $T_{c0}(0) = 18$ K.

The formula for calculating the critical current density is

$$J_c(B, T, \varepsilon) = \frac{C(\varepsilon)}{\sqrt{B}} \left[1 - \frac{B}{B_{c2}(T, \varepsilon)} \right]^2 \left[1 - \left(\frac{T}{T_{c0}(\varepsilon)} \right)^2 \right]^2, \quad C(\varepsilon) = C(0) \left[1 - \alpha |\varepsilon|^{1.7} \right]^{1/2}$$

where $C(0)$ is a scale parameter.

To calculate the critical parameters of HTS, there are also corresponding approximate formulae. Let us write the most common of them.

Bottura (Bottura 2002b), summarizing the results presented in (van der Laan et al. 2001, Wesche 1995) introduced the expression describing the critical current density of Bi-based superconductors as a function of magnetic field and temperature as follows:

$$J_c(T, B) = J_0 \left(1 - \frac{T}{T_{c0}} \right)^\gamma \left[(1 - \chi) \frac{B_0}{B_0 + B} + \chi \exp \left(-\frac{\beta B}{B_{c0} \exp(-\alpha T / T_{c0})} \right) \right] \tag{1.8}$$

Here, α, β, γ, χ and B_0 are fitting constants, T_{c0} is the critical temperature of superconductor in zero magnetic fields, J_0 and B_{c0} are the approximation of the critical current density and the critical magnetic field at zero temperature, respectively.

This formula considers the huge flux-creep states of Bi-based superconductors in high magnetic fields which lead to the strong temperature degradation of the critical current. The Formula (1.8) may be also used to calculate the effective values of J_{c0} and T_{cB} by the linear fitting to the corresponding nonlinear Curves $J_c(T, B)$. As for illustrations, Figure 1.1 and Figure 1.2 show the corresponding comparisons between experimental data and calculations which were made for Ag/Bi2212 tape discussed in (Romanovskii et al. 2006). The next constants,

$$T_{c0} = 87.1 \text{ K}, \alpha = 10.3, \beta = 3.3, \gamma = 1.73, \chi = 0.27, B_{c0} = 465 \text{ T},$$
$$B_0 = 75 \times 10^{-3} \text{ T}, J_0 = 1.1 \times 10^{10} \text{ A/m}^2 \tag{1.9}$$

were used.

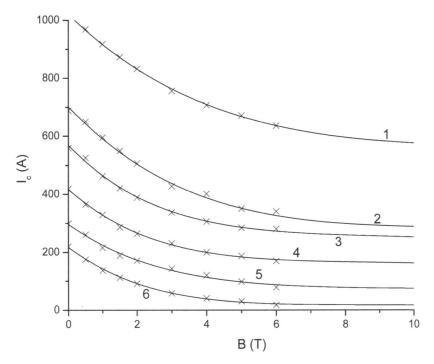

FIGURE 1.1 Critical currents I_c of the Ag-sheathed $Bi_2Sr_2CaCu_2O_8$ superconductor versus applied magnetic field B at various operating temperatures T_0: (×) - experiment, (———) - fit calculations, 1 - T_0 = 4.2 K, 2 - T_0 = 10 K, 3 - T_0 = 15 K, 4 - T_0 = 20 K, 5 - T_0 = 25 K, 6 - T_0 = 30 K.

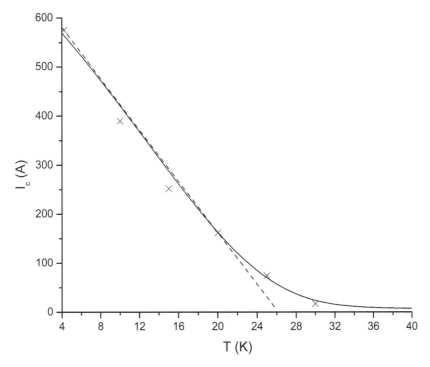

FIGURE 1.2 Temperature dependence of the critical current of Ag-sheathed $Bi_2Sr_2CaCu_2O_8$ at B = 10 T: (×) - experiment, (———) - fit calculation, (- - - -) - linear approximation.

In order to calculate the critical current density of YBCO in a wide range of temperature and magnetic field variations, different models may be used. In particular, the $J_c(T, B)$ dependence and the corresponding parameters are given in (Kiss T. et al. 2003, Wesche 2011). In these approximations, it has the following scaling form

$$J_c(T, B) = \frac{A}{B} B_{\mathrm{irr}}^r(T) \left(\frac{B}{B_{\mathrm{irr}}}\right)^p \left(1 - \frac{B}{B_{\mathrm{irr}}}\right)^q \tag{1.10}$$

Here, $B_{\mathrm{irr}} = B_{\mathrm{irr}}(0)(1 - T/T_c)^\alpha$ according to Wesche, T_c is the critical temperature and $B_{\mathrm{irr}}(0)$, A, p, r, q, β are the fitting parameters.

Another variant of such approximation was proposed by Awaji et al. (1996). Accordingly,

$$J_c(T, B) = J_0 \left(1 - \frac{T}{T_c}\right)^\alpha B^p \left(1 - \frac{B}{B_{\mathrm{irr}}}\right)^q \tag{1.11}$$

where $B_{\mathrm{irr}} = B_{\mathrm{irr}}(0)\exp(-\beta T)$ with fitting parameters $B_{\mathrm{irr}}(0)$, J_0, p, q, α, β.

1.3   THERMO-ELECTRODYNAMIC FEATURES OF MACROSCOPIC PHENOMENA IN TECHNICAL SUPERCONDUCTORS

As discussed above, the dynamics of the magnetic flux and electric field generated by the vortices motion at $B \gg B_{c1}$ can be studied in the macroscopic approximation using the Maxwell equations

$$\mathrm{rot}\, \boldsymbol{E} = -\partial \boldsymbol{B}/\partial t, \; \mathrm{rot}\, \boldsymbol{B} = \mu_0 \boldsymbol{J} \tag{1.12}$$

In this case, the voltage-current characteristics (1.1)–(1.4) are the material equations of technical superconductors, taking into account of the formation features of their resistive states in the averaged form. Since energy losses occur during the vortices motion, the analysis of their diffusion should allow for the change in the initial thermal state of the superconductor, taking into account its possible transition to a normal state, even in any of its local area. The description of dissipative phenomena in the superconductors is based on the solution of the quasilinear heat equation, which generally has the form

$$C \frac{\partial T}{\partial t} = \mathrm{div}[\lambda \, \mathrm{grad}\, T] + G + P \tag{1.13}$$

Here, C and λ are the volumetric heat capacity and the thermal conductivity coefficient of the superconducting material, respectively, depending on the spatial coordinates and temperature; $G = JE$ is the density of the Joule heat dissipation; P is the volumetric density of external heat sources. The remaining parameters were defined above.

The system of equations (1.12) and (1.33) supplemented by the corresponding initial and boundary conditions as well as the V-I characteristic facilitate the studying of the macroscopic phenomena in low- and high-temperature superconducting materials. Moreover, their solutions also allow one to perform the analysis of operating modes of superconducting magnets in the appropriate setting. The problems that arise in this case but in more simplified formulations making it possible to understand their physical features are discussed below in detail.

The system of equations (1.12) and (1.13) may be written in a scalar form as follows

$$\mu_0 \frac{\partial J}{\partial t} = \Delta E, \tag{1.14}$$

$$C(T) \frac{\partial T}{\partial t} = \nabla[\lambda(T)\nabla T] + EJ + P \tag{1.15}$$

for most practically important cases. Here, Δ is the Laplace operator and ∇ is the nabla operator.

Let us assume that the V-I characteristic of a technical superconductor can be represented as $J(T, B) = J_c(T, B)f(E/E_c)$, where f is a certain dimensionless function, describing the features of the increase of V-I characteristic of a superconductor. For example, for the V-I characteristic of power and exponential types, it is described by the relations $f(E/E_c) = (E/E_c)^{1/n}$ and $f(E/E_c) = 1 + \delta \ln(E/E_c)$, respectively. Since

$$\frac{\partial J}{\partial t} = \frac{\partial J}{\partial E}\frac{\partial E}{\partial t} + \frac{\partial J}{\partial T}\frac{\partial T}{\partial t} + \frac{\partial J}{\partial B}\frac{\partial B}{\partial t},$$

equation (1.14) is transformed as follows

$$\mu_0 \sigma \frac{\partial E}{\partial t} = \Delta E - \mu_0 \left(\frac{\partial J}{\partial T}\frac{\partial T}{\partial t} + \frac{\partial J}{\partial B}\frac{\partial B}{\partial t} \right) = \Delta E + \mu_0 \left(\left|\frac{\partial J_c}{\partial T}\right|\frac{\partial T}{\partial t} + \left|\frac{\partial J_c}{\partial B}\right|\frac{\partial B}{\partial t} \right) f\left(\frac{E}{E_c}\right).$$

Here, it was taken into account that $\sigma = \partial J/\partial E$ is the differential conductivity of the superconductor, and J_c is the monotonically falling function of the magnetic field and temperature. Considering the equation of thermal conductivity (1.15), the above-equation is reduced to the form

$$\mu_0 \sigma \frac{\partial E}{\partial t} = \Delta E + \gamma E + Q \tag{1.16}$$

where

$$\gamma = \frac{\mu_0 J_c(T, B)}{C(T)}\left|\frac{\partial J_c}{\partial T}\right|f^2, \quad Q = \mu_0 f\left[\frac{1}{C(T)}\left|\frac{\partial J_c}{\partial T}\right|(\nabla\lambda\nabla T + P)f + \left|\frac{\partial J_c}{\partial B}\right|\frac{\partial B}{\partial t}\right]$$

Equations of such types describe diffusion processes in so-called bulk multiplication media (Murray 2009). Typical such phenomenon is a chain reaction of neutron fission in the active medium with a multiplication factor γ and an additional volume density of their source Q. First, at $\gamma > 0$ and $Q = 0$, the bulk self-generation of neutrons prevails over their absorption and an uncontrolled chain reaction occurs at a certain γ value. According to this analogy, a spontaneous avalanche-like increase in the electric field (hence, the magnetic field and temperature) is possible in superconductors leading to the destruction of the initial superconducting state even at $Q = 0$. Such processes are called *instabilities*. They occur because the critical current density of technical superconductors decreases with temperature and magnetic induction. In addition, $Q > 0$ even at a uniform temperature distribution in the superconductor, if the critical current density of superconductor has the magnetic induction dependence or the superconductor is affected by external sources of heat $(P > 0)$. These two factors increase the probability of spontaneous transition of the superconductor to the normal state. As will be shown below there is a critical heat release at $P > 0$, after exceeding which the superconductivity is destroyed. Consequently, the stability of the superconducting state and superconducting magnets can be destroyed even when their operating parameters are in the subcritical region. Note that if in equation (1.16) the second and third terms on its right part are small and may be neglected, then it describes stable processes occurring in technical superconductors in the isothermal approximation. As a rule, the electrodynamics of technical superconductors is investigated under the isothermal approximation.

Let us discuss other causes of the superconducting state instabilities. First, it is necessary to note the role of the thermal state in the stability of superconducting states. As the temperature increases, the critical current density $J_c(T, B)$ decreases and the heat capacity $C(T)$ increases. This leads to a corresponding decrease in γ and Q, which means more stable superconducting states, and even to the possible emergence of new stable states at high temperatures. The temperature effect on the electrodynamic states of superconducting materials is discussed in detail in Chapters 4-6. As a result, instabilities may not be catastrophically avalanche-like. On the other hand, the conditions for instabilities depend on the temperature and the magnetic field. Therefore, in superconducting materials with high values of the critical current density, the onset of unstable states in them is most likely and their development will have a rapidly increasing character at all other parameters being equal. In addition, the nature of the formation of stable superconducting states and the onset of instabilities depend on the temperature distribution in the superconductor, as it follows from equation (1.26), and also depend on the magnitude and sign of the term div $\lambda(T)$ grad T. The latter can be either positive or negative. The sign depends on the cooling conditions, the transverse size of the superconducting filaments and the composite as the whole and their thermal conductivity coefficients which in turn affect the stability conditions of the superconducting states.

It should be noted the influence of the intensity of the V-I characteristic increases on the conditions of occurrence and development of instabilities. This depends on the type of function $f(E/E_c)$ and the changing character of σ. In particular, as will be shown in Chapter 4 for power V-I characteristic, the values of γ and Q are proportional to the change of multiplier $(E/E_c)^{1/n}$. Therefore, if the n-value increases or δ decreases, then the V-I characteristic is steeper. As a result, it is more likely to cause instability and its rapid development. Thereby, the instabilities in the LTS will occur almost at zero background electric field, and their development will have sharply increasing character. At the same time, the development of the electrodynamic states will be smoother and occur against the background of high electric fields in the HTS. (In Chapters 4-8, the chain formation of thermo-electrodynamic states in superconductors with exponential and power V-I characteristic is discussed in more detail).

The occurrence of instabilities in superconductors is associated with the falling dependence of the critical current density of a superconductor on the temperature and magnetic field. According to equation (1.16), this leads to additional heat. Indeed, an increase in the temperature or the magnetic field inside composite will lead to a decrease in its critical current density. At a certain size of the heat release or heat transfer conditions, this process becomes avalanche-like and the superconducting state loses stability.

Thus, the macroscopic states of technical superconductors are metastable, and their irreversible destruction can be avalanche-like due to the development of instabilities of different nature. Collective thermal and electrodynamic processes that are weakly dependent on the microscopic properties of a superconductor play a key role in their occurrence. As a result, the stability criteria of the superconducting state can be formulated in the framework of macroscopic models. At the same time, the consideration of temperature changes in technical superconductors before instability onset is important for the correct description of the processes occurring in them due to the dependence of C, J_c and $\partial J_c/\partial T$ on the temperature. The temperature of superconductors also depends on cooling conditions, transverse dimensions of both superconducting filaments and composite as the whole. These features are taken into account in the chapters discussed below.

1.4   INSTABILITIES OF THE SUPERCONDUCTING STATES

In spite of the success of the technology development of superconductors, their application in the form of monofilament wires or massive tapes is unsuccessful. Superconducting magnets based on such current-carrying elements may have the transition into the normal state at currents smaller than the critical current of a superconductor. Let us discuss the main reasons for such superconducting state instabilities.

During the manufacture of the first superconducting magnets, the problem of *degradation* of the current-carrying capacity of superconductors, used in their winding, appeared almost immediately (Wolgast et al. 1963). It was found that the values of the critical currents obtained on short samples were not reproduced on solenoids made on their basis. Consequently, the fracture currents of their superconducting state were unpredictably smaller than the corresponding critical currents obtained by measurements on short samples. Moreover, the degree of degradation could increase with the increase in the size of the transverse dimensions of the superconducting wire or solenoid. As a result, a premature transition of the superconducting winding to the normal state may occur at currents lower than the critical current of the superconducting wire. This is the uncontrolled transition of a superconducting current-carrying element to the non-superconducting state, named as *quenching phenomenon*. It may lead to its burnout or electrical breakdown with subsequent evaporation of helium and an increase in pressure in the cryostat up to the explosion.

Later it turned out that the degradation phenomenon is associated with another effect, which was called *training* effect. It is characterized by a gradual increase in the currents of the premature transition of the superconducting winding as a result of its successive transitions to the normal state.

Thus, premature destruction of superconductivity may be observed even when the superconductor is in the subcritical state (operating values of current, temperature and magnetic field do not exceed the corresponding critical values) because the superconducting state is metastable according to equation (1.16). Therefore, under specified conditions, it can be destroyed as a result of any external disturbance. They can lead to an irreversible transition of the entire superconducting magnetic system to the normal state.

As it was apparent, the superconducting magnets are subject to many external disturbances, which can initiate local destruction of a superconducting state of filaments or their current-carrying elements. Indeed, varying external magnetic field or transport current lead to a temperature increase of superconductor. Correspondingly, the critical current density of superconductor decreases because its $J_c(T)$ dependence, as a rule, has a negative slope. As a result, the magnetization of a superconductor also decreases leading to further penetration of the external magnetic field into a superconductor. At specific conditions, the energy loss caused by the decrease of the magnetization of superconductor can be sufficient to initiate its transition into a resistive state or even to a normal state. From the thermodynamics viewpoint, this process is due to the metastable nature of the vortex penetration. It leads to the so-called *electrodynamic instabilities* (*magnetic* or *current*) which are the reasons why the superconducting magnets may not be workable. They exist in LTS and HTS. Therefore, the instability problems of the superconducting state are one of the key problems arising at the designing of the superconducting magnet. Their studies are of interest not only in light of potential applications of superconductors but from a basic physical viewpoint. As a result, they are one of the peculiar phenomena of research interest in both LTS and HTS as their investigation allows one to understand the complexity of the vortex matter formation in the mixed-phase of these superconductors.

Magnetic instability is the first instability that was observed experimentally (Evetts et al. 1964, Hancox 1965, Wipf 1965). This phenomenon is also known as a *flux jump* which was the reason why the first superconducting magnets lost their superconducting state. It leads to avalanche flux redistribution toward a new equilibrium state, which is characterized by the corresponding penetration of magnetic flux into the superconductor. As a result, the corresponding sharp increase in its temperature happens. So, the screening currents may decrease to zero during flux jump.

Since the flux jump is a diffusion process of the redistribution of the magnetic field and heat flux in the superconductor, then it depends on the relation between magnetic ($D_m = \rho_f/\mu_0$) and thermal ($D_t = \lambda_s/C_s$) diffusivity coefficients. Here, C_s is the volumetric heat capacity of a superconductor, λ_s is the coefficient of its heat conductivity, ρ_f is the superconductor resistivity and μ_0 is the magnetic permeability of a vacuum. Quantitative analysis of the flux jump involves two approximations. First (so-called adiabatic approximation), the flux penetration is accompanied by practically adiabatic heating of the superconductor at $\Lambda = D_t/D_m = \mu_0\lambda_s(T_0)/C_s(T_0)\rho_f \ll 1$ since its thermal diffusivity is much lower than the magnetic diffusivity. Second, it is the so-called dynamic approximation when the spatial redistribution of flux remains practically fixed during the sharp heating at $\Lambda \gg 1$. These approximations are due to the difference between relatively high resistivity of the superconductor in the normal state and low resistivity of the nonsuperconducting materials such as copper or silver used as stabilizers.

Two different types of flux jumps can be observed, namely, global and partial. A global flux jump involves vortices into motion in the entire volume of the superconductor. A partial flux jump occurs in a small volume of the superconductor. The first turns the superconductor to the normal state. The second self-terminates when the temperature is still less than the critical temperature of the superconductor.

Usually, magnetic instabilities of nonideal type-II superconductors are investigated in the framework of the critical state model (Evetts et al. 1964, Hancox 1965, Wipf and Lubell 1965, Kremlev 1974, Duchateau and Turk 1975, Mints and Rakhmanov 1975 Mints and Rakhmanov 1982, Akachi et al. 1981, Gerber et al. 1993, Legrand et al. 1993, Muller and Andrikidis 1994, Milner 2001, Fisher et al. 2001, Chabanenko et al. 2002). In the framework of the developed theory, it is assumed that the flux jump is initiated by a small fluctuation of the temperature or external magnetic field. Then a stable magnetic field distribution in the superconducting slab exists, if the so-called adiabatic condition

$$\beta = \mu_0 a^2 J_c^2(T_0, B_a)/[C_s(T_0)(T_{cB} - T_0)] < 3 \tag{1.17}$$

takes place (Evetts et al. 1964, Hancox 1965, Wipf and Lubell 1965). It was obtained under the assumption that $\Lambda \to 0$ while supposing that the temperature of the superconductor is equal to coolant temperature before instability onset. Here, a is the half-thickness of the slab. The corresponding instability field for the first flux jumps after cooling of the superconductor in a zero magnetic field is equal to

$$B_f = \sqrt{3\mu_0 C_s(T_0)(T_{cB} - T_0)} \tag{1.18}$$

According to Wilson (1983), the allowance for the finite value of Λ leads to the corresponding correction in the right side of the criterion (1.17). Namely, it can be written as

$$\beta < \pi^2(1 + 2\sqrt{\Lambda})/4 \tag{1.19}$$

There should be a special emphasis on the fact that the isothermal approximation was used to formulate the magnetic instability conditions (1.17)-(1.19). Indeed, the critical current density, the heat capacity and the heat conductivity coefficient are determined at coolant temperature T_0. In other words, these criteria describe the stability conditions of the thermally isolated superconductor whose temperature is equal to the coolant temperature upon adiabatic penetration of magnetic flux in it.

Analysis of the magnetic instability conditions in the nonideal type-II superconductors taking into account of the flux-creep states described by the exponential equation of the V-I characteristic, written in the form

$$E = E_c \exp[J/J_\delta + (T - T_{cB})/T_\delta] \tag{1.20}$$

has been performed in (Klimenko and Martovetsky 1992; Zhou and Yang 2006) both for the LTS and HTS. In particular, it was proposed as the model in whose framework the instability criterion is written as follows

$$\frac{1}{S}\int_S E ds > E_m = \frac{hp}{S}\frac{J_\delta}{J_c|\partial J_c/\partial T|} \tag{1.21}$$

Here, h is the heat transfer coefficient, p is the cooled perimeter of the conductor, S is its cross-sectional area and T_δ and J_δ are the temperature and current creep parameters of the V-I characteristic (1.20). Accordingly, the temperature of superconductor T_i before instability is small and equals

$$T_i = T_0 + J_\delta/|\partial J_c/\partial T| \tag{1.22}$$

where the current density J is constant over the cross-sectional area of the conductor.

The physical sense of condition (1.21) is obvious. The superconducting state is stable when the average value of the electric field over its cross-sectional area is lower than the characteristic value E_m. However, it is easy to find that criterion (1.21), which determines the first flux jump field in superconductors during flux-creep, do not satisfy the limiting transition to criterion (1.17) at $h \to 0$ or T_δ and $J_\delta \to 0$ because the flux jump field will be equal to zero in these transitions. In other words, the limiting transition to the critical state model ($J_\delta \to 0$ or $T_\delta \to 0$) is not possible in the framework of such approximations.

Criteria (1.17)-(1.19) show parameters which affect the magnetic stability of the critical state. These are the superconductor thickness, its heat capacity, the critical current density and the so-called temperature margin that is equal to T_{cB}-T_0. It is seen that the flux jumps can be avoided by reducing the thickness of a superconductor. According to these criteria, the typical half-thickness of the superconductor is of the order of 10-100 μm at 4.2 K. Therefore, superconducting wires must be multifilament wires with filaments each of which do not lead to the flux jump. Accordingly, technical superconductors are produced as conductors in which superconducting core are located in a normal conducting matrix or coated by a normal metal. It is the so-called intrinsic method of stabilization of the superconducting state.

At the same time, the probability of the magnetic instability occurrence decreases with increasing temperature margin but it increases with increasing the critical current density. Besides, the higher the heat capacity, lesser is the instability probability. As a rule, magnetic instability of LTS and HTS is most probably in smaller magnetic fields and at operating temperatures close to liquid helium. In general, HTS is more stable to flux jumps than LTS due to the high-temperature margin.

In addition, there is a magnetic bond between the superconducting filaments located in the electro-conductive matrix. It negates the advantages of multifilament technics. As a result, multifilament technical superconductors may be subject to flux jumps as well as the monolithic superconductor. The magnetic bond between the filaments is avoided by twisting them so that they form screw trajectories along the longitudinal axis of the composite.

The simple stability criteria, described above, allow one to explain the influence of the properties of the superconductor on the magnetic instability onset. However, in general cases, it is not possible to write down the stability criteria in the analytical form even when the solution for the problem of magnetic instabilities in a cooled superconductor is reduced to the study of only the initial state development under the action of infinitely small perturbations. Activities undertaken in this analysis is based on the numerical analysis of the spectrum of the eigenvalues to determine such boundary parameter β_c when infinitely small perturbations are damped at $\beta < \beta_c$ and when the distribution of the shielding currents is already unstable to infinitely small perturbations at $\beta > \beta_c$ as formulated in (Wilson 1983, Gurevich et al. 1997c). This concept allowed one to formulate a general theoretical approach to the definition of the stability criteria of the critical state with respect to infinitely small perturbations. It is widely used in the study of flux jumps in both LTS (Wipf and Lubell 1965, Wipf 1967, Wilson et al. 1970, Kremlev 1974, Duchateau and Turk 1975, Mints and Rakhmanov 1975a, Akachi T. et al. 1981) and HTS (Legrand et al. 1993, Muller and Andrikidis 1994, Legrand et al. 1996, Khene and Barbara 1999, Milner 2001) and also allows one to explain the instabilities which are characteristic only for HTS (Fisher et al. 2001). (The features of the onset and development of magnetic instabilities are discussed in more details in Chapters 4 and 5).

The instability of the superconducting state may be initiated by changing transport current in superconductor or superconducting composite which is being supplied by the external power supply. This phenomenon is known as the *current instability*.

In this case, the problem of determining the conditions for the absence of unstable states includes the analysis of interrelated thermal and electrodynamics phenomena occurring in superconducting composites, as a result of which the maximum permissible values of currents and the corresponding values of the electric field and the temperature preceding the onset of current instability are determined. For the first time, the current instability conditions were formulated (Polak et al. 1973) under the assumption of an infinitely slow current charging into the Nb-Ti composite. It was found that the maximum current flowing steadily in a superconductor depends on the nonlinear part of the V-I characteristic of superconductor. In subsequent works, the corresponding estimates of the maximum permissible currents depending on external conditions were written out. For example, in Andrianov et al. (1983) equation for determining the current carrying capacity of a superconducting composite under different current charging conditions is obtained. In particular, the stable value of the current I_m charging at a constant rate dI/dt follows from the solution of the equation

$$\frac{I_m}{I_c(T_0, B_a)} + \ln\left(1 - \frac{I_m}{I_c(T_0, B_a)}\right) + \frac{1}{\alpha_{\text{eff}}} = 0, \quad \alpha_{\text{eff}} = \frac{\mu_0 \eta J_c(T_0, B_a)S}{4\pi h p J_\delta} \frac{dI}{dt}\left|\frac{dJ_c}{dT}\right|_{T_0, B_a} \tag{1.23}$$

for a round wire with the exponential equation of the voltage-current characteristic (1.4) placed in a constant external magnetic field. Here, η is the volume fraction of the superconductor in the composite. The physical meaning of the other parameters was determined earlier. The calculation was made under the assumption that the increase in the temperature of the wire relative to the temperature of the coolant before the current instability onset was negligible, and their own magnetic field is much smaller than the external field. It leads to a falling dependence of $I_m(dI/dt)$ which is observed in experiments.

The main reason for the current instability is the thermal imbalance between the heat release induced by transport current and heat removal to the coolant. It may occur in the partially or fully penetrated currents. In the first case, the current-carrying capacity of superconducting wires may be essentially smaller than their critical current. The characteristic quantities of the current instability phenomenon as a function of operating parameters are discussed in Chapter 6.

The effects of degradation and training can be also caused by high mechanical forces arising in the winding that carry a high-density current and induce a high magnetic field. Under their action, in the winding can be the movement of turns, their friction with each other, cracking of the compound and the slip of the winding relative to the frame. In addition, high mechanical stresses may be accompanied by corresponding plastic deformations of the wire which are intermittent at low temperatures. All of these occurrences lead to additional heat generation in the volume of the magnet that can be captured by any perturbation.

Due to these pulsed heat releases, local destruction of the superconducting properties of the wire is possible which leads to the so-called *thermal instability*. The analysis of their permissible level led to the concept of the so-called *critical energy of thermal perturbation or the minimum quench energy* (Wipf 1978). If the energy of heat release is larger than the critical energy, then it causes a thermal runaway (quench) while the energy of heat release that is smaller than the critical energy leads to a recovery of superconducting properties of the composite. In general, to find the critical energy, it is required to have a numerical solution of the thermal diffusion equation with highly nonlinear thermo-electric properties and temperature-dependent heat generation term (Chen and Purcell 1978, Schmidt 1978, Nick et al. 1979, Ishibashi et al. 1979, Anashkin et al. 1979, Keilin and Romanovsky 1982). They show that for a composite superconductor with DC current, the critical energy depends primarily on the cooling conditions and the length and duration of the thermal disturbance. It was obtained that the critical energy, as a role, increases with increasing values of the length and duration of the thermal disturbance (Keilin and Romanovsky 1982). It was shown that the most dangerous aspect is the short local heat release. At the same time, unlike the magnetic or current disturbances, for which measures to prevent them are clear, the unpredictable onset of thermal instabilities is possible in superconducting magnets. They arise due to the inevitable technological imperfection of the process of the superconducting magnet manufacturing, for example, when it is wound that may lead to its degradation or/and training. It follows from the above that measures to combat degradation and training should consist of reducing the energy of numerous mechanical disturbances. To do this, the winding should be as rigid as possible while securely fastening the entire winding volume during the winding process.

In order to avoid degradation, Stekly proposed a method of *cryogenic stabilization* (Kantrowitz and Stekly 1965, Stekly and Zar 1965). It stated that even when a current, which is equal to the critical current, flows through the cooled superconducting composite, its heating would not exceed the critical temperature of the superconductor. This state of stability is achieved by selecting the appropriate amount of stabilizing matrix and intensive cooling conditions. Then the nonsuperconducting area (the so-called *normal zone*) arising because of any perturbation will not propagate throughout the magnet since the presence of an appropriate amount of the stabilizing matrix and effective heat transfer can contribute to the subsequent return of the current-carrying element to the superconducting state. To estimate the degree of stabilization of a thin composite superconducting composite, a dimensionless parameter

$$\alpha = \frac{I_c^2(T_0, B_a)\rho_m}{hp[T_{cB}(B_a) - T_0]S(1-\eta)}$$

(1.24)

maybe used according to Stekly's idea. It is equal to the ratio of the Joule heat dissipation in the matrix at the critical current flowing through it and the heat flow to the coolant from the surface of the composite when its temperature is equal to the critical temperature of the superconductor. Here, ρ_m is the electrical resistivity of the matrix. If $\alpha \leq 1$, then the superconducting composite is *cryogenically* stabilized since at any current less than the critical one all Joule heat release in the composite is discharged into the coolant.

The method of cryogenic stabilization led to a significant increase in the reliability of superconducting magnets. However, in this case, the superconductor is a small percentage of the cross-sectional area of the current-carrying element (typically 5 to 15% depending on the size of the magnet). Nevertheless, the operating modes of large superconducting systems are still being designed based on the conditions of cryogenic stabilization. For the first time, the method of cryogenic stabilization was used by Morpurgo to create superconducting magnets of the bubble chamber in CERN.

The cryogenic stabilization was generalized by Maddock and his co-authors (Maddock et al. 1969). They took into account of the effect of the thermal conductivity $\lambda_k(T)$ of a thin wire with a current I smaller than the critical I_c—on the equilibrium condition in it of the heat release $G(T)$ and the heat transfer to the coolant $W(T)$—when λ_k, G and W are the functions of temperature T. It was shown that under the condition

$$\int_{T_1}^{T_2} \lambda_k(T)(W - G)dT = 0$$

(1.25)

when one end of the superconducting composite is at a temperature T_1 and the other end is at a temperature T_2, the equilibrium temperature distribution will be established inside the conductor regardless of the nature of the thermal disturbance. In the approximation, when the thermal conductivity does not depend on the temperature, this condition leads to the so-called *theorem of equal areas*. This name is due to the fact that in the temperature range from T_1 to T_2, the equilibrium state in the composite superconductor occurs if the areas described by the temperature-dependent Curves $G(T)$ and $W(T)$ are equal to each other. This theorem allows one to formulate the condition of the full thermal stabilization of a composite superconductor. In particular, in the quasilinear approximation (when the thermo-electric properties of the composite and the heat transfer coefficient do not depend on the temperature), it can be written in the form

$$\alpha i^2 - 2i + 1 = 0$$

(1.26)

If $i = I/I_c(T_0) = 1$, then Maddock's condition (1.26) passes into the Stekly condition. From the existence of the critical energies, the conditions (1.25) and (1.26) describe operating modes that are stabilized to arbitrary thermal disturbances. This conclusion is proved in Chapter 7.

Thus, thermal modes of the superconducting composites are characterized by stable and unstable regimes. They follow from the existence of the avalanche-like solution of equation (1.16). There exist different models allowing one to describe them. In this connection, it should be mentioned the so-called concept of the *minimum propagating zone* (the *MPZ concept*) which was suggested by Martinelli and Wipf (Martinelli and Wipf 1972) and later generalized by Wilson and Iwasa (Wilson and Iwasa 1978). It was one of the first thermal stability theories that used to explain the thermal instability of superconducting devices and to estimate their energy margin against short instantaneous thermal perturbations.

The MPZ concept is based on the existence of the steady nonuniform temperature distribution $T_{MPZ}(x)$ along with the superconducting composite with maximum temperature T_{max} in a central point of composite determined by the equality

$$\int_{T_0}^{T_{max}} \lambda_k(T)(W - G)dT = 0$$

which is similar to the equal-area condition (1.35) and obeys the boundary conditions $T|_{x \to \pm \infty} = T_0$. According to the MPZ concept, such temperature profile shows that any normal zone with an initial temperature distribution below the MPZ profile obtained for a given heat generation G and cooling heat flux W, that is at $T(x,t)|_{t=0} < T_{MPZ}(x)$ will not collapse since the cooling exceeds the heat generation. As a result, the superconductivity will be maintained. When an initial temperature profile is above that MPZ temperature distribution, that is if $T(x,t)|_{t=0} > T_{MPZ}(x)$, then the normal zone will grow in time and lead to the quenching. In other words, it is asserted that if a normal zone during thermal perturbation has an initial length smaller than the so-called MPZ length, then the temperature profile evolves quickly toward the steady profile. At the same time, the resistive area greater at initial time than the MPZ length will grow without limit and lead to the quench of a magnet. Thus, it was postulated that the energy margin can be estimated as the energy necessary to instantaneously create the MPZ temperature distribution. Its value for an infinitely long superconducting composite having temperature-dependent volumetric heat capacity $C(T)$ is determined as follows

$$E_{MPZ} = S \int_{-\infty}^{+\infty} dx \int_{T_0}^{T_{max}} C(T)dT$$

(Wipf 1978, Elrod et al. 1981). In the framework of the MPZ concept, this energy instantaneously deposited at a point which is regarded as the minimum energy and is necessary *in all conditions* to quench the composite superconductor. Thereby, this concept was called the Minimum Propagating Zone. The legality of such approximation is discussed in Chapter 7.

If the magnetic system is partially stabilized, that is, $\alpha > 1$ and therefore does not satisfy the cryostabilization condition then despite the taken measures of the thermal stabilization, its emergency transition to a normal state is possible when the energy of thermal perturbation exceeds the critical value. In this case, the nonsuperconducting area (normal zone) propagates inside the magnet in both longitudinal and transverse directions while increasing its temperature. Its maximum value depends on the stored energy, the material of the superconductor, the rate of the current attenuation in the magnet-power supply circuit and the time of its disconnection. If the released energy exceeds hundreds of kilojoules, then the central part of the nonsuperconducting area can be heated to high temperatures or high electrical voltage can occur (Smith 1963). As a result, the superconducting magnet can irreversibly collapse. To avoid these phenomena, special protection measures have been developed. Conventionally, they can be divided into two methods. First, it is necessary to create conditions under which the stored energy is evenly distributed throughout the entire volume of the magnet. In this case, measures should increase the propagation velocity of a normal zone to the released stored energy that does not lead to a localized burn. However, it is necessary to ensure that liquid helium is removed from the magnet to avoid the explosion of the cryostat. Second, it is necessary to ensure that the operating current is removed from the magnetic system as soon as possible. One of the effective methods of such protection is shunting the parts of the winding by external resistances. In the case of an irreversible transition of the magnet to the normal state, a part of the stored energy is dissipated on them. Both methods of protection must take into account the propagation velocity of the normal zone inside the winding. In any case, however, it is necessary to disconnect the power supply as soon as possible.

To summarize, the instability of the superconducting state shows that superconducting current-carrying elements of the magnet system are subjected to many disturbances of different nature. They depend on their size, shape and operating modes. Therefore, these peculiarities must be taken into account during the development of large-scale superconducting magnet systems, primarily with poorly cooled, massive current-carrying elements immersed in rapidly varying magnetic fields or carrying AC transport current. The corresponding stability conditions are discussed in detail in Chapters 4-8.

1.5   CONCLUSIONS

In this Chapter, the main results of the theories of fundamental and applied superconductivity were briefly discussed. The basic electrodynamic and thermal stability criteria of the superconducting state are written down. They permit to estimate the boundary of the stable operability of superconducting devices subjected to many external perturbations. However, they are formulated within the framework of the models that a priori set the temperature of the superconductor before the instability onset. At the same time, according to equation (1.16), the investigation of the electrodynamics states formation of low- and high-temperature superconductors must take into account their thermal prehistory. It is this general approximation that will allow to describe correctly essentially nonlinear phenomena in superconductors, in particular, to explain the main reasons of the onset and an avalanche-like development of the instabilities of different nature. The results of such analysis extend understanding of the character of macroscopic phenomena developing in the low- and high-temperature superconductors. Therefore, in Chapters 4-6 and sections 8.5 and 8.6, the formation mechanisms of the stable and unstable electrodynamics states of low- and high-temperature superconductors are discussed considering their thermal prehistory.

Since superconductors are subjected to instabilities, the superconducting magnet operating regimes can be destroyed if it is not correctly designed. Due to the large cost of the full-scale prototypes and test equipment, a possible strategy is to develop the theoretical models to understand the behavior of small-scale samples. It will allow not only the understanding of the stability mechanisms of superconducting state but enhance its stable operating conditions for superconducting devices. Therefore, the results presented below are focused on the simulation of the macroscopic electromagnetic and thermal phenomena in LTS and HTS. They may be used in developing the superconducting magnets and during analysis of conditions saving the superconductivity within a wide range of the change in their operating parameters at external perturbations different nature.

2

Macroscopic Electrodynamics of NonIdeal Type-II Superconductors: Isothermal Models

To optimize the design of superconducting magnet systems, it is necessary to predict the electromagnetic behavior of their superconducting current-carrying elements to analyze the field and current distributions and to calulate the electromagnetic losses and forces. Using various types of V-I characteristics, the superconducting state forming can be investigated in terms of the macroscopic models obtained by averaging the microscopic currents over Abrikosov's vortices as mentioned above. For this aim, it is convenient to present the Maxwell equations as equations for the local electric field $E = v \times B$, which is caused by the motion of vortices with velocity v. Here, B is the average magnetic induction (flux density) in a volume element. This approximation is valid at $B \gg B_{c1}$. Let us consider the possible models, which can be used to describe the macroscopic electrodynamic states of nonideal type-II superconductors.

2.1 BEAN APPROXIMATION OF THE MAGNETIC FLUX DIFFUSION IN SUPERCONDUCTORS ($\rho_f \to \infty$)

As above-mentioned, Bean made the application of the macroscopic theory to explain the hysteresis phenomenon observed in hard superconductors. It is a quasi-static field theory based on the solution of equations (1.12) with an ideal V-I characteristic (1.1). According to Bean's assumption, a superconductor is considered as an isothermal isotropic slab and the screening current density is either zero or equal to the critical current density J_{c0} = const at operating temperature T_0 and external magnetic field B_a. Their orientations follow from the right-hand law. Magnetic flux penetrates the superconductor while changing consequently the magnetic field in it. The Meissner phase and the dependence of the critical current density on the magnetic induction are not considered. In this case, it is said that the superconductor is in the critical state since the current with the critical density flows in a superconductor. Correspondingly, the model is called the critical state model (CSM).

To understand the features of the magnetic field and current distributions in a superconductor defined by the Bean approximation, let us consider an initially nonmagnetized superconducting slab without transport current having thickness $2a$ in the x-axis (Figure 2.1a). Assume that it is placed in a uniform external magnetic field B_a that is increasing with a constant sweep rate $\dot{B}_a = dB_a/dt$ from zero. Suppose that the y-z plane extends infinitely and the external magnetic field is along the z-axis. Therefore, the induced magnetic field in the superconductor is also along the z-axis, and its distribution is symmetrical. However, induced currents and electric fields are bipolar (Figure 2.1b). Then the system of equation (1.12) converts into the following system

$$\frac{\partial B}{\partial x} = \pm \mu_0 J_{c0}, \quad \frac{\partial E}{\partial x} = -\frac{\partial B}{\partial t}, \quad (-a < x < a, -\infty < y < \infty, -\infty < z < \infty) \qquad (2.1)$$

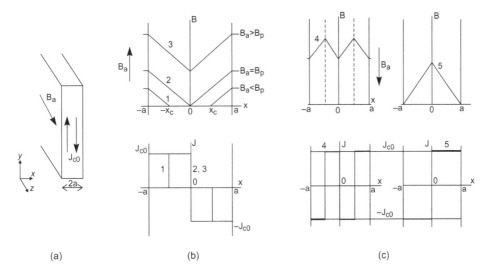

(a)   (b)   (c)

FIGURE 2.1 Distributions of a magnetic field and screening current in a slab (a) according to the Bean model during increasing (b) and decreasing (c) external magnetic field.

One-dimensional (1D) isothermal isotropic models like (2.1) are often used to describe the macroscopic electrodynamic properties of nonideal type-II superconductors. Such simple approximation is used to investigate the dynamics of vortices in superconductors (for example, van der Beek et al. 1991, Gilchrist and van der Beek 1994, Gurevich 1995a, Gurevich and Brandt 1997b, Rhyner 1993, Vinokur et al. 1991) for the magnetization studying of superconductors (Ding et al. 2000, Gurevich 1995a, Gurevich and Brandt 1997b, Yamafuji and Mawatari 1992) and during analysis of their magnetic susceptibility (Gurevich and Brandt 1997b, Qin and Yao 1996). To study the electromagnetic properties of superconductors of arbitrary shape, the general simulation algorithms were proposed by Brandt (1996) and Labusch and Doyle (1997). The magnetic properties of superconductors for various geometries (for instance, strips, discs or rings) and field orientation (parallel or perpendicular) may be computed by these methods. Moreover, to investigate the magnetization of superconductors in 2D or 3D approximations, the finite difference, finite element and boundary element methods including commercial software may be used (see examples, Amemiya et al. 1998, Lahtinen et al. 1996, Vinot et al. 2000, Wakuda et al. 1997, Wetzko et al. 1995). However, in many practical cases, one-dimensional models may be used. As a result, 1D consideration allows one to understand the basic physical features of the electrodynamic phenomena solving simplified equations without a large volume of computations. From this point of view, the possible 1D solution in the closed-form is convenient to evaluate the experiments even for the realistic geometry of superconductors. Therefore, the flux-creep features of the nonlinear diffusion of the electromagnetic field are discussed below for the parallel geometries by taking different models of V-I characteristics into consideration.

The isothermal approximations are also usually used in the study of the energy losses in superconductors (for example, Carr 1983, Amemiya et al. 1998, Nibbio et al. 2001, Paasi and Lahtinen 1998, Rhyner 1993, Stavrev et al. 2001, Stavrev et al. 2003). Note that during similar simulations it is assumed that the permissible level of the heat release, at which superconductivity is maintained, should not lead to the heating of the superconductor to a temperature exceeding its critical value, regardless of the nature of the change in the external magnetic field and/or transport current. The isothermal approximation is true at intense heat transfer conditions, relatively low rates of the change of external magnetic field or charging current and small transverse dimensions of a superconductor. However, its use should be strictly justified, that is it is necessary to give a correct answer about the range of the stable temperature increase of the superconductor.

According to equations (2.1) and the boundary condition $B(a) = B_a(t) = \dot{B}_a t$, the distributions of the magnetic and electric fields in the interior of the slab are given by

$$B(x,t) = \begin{cases} B_a(t) - \mu_0 J_{c0}(a-x), & 0 < x_c(t) < x < a \\ 0, & -x_c(t) < x < x_c(t) \\ B_a(t) - \mu_0 J_{c0}(a+x), & -a < x < -x_c(t) < 0 \end{cases}$$

$$E(x,t) = \begin{cases} \dot{B}_a(t)(x_c(t) - x), & 0 < x_c(t) < x < a \\ 0, & |x| \le x_c(t) \\ -\dot{B}_a(t)(x + x_c(t)), & -a < x < -x_c(t) < 0 \end{cases} \tag{2.2}$$

during partially penetrated states (lines 1). Here, the penetration coordinates of the external magnetic field x_c are defined as follows $x_c(t) = a - B_a(t)/(\mu_0 J_{c0})$. It may be also rewritten in the form $x_c(t) = a - x_0(t)$, where $x_0(t) = B_a(t)/(\mu_0 J_{c0})$ is the penetration depth of an external magnetic field.

As shown in (2.2), there are two regions ($-a < x < -x_c$ and $x_c < x < a$) in which the magnetic field and current exist in the partially penetrated modes. Accordingly, the magnetic field and the screening current are zero in the region $-x_c(t) < x < x_c(t)$.

At a certain moment, the induced magnetic field reaches the center of a slab (lines 2 in Figure 2.1b). This magnetic field is called the fully penetrated magnetic field B_p, and it is equal to $B_p = \mu_0 J_{c0} a$ according to the solution (2.2). The magnetic field occupies the whole volume of the slab, and the current density inside the slab is equal to the critical current density $J = \pm J_{c0}$ (lines 2 in Figure 2.1b).

If the external magnetic field B_a increases further, then the magnetic field in the slab linearly increases at each point (lines 3 in Figure 2.1b). At the same time, the screening current density in the slab is constant according to the Bean model and it equals $\pm J_{c0}$.

Thereby, the evolution of the magnetic moment of the slab during sweep-rate mode is described as follows

$$M_c(t) = \frac{1}{\mu_0}\left[\frac{1}{2a}\int_{-a}^{a} B(x,t)dx - B_a(t) \right] =$$

$$= \begin{cases} -\left\{ \frac{1}{\mu_0}\left[(B_a - B_p)\left(1 - \frac{x_c}{a}\right) + \frac{B_p}{2}\left(1 - \frac{x_c^2}{a^2}\right) - B_a \right] = \frac{B_a}{\mu_0} - \frac{B_a^2}{2\mu_0 B_p}, & B_a(t) \le B_p \right. \\ J_{c0}a/2, & B_p < B_a < B_{c2} \end{cases} \tag{2.3}$$

The presented solution allows one to find important characteristics of the stable superconducting states taking place during the magnetic field penetration in the nonideal type-II superconductors, namely, the velocity of the flux penetration $V_c = dx_c/dt$ and hysteresis losses $G_c = \int_0^t \int_{x_c}^a EJ_{c0}dx\,dt$. Using (2.2), it is easy to find the following relationships

$$|V_c| = \frac{\dot{B}_a}{\mu_0 J_{c0}}, \quad G_c = \begin{cases} \dfrac{B_a^3}{6\mu_0 B_p}, & B_a \le B_p = \mu_0 J_{c0} a \\ \dfrac{B_p}{2\mu_0}\left(B_a - \dfrac{2}{3}B_p \right), & B_a > B_p \end{cases} \tag{2.4}$$

If an external magnetic field starts to decrease, then new bipolar currents of opposite polarity are induced. Such states are described first by lines 4 and then by lines 5 in Figure 2.1c. Here, the state described by the Curves 5 appears according to the hysteresis phenomenon in nonideal type-II superconductors when a nonzero magnetic field exists in a superconductor at a zero external magnetic field.

The written expressions describe modes of the superconductor in the simplest cases of the monotonic increase/decrease regimes of the external magnetic field. Using the CSM, it is easy to find energy losses for other modes. For example, if B_a increases with a constant sweep rate from zero to B_k that is less than the full penetration field B_p and then decreases to zero. For this mode, the energy losses are described by the formulae

$$G_c = \begin{cases} \dfrac{B_k^3}{6\mu_0 B_p}, \ \dot{B}_a > 0 \\ \dfrac{B_k^3}{24\mu_0 B_p}, \ \dot{B}_a < 0 \end{cases}$$

If the external magnetic field changes periodically from $B_a = B_0 - B_k$ to $B_a = B_0 + B_k$, the volume density of losses for the period of this cycle is determined by the relations

$$G_c = \begin{cases} \dfrac{2B_k^3}{3\mu_0 B_p}, \quad B_k \leq B_p \\ \dfrac{2B_p}{\mu_0}\left(B_k - \dfrac{2}{3}B_p\right), \quad B_k > B_p \end{cases}$$

These formulae show that the losses arising in superconductors placed in variable external magnetic fields are hysteresis that is they depend on an initial distribution of the magnetic fields in a superconductor.

The Bean approximation also allows one to describe many other practically important electrodynamic states of nonideal type-II superconductors. For example, let us discuss the states when the slab is charged by transport current $I(t)$ without an external magnetic field ($B_a = 0$) which called the self-field mode. Let us assume that current flows along the z-axis (Figure 2.2a). In this case, the system (1.12) also leads to 1D differential equations, like (2.1) which can be easily solved. Figure 2.2 demonstrates the corresponding distribution of the magnetic field and the transport current.

During these operating modes, the transport current begins to penetrate first from the slab surface (partially penetrated mode) with increasing value. The induced magnetic field correspondingly increases. As shown in Figure 2.2b, the current in the slab distributes symmetrically in the cross-sectional of a superconductor (lines 1). However, the induced magnetic field distribution inside the slab is antisymmetric. Correspondingly, positive and negative values correspond to the magnetic field distributions at the right and left-hand sides of the slab, respectively. The penetrated current density is equal to its critical value J_{c0} in each region of $-a < x < -x_c(t)$ and $x_c(t) < x < a$ through which the magnetic field penetrates, respectively. The transport current and magnetic field are zero in the region $-x_c < x < x_c$. The transport current and magnetic field may increase to a certain value which was defined above as a fully penetrated state. This limiting state is shown by lines 2 in Figure 2.2b at which the current and magnetic field reaches simultaneously to the slab center from both left and right sides. At this moment, the transport current is equal to its critical current $I_c = 2J_{c0}a$. Correspondingly, the magnetic field $B_p = \mu_0 J_{c0}a$ produced by this current is the fully penetrated field. It should be underlined that the critical current I_c and the fully penetrated field B_p are the maximum allowed values, which characterize such limiting electrodynamic states of the nonideal type-II superconductor in the framework of the CSM.

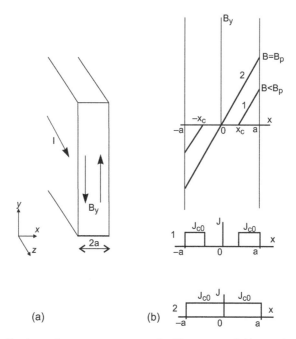

FIGURE 2.2   Distributions of transport current and self-magnetic field according to the Bean model.

Thus, the distributions of the current and magnetic field in the nonideal type-II superconductor can be simply described by the Bean model. As a result, it significantly simplifies the mathematical simulations, and it is often used to study the electromagnetic properties of nonideal type-II superconductors. At the same time, the critical current density depends on the magnetic field in practical cases. Therefore, other critical state models are used. As above-mentioned, one of them is the Anderson-Kim model. Although there are some discrepancies between experimental results and predictions leading from the Bean and Anderson-Kim models, they are still most widely used because of their simplicity and intuitiveness allowing easy understanding of the basic physical features of the electromagnetic phenomena in the nonideal type-II superconductors.

2.2   THE SWEEP-RATE PROBLEM FOR SUPERCONDUCTORS WITH DIFFERENT TYPES OF NONIDEAL VOLTAGE-CURRENT CHARACTERISTICS. THE SPATIAL-TEMPORAL CHARACTERISTICS OF THE PARTIALLY PENETRATED MODES

In the previous paragraph, the forming dynamics of the stable superconducting states in the case of the ideal V-I characteristic was considered, which is based on a jump change in the differential resistivity of the superconductor from zero to infinitely large after reaching the critical current. Let us discuss the physical features of the magnetic flux diffusion in nonideal type-II superconductors during partially penetration mode when their differential resistivity depends on the type of the nonlinearity of V-I characteristics.

Consider an ideally cooled half-infinite superconductor ($0 < x < \infty$, $-\infty < y < \infty$, $-\infty < z < \infty$) placed in the external magnetic field B_a ($B_a \gg B_{c1}$) which is parallel to its surface in the z-direction. Formally, this mode corresponds to the case when the center of the coordinate system is located on the left side of the slab as shown in Figure 2.1a with infinitely large geometric dimensions in y and z directions and half-infinite in x direction. As above-mentioned, let us assume that the external magnetic field on the surface is zero at the initial time and then it increases with a constant sweep rate $\dot{B}_a$. Let us suppose for simplicity that the critical current density does not depend on the magnetic induction. In this case, the Maxwell equations (2.1) are easily reduced to the following equation

$$\mu_0 \partial E/\partial t = \rho_d \partial^2 E/\partial x^2 \tag{2.5}$$

that describes the one-dimensional evolution of the induced electric field $E = E_y(x, t)$. Here, ρ_d is the differential resistivity of the superconductor which is equal to

$$\rho_d = \begin{cases} \rho_f \\ n(E_c/E)^{1/n} E/J_{c0} \\ E/J_\delta \end{cases} \tag{2.6}$$

for the flux-flow and flux-creep states described by the power and exponential equations, respectively.

Taking into account of the character of the magnetic field variation, the necessary boundary and initial conditions describing the sweep-rate modes are written as follows

$$\partial E/\partial x(0, t) = -\dot{B}_a, \quad E(\infty, t) = 0, \quad E(x, 0) = 0 . \tag{2.7}$$

The first boundary condition takes into account of the rise of the external magnetic field growing at a constant rate, and the second one is the apparent condition of the limited solution for a medium with unlimited geometric dimensions.

For the correct analysis the extra condition

$$\mu_0 \int_0^\infty J(x, t) \, dx = B_a(t), \tag{2.8}$$

which follows from Ampere's law is also needed. It is written for the total volume of the superconductor in the x-direction ($0 < x < \infty$). Here, the quantity of the current density $J(x, t)$ in the superconductor comes from the corresponding equations of the V-I characteristics.

Note that the conditions (2.7) and (2.8) are initially written under the assumption that the screening current flows through the entire volume of the superconductor. From the formal viewpoint, this assumption is justified since the finite electric field in the creep arises in the superconductor long before the critical current is reached and leads to the appearance of the resistance in it. Therefore, it is assumed that the stable electric field arising in the superconductor leads to an instantaneous penetration of the external electromagnetic field into it.

The phenomena described by equation (2.5) with conditions (2.6)-(2.8) are characterized by the space-temporal scales of their development like all typical diffusion phenomena. Besides, the shape of the V-I characteristic also affects the features of the macroscopic magnetic flux dynamics. The corresponding features are discussed below in detail. However, some qualitative aspects of this phenomenon may be understood using simple dimensional analysis. Let us discuss them for the problem under consideration.

As it is known, the change of the electromagnetic field in the conductor is characterized by the diffusion time of the magnetic flux which determines the boundary between the transient and quasi-steady states of the diffusion phenomenon considered. This quantity is defined as $t_x = \mu_0 L_x^2/\rho_x$, where L_x and ρ_x are the characteristic values of the length and resistivity of the conductor. Their choice depends on the features of the phenomenon under consideration. It is convenient to choose the following L_x value

$$L_x = E_x/\dot{B}_a \tag{2.9}$$

in the case of diffusion of magnetic flux in a semi-infinite superconductor, which is placed in a constantly increasing external magnetic field, that is during its partially penetrated state. It follows from the first boundary condition (2.7). Here, E_x is the characteristic value of the electric field. Then, the corresponding characteristic time is equal to

$$t_x = \mu_0 E_x^2 / (\rho_x \dot{B}_a^2) \tag{2.10}$$

The given constant permits to estimate the evolution character of the sweep-rate states. Namely, the influence of the flux-creep on the electric field distribution in a superconductor will be slight at $t \gg t_x$. Under this condition, equation (2.5) has the quasi-static form $\partial^2 E / \partial x^2 \approx 0$. In particular, this approximation will be more suitable at relatively high sweep rates according to (2.10). However, the quasi-steady Bean approximation may lead to inadequate analysis of the electrodynamic states of superconductors located in weakly variable magnetic fields. Besides, this formula depicts the effect of the resistivity on superconductor. Namely, if a superconductor has relatively large resistivity during flux-creep, then the electric field distribution does not practically depend on its value. From the formal point of view, the electromagnetic states of such superconductors are close to the critical state. However, the electrodynamic phenomena in the superconductor will have a nonlinear character at the relatively small value of the resistivity. These conclusions show the nontrivial influence of the differential resistivity of superconductors on the character of the electromagnetic field diffusion during flux-creep because of the flux creep resistivity of a superconductor is a function of the electric field. Indeed, the electric field is ultra-low at the initial sweep-rate stage. Therefore, the differential resistivity of a superconductor is also small, and the magnetic diffusion phenomenon may have nonlinear character unlike the critical state model. However, the differential resistivity of a superconductor increases with the time. Then, the quasi-static regime may be reached after ample time. Formally, it means that the critical state could be applied during this stage. At the same time, to describe correctly the flux diffusion, taking into consideration of the possible initial nonlinear sweep-rate stage, the extended critical state model must be used. This possibility is discussed in detail below for the different models of the flux-creep when the nonideal type-II superconductors are in the applied magnetic field or under the varying transport current.

2.3　SCALING DYNAMICS OF THE ELECTROMAGNETIC FIELD DURING VISCOUS FLUX-FLOW MODE (ρ_f = const)

Let us discuss the features of the magnetic flux penetration into a superconductor with the V-I characteristic described by the viscous flux-flow model (1.2). Let us introduce the dimensionless variables $e = E/E_x$, $X = x/L_x$, $\tau = t/t_x$. In this case, it is convenient to select the following characteristic values $\rho_x = \rho_f$ and $E_x = J_{c0}\rho_f$. Then, $L_x = J_{c0}\rho_f/\dot{B}_a$, $t_x = \mu_0 J_{c0}^2 \rho_f / \dot{B}_a^2$. The physical meaning of L_x and t_x is discussed below. As a result, the initial-boundary problem (2.5)-(2.8) is transformed into a dimensionless form

$$\frac{\partial e}{\partial \tau} = \frac{\partial^2 e}{\partial X^2} \tag{2.11}$$

$$\frac{\partial e}{\partial X}(0, \tau) = -1, \qquad e(\infty, \tau) = 0, \qquad e(X, 0) = 0 \tag{2.12}$$

$$\int_0^\infty (1 + e)\, dX = \tau \tag{2.13}$$

Let us find the solution to the problem (2.11)-(2.13) in the framework of the so-called scaling approximation using the Boltzmann transformation. To do this, introduce the invariants

$$V(Z) = \partial e / \partial X, \quad Z = X / \sqrt{\tau} \tag{2.14}$$

for which the partial differential equation (2.11) is converted into an ordinary differential equation

$$\frac{d^2 V}{dZ^2} + \frac{Z}{2} \frac{dV}{dZ} = 0$$

with boundary conditions $V(0) = -1$ and $V(\infty) = 0$. The following function

$$V(Z) = C_0 \, erf(Z/2) - 1$$

is its solution. Here,

$$erf(y) = \frac{2}{\sqrt{\pi}} \int_0^y \exp(-t^2) \, dt$$

is the integral of errors, C_0 is the positive constant with a value less than one. It was not written due to its bulkiness.

It can be proved that $V(Z) < 0$. Therefore, $\partial e / \partial X < 0$. As a result, the length of the region with the positive values of $e(X, \tau)$ is finite. Thus, there is a moving boundary (front) of magnetization $x_0 \, (0 < x < x_0)$ during the viscous flux-flow of the magnetic flux. This conclusion should be taken into account in the condition (2.13). As a result, the distribution of the electric field in a superconductor in the dimensional form is described by the following expression

$$E(x, t) = \dot{B}_a (x_0 - x) + \dot{B}_a x_0 \left[\Psi(\xi, \omega) - \Psi(1, \omega) \right], \quad 0 < x < x_0(t)$$

Here,

$$\xi = x/x_0, \, \omega = 0.5 x_0(t) \sqrt{\mu_0/(\rho_f t)}, \, \Psi(\xi, \omega) = \xi \, erf(\xi\omega) + \exp(-\xi^2\omega^2)/\omega\sqrt{\pi}$$

and the coordinate of the moving front $x_0(t)$ is a solution of the equation

$$\frac{\mu_0 x_0 J_{c0}}{\dot{B}_a} + \frac{\mu_0 x_0^2}{2\rho_f} - \frac{\mu_0 x_0^2}{\rho_f} \left[\Psi(1, \omega) - \int_0^1 \Psi(Z, \omega) \, dZ \right] = t \tag{2.15}$$

It may be proved that for any finite value of ρ_f the distribution of the electric field induced in the superconductor during the viscous flow of the magnetic flux is close to the Bean distribution of the electric field due to small values of Ψ. In this case, equation (2.15) can be rewritten in a simplified form

$$\frac{\mu_0 x_0 J_{c0}}{\dot{B}_a} + \frac{\mu_0 x_0^2}{2\rho_f} = t$$

Using the characteristic values of the length and time introduced above, it takes the form

$$\frac{x_0}{L_x} + \frac{x_0^2}{2L_x^2} = \frac{t}{t_x} \tag{2.16}$$

The solution of equation (2.16) leads to the following nonlinear moving law of the magnetization boundary

$$x_0(t) = L_x \left(\sqrt{1 + 2t/t_x} - 1 \right)$$

Accordingly, its propagation velocity is equal to

$$V_f = \frac{dx_0}{dt} = \frac{|V_c|}{\sqrt{1 + 2t/t_x}}$$

These expressions show that the conditions $x_0 \ll L_x$ and $V_f \approx |V_c|$ are satisfied at $t \ll t_x$, that is the coordinate of the magnetization boundary changes as $x_0(t) \approx |V_c| t$ and the magnetization front moves with a velocity of $V_f \approx \dot{B}_a / \mu_0 J_{c0}$. Therefore, the electrodynamic states of superconductors with the viscous flux-flow V-I characteristic are close to the states following from the CSM at $t \ll t_x$. In the other hand, the conditions $x_0 \gg L_x$ and $V_f < |V_c|$ are satisfied at $t \gg t_x$ and the following approximate expression can be used

$$x_0(t) \approx \sqrt{2\frac{\rho_f t}{\mu_0} - \frac{\rho_f J_{c0}}{\dot{B}_a}},$$

to define x_0, that is the Bean approximation is not performed at $t \gg t_x$. In general, the values V_f decrease monotonically with time and are always less then $|V_c|$. Their equality is satisfied only at $t = 0$. Thus, the quantities L_x and t_x are the characteristic spatial-temporal scales separating the linear (Bean) and nonlinear (non-Bean) phases of the magnetic flux penetration in the framework of the viscous flux-flow model.

From the written solution it is easy to find the field of the fully penetrated field $B_{f,p}$ and time $t_{f,p}$ over which it is reached at the viscous flux-flow state. For a superconducting slab with a thickness of $2a$, they are equal to

$$B_{f,p} = B_p\left(1 + \frac{\mu_0 a^2 \dot{B}_a}{2\rho_f B_p}\right), \quad t_{f,p} = \frac{\mu_0 a^2}{2\rho_f} + t_p$$

where B_p and $t_p = \mu_0 J_{c0} a/(dB_a/dt)$ are the magnetic field and time of the full penetration, respectively, in the framework of the CSM. It is seen that the finite value of the differential resistivity of a superconductor leads to an increase in the values of the magnetic field of full penetration and the forming time of this mode due to the lower value of the penetration velocity of magnetization front. That is why this mode is called the viscous flux-flow.

Overall, the viscous flux-flow model demonstrates the basic features of the diffusion phenomena occurring in superconductors with the finite differential resistivity. First, if this quantity is high, then the forming of the superconducting states is almost close to the states described by the Bean approximation. However, the flux-flow model leads to the nonlinear nature of the superconducting states' formation at small values of ρ_f as follows from equation (2.15). The corresponding influence of the differential resistivity on the development of the electrodynamic states of a superconductor will be also observed at the flux-creep by taking into account that ρ_f is small at the initial stage, but it increases with increasing of the electric field. As follows from equation (2.16), the forming of the superconducting states will also depend on the sweep rate of the external magnetic field and the nonlinearity character of the V-I characteristic. As a consequence of the dependence of ρ_f on the induced electric field, the Bean approximation describes the electrodynamics of the LTS more accurately than the same by the HTS. It is explained by the fact that the V-I characteristics of the LTS have sharper rise than that of the HTS.

2.4 DIFFUSION FEATURES OF THE SCREENING CURRENTS AT THE FLUX-CREEP AND LOCAL CHANGE OF THE DIFFERENTIAL RESISTIVITY OF SUPERCONDUCTORS WITH THE POWER OR EXPONENTIAL VOLTAGE-CURRENT CHARACTERISTICS

To analyze the flux-creep states considering equation (2.5), the second and third terms of the relations (2.6) and conditions (2.7) and (2.8), let us use the dimensionless variables $e = E/E_x$, $X = x/L_x$ and $\tau = t/t_x$. Here, $E_x = E_c$ and $\rho_x = E_c/J_{c0}$ are the characteristic quantities for a superconductor with the power V-I characteristic and $E_x = E_c$, $\rho_x = E_c/J_\delta$ for a superconductor with the exponential V-I characteristic. Then, it is easy to get $L_x = E_c/\dot{B}_a$ and $t_x = \mu_0 J_{c0} E_c/\dot{B}_a^2$ for the power V-I characteristic and $L_x = E_c/\dot{B}_a$, $t_x = \mu_0 J_\delta E_c/\dot{B}_a^2$ for the exponential V-I characteristic. Accordingly, the variation of the electric field in space and time is the solution of the following dimensionless initial-boundary problems

$$\frac{\partial^2 e}{\partial X^2} = \begin{cases} \dfrac{e^{(1-n)/n}}{n}\dfrac{\partial e}{\partial \tau} & (2.17) \\[2ex] \dfrac{1}{e}\dfrac{\partial e}{\partial \tau} & (2.18) \end{cases}$$

$$\partial e/\partial X(0,\tau) = -1, \quad e(\infty,\tau)0, \quad e(X,0)=0 \tag{2.19}$$

with extra conditions

$$\int\limits_0^\infty e^{1/n}dX = \tau \tag{2.20}$$

$$\int\limits_0^\infty (1+\delta\ln e)\,dX = \delta\tau, \quad \delta = J_\delta/J_{c0} \tag{2.21}$$

for the V-I characteristics of the power and exponential types, respectively.

Let us find the dimensionless distribution of the electric field $e(X,\tau)$ in the class of the scaling approximations using the group properties of equations (2.17) or (2.18).

The scaling solution for superconductors with the power V-I characteristic is convenient to find in the form $V(Z) = e/\tau^p$ where $Z = X/\tau^p$, $p = n/(n+1)$. In this case, equation (2.17) leads to the integration of the following ordinary differential equation

$$(n+1)d^2V/dZ^2 + ZV^{(1-n)/n}dV/dZ - V^{1/n} = 0 \tag{2.22}$$

that must satisfy the conditions

$$dV/dZ\big|_{Z=0} = -1, \quad V(\infty) = 0 \tag{2.23}$$

$$\int\limits_0^\infty V^{1/n}dZ = 1 \tag{2.24}$$

The family of the phase trajectories of equation (2.22) is shown in Figure 2.3. In this case, instead of the boundary problem (2.22) and (2.23), the Cauchy problem was numerically solved. In this approximation, the initial value of the dV/dZ was specified at the point $Z = 0$ according to the condition (2.23) and the value $V(0)$ was varied. In addition, the condition (2.24) was also checked. The curves presented show the features of the solution of equation (2.22). It is described only by Curve 3. Indeed, Curves 1 and 2 do not satisfy the condition (2.24), and Curves 4 and 5 have positive values dV/dZ. Correspondingly, these values of $V(Z)$ do not satisfy the limiting condition at $Z \to \infty$. Therefore, according to Figure 2.3, the invariant $V(Z)$ has a negative slope and as a result it exists in the finite region $0 \le Z \le Z_0$. So, the second boundary condition (2.23) and condition (2.24) should be rewritten as follows

$$V(Z_0) = 0 \tag{2.25}$$

$$\int\limits_0^{Z_0} V^{1/n}dZ = 1 \tag{2.26}$$

Equality (2.26) not only allows one to find the value Z_0 but is also the unambiguity condition of the solution of equation (2.22).

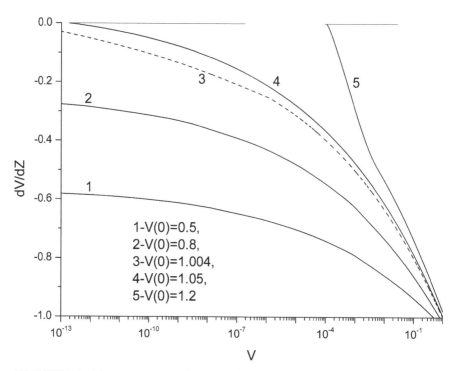

FIGURE 2.3   Phase trajectories of equation (2.22) at $n = 10$ and different values of $V(0)$.

The existence of the finite value Z_0 at which $V(Z_0) = 0$ determines the specifics of the magnetic flux penetration into the superconductors with a power V-I characteristic. First, the screening currents originating in the superconductor by any variation of the external magnetic field will exist in the finite region $0 \le X \le X_0(\tau) = Z_0 \tau^p$ with a moving boundary $X_0(\tau)$ until the full penetration of magnetic flux is achieved. Second, the diffusion of the screening currents in a superconductor with a power V-I characteristic occurs with a finite rate as in the CSM. In this case, not only $e(X_0, \tau) = 0$ but the other conditions of the variation of $e(X_0, \tau)$ exist on the moving boundary of the magnetization region. Let us formulate them. To do this, let us differentiate equation (2.20) by time in which the upper limit is replaced by $X_0(\tau)$ in accordance with the above-mentioned feature. As a result, it is easy to get $e^{1/n} dX_0/d\tau + \partial e/\partial X \big|_{X=X_0} = 0$. Since $e(X_0, \tau) = 0$, then $\partial e/\partial X(X_0, \tau) = 0$. Therefore, $dV/dZ = 0$ at $Z = Z_0$. Then according to (2.22), one may find that $d^k V/dZ^k$ ($k = 2, 3\ldots$) are also equal to zero at $Z = Z_0$. Therefore, the invariant $V(Z)$ at $Z \ge Z_0$ and thus $e(X_0, \tau)$ at $X \ge X_0$ continuing to be the identical zero, that is the electromagnetic field penetration in the superconductor with the power V-I characteristic occurs against the background of an unperturbed state.

Thus, the distributions of the electric and magnetic fields as well as the density of the screening current and the differential resistivity of a superconductor in the scaling approximation are described by the expressions

$$E(x, t) = E_a(t) V(Z), \quad B(x, t) = B_a(t) \left[1 - \int_0^Z V^{1/n} dy \right],$$

$$J(x, t) = J_{c0} \left[\frac{V(Z) E_a(t)}{E_c} \right]^{1/n}, \quad \rho_d(x, t) = \frac{n E_a(t)}{J_{c0}} V(Z)^{(n-1)/n}$$

where $V(Z)$ is the solution to the problem (2.22)-(2.26) and $Z = (x/L_x)(t_x/t)^{n/(n+1)} = (x/L_x) E_c/E_a(t)$, $x_0(t) = Z_0 L_x (t/t_x)^{n/(n+1)} = Z_0 L_x E_a(t)/E_c$. Here $E_a(t) = E_c (t/t_x)^{n/(n+1)}$.

This solution exists in the area $0 < x < x_0(t)$ with a moving front $x_0(t)$ penetrating the superconductor at the velocity

$$\frac{dx_0}{dt} = \frac{Z_0}{\dot{B}_a} \frac{dE_a}{dt}.$$

The written relations show that the electrodynamic phenomena occurring in superconductors during flux-creep depend on the variation of the electric field on its surface E_a. Moreover, the specifics conditions $E = 0$, $B = 0$, $\partial^k E/\partial x^k = 0$, $\partial^k B/\partial x^k = 0$ $(k = 1, 2...)$ take place at the moving front.

Let us estimate the asymptotic character of the $V(Z)$ change in the vicinity of the point Z_0. As follows from Figure 2.3, the values $V(Z)$ are small near Z_0. Then, equation (2.22) can be written as follows

$$d^2V/dZ^2 + \frac{n}{n+1}\frac{d(ZV^{1/n})}{dZ} \approx 0$$

This equation with the conditions $V(Z_0) = 0$ and $dV/dZ(Z_0) = 0$ has the following solution

$$V(Z) = \left[\frac{n-1}{2(n+1)}(Z_0^2 - Z^2)\right]^{n/(n+1)}$$

which demonstrates the qualitative character of the $V(Z)$ change near Z_0. As a result, when $n > 10$, the size of the interface—where $V(Z)$ and all its derivatives approach zero—is very small. This is explained by the fact that if $n > 10$, then the power V-I characteristic asymptotically approaches the ideal V-I characteristic.

The effect of the n-value on the scaling distributions of the electric $(E \sim V)$ and magnetic fields, the screening current density $(J \sim V^{1/n})$ and the differential resistivity $(\rho_d \sim V^{(n-1)/n})$ in the superconductor are shown in Figure 2.4. The corresponding invariants $V(Z)$ were determined based on the numerical solution of the problem described by equation (2.22) with conditions (2.25) and (2.26).

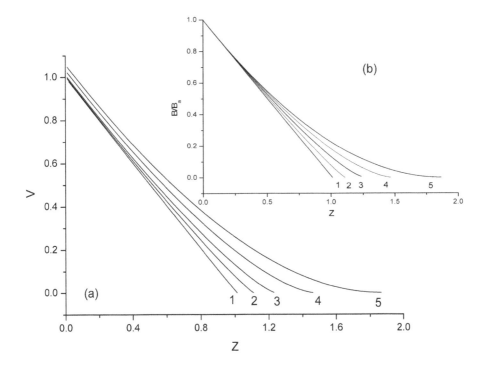

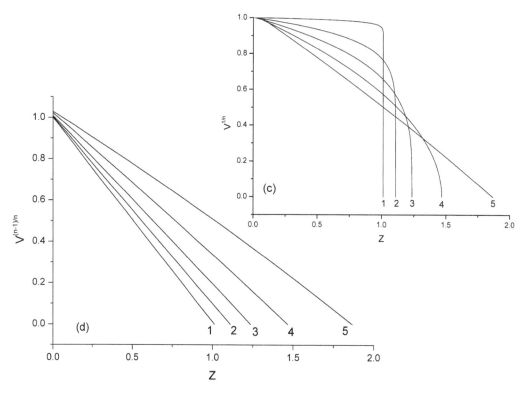

FIGURE 2.4 Scaling distributions of the electric field (a), magnetic field (b), screening current density (c) and differential resistivity (d) for different n-values: 1 - $n = 100$, 2 - $n = 10$, 3 - $n = 5$, 4 - $n = 3$, 5 - $n = 2$.

The curves presented show that for $n < 10$ the distribution of the electric and magnetic fields in the superconductor is essentially nonlinear. At the same time, their distributions approach the Bean dependencies when $n > 10$. This regularity allows one to introduce the following definition of the flux-creep states: strong creep at $n < 10$ and weak creep at $n > 10$.

Using the scaling approximation, let us find the hysteresis losses in the superconducting slab with half-thickness a during partial penetration of the magnetic flux, induced by a constantly increasing external magnetic field. After simple manipulations, the volume density of losses may be written as follows

$$G(t) = G_c \frac{6(n+1)}{3n+2} \left(\frac{E_c}{E_a(t)} \right)^{1/n} \int_0^{Z_0} V^{(n+1)/n} dZ$$

where G_c are the losses calculated in the framework of the CSM and written in paragraph 2.1. The dimensionless factor in the right side of this expression, which is a function of n-value, allows one to estimate the effect of the flux-creep on dissipative phenomena in superconductors. In particular, it implies the losses dependence on the sweep rate, which is absent in the CSM, but is observed in experiments. This dependence becomes stronger if the n-value decreases.

The physical meaning of the characteristic values of L_x and t_x for superconductors with the power V-I characteristic follows from the scaling solution. At $t \ll t_x$, the electrodynamic states formation occurs in the region $0 < x < x_0(t) \ll Z_0 L_x$. Here, $Z_0 \sim 1$, as follows from Figure 2.4a. Thus, these penetration states have a nonlinear formation character due to the nonlinear dependence of x_0 and E_a on time. Next, the electric field on the surface of the superconductor exceeds the value E_c at $t \gg t_x$. Then the size of the region occupied by the screening currents will satisfy the condition $x_0(t) \gg Z_0 L_x$ at $t \gg t_x$. In this case, the diffusion of the electromagnetic field, as in the CSM, will occur almost in a quasi-stationary regime. Since $L_x \sim 1/\dot{B}_a$, $t_x \sim 1/\dot{B}_a^2$, then faster the Bean modes will be established, the higher the sweep rate of the external magnetic field as shown in the dimensional analysis made above.

The special character of the electrodynamic state formation in a superconductor with the power V-I characteristic is also characterized by a nontrivial spatial-temporal change in its differential resistivity unlike the distribution of the differential resistivity of superconductor in the frameworks of the CSM or the viscous flux-flow model. According to the scaling approximation, the differential resistivity of a superconductor (written above) not only depends on the sweep rate and n-value but it also monotonously decreases to zero in the direction of the moving magnetization boundary. This feature is not taken into account in both the CSM and viscous flux-flow model. At the same time, the corresponding conjugation conditions also take place on the moving front: $\rho_d = 0$, $\partial^k \rho_d/\partial x^k = 0$ ($k = 1, 2, 3\ldots$). Due to the uneven distribution ρ_d, the increase of the electric field in the magnetization region occurs with an uneven rate. In particular, its largest value is reached on the surface of the superconductor and is at a minimum on the moving penetration boundary. As a result, the inhomogeneous distribution of the differential resistivity of a superconductor in the magnetization region leads to a substantially nonlinear character of the electromagnetic field penetration into the superconductor at the flux-creep which differs from the electrodynamic states described by the CSM and first at $t < t_x$.

Let us discuss the validity use of the CSM in more detail using the noted properties of the invariant $V(Z)$. To do this, transform equation (2.22) to an equivalent integral equation of the second kind integrating it with respect to Z between the limits from 0 to Z and then from Z_0 to Z. As a result, one obtains

$$V(Z) = Z_0 - Z + \int_{Z_0}^{Z} dx \int_0^x V^{1/n}(y)\,dy - \frac{n}{n+1} \int_{Z_0}^{Z} V^{1/n}(y)y\,dy$$

The solution of this equation is sought by the method of successive approximations. In this case, $V_0(Z) = Z_0 - Z$ is zero approximation, where $Z_0 = [(n + 1)/n]^{n/(n+1)}$ according to the condition (2.26). Substituting zero approximation into the right-hand side of this integral equation, one finds the following approximation $V_1(Z) = V_0(Z) - pZ_0^{1/p} V_0(Z) + p^2 Z_0 V_0^{1/p}(Z)$ and so on. At the same time, the nonzero approximations make a small contribution to the overall solution at $n > 10$ as follows from Figure 2.4. From zero solution $V_0(Z)$ follows the linear distribution of the electric field in the superconductor. Overall, this tendency will be observed at large sweep rates and n-values. It is easy to understand this regularity. According to equation (2.17), the spatial distribution of the electric field is almost linear in the range of values $E/E_c > 1/n$.

Thus, the approximate distribution of the electric field in the magnetization region may be written as follows

$$E(x, t) = \dot{B}_a[x_0(t) - x], \ 0 \le x \le x_0(t), \ t > 0 \tag{2.27}$$

according to the zero scaling solution. This formula completely coincides with the expression that follows from the CSM and the approximate solution obtained in the framework of the viscous flux-flow model. Consequently, the Bean distribution of the electric field is zero solution of the Maxwell equations for superconductors with the power V-I characteristic. Under this approximation, the magnetic field distribution, equation of the magnetization front motion and evolution of the differential resistivity of a superconductor are described by the relations

$$B(x, t) = B_a(t) + \mu_0 J_{c0}(\dot{B}_a/E_c)^{1/n}[(x_0(t) - x)^{1+1/n} - x_0(t)^{1+1/n}]n/(n+1),$$

$$x_0(t) = [t(n+1)/n]^{n/(n+1)} \dot{B}_a^{\frac{n-1}{n+1}} [E_c/(\mu_0^n J_{c0}^n)]^{1/(n+1)} \tag{2.28}$$

$$\rho_d(x, t) = \frac{nE_c^{1/n}}{J_{c0}} [\dot{B}_a(x_0(t) - x)]^{(n-1)/n}$$

Correspondingly, the volume density of the hysteresis losses and magnetic moment of the superconducting slab with a half-thickness a change with time as follows

$$G(t) = G_c \frac{6n(n+1)J_{c0}}{(2n+1)(3n+2)} \left(\frac{n+1}{n}\right)^{\frac{2n+1}{n+1}} \left(\frac{E_c}{\dot{B}_a^2 t \mu_0 J_{c0}^n}\right)^{1/(n+1)},$$

$$M(t) = M_c(t) + \frac{J_{c0} x_0^2}{2a} \left[1 - \frac{2n}{2n+1}\left(\frac{x_0 \dot{B}_a}{E_c}\right)^{1/n}\right]$$

(2.29)

at partially penetrated mode. Here, $M_c(t) = B_a(t)(x_0/a - 1)/\mu_0 - J_{c0} x_0^2/2a$ is the magnetic moment of the slab which is calculated above in the CSM approximation for the partially penetrated mode.

Let us use zero approximation to find the penetration velocity of the magnetization front. According to (2.28), it is easy to get

$$V_n = \frac{dx_0}{dt} = \dot{B}_a^{\frac{n-1}{n+1}} \left(\frac{n}{n+1} \frac{E_c}{\mu_0^n J_{c0}^n t}\right)^{1/(n+1)} = V_c \left(\frac{n}{n+1} \frac{t_x}{t}\right)^{1/(n+1)}$$

The following estimates follow from these formulae, namely, $V_n > V_c$ at $t < t_v$ and $V_n < V_c$ at $t \geq t_v$, where $t_v = n\mu_0 J_{c0} E_c/[(n+1)\dot{B}_a]$. Replacing t_v in the formulae written above for x_0 and E_a, one can obtain that $x_0 = L_x$ and $E_a = E_c$ at $t = t_v$. Therefore, t_v is the boundary time between the subcritical and supercritical states of the superconductor. This quantity depends on the parameters of the V-I characteristic and the sweep rate.

The written solution demonstrates the nonlinear features of the diffusion phenomena occurring during flux-creep and their difference from the Bean approximation. Indeed, the magnetic flux penetration occurs at a constant velocity in a CSM based on an ideal V-I characteristic. However, it is uneven and decreases with time during flux-creep described by a power V-I characteristic as in the viscous flux-flow model. However, in contrast to this model, the velocity V_n of the magnetization front at $t < t_v$ is higher than the corresponding value V_c calculated according to the CSM. This difference increases with a decrease in the n-value. The difference in the values V_n and V_c will be more noticeable at a strong creep. Besides, the difference in the calculations carried out in the frameworks of the CSM and the model with the power V-I characteristic increases with time at $t \geq t_v$. As a result, there is no, strictly speaking, physically valid method for determining the critical current density using the real V-I characteristic to do calculations in the framework of the CSM. To fit the calculations performed within the CSM with the experiment data, the current, which separates the partially and fully penetrated modes, is often used as the 'critical current'.

The proposed scaling solution of the diffusion phenomena in superconductors with a power V-I characteristic allows one to investigate asymptotic states in its physical sense that are observed when a large time window has passed. However, as it will be shown below, comparing the written scaling solution with the numerical one of the corresponding problems in the total formulations, the scaling approximation can be used to describe the formation of the electrodynamic states in LTS, HTS and composites based on them with a satisfactory degree of accuracy throughout the time of the existence of partially penetrated mode.

Let us investigate the formation features of the electrodynamic states in superconductors with the exponential V-I characteristic (1.4) in the scaling approximation. Using invariants $V(Z) = e/\tau, Z = X/\tau$ in which the quantity X and τ are the dimensionless variables introduced above, the initial-boundary problem (2.18), (2.19) lead to the integration of equation

$$d^2V/dZ^2 + ZdV/dZ - V = 0$$

(2.30)

with the boundary conditions (2.23) and the extra condition (2.24).

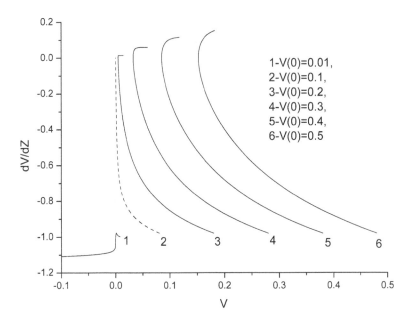

FIGURE 2.5 Phase trajectories of equation (2.30).

The phase trajectories of equation (2.30) numerically calculated the equivalent Cauchy problem for different values $V(0)$ are shown in Figure 2.5. As follows from this figure, the desired invariant $V(Z)$ also exists in a finite area $0 \leq Z \leq Z_0$ since dV/dZ should not have positive values that are observed in Curves 3-6. However, Curve 1 does not satisfy the condition (2.24). Only the Curve 2 corresponds to this condition. Since $V(Z_0) = 0$, which leads to the front-like solution, then it is easy to prove the fulfillment of the conditions $d^k V/dZ^k = 0$ ($k = 1, 2, 3...$) at $Z = Z_0$, that is $V(Z_0) \equiv 0$ at $Z \geq Z_0$. So, the propagation of an electromagnetic field in a superconductor with the exponential V-I characteristic also occurs at a finite rate. In this case, on the moving boundary of the magnetization region, special conjugation conditions take place according to which the electromagnetic field induced by a changing external magnetic field penetrates the superconductor against the background of unperturbed initial states.

To analyze the functional dependence of the invariants $V(Z)$, let us transit from the differential equation (2.30) with the conditions (2.23) and (2.25) to the equivalent integral equation. Integrating as above equation (2.30) twice in the spatial coordinate, it is easy to obtain

$$V(Z) = Z_0 - Z + \frac{Z^2 - Z_0^2}{2} - \int_{Z_0}^{Z} dx \int_0^X \frac{y}{V} \frac{dV}{dy} dy$$

As follows from Figure 2.5, the desired invariant $V(Z)$ satisfies the condition $-1 < dV/dZ < 0$. Therefore, it suffices to accept a linear relationship $V_0(Z) = Z_0 - Z$ as zero approximation. Consequently, the distribution of the electric field in the magnetization region of a superconductor with the exponential V-I characteristic also coincides with the Bean approximation (2.27) in zero approximation. In this approximation, the distribution of the magnetic field and differential resistivity of the superconductor in the magnetization region is described by the expressions

$$B(x, t) = B_a(t) - \mu_0 J_{c0}(1 - \delta)x - \mu_0 J_\delta x_0 \ln(x_0/L_x) + \mu_0 J_\delta(x_0 - x)\ln(x_0 - x)/L_x,$$

$$\rho_d = \dot{B}_a[x_0(t) - x]/J_\delta$$

The moving coordinate of the magnetization front and its velocity are the solutions of equations

$$B_a(t) = \mu_0 J_{c0} x_0(t) + \mu_0 J_\delta x_0(t) \left[\ln \frac{x_0(t)}{L_x} - 1 \right], \quad \frac{dx_0}{dt} = \frac{V_c}{1 + \delta \ln(x_0/L_x)}$$

All of these expressions satisfy the limit transition to the CSM at $\delta \to 0$.

The written relations show that the electromagnetic field penetrates a superconductor with the exponential V-I characteristic with uneven velocity as in the case of a superconductor with a power V-I characteristic. This velocity decreases with time depending on the factor $\ln(x_0(t)/L_x) = \ln(E_a(t)/E_c)$ in zero approximation, where $E_a = E(0,t) = x_0(t)\dot{B}_a$ is the electric field on the surface of the superconductor. Therefore, the size of the magnetization region and the velocity of its front will exceed the corresponding values that follow from the CSM at $x_0(t) < L_x$, that is at $E_a < E_c$. In particular, there is the following estimate $v \sim \dot{B}_a/[\mu_0(J_{c0} - 2J_\delta)] > V_c$ for the penetration velocity of the magnetic flux at $x_0 \ll E_c/\dot{B}_a$. At the same time, the velocity of the magnetization front will be less than V_c at $x_0(t) > L_x$. On the basis of the noted nonlinear dynamics of superconducting states lies not only in the finite value of the differential resistivity of superconductor but also in its uneven distribution in the magnetization region. This distribution will appropriately affect the forming nature of the magnetization region as follows from the viscous flux-flow model.

A direct result of the discussed regularities is a dependence of the hysteresis losses and magnetic moment of superconductor on the sweep rate and therefore the frequency dependence exists. For a superconducting slab with a half-thickness a, they are described by the expressions

$$G(t) = G_c \left[1 - \frac{7}{6}\delta + \frac{7}{18}\delta^2 + \left(2 - \frac{5}{2}\delta \right) \delta \ln \frac{x_0(t)}{L_x} + \delta^2 \left(\ln \frac{x_0(t)}{L_x} \right)^2 \right],$$

$$M(t) = M_c(t) + J_{c0} x_0^2 \delta (0.5 - \ln(x_0/L_x))/2a$$

the logarithmic variation character of which depends on the factor $\ln(x_0/L_x) = \ln(E_a/E_c)$. Note that these expressions satisfy the limit transition to the CSM at $\delta \to 0$.

Thus, the magnetic flux penetration occurs with finite velocity in superconductors with the power and exponential types of V-I characteristics. These diffusion phenomena are characterized by the dependence on the hysteresis losses and magnetic moment on the sweep rate which is observed in the experiment. The distribution of the electric field in the superconductors satisfactorily describes the Bean approximation in zero approximation. However, the change in time of the penetration front velocity differs on the change in time of the front velocity described by the CSM and viscous flux-flow model. Besides, the special conditions for the coupling of the perturbed and unperturbed states are satisfied on the moving boundary of the magnetization region. In addition, there is a nontrivial distribution of the differential resistivity in the superconductor. This quantity is not only finite but also monotonously decreases in the direction from the surface of the superconductor to the moving front of the magnetic flux penetration while taking zero value on the front. The differential resistivity depends not only on the properties of the superconductor but also on the sweep rate of the external magnetic field that is the size of the magnetization region. The written zero scaling solutions also show that the use of the power and exponential types of V-I characteristics for describing diffusion phenomena in superconductors will lead to almost identical results in the region of the weak creep, that is at $n > 10$ or $\delta < 0.1$.

The above-discussed results indicate that such electrodynamics behavior of the nonideal type-II superconductors is the result of the corresponding variation of the superconductor differential resistivity in the case of the partial flux penetration. Indeed, integrating equation (2.5) over x from 0 to x and then from x to $x_0(t)$, one can obtain the following equation

$$E(x,t) = \dot{B}_a(x_0 - x) - \mu_0 \int_x^{x_0} dz \int_0^z \frac{1}{\rho_d(E)} \frac{\partial E}{\partial t} dy$$

using the above-formulated conditions at $x = x_0(t)$. This relation allows one to estimate the influence of the differential resistivity on the evolution of the macroscopic electrodynamic states of superconductors. In particular, if the I-V characteristic is steep, then the differential resistivity in the magnetization region of the superconductor rises quickly from 0 up to a high value. Moreover, the differential resistivity of superconductors increases with time. Therefore, the spatial distribution of the electric field in the partially penetrated mode is practically linear at the sweep-rate stage for all steep equations of the I-V characteristics. Contrariwise, when the differential resistivity of the superconductor is small, then the influence of the nonlinear part of the current-voltage characteristics on the evolution of the electrodynamics states will become essential.

It should be also noted that the invariance of the formulae obtained concerning the choice of the conditionally given critical values J_{c0} and E_c. It is easy to prove that the obtained results remain unchanged when moving from one pair of values $\{J_{c0}, E_c\}$ to another (unlike the CSM or viscous flux-flow model) for continuously increasing V-I characteristic of the power and exponential types. This greatly facilitates the selection of conditional critical values. However, it should be taken into account that the invariance of the $\{J_{c0}, E_c\}$-choice takes place if the measurements comply with the isothermal conditions. Otherwise, the correctness of the analysis will be violated even at the stage of the stable states due to a possible increase in the temperature of the superconductor, as will be shown below.

2.5   PARTIALLY PENETRATED SWEEP-RATE STATES OF SUPERCONDUCTORS WITH ARBITRARY VOLTAGE-CURRENT CHARACTERISTICS

Let us use the above-formulated penetration features of the electromagnetic field during flux-creep and define the distribution of the electric field in a superconductor with an arbitrary V-I characteristic for the Cartesian coordinate system. Integrating equation (1.14) with respect to x in the framework of a one-dimensional model, it is easy to get for a slab the following relation

$$\mu_0 \int_0^x \frac{\partial J}{\partial t} dx = \frac{\partial E}{\partial x} + \frac{dB_a}{dt} \tag{2.31}$$

taking into account the boundary condition (2.7) on the surface of the superconductor. Let us replace the variable

$$\frac{\partial}{\partial t} = \frac{\partial}{\partial x_0} \frac{dx_0}{dt} \tag{2.32}$$

Using the condition $\partial E/\partial x = 0$ at $x = x_0$, one can get

$$\frac{dx_0}{dt} = \frac{\dot{B}_a}{\mu_0 \int_0^{x_0} \frac{\partial J}{\partial x_0} dy} \tag{2.33}$$

Integrating (2.31) over x from x to x_0 and taking into account of the equations (2.32) and (2.33), it is easy to write

$$E(x, t) = \frac{dB_a}{dt} \left[x_0 - x - \frac{\int_x^{x_0} dz \int_0^z \frac{\partial J}{\partial x_0} dy}{J\big|_{x=x_0} + \int_0^{x_0} \frac{\partial J}{\partial x_0} dy} \right]$$

Let us rewrite this formula using the relation

$$\frac{\partial J}{\partial x_0} = \frac{\partial J}{\partial E}\frac{\partial E}{\partial x_0} + \frac{\partial J}{\partial B}\frac{\partial B}{\partial x_0} = \frac{1}{\rho_d}\frac{\partial E}{\partial x_0} + \frac{\partial J}{\partial B}\frac{\partial B}{\partial x_0}$$

The latter corresponds to the case when the current is the function of the electric and magnetic fields. Then, the electric field distribution is the solution of equation

$$E(x,t) = \frac{dB_a}{dt}\left[x_0 - x - \int_x^{x_0} dz \int_0^z \left(\frac{1}{\rho_d}\frac{\partial E}{\partial x_0} + \frac{\partial J}{\partial B}\frac{\partial B}{\partial x_0}\right)dy \Big/ \left[J\big|_{x=x_0} + \int_0^{x_0}\left(\frac{1}{\rho_d}\frac{\partial E}{\partial x_0} + \frac{\partial J}{\partial B}\frac{\partial B}{\partial x_0}\right)dy\right]\right] \quad (2.34)$$

It may be solved according to the method of the successive approximation. Then, zero solution is described by the Bean approximation (2.27). Inserting this formula into equations (2.33) and (2.34), one can find the following approximations and so on.

Let us consider some practically interesting cases using the above-presented solution.

For the superconductor with the flux-flow V-I characteristic (1.2) (ρ_f = const), it is easy to find the following conditions: $\partial E^{(0)}/\partial x_0 = \dot{B}_a$, $J\big|_{x_0} = J_{c0}$. Then, the next approximation is described by the formulae

$$E^{(1)}(x,t) = \dot{B}_a(x_0 - x)\frac{2J_{c0}\rho_f + \dot{B}_a(x_0 - x)}{2J_{c0}\rho_f + 2\dot{B}_a x_0}, \quad x_0(t) = \frac{\rho_f J_{c0}}{\dot{B}_a}\left(\sqrt{1 + \frac{2\dot{B}_a B_a}{\mu_0 \rho_f J_{c0}^2}} - 1\right)$$

The written expression for $x_0(t)$ with a good degree of accuracy coincides with the exact solution of the flux-flow problem as discussed in section 2.3.

For the superconductor with the power V-I characteristic and $J_c = J_c(B, T_0)$, zero approximation of the magnetic field distribution inside the magnetization region is described by the formula

$$\int_B^{B_a}\frac{dB}{J_c(B,T_0)} = \mu_0\left(\frac{\dot{B}_a}{E_c}\right)^{1/n}\frac{n}{n+1}[(x_0(t)-x)^{1+1/n} - x_0(t)^{1+1/n}]$$

and the moving boundary $x_0(t)$ obeys the equation

$$\int_0^{B_a}\frac{dB}{J_c(B,T_0)} = \mu_0\left(\frac{\dot{B}_a}{E_c}\right)^{1/n}\frac{n}{n+1}x_0(t)^{1+1/n}$$

The next approximation is too cumbersome to write here. However, in the case $J_c(B) \approx$ const, it may be written as

$$E^{(1)}(x,t) = \dot{B}_a(x_0 - x)\left[2 - \frac{n}{n+1}\left(\frac{x_0 - x}{x_0}\right)^{1/n}\right],$$

$$B^{(1)}(x,t) = B_a(t) - \mu_0 J_{c0}\left(\frac{\dot{B}_a}{E_c}\right)^{1/n}x_0^{1+1/n}\int_0^{x/x_0}(1-y)^{1/n}\left[2 - \frac{n}{n+1}(1-y)^{1/n}\right]^{1/n} dy$$

$$x_0(t) = \left\{\frac{B_a(t)}{\mu_0 J_c\left(\frac{\dot{B}_a}{E_c}\right)^{1/n}\int_0^1(1-y)^{1/n}\left[2 - \frac{n}{n+1}(1-y)^{1/n}\right]^{1/n} dy}\right\}^{\frac{n}{n+1}}$$

As follows from (1.4), the electric field is equal to $E_0 = E_c \exp(-J_{c0}/J_\delta)$ at $J = 0$. To analyze the electrodynamic states in the region of ultra-low values of the electric field ($E < E_0$), the V-I characteristic (1.4) should be written in the form $E = E_c \exp[(J - J_{c0})/J_\delta] - E_0$. Then, the corresponding distribution of the electric field in the first approximation is written as follows

$$E^{(1)}(x,t) = \dot{B}_a(x_0 - x)\left\{1 - \frac{\left[1 + \dfrac{E_0}{\dot{B}_a(x_0-x)}\ln\dfrac{E_0}{E_c} + \ln\left[\dfrac{E_0 + x_0\dot{B}_a}{E_c}\right]\right] - \left(\dfrac{E_0}{\dot{B}_a(x_0-x)} + 1\right)\ln\left[\dfrac{E_0 + (x_0-x)\dot{B}_a}{E_c}\right]}{\ln[(E_0 + x_0\dot{B}_a)/E_0]}\right\}$$

This approximation takes into consideration of the quantity E_0 which, however, may be arbitrary in the general case.

Finally, let us write the relevant approximation of the electric field distribution in a superconductor with the V-I characteristic describing the collective flux-creep states when $E = E_c \exp\{[1 - (J_{c0}/J)^\beta]/\gamma\}$, ($\beta$, γ = const). In this case, the first approximation is given by

$$E^{(1)}(x,t) = \dot{B}_a\left(1 - \gamma\ln\frac{\dot{B}_a x_0}{E_c}\right)^{1/\beta}\int_x^{x_0}\left[1 - \gamma\ln\frac{\dot{B}_a(x_0-y)}{E_c}\right]^{-1/\beta} dy$$

Thus, the written solutions show that the critical state model is the linear (zero) approximation of the flux-creep sweep-rate states of a superconductor with arbitrary V-I characteristics. The nonlinear effects occur in its background. Such behavior is the result of the corresponding variation of a differential resistivity of a superconductor, as discussed above.

2.6   FEATURES OF TRANSPORT CURRENT PENETRATION IN COMPOSITE SUPERCONDUCTORS DURING FLUX-CREEP

The above-presented solutions demonstrate the physical features of the magnetic flux penetration in nonideal type-II superconductor induced by a changing external magnetic field. For practical applications, the problem of the transport current charging in a composite superconductor is also relevant. In this regard, let us analyze the current penetration into the superconducting composite when it is charged with a finite rate and its value in the matrix is negligible compared to the current in the superconducting core.

First, let us investigate the partially penetrated states of *a superconducting composite slab* (Figure 2.2).

Consider an ideally cooled composite with slab geometry ($-a < x < a, -b < y < b, -\infty < z < \infty$, $a \ll b$). To simplify the analysis, suppose that

- a superconductor is uniformly distributed over the composite cross-section with the filling coefficient η, so the use of a continuum medium model is rightful;
- there is no current at the initial time, and then it rises at the constant rate dI/dt where $I = JS = (dI/dt)t$ is the total current flowing in the slab along z-direction;
- the background magnetic field is fixed;
- the critical current density does not depend on the magnetic induction;
- the current in the matrix is negligibly low to compare with one in the superconductor, that is, the relation $J \approx \eta J_s$ takes places. Here, J_s is the current density in the superconductor.

For that planar case, the system of Maxwell's equations is reduced to a one-dimensional partial differential equation relative to the single z-component of the electric field $E = E_z(x, t)$. As a result, the dynamics of the transport current is described by equation (2.5) where

$$\rho_d = \begin{cases} n(E_c/E)^{1/n} E/(\eta J_{c0}) \\ E/(\eta J_\delta) \end{cases} \tag{2.35}$$

for the composite superconductor with the power and exponential V-I characteristics, respectively. The necessary initial and boundary conditions are given by

$$E(x,0) = 0, \quad \frac{\partial E}{\partial x}(a,t) = \frac{\mu_0}{4b}\frac{dI}{dt}, \quad \frac{\partial E}{\partial x}(0,t) = 0 \tag{2.36}$$

Here, the boundary condition on the surface of the superconductor follows from Ampere's current law, and the boundary condition at $x = 0$ takes into account of the symmetric distribution of the electric field in the composite. The extra condition must be also taken into consideration

$$4b\int_0^a J(x,t)\,dx = \frac{dI}{dt}t \tag{2.37}$$

describing the conservation law of the charged current. It is written for the entire cross-section of the composite.

Introduce the scaling variables $V(Z) = \gamma_1 E/t^p$, $Z = \gamma_2 x/t^p$, to find scaling solutions describing the temporal and spatial evolution of the electric field. Here, $V(Z)$ is the unknown function to be determined; p, γ_1, γ_2 are the constants defined as follows

$$p = n/(n+1), \quad \gamma_1 = \frac{1}{E_c}\left[\eta J_{c0}E_c/\mu_0\left(\frac{1}{4b}\frac{dI}{dt}\right)^2\right]^p, \quad \gamma_2 = \gamma_1\frac{\mu_0}{4b}\frac{dI}{dt},$$

for a superconducting composite with the power V-I characteristic and the corresponding values are given by

$$p = 1, \quad \gamma_1 = \eta J_\delta/\mu_0\left(\frac{1}{4b}\frac{dI}{dt}\right)^2, \quad \gamma_2 = \gamma_1\frac{\mu_0}{4b}\frac{dI}{dt}$$

for a superconducting composite with the exponential V-I characteristic.

Using these variables, the partial differential equation (2.5) is reduced to the ordinary differential equations (2.22) or (2.30) for the superconducting composite with the power and exponential V-I characteristics, respectively. Then, as discussed above, the scaling functions $V(Z)$ are equal to zero at a certain point Z_0. Besides, $d^k V/dZ^k = 0$, $k = 1, 2, 3, \ldots$ at $Z = Z_0$. Hence, the functions $V(Z)$ approach smoothly to zero value at $Z = Z_0$. Respectively, the transport current propagates as a characteristic wave with the finite velocity when the values of the current, electric field, magnetic induction and differential resistivity of the composite, as well as all corresponding spatial derivatives, are equal to zero at the moving current penetration boundary $x_p(t)$.

According to the corresponding zero solutions of equations (2.22) and (2.30), the electric field induced in the superconducting composite slab by the current charging can be described by the approximate formulae

$$E(x,t) = \begin{cases} 0, \ 0 \leq x \leq x_p \\ \frac{\mu_0}{4b}\frac{dI}{dt}(x - x_p), \ x_p \leq x \leq a \end{cases} \tag{2.38}$$

where the coordinate of the moving current front is given by

$$x_p(t) = a - \left(\frac{n+1}{n}t\right)^{\frac{n}{n+1}}\left(\frac{\mu_0}{4b}\frac{dI}{dt}\right)^{\frac{n-1}{n+1}}\left[\frac{E_c}{\mu_0^n\eta^n J_{c0}^n}\right]^{\frac{1}{n+1}} \tag{2.39}$$

for the superconducting composite with the power V-I characteristic and by equation

$$\eta J_{c0}(a - x_p) + \eta J_\delta(a - x_p)\left[\ln\left(\frac{a - x_p}{E_c}\frac{\mu_0}{4b}\frac{dI}{dt}\right) - 1\right] = \frac{1}{4b}\frac{dI}{dt}t \tag{2.40}$$

for the superconducting composite with the exponential V-I characteristic.

Then, it is easy to find the distributions of the magnetic induction $B(x, t)$ and differential resistivity ρ_d induced in the superconducting composite by the current charging in the region $x_p \leq x \leq a$. For example, the respective values are equal to

$$B(x, t) = \frac{\mu_0}{4b}\frac{dI}{dt}t + \mu_0 \eta J_{c0}\left(\frac{\mu_0}{4bE_c}\frac{dI}{dt}\right)^{1/n}\frac{n}{n+1}[(x_p(t) - x)^{1+1/n} - x_p(t)^{1+1/n}],$$

$$\rho_d = \frac{nE_c}{\eta J_{c0}}\left(\frac{\mu_0}{4bE_c}\frac{dI}{dt}[x_p(t) - x]\right)^{(n-1)/n}$$

for the superconducting composite slab with the power V-I characteristic.

Using the obtained solutions, one may find the time t_f during which the transport current fills the cross-section of the composite. In this case $x_p = 0$. As an example, consider a superconductor with the power V-I characteristic. Then according to (2.39), this value is equal to

$$t_f = \frac{n}{n+1}\frac{\mu_0 \eta J_{c0}}{E_c^{1/n}}a^{(n+1)/n}\left(\frac{\mu_0}{4b}\frac{dI}{dt}\right)^{(1-n)/n} \tag{2.41}$$

Consequently, the partially penetrated mode exists for $t \leq t_f$. During this time, the distribution of the electric field and current front evolution can be described by the formulae (2.38) and (2.39). If $t > t_f$, then they have no physical meaning. The simplified model that can be used to describe fully penetrated current modes at $t > t_f$ is discussed in Chapter 6 in detail.

The solutions presented above describe the stable penetration of the transport current into the superconducting slab. However, for many practical applications, the analysis of the current diffusion in a cylindrical superconducting composite is of undoubted interest. Let us find the solution to the problem of the current charging into the infinitely long round composite superconductor, which is in the magnetic field of its current. For this case, the electric field distribution $E = E_z(r, t)$ in the cylindrical coordinates (r, z, φ) is independent of z and φ directions, and it is described by the nonlinear differential equations

$$\frac{1}{r}\frac{\partial}{\partial r}\left(r\frac{\partial E}{\partial r}\right) = \mu_0 \eta \frac{J_{c0}}{nE}\left(\frac{E}{E_c}\right)^{1/n}\frac{\partial E}{\partial t}$$

for the composite with the power V-I characteristic and

$$\frac{1}{r}\frac{\partial}{\partial r}\left(r\frac{\partial E}{\partial r}\right) = \mu_0 \eta \frac{J_\delta}{E}\frac{\partial E}{\partial t}$$

for the composite with the exponential V-I characteristic.

The corresponding current conservation law, initial and boundary conditions have the following forms

$$2\pi\int_{r_p}^{a} Jr\,dr = I_t = \frac{dI}{dt}t, \quad E(x, 0) = 0, \quad \frac{\partial E}{\partial r}(a, t) = \frac{\mu_0}{2\pi a}\frac{dI}{dt}$$

Here, a is the radius of composite, $r_p(t)$ is the time-varying penetration boundary of the transport current.

To find the solution to the given problem, let us transform the divergent part of the initial equations by replacing variable $R = 1 - r/a$ using the series expansion $(1 - R)^{-1} \sim 1 - R + \ldots$ for $R < 1$. Then the spatial differential operator relatively to a new variable $y = \ln(1 - R)$ is transformed as follows

$$\frac{1}{r}\frac{\partial}{\partial r}\left(r\frac{\partial E}{\partial r}\right) = \frac{1}{a^2}\frac{\partial^2 E}{\partial y^2}$$

and the boundary condition is written like this

$$\frac{\partial E}{\partial y}(0, t) = \frac{\mu_0}{2\pi}\frac{dI}{dt}$$

To solve this problem, let us use the dimensionless variables introduced above for the power and exponential V-I characteristics. Then, one can get the following initial-boundary problem

$$\frac{\partial^2 e}{\partial y^2} = \begin{cases} \dfrac{e^{(1-n)/n}}{n}\dfrac{\partial e}{\partial \tau} \\[2ex] \dfrac{1}{e}\dfrac{\partial e}{\partial \tau} \end{cases}$$

$$\frac{\partial e}{\partial y}(0, \tau) = q = \frac{\mu_0}{2\pi E_c}\frac{dI}{dt}, \quad e(y, 0) = 0$$

It is easy to find the scaling solutions of these transformed equations using the above-introduced procedures. Then, the formula describing in zero approximation the distribution of the electric field in the round composite superconductor has the form similar to that of the critical state model (Gurevich et al. 1997c), namely

$$E(r, t) = \frac{\mu_0}{2\pi}\frac{dI}{dt}\ln\frac{r}{r_p(t)} \tag{2.42}$$

where according to the condition of the current conservation law, the moving current front $r_p(t)$ is the solution of the equation

$$\frac{dI}{dt}t = 2\pi\eta J_{c0}\left(\frac{\mu_0}{2\pi E_c}\frac{dI}{dt}\right)^{1/n}\int_{r_p}^{a}\left(\ln\frac{r}{r_p(t)}\right)^{1/n}r\,dr$$

for the composite with the power V-I characteristic and

$$\frac{dI}{dt}t = \pi\eta J_{c0}(a^2 - r_p^2) + 2\pi\eta J_\delta\int_{r_p}^{a}\ln\left(\frac{\mu_0}{2\pi E_c}\frac{dI}{dt}\ln\frac{r}{r_p(t)}\right)r\,dr$$

for the composite with the exponential V-I characteristic.

Thus, according to the written above solutions, the nonlinear current diffusion in the composite superconductors may be studied in a simple form taking into account of the essentially nonlinear nature of V-I characteristics. To illustrate the legitimacy of the proposed scaling solutions in the analysis of electrodynamic states in a round superconducting composite, the dynamics of the current penetration boundary and the increase in the electric field on the surface of the perfectly cooled multifilament Nb-Ti composite with different V-I characteristics are shown in Figure 2.6. The parameters of the composite were taken equal to $a = 5 \times 10^{-4}$ m, $J_{c0} = 4 \times 10^9$ A/m^2, $E_c = 10^{-4}$ V/m, $\rho_m = 2 \times 10^{-10}$ $\Omega \times$ m and $\eta = 0.5$. The parameters of the V-I characteristics satisfied the condition

$$n = J_{c0}/J_\delta \tag{2.43}$$

at $E = E_c$. In this case, the power and exponential V-I characteristics have a common tangency point at $E = E_c$. Under this condition, the calculated values of the electric fields for both V-I characteristics do not considerably differ from each other near E_c. The length of this nearness region depends on the values n, J_{c0}, and E_c. Accordingly, the value $n = 100$ was used for the power V-I characteristic and $J_\delta = 4 \times 10^7$ A/m^2 was used for the exponential V-I characteristic.

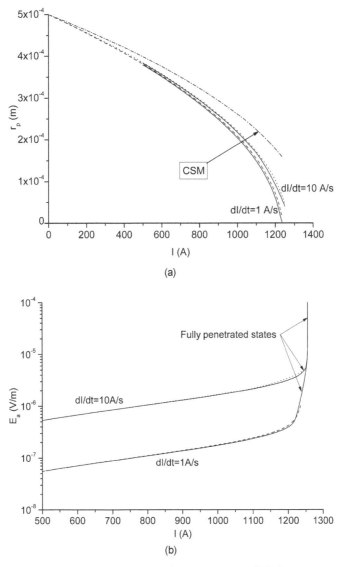

FIGURE 2.6   Comparison of zero scaling solution (- - - - the power V-I characteristic (1.3), · · · · · · - the exponential V-I characteristic (1.4)) and numerical solution (——— - the exponential V-I characteristic (1.4)) describing the current penetration boundary (a) and the electric field on the surface of a round composite superconductor (b) at different current charging rates.

The curves presented in Figure 2.6 were obtained based on the corresponding zero scaling approximations for composites with V-I characteristic of the power and exponential types. For a composite superconductor with the exponential V-I characteristic, the system of Maxwell's equations, which takes into account the so-called current sharing mechanism between the superconducting core and the matrix, was also numerically solved. The formulation of the current diffusion problems in composite conductors is discussed in Chapters 5 and 6. In this case, in cylindrical coordinates, the electric field $E = E_z(r, t)$ and currents in the superconductor J_s and matrix J_m are the solutions to

$$\mu_0 \frac{\partial J}{\partial t} = \frac{1}{r}\frac{\partial}{\partial r}\left(r\frac{\partial E}{\partial r}\right), \quad J = \eta J_s + (1-\eta)J_m, \quad E = E_c \exp\left(\frac{J_s - J_{c0}}{J_\delta}\right) = J_m \rho_m$$

with the following initial, boundary and conservation current conditions

$$E(r,0) = 0, \quad \left.\frac{\partial E}{\partial r}\right|_{r=a} = \frac{\mu_0}{2\pi a}\frac{dI}{dt}, \quad E(r_p, t) = 0, \quad 2\pi \int_{r_p}^{a} J(r,t)r\,dr = \frac{dI}{dt}t$$

taking into account the current charging at a constant rate dI/dt and the existence of a moving current penetration front r_p.

Figure 2.6 demonstrates that the dynamics of the current in a composite superconductor can be described by zero scaling approximation with a good degree of accuracy during the partially penetrated mode. (The fully penetrated states, which follow from the numerical simulations, are depicted in Figure 2.6). As mentioned above, the distribution of the electric field in the superconducting composite may be described by the CSM in zero approximation. However, the moving boundary of the current front must be determined to take into account the corresponding V-I characteristic, as already noted. Otherwise, the direct calculations in the framework of the Bean approximation will lead to incorrect determination of the current penetration boundary and hence to the distortion of the results obtained, as follows from Figure 2.6a.

Thus, the finite character of the electromagnetic field penetration specifies the features of the macroscopic electrodynamic phenomena in superconducting materials with arbitrary V-I characteristics. The spatial-time change of the electric field is described by the Bean model in zero approximation, which allows analyzing the electromagnetic phenomena in superconductors in a close form. However, the dynamics of the penetration front must take into account the features of variation in the differential resistivity of a superconductor. As a result, the above written solutions lead to the logarithmic losses dependence on the sweep rate of the external magnetic field, as the experiments demonstrate. It cannot be obtained directly within the framework of the CSM.

It should be emphasized that the essentially nonlinear character of the V-I characteristic is inherent for the HTS due to the physical inhomogeneity of their properties and the complexity of the phenomena occurring at the boundaries between the grains. Therefore, the nonlinear features of the diffusion phenomena in superconductors formulated above are valid for HTS. They must be considered when describing the phenomena occurring in them.

2.7 INFLUENCE OF VOLTAGE-CURRENT CHARACTERISTIC NONLINEARITY ON THE ELECTRODYNAMIC STATES OF NONIDEAL TYPE-II SUPERCONDUCTORS

Let us analyze the effect of V-I characteristic on the dynamics of the flux penetration considering a superconducting slab ($0 < x < 2a$) which is in a longitudinal magnetic field. To illustrate the above-discussed features of the electrodynamic phenomena in superconductors and the possibility of using approximate solutions, the results of the numerical simulations based on the system of Maxwell's equations and calculations performed in the framework of zero scaling approximation are compared below. In the first case, equation (2.5) was solved numerically with conditions (2.7) and (2.8) for the V-I characteristics of the power and exponential types, taking into account the finite size of the magnetization region with a moving front. (A numerical method for solving similar problems is discussed in Chapter 4). In the second case, the transition to the slab geometry is obvious.

The diffusion of the electric field in the superconducting slab with a half-thickness $a = 10^{-5}$ m and the increase in its differential resistivity demonstrate Figures 2.7 and 2.8, respectively. The critical parameters of the superconductor were set as follows: $E_c = 10^{-4}$ V/m, $J_{c0} = 4 \times 10^9$ A/m^2. The results presented in Figure 2.7 correspond to the LTS with different types of V-I characteristics. Namely, their rise parameters were assumed to be $n = 80$ for the power V-I characteristic and J_δ

$= 4 \times 10^7$ A/m^2 for the exponential V-I characteristic. Curves depicted in Figure 2.8 describe an increase in the differential resistivity of superconductors at different sweep rates. Namely, LTS was assumed to be in an external magnetic field, increasing at the sweep rate $dB_a/dt = 1$ T/s (Figure 2.8a), and HTS at the sweep rate $dB_a/dt = 10^{-2}$ T/s (Figure 2.8b). The V-I characteristics of both superconductors were described by the power equation $E_c = 10^{-4}$ V/m, $J_{c0} = 4 \times 10^9$ A/m^2.

The presented results show a satisfactory coincidence of the numerical simulations and calculations made within the framework of zero scaling approximation, proving its operability. They confirm the conclusions formulated above. First, there is a moving boundary between perturbed and unperturbed states for both LTS and HTS, which have continuously increasing V-I characteristics. The current front moves with a finite velocity. Second, the differential resistivity of a superconductor in the magnetization region is not a fixed value. The quantity ρ_d constantly increases with an increase in the external magnetic field and decreases monotonously as it approaches the moving current front where it takes zero value. Although the CSM does not take into account the true character of the differential resistivity variation of a superconductor, this model may be used to simulate the electrodynamic states both LTS and HTS (Figures 2.7 and 2.8), as discussed above. In this approximation, it allows one to take into account the real character of the differential resistivity variation according to the scaling approximation. Third, the nonlinear character of the electromagnetic field penetration becomes more noticeable with decreasing of n (or $1/\delta$) values. Therefore, the deviation of the modes described by the linear scaling approximations from real ones will be more noticeable in HTS.

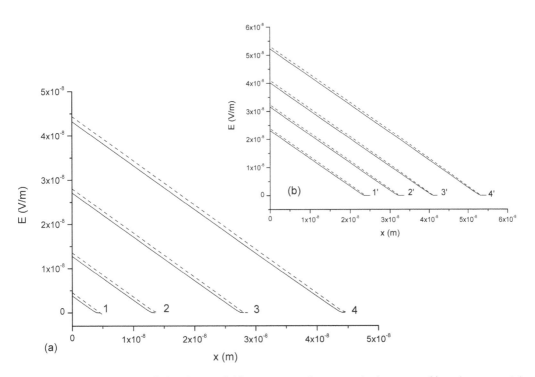

FIGURE 2.7 Dynamics of the electric field in superconductors with the power (a) and exponential (b) V-I characteristics at $dB_a/dt = 1$ T/s (——— numerical calculation, - - - - - zero scaling approximation): 1 - 0.2×10^{-4} s, 2 - 0.6×10^{-4} s, 3 - 1.3×10^{-4}, 4 - 2×10^{-4} s; 1' - 0.9×10^{-4} s, 2' - 1.2×10^{-4} s, 3' - 1.5×10^{-4}, 4' - 2×10^{-4} s.

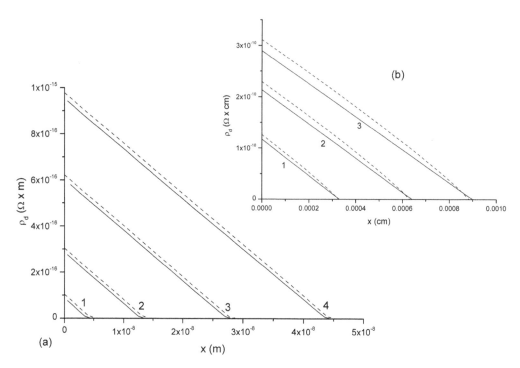

FIGURE 2.8 Constant increase of the differential resistivity of superconductors with the power V-I characteristic (a - n = 100, dB_a/dt = 1 T/s, b - n = 10, dB_a/dt = 10^{-2} T/s) in the increasing external magnetic field (——— numerical calculation, - - - - - zero scaling approximation): 1 - 0.2 × 10⁻⁴ s, 2 - 0.6 × 10⁻⁴ s, 3 - 1.3 × 10⁻⁴ s, 4 - 2 × 10⁻⁴ s, 1′ - 10⁻⁴ s, 2′ - 2 × 10⁻⁴ s, 3′ - 3 × 10⁻⁴ s.

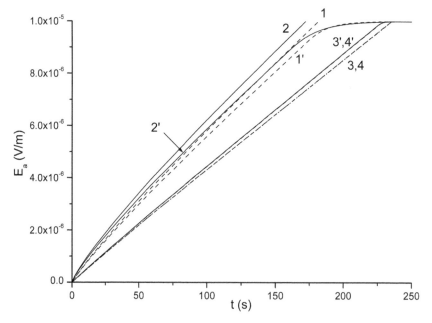

FIGURE 2.9 Electric field on the surface of superconductors as a function of time at dB_a/dt = 0.01 T/s. Here, (1, 1′ - n = 10; 3, 3′ - n = 50) – for the power V-I characteristic; (2, 2′, 4, 4′) – for the exponential V-I characteristic under the condition (2.43). Calculation models: (1-4) – zero scaling approximations, (1′-4′) – numerical computations.

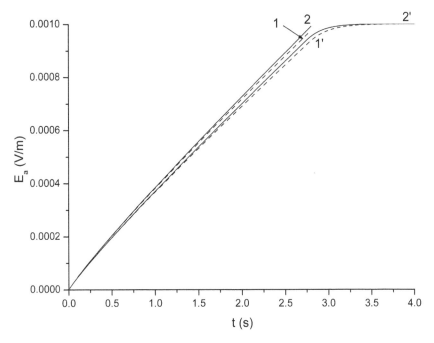

FIGURE 2.10   Time dependence of the electric field on the surface of superconductors under the condition (2.43) at $dB_a/dt = 1$ T/s for the power V-I characteristic (1, 1' - $n = 10$) and the exponential V-I characteristic (2, 2'). Calculation models: (1, 2) – zero scaling approximations, (1', 2') – numerical simulations.

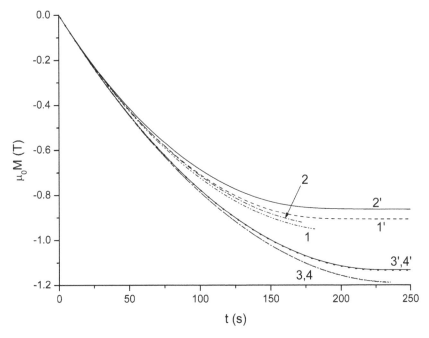

FIGURE 2.11   Magnetization of superconductors at $dB/dt = 0.01$ T/s: 1, 1' - $n = 10$; 3, 3' - $n = 50$; 2, 2', 4, 4' – exponential V-I characteristic under the condition (2.43). Calculation models: (1-4) – zero scaling approximations, (1'-4') – numerical computations.

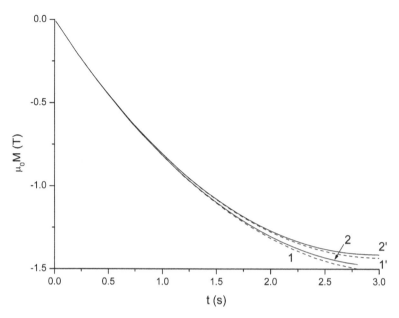

FIGURE 2.12 Magnetization of superconductors at $dB/dt = 1$ T/s: 1, 1' - power V-I characteristic at $n = 10$; 2, 2' – exponential V-I characteristic under the condition (2.43). Calculation models: (1, 2) – zero scaling approximation, (1', 2') – numerical simulations.

In Figures 2.9 and 2.10, the electric field evolution on the surface of the superconducting slab is shown for relatively small and high values of the sweep rate. Figures 2.11 and 2.12 present the corresponding time-variation of the magnetic moment, which occurs in accordance with the dynamics of the electric field on the surface of the superconductors (Figures 2.9 and 2.10). The parameters of the V-I characteristics were set according to the condition (2.43). In the simulations, the parameters of the power V-I characteristic were specified as follows: $E_c = 10^{-4}$ V/m, $J_{c0} = 1.91 \times 10^9$ A/m^2. For the two parameters $n = 10$ and $n = 50$ used in calculations, the parameters of the exponential V-I characteristic equal $J_\delta = 1.91 \times 10^8$ A/m^2 and $J_\delta = 3.82 \times 10^7$ A/m^2, respectively. As above-mentioned, the calculations were carried out based on a numerical solution and zero scaling approximation. The results lead to the following conclusions.

First, zero scaling approximation satisfactorily describes the electrodynamic states of superconductors with a real V-I characteristic during partially penetrated states, as already noted. Correspondingly, the sharper the V-I characteristic increases, the higher the simulation accuracy. Besides, the difference between the electrodynamic behavior of superconductors with different types of V-I characteristics decreases at high values of n or small J_δ since they approach ideal ones. This difference increases at a relatively small sweep rate for the given value of the n-value. As a result, the influence of the nonlinear type of V-I characteristic on the flux-creep sweep-rate dynamics is not noticeable for superconductors with the weak creep (n, $1/\delta > 10$).

Second, the screening current penetration into the superconductor with the exponential V-I characteristic occurs more intensively than their diffusion in a superconductor with the power V-I characteristic. This difference increases with a decrease in the sweep rate. Conversely, the electrodynamic states of technical superconductors will become close to ideal ones at high values of dB_a/dt. As simulations show for the modes under consideration, the penetration velocity of the magnetization front in superconductors with the real V-I characteristics is higher than the corresponding penetration velocity $|V_c|$ for the superconductor with an ideal V-I characteristic which was defined in the framework of CSM and is estimated by (2.4). Moreover, the more smoothed character of the V-I characteristic (lower n or higher J_δ), the higher the difference between results obtained in the framework of the ideal and real V-I characteristics.

Third, the scaling approximation does not allow one to describe superconducting states formation during full penetration of the magnetic flux for obvious reasons. In Figure 2.9, these states represent Curves 1′-4′ obtained based on a numerical solution of the system of Maxwell's equations. As follows from Figure 2.9, after reaching the full penetration state, dE_a/dt value monotonously decreases to zero, that is the distribution of the electric field approaches the stationary distribution. However, the scaling approximation makes it possible to find the occurrence time of the full penetration regime t_f in superconductors with the real V-I characteristics. The value t_f was determined above for the operating modes of the superconducting composite slab when a transport current is charged into it and is described by the Formula (2.41).

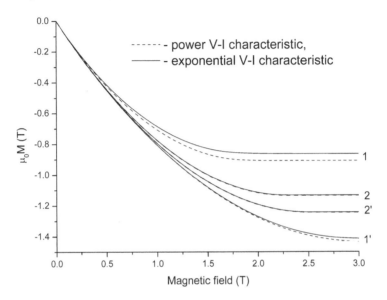

FIGURE 2.13   Magnetization of superconductors with the smoothness parameters obeying the condition (2.43) at $dB_n/dt = 0.01$ T/s (1 - n = 10, 2 - n = 50) and $dB_a/dt = 1$ T/s (1′ - n = 10, 2′ - n = 50).

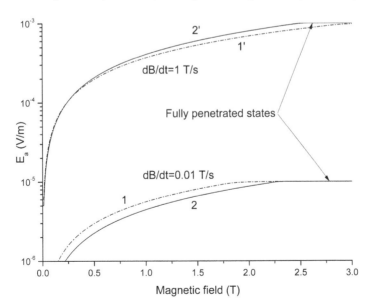

FIGURE 2.14   Dynamics of the electric field on the surface of superconductors with the power V-I characteristic: 1, 1′ - n = 10; 2, 2′ - n = 50.

The magnetic moment as the function of the smoothness parameters of V-I characteristics and sweep rate is shown in Figure 2.13. The results of the numerical solution of Maxwell's equations depict the features of the sweep-rate dependence of magnetization curves, which follows from the above-discussed investigation. It is seen that the magnetic moment increases with n or when J_δ decreases at a relatively small value of the sweep rate. However, at relatively high value of the sweep rate, this regularity has a reverse tendency: the magnetic moment of a superconductor decreases with J_δ or when the n-value increases. This feature is based on the relevant change of the electric field on the surface of a superconductor (Figure 2.14), which defines the electrodynamics states of superconductors, as follows from the results discussed in section 2.4. Here, the constant values of the electric field correspond to the fully penetrated states.

Thus, the above-mentioned analysis shows that the equivalence in the description of the superconductor electromagnetic states during flux-creep, which is described by the power or exponential V-I characteristics, depends on the smoothness parameters of V-I characteristics. First of all, the qualitative and quantitative differences between the obtained results increase if n decreases or δ increases. The difference will also increase when the values $\{J_{c0}, E_c\}$ decreases. The qualitative difference between the flux diffusion in superconductors with the various types of V-I characteristics is related to the time-variation of the electric field on the surface of the superconductor. This quantity directly affects the penetration character of the moving boundary of the magnetization region and depends strongly on the distribution of the differential resistivity of a superconductor in its interior. Along with these features, the above obtained results indicate that the exponential and power equations of V-I characteristics with relatively high values of the smoothness parameters lead to practically equivalent results.

Besides, these solutions demonstrate the possibility of the CSM in the investigation of the flux-creep sweep-rate states of superconductors. According to the performed analysis, the critical state model defines a linear approximation of the general solution that describes the electric field distribution in the superconductors during the sweep-rate stage. As a result, the Bean model may be directly used to investigate the sweep-rate diffusion phenomena taking place in the superconductors with weak creep ($n > 10$ or $\delta < 0.1$), selecting the corresponding effective value of the critical current density. However, to describe more precisely the flux-creep state evolution, the equation of the moving magnetization boundary must consider the nonlinear character of V-I characteristics. This electric field approximation is right and leads to the dependence of the front slope of the magnetic field penetration on its sweep rate.

The results presented in Figures. 2.7-2.14 were calculated under the assumption that the critical current density is constant. Let us estimate the effect of the magnetic field dependence of J_c on the forming of the electrodynamics states of superconductors considering the Kim approximation using the following relation

$$J_c(B) = J_{c0} \frac{B_0}{B + B_0}, \quad J_{c0}, B_0 = \text{const}$$

For this case, the condition allowing to connect the smoothness parameters of the power and exponential V-I characteristics has the form

$$n = \frac{J_{c0}}{J_\delta} \frac{B_0}{B + B_0} \tag{2.44}$$

Figures 2.15 and 2.16 show the curves of the superconducting slab magnetization ($a = 10^{-3}$ m) for both the partially and fully penetrated states at $B_0 = 8$ T obtained by numerical solution of Maxwell's equations. The parameters of the V-I characteristics were taken as it was written above. The quantity J_δ for the superconductor with the exponential V-I characteristic was defined according to the condition (2.44). Figure 2.15 depicts the influence of the nonlinearity of V-I characteristics on the magnetization curves for various sweep rates. It shows that the operating states, which will be observed in superconductors with the power and exponential V-I characteristics, have practically similar behavior when the condition (2.44) is satisfied. However, the magnetic flux penetration in a superconductor with the exponential V-I characteristic is more intensive than that in a superconductor with the power V-I characteristic, as discussed above. The effect of the smoothness parameter of the V-I characteristic and sweep rate on the magnetic moment is presented in Figure 2.16 for partial and full penetration modes.

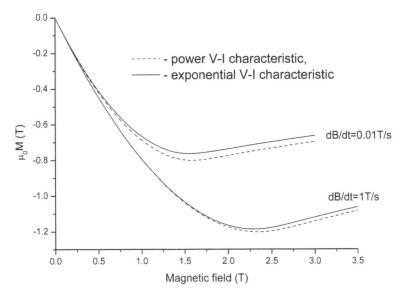

FIGURE 2.15   Sweep rate effect on the stable magnetic moment variation of the superconductors with the power (n = 10) and exponential V-I characteristics which obey the condition (2.44).

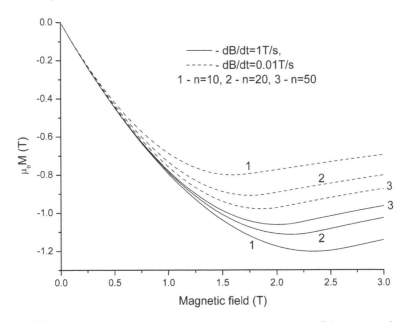

FIGURE 2.16   Effect of the sweep rate and n-value on the magnetization of the superconductor with the power V-I characteristic.

The results depict that the magnetic field dependence of the critical current of superconductor influence essentially on its magnetization during fully penetration modes. In these modes, the magnetic moment decreases when the external magnetic field increases. Overall, comparing the calculation results presented in Figures. 2.11-2.13, 2.15 and 2.16, it is seen that the magnetic field dependence of the critical current may quantitatively change the calculation results of the magnetic moment as the magnetization region increases.

2.8　MAGNETIC RELAXATION OF THE PARTIALLY PENETRATED SCREENING CURRENT IN SUPERCONDUCTORS

The study of the magnetic relaxation of superconductors is important investigations from applications as well as scientific point of view. The investigations of these nonequilibrium phenomena in the macroscopic approximation became a useful tool for understanding the microscopic mechanisms of pinning. That is why considerable attention is paid to the study of the magnetic relaxation problems both experimentally and theoretically. Let us find the scaling solutions for Maxwell's equations describing the magnetic relaxation of the partially penetrated screening current in terms of the power V-I characteristic described by equation (1.3).

Consider an ideally cooled half-infinite superconductor placed in the homogeneous external magnetic field B_a, which is parallel to its surface in the z-direction at the initial time. Assume that the external magnetic field after any field disturbance takes the constant value $B_{0,i}$ at $t = t_i$, and it does not change the constants J_{c0}, J_δ and n. This perturbation produces the extra magnetic induction and induces the screening current in the y-direction that will diffuse into the superconductor. In this case, Maxwell's equations describing the macroscopic decay of the induced electric field $E = E_y(x, t)$ are given by one-dimensional unsteady equation (2.5) with the corresponding value of ρ_d according to the relation (2.6), initial-boundary conditions

$$E(x, t_i) = E_0(x),$$

$$\frac{\partial E}{\partial x}(0, t) = 0, \quad E(\infty, t) = 0, \quad t \geq t_i \tag{2.45}$$

and the extra condition

$$\mu_0 \int_0^\infty J(x, t)\, dx = B_{0,i} \quad t \geq t_i \tag{2.46}$$

which follows from the conservation law of the induced screening current. Here, $E_0(x)$ is the distribution of the electric field at the instant when the external magnetic field ceases to rise. Second boundary and extra condition are based on the assumption that the screening current flows over the total volume of the superconductor, like in the computer simulations (for example, Brandt 1996).

Introduce the dimensionless variables $e = E/E_c$, $X = x/L_x$, $\tau = t/t_x$, where $L_x = B_{0,i}/\mu_0 J_{c0}$, $t_x = B_{0,i}^2/(\mu_0 J_{c0} E_c)$. The scaling solution of the problem under consideration can be found in the form

$$e = (\tau + \tau_0)^q\, W(Z),\ X = (\tau + \tau_0)^p Z,\ q = -n/(n + 1),\ p = 1/(n + 1)$$

where τ_0 is the constant to be determined. Then the partial differential equation is transformed to the ordinary differential equation

$$(n+1)\frac{d^2 W}{dZ^2} + \frac{Z}{n} W^{(1-n)/n} \frac{dW}{dZ} + W^{1/n} = 0 \tag{2.47}$$

with the boundary conditions

$$\frac{dW}{dZ}(0) = 0, \quad W(\infty) = 0 \tag{2.48}$$

Similarly to the sweep-rate problems investigated above, it is easy to show that both the nontrivial value of the scaling function $W(Z)$ and all the derivatives over Z equal zero at a certain point $Z = Z_0$. Then the value Z_0 is given by

$$\int_0^{Z_0} W^{1/n} dy = 1 \tag{2.49}$$

Equation (2.47) with conditions (2.48) and (2.49) has the analytical solution

$$W(Z) = \left[\frac{n-1}{2n(n+1)} (Z_0^2 - Z^2) \right]^{n/(n-1)}$$

where

$$Z_0 = \left[\frac{2n(n+1)}{n-1} \right]^{1/(n+1)} \left(\frac{1}{\Psi_1} \right)^{(n-1)/(n+1)} \quad , \quad \Psi_1 = \int_0^1 (1-y^2)^{1/(n-1)} dy$$

Therefore, relative to the dimensional variables the decay of the electromagnetic field in the superconductor is described by the following expressions

$$E(x, t) = E_a(t)(1 - x^2/x_0^2)^{n/(n-1)} \tag{2.50}$$

$$B(x, t) = B_{0,i} \left[1 - \frac{1}{\Psi_1} \int_0^{x/x_0} (1-y^2)^{1/(n+1)} dy \right] \tag{2.51}$$

Here,

$$E_a(t) = E_c \left(\frac{t_n}{t + t_0} \right)^{n/(n+1)} \quad , \quad t_n = \frac{n-1}{2n(n+1)} \frac{t_x}{\Psi_1^2} \tag{2.52}$$

The moving boundary of the magnetization region and the differential resistivity of the superconductor are given by

$$x_0(t) = \frac{B_{0,i}}{\mu_0 J_{c0} \Psi_1} \left(\frac{E_c}{E_a(t)} \right)^{1/n} , \tag{2.53}$$

$$\rho_d(t) = n \frac{E_c}{J_{c0}} \left(\frac{E_a(t)}{E_c} \right)^{(n-1)/n} \left(1 - \frac{x^2}{x_0^2} \right) \tag{2.54}$$

In the given exact solution constant t_0 is unknown. To find it, let us use the value of the electric field $E_{0,i} = E_0(0)$ on the surface of the superconductor at $t = t_i$. This leads to the following value: $t_0 = t_e - t_i$, where $t_e = t_n (E_c/E_{0,i})^{(n+1)/n}$ is the characteristic time of the magnetic flux relaxation during the partially penetrated mode, according to (2.52).

The curves in Figure 2.17 demonstrate the influence of the n-value on the distribution of the relaxing magnetic field in the magnetization region. It is easy to see that similarly to the sweep-rate stage, the value $n = 10$ describes satisfactorily the boundary value above which the distribution of the magnetic induction inside the superconductor is nearly linear similar to the sweep-rate stage. This estimation gives reasons for the usage of simplified methods of the relaxation analysis in superconductors as it was done, for example, in Schnack and Griessen (1992).

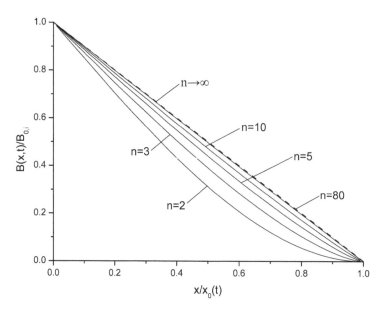

FIGURE 2.17 Scaling distribution of the magnetic field in a superconductor with various n-value during relaxation mode.

Thus, as in the sweep-rate modes, the relaxation modes are accompanied by the formation of the special states. The screening currents exist only in the limited part of the superconductor and penetrate into it with the finite rate

$$\frac{dx_0}{dt} = \frac{2n\Psi_1}{n-1}\frac{E_a(t)}{B_{0,i}}$$

The smooth transition of the disturbed values of the electromagnetic field to undisturbed ones takes place on the moving boundary $x_0(t)$: $E = 0$, $B = 0$, $\partial^k E/\partial x^k = 0$, $\partial^k B/\partial x^k = 0$, $k = 1, 2, 3....$ They lead to a nontrivial change in the differential resistivity of superconductor. It monotonously decreases and $\rho_d \equiv 0$ at $x_0(t) \geq 0$, as follows from (2.54).

Let us use the solution formulated above to define the magnetic moment of the superconducting slab of the half-thickness a before the full penetration of the screening current into its cross-section. According to (2.51), the result is

$$-\mu_0 M(t) = B_{0,i}[1 - \phi_n x_0(t)/a]$$

for all $x_0(t) < a$. Here,

$$\phi_n = 1 - \frac{1}{\Psi_1}\int_0^1 \psi(\eta)d\eta, \quad \psi(\eta) = \int_0^\eta (1 - y^2)^{1/(n-1)}dy$$

This formula can be reduced as follows

$$M(t) = M(t_i) + M_1\left(\frac{E_c}{E_{0,i}}\right)^{1/n}\left[\left(\frac{t - t_i + t_e}{t_e}\right)^{1/(n+1)} - 1\right]\phi_n, \quad t \geq t_i$$

where $M(t_i)$ is the magnetic moment of the slab at $t = t_i$ and $M_1 = B_{0,i}^2/(\mu_0^2 aJ_{c0}\Psi_1)$.

Differentiating $M(t)$ over time, it is easy to find

$$\frac{dM}{dt} = \frac{M_1\phi_n}{(n+1)(t - t_i + t_e)}\left(\frac{t - t_i + t_e}{t_n}\right)^{1/(n+1)}$$

As follows from this formula, there are two characteristic rates in the relaxation mode. If $t - t_i \ll t_e$, the rate of the magnetic moment relaxation is almost constant and can be estimated as follows

$$\frac{dM}{dt} \approx \frac{M_1 \phi_n}{(n+1)t_e} \left(\frac{t_e}{t_n} \right)^{1/(n+1)}$$

During long-time relaxation, that is at $t - t_i \gg t_e$, the decreasing relaxation rate equals

$$\frac{dM}{dt} \approx \frac{M_1 \phi_n}{(n+1)(t-t_i)} \left(\frac{t-t_i}{t_n} \right)^{1/(n+1)} , \quad t - t_i \gg t_e$$

In the logarithmic time scale, this formula may be written as follows

$$\frac{dM}{d \ln t} \approx \frac{M_1 \phi_n}{n+1} \frac{t}{t-t_i} \left(\frac{t-t_i}{t_n} \right)^{1/(n+1)}$$

The given analytical solution of the relaxation problem of magnetic flux during sweep-rate mode demonstrates the qualitative features of the phenomena occurring in this case. According to the Formulae (2.50)-(2.54), the relaxation of the partially penetrated screening current is determined by the decay of the electric field on the surface of the superconductor as during sweep-rate modes. This feature is characterized by the universal electric field distribution of the form (2.50). As a result, when all other things are being equal, the increase in the n-value will be accompanied by a decrease in the time decay of $E_a(t)$. Therefore, the penetration rate of the screening current in the superconductor and the relaxation rate of its magnetic moment will also decrease since $dM/dt \sim dx_0/dt \sim E_c/(n+1)$. In the limiting case at $n \to \infty$, the following relations take place $E_a \to 0$, $dx_0/dt \to 0$, $dM/dt \to 0$, $x_0 \to B_{0,i}/\mu_0 J_{c0}$. They are a direct consequence of a corresponding increase in the differential resistance of a superconductor when $\rho_d \to \infty$ at $n \to \infty$. As known, this reason underlies the explanation of the fundamental difference in the relaxation times in low- and high-temperature superconductors.

To illustrate the scaling solution possibility, the simulation results of the corresponding electrodynamic states are shown in Figures 2.18-2.21.

Figure 2.18 depicts the initial relaxation stage in the LTS ($n = 80$, $E_c = 10^{-4}$ V/m, $J_{c0} = 4 \times 10^9$ A/m^2). The electric field distribution $E_0(x)$ specified at $t = t_i$ was taken according to the numerical simulation of the sweep-rate stage. Here and below, the sweep-rate simulations are based on the numerical solution of equations (2.5), (2.6) under the conditions

$$\partial E/\partial x(0, t) = -\dot{B}_a, \quad E(x_0, t) = 0, \quad E(x, 0) = 0, \quad \mu_0 \int_0^{x_0} J(x, t) dx = B_a(t)$$

Curve 1 corresponds to the electric field distribution at $t_i = 2 \times 10^{-4}$ s when the external magnetic field was increased with the sweep rate $dB_a/dt = 1$ T/s and then it is fixed at $t \geq 2 \times 10^{-4}$ s. Solid curves are obtained using the numerical solution of the corresponding magnetic relaxation problem. Dashed Curves 4-6 are calculated according to the Formulae (2.50)-(2.53). Solid curves in Figure 2.19 correspond to the numerical solution to the problem of the sweep and relaxation stages, and dashed curves show the scaling solution during relaxation stages.

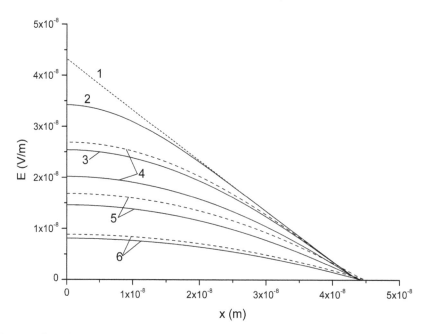

FIGURE 2.18 Distribution of the electric field in Nb-Ti at the relaxation stage (——— - numerical solution, - - - - scaling solution): 1 - $t = 2 \times 10^{-4}$ s, 2 - $t = 2.001 \times 10^{-4}$ s, 3 - $t = 2.005 \times 10^{-4}$ s, 4 - $t = 2.01 \times 10^{-4}$ s, 5 - $t = 2.02 \times 10^{-4}$ s, 6 - $t = 2.05 \times 10^{-4}$ s.

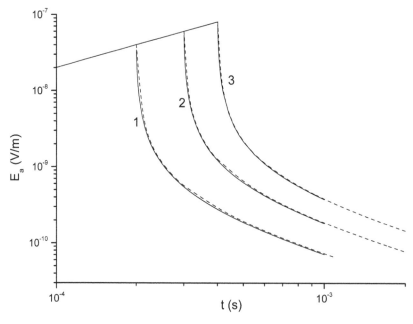

FIGURE 2.19 Time variation of the electric field on the surface of Nb-Ti: 1 - $t_i = 2 \times 10^{-4}$ s, 2 - $t_i = 3 \times 10^{-4}$ s, 3 - $t_i = 4 \times 10^{-4}$ s: ——— - numerical solution, - - - - scaling solution.

Figure 2.20 shows the curves describing the evolution of the moving boundary of the magnetization region in the HTS with parameters defined as follows $n = 23$, $E_c = 1.78 \times 10^{-4}$ V/cm, $J_{c0} = 2 \times 10^5$ A/cm². These curves were obtained under the assumption that the external magnetic field was increased with the sweep rate $dB_a/dt = 7.5 \times 10^{-3}$ T/s and it is fixed at $t \geq 2 \times 10^{-4}$ s.

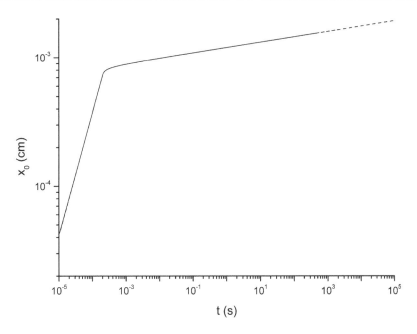

FIGURE 2.20   Dynamics of the moving boundary in YBCO at the sweep-rate and relaxation stages: ——— - numerical solution, - - - - - scaling solution.

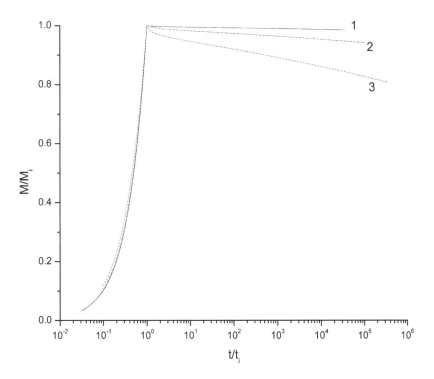

FIGURE 2.21   Time dependence of the magnetic moment of YBCO at the sweep-rate and relaxation stages: 1 - $dB_a/dt = 10^{-4}$ T/s, $t_i = 3 \times 10^{-4}$ s; 2 - $dB_a/dt = 10^{-4}$ T/s, $t_i = 10^{-3}$ s, 3 - $dB_a/dt = 10^{-3}$ T/s, $t_i = 3 \times 10^{-4}$ s.

The results in Figures 2.18-2.20 show that the scaling solution is a good approximation which can be used to describe exactly enough the relaxation modes. To demonstrate this capability, the magnetic moment evolution of the above-mentioned HTS is shown in Figure 2.21. The calculations of the sweep-rate stage were based on the numerical solution which is used to calculate relaxation stages.

Observation of the macroscopic quantities of superconductors is usually carried out by the study of the fully penetrated electromagnetic field. This analysis is a useful tool to estimate the critical current density, pinning potential, losses, etc. As a result, accomplished investigations allow one to get the macroscopic description explicating exactly enough the basic electromagnetic properties of superconductors placed in the changing field or carrying the transport current. Below, the macroscopic evolution peculiarities of the flux-creep states are discussed which are usually observed during voltage relaxation of the fully penetrated mode.

Relaxation of the electric field in a superconductor with the power V-I characteristic may be defined according to Vinokur et al. (1991), finding the scaling solution of equations (2.5) with conditions (2.45) in the form of a product of x- and t-dependent functions. So, the decay of the electric field and magnetic moment are given by the formulae

$$E(x, t) = E_c X(x) \left[\frac{t_{x,f}}{(n-1)(t-t_i+t_e)} \right]^{n/(n-1)} , \quad M(t) = -J_{c0} \xi(a, n) \left[\frac{t_{x,f}}{(n-1)(t-t_i+t_e)} \right]^{n/(n-1)}$$

where $t_{x,f} = \mu_0 J_{c0} a^2 / E_c$, the implicit function X obeys the integral equation

$$\int_0^X \frac{dy}{\sqrt{X_0^{(n+1)/n} - X^{(n+1)/n}}} = \sqrt{\frac{2n}{n+1}} \left(1 - \frac{x}{a} \right), \quad X_0 = \left(\frac{2n}{n+1} \right)^{n/(n-1)} \left(\int_0^1 \frac{dx}{\sqrt{1-x^{(n+1)/n}}} \right)^{2n/(1-n)}$$

and the coefficient ξ is equal to

$$\xi(a, n) = \frac{1}{a} \int_0^a dy \int_0^y X^{1/n} dx$$

which depends on the n-value and transverse dimension of a slab (for instance, $\xi \sim a/2$ at $n \gg 1$).

The characteristic time decay should be written as follows

$$t_e = \frac{\mu_0 J_{c0} a^2}{(n-1)E_c^{1/n}} \left(\frac{X_0}{E_{0,i}} \right)^{(n-1)/n}$$

if one will take into consideration the magnitude of the electric field on the surface of superconductor $E_{0,i}$ at $t = t_i$.

It is seen that the quantity t_e depends on the macroscopic parameters of the superconductor. In particular, it may be shown that the magnitude of the characteristic time is large at $n \sim 1$ and small at $n \gg 1$. Therefore, the initial stage of the magnetic moment relaxation in a superconductor, which has strong creep ($n < 10$), will be more detectable than that in a superconductor with the weak creep ($n > 10$). During this stage, the rate of the magnetic moment relaxation can be estimated as follows

$$\frac{dM}{dt} \approx \frac{J_{c0} \xi(a, n)}{t_e} \left[\frac{t_{x,f}}{(n-1)t_e} \right]^{1/(n-1)}$$

at $t - t_i \ll t_e$, that is it is practically constant and depends on the initial state of a superconductor and its macroscopic parameters. In particular, for the superconductors with the strong creep the values dM/dt is smaller than ones for the superconductors with the weak creep.

The long-time relaxation $(t - t_i \gg t_e)$ is characterized by the decreasing rate. It is equal to

$$\frac{dM}{dt} \approx \frac{J_{c0}\xi(a, n)}{t - t_i}\left[\frac{t_{x,f}}{(n-1)(t - t_i)}\right]^{1/(n-1)}$$

and does not depend on the initial distribution of the electric field. In accordance with this formula, the relaxation is also close to the logarithmic time-variation.

Thus, the above theoretical results indicate the existence of the magnetic relaxation peculiarities. The relaxation phenomena are characterized by the formation of the electromagnetic wave propagating in the superconductor with a finite rate. In this case, the electric and magnetic fields induced in a superconductor by the external perturbations smoothly approach their undisturbed values on the moving penetration boundary. The considerable influence of the flux-creep will be seen at $n < 10$ when the creep strongly affects the electromagnetic field distribution. The proposed analytical solutions also show that the main qualitative difference is in the dynamics of the electric field on the surface of the superconductor which determines the character of the magnetic relaxation.

2.9 CONCLUSIONS

The electrodynamic phenomena in the nonideal type-II superconductors have intrinsic features. The obtained solutions and the performed analysis of the magnetic flux penetration in superconductors with different nonlinearity types of the V-I characteristics show that the nontrivial distribution of the electromagnetic field is originated in the superconductor during flux-creep in response to electromagnetic perturbations.

It is shown that the critical state model based on the ideal V-I characteristic is zero approximation of the general solution of the Maxwell equations that describes macroscopic distribution of the induced electric field in the superconductors with real V-I characteristics, in particular with the power $(n > 10)$ and exponential $(J_{c0}/J_\delta < 0.1)$ V-I characteristics. As a result, similar to the Bean approximation, the flux-creep phenomena have a local penetration nature even though the steady voltage in the superconductor arises well before the disruption of the stability conditions of the superconducting state. Therefore, the flux-creep modes are characterized by the forming of the electromagnetic wave which propagates in a superconductor with a finite velocity. However, unlike the critical state model, special conditions are observed on the moving boundary of the magnetic flux leading to the existence of the nontrivial states. In contrast to the critical state model, all derivatives of the electrodynamic quantities of superconductors with respect to the spatial coordinate are equal to zero at the moving boundary of the magnetization region. Thereby, near the magnetic flux front, a smooth transition takes place from the value of the electric field induced by the external electromagnetic disturbance to the corresponding nonperturbed value. Besides, the nontrivial distribution of the differential resistivity in the superconductor takes place. It monotonically decreases toward the moving boundary of the magnetization region and depends on the magnetic field sweep rate or current charging. As a result, the losses and magnetic moment change in accordance with this local formation of the flux-creep state. At the same time, to correctly describe the electrodynamic states of nonideal type-II superconductors using the Bean approximation, the nonlinear relations—which describe the spatial-time distributions of the magnetic field, differential resistivity and the moving boundary—should be written by taking into account the corresponding nonlinear character of V-I characteristic.

The performed analysis also reveals that the use of the power and exponential V-I characteristics for a description of the flux-creep states may give practically similar results. However, the observance in the equivalence of the flux-creep states depends on n or δ values. The smaller n or higher δ, the greater the differences between the obtained results. A considerable difference will be seen at strong flux-creep states when $n < 10$ or $\delta > 0.1$. This conclusion should be taken into account when the equation of the I-V characteristics is chosen to describe the flux-creep states of HTS.

3

Physical Features of the Dissipative Phenomena in Superconductors in the Self-Field Modes

Knowledge of losses in superconductors is crucial for designing of superconducting devices. They lead to the heat release and may be characterized by a finite temperature rise at stable modes. Therefore, the loss reduction is an important topic in applied superconductivity.

The Bean model and its modifications including the field-dependent critical current density are often used to simulate the losses in nonideal type-II superconductors. They allow one to investigate the AC losses or the losses in superconductors placed in the changing parallel or transverse external magnetic fields (for example: Norris 1970, Prigozhin 1997, Däumling 1998, Carr 1983, Hong et al. 2010, Wilson 1983). However, as discussed above, the transport current in a superconductor cannot exceed the critical value in the CSM approximation. It is a result of the jump-like V-I characteristic used in the framework of the CSM. Moreover, the CSM does not allow one to investigate the losses as a function of the current charging rate (in general case current frequency), as follows from the Formula (2.4). Using general models (for example: Brandt 1996, Amemiya et al. 1998, Stavrev et al. 2001), which take into account the smooth nonlinear character of V-I characteristics, investigations of the losses may be made over more wide changing ranges of applied currents. At the same time, even in the framework of these simulating models, it is often assumed that the transport current cannot exceed a priory-defined critical value during current charging in the fully penetrated states. However, the smooth increase of the V-I characteristic leads to the existence of the stable supercritical fully penetrated states both in the low-T_c and high-T_c superconductors (Polak et al. 1973, Romanovskii and Watanabe 2005a). Many experiments prove this peculiarity. Therefore, let us discuss the basic physical features of the self-field dissipative phenomena in nonideal type-II superconductors with the continuously increasing V-I characteristic.

3.1 DISSIPATIVE PHENOMENA DURING CURRENT CHARGING WITH A CONSTANT RATE

As above-made, consider a perfectly cooled superconducting slab (Figure 2.2a) with a cross-sectional area $S = 2a \times 2b$ ($-a < x < a$, $-b < y < b$, $-\infty < z < \infty$, $b \gg a$) and the V-I characteristic is described by the power equation (1.3). Suppose that the current in the slab is zero at the initial time, and then it begins to increase with a constant rate dI/dt. Accordingly, $I(t) = dI/dt \times t$. In the framework of the CSM, the full penetration state is the maximum permissible since the superconductor is completely in the critical state (Figure 2.2b). In a superconductor with a power V-I characteristic, the fully penetrated current may further increase even over the critical value (Figure 2.6b). (The conditions defining the existence of the stable sub- and supercritical currents are discussed in Chapter 6 in detail).

Similar to the results obtained in section 2.1, it is easy to find the distribution of the electric field $E(x, t)$, current density $J(x, t)$, the moving current penetration front $x_c(t)$, the time of the fully penetrated state t_c and the averaged volume density of the heat release $G_c(t)$ in the framework of the CSM during current charging. They are described by the expressions

$$E(x, t) = \begin{cases} 0, & 0 \le x \le x_c \\ \dfrac{\mu_0}{4b}\dfrac{dI}{dt}(x - x_c), & x_c \le x \le a \end{cases}, \quad J(x, t) = J_{c0}[E(x, t)/E_c]^{1/n} \tag{3.1}$$

$$x_c(t) = a - \frac{1}{4b}\frac{I(t)}{J_{c0}}, \quad t_c = I_c(dI/dt)^{-1} \tag{3.2}$$

$$G_c(t) = \frac{1}{a}\int_{x_p}^{a} dx \int_{0}^{t} EJ_{c0} dt = \frac{\mu_0 a^2 J_{c0}^2}{6}\left[\frac{I(t)}{I_c}\right]^3 \tag{3.3}$$

when a current is charged with a constant charging rate dI/dt ($I(t) = \frac{dI}{dt}t$). The relations (3.1)-(3.3) are met when $I \le I_c = J_{c0}S$.

According to zero scaling solution obtained in 2.6, the corresponding time dependencies of the electric field, coordinate of the moving current penetration front and heat release in the superconductor ($\eta = 1$) are defined as follows

$$E(x, t) = \begin{cases} 0, & 0 \le x \le x_p \\ \dfrac{\mu_0}{4b}\dfrac{dI}{dt}(x - x_p), & x_p \le x \le a \end{cases} \tag{3.4}$$

$$x_p(t) = a - \left(\frac{n+1}{n}t\right)^{\frac{n}{n+1}}\left(\frac{\mu_0}{4b}\frac{dI}{dt}\right)^{\frac{n-1}{n+1}}\left[\frac{E_c}{\mu_0^n J_{c0}^n}\right]^{\frac{1}{n+1}} \tag{3.5}$$

$$G_p(t) = \frac{\mu_0 I_c^2}{S}\frac{n(n+1)}{(2n+1)(3n+2)}\left(\frac{n+1}{n}\right)^{\frac{2n+1}{n+1}}\left(\frac{a}{4b}\right)^{\frac{n}{n+1}}\left(\frac{I(t)}{I_c}\right)^{\frac{3n+2}{n+1}}\left[\frac{E_c}{\mu_0 \frac{dI}{dt}}\right]^{\frac{1}{n+1}} \tag{3.6}$$

when the charging current increases with a constant rate. These formulae describe the partially penetrated modes ($0 < x_p < a$) in the time window

$$0 < t \le t_f = \frac{n}{n+1}\frac{\mu_0 J_{c0}}{E_c^{1/n}}a^{(n+1)/n}\left(\frac{\mu_0}{4b}\frac{dI}{dt}\right)^{(1-n)/n} $$

according to the Formula (2.41). Then, the charged current I_f and electric field on the surface E_f is equal to

$$I_f = \frac{n}{n+1}\frac{\mu_0 J_{c0}}{E_c^{1/n}}a^{\frac{n+1}{n}}\left(\frac{\mu_0}{4b}\frac{dI}{dt}\right)^{\frac{1-n}{n}}\frac{dI}{dt}, \quad E_f = \frac{a\mu_0}{4b}\frac{dI}{dt} \tag{3.7}$$

just after the full penetration mode ($x_p = 0$). Note that the expressions (3.4)-(3.7) comply with the limiting transition to the CSM formulated in the form (3.1)-(3.3) since $x_p \to x_c$, $G_p \to G_c$, $I_f \to I_c$ at $n \to \infty$.

To describe the electrodynamic states during full penetration modes, let us use the so-called zero-dimensional model. It is based on the assumption of a uniform distribution of the charged current over the cross-section of a superconductor. Accordingly, the time variation of the electric field in the superconductor with the power V-I characteristic (1.3) is written as follows

$$E(t) = E_c[I(t)/I_c]^n, \quad I(t) = \frac{dI}{dt}t \tag{3.8}$$

Then, the volume density of the self-field losses is equal to

$$G_f = \frac{1}{a}\int_0^a dx \int_0^t EJdt = \frac{\mu_0 I_c^2}{4ab(n+2)}\left[\frac{I(t)}{I_c}\right]^{n+2} \frac{E_c}{\mu_0 \frac{dI}{dt}} \tag{3.9}$$

For a general analysis of the electrodynamic states of the superconductor at the partial and full penetration modes, the system of Maxwell's equations should be used with the corresponding initial and boundary conditions. In the case under consideration, according to Equation (1.14) and results discussed in section 2.4, let us solve the following one-dimensional initial-boundary problem of the form

$$\mu_0 \frac{\partial J}{\partial t} = \frac{\partial^2 E}{\partial x^2}, \quad t > 0, \quad 0 \le x_p < x < a$$

$$E(x, 0) = 0,$$

$$E(x_p, t) = 0, \quad x_p > 0, \quad \frac{\partial E}{\partial x}(0, t) = 0, \quad x_p = 0 \tag{3.10}$$

$$\frac{\partial E}{\partial x}(a, t) = \frac{\mu_0}{4b}\frac{dI}{dt}$$

$$4b \int_{x_p}^a J(x, t)dx = I(t)$$

taking into account the power V-I characteristic (1.3). In this case, the average density of the heat release on the right side of the superconducting slab is determined numerically as follows:

$$G = \frac{1}{a}\int_0^a dx \int_0^t EJdt.$$

In the formulated problem (3.10), it was taken into account that the electric field in the superconductor is absent at the initial time, the electric field is zero at the moving current penetration boundary and the electric field distribution in the superconductor is symmetric in the full penetration mode.

Using these models, one can discuss the physical features of the electrodynamic state formation in a superconductor during partial and full penetration modes when the current is charged with a constant rate from zero value.

The typical time dependencies of the electric field (Figure 3.1a), current density (Figure 3.1b) and coordinate of the moving current front (Figure 3.1c) during current charging into the superconductor with different n-values are shown in Figure 3.1. The Curves 4, 4' and 5, 5' describe the corresponding values on the surface and in the center of the superconductor in Figures 3.1a and 3.1b. The fully penetrated time t_f and the corresponding value of the electric field E_f are depicted in Figure 3.1 as well as the time at which a priory defined values E_c and J_{c0} are achieved. The change of the electric field on the surface of superconductor and self-field losses at $n = 10$ and different current charging rates are shown in Figure 3.2. In both cases, the simulations were made at $B = 10$ T, $E_c = 10^{-6}$ V/cm, $J_{c0} = 1.52 \times 10^4$ A/cm^2, $a = 10^{-3}$ cm, $b = 10^{-2}$ cm.

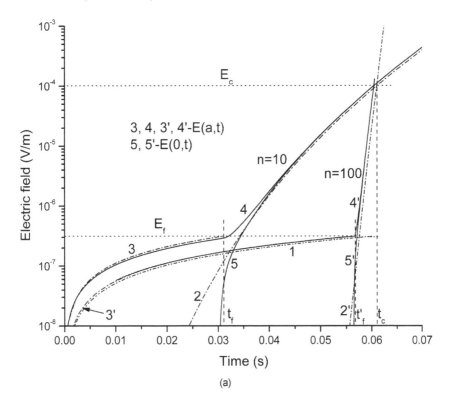

(a)

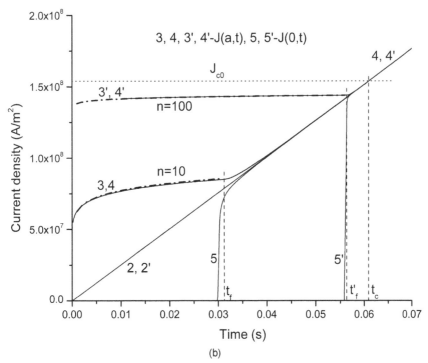

(b)

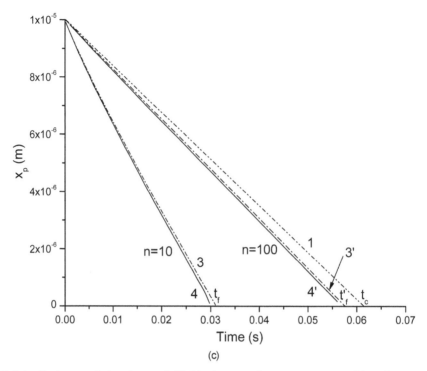

(c)

FIGURE 3.1 Evolution of the electric field (a), density of transport current (b) and moving current penetration front (c) in superconducting slab at dI/dt = 10 A/s simulated within the framework of various approximations: 1 – the critical state model (3.1), (3.2); 2, 2′ – zero-dimensional model (3.8); 3, 3′ – scaling approximation (3.4), (3.5); 4, 5, 4′, 5′ – numerical solution of the problem (3.10).

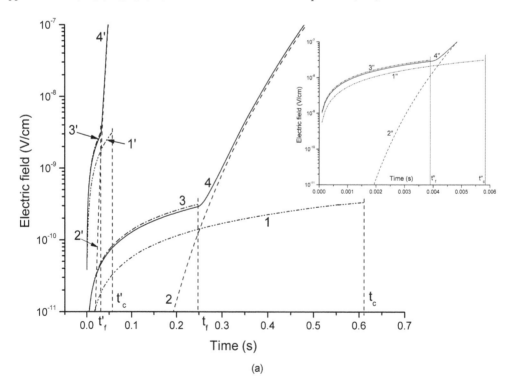

(a)

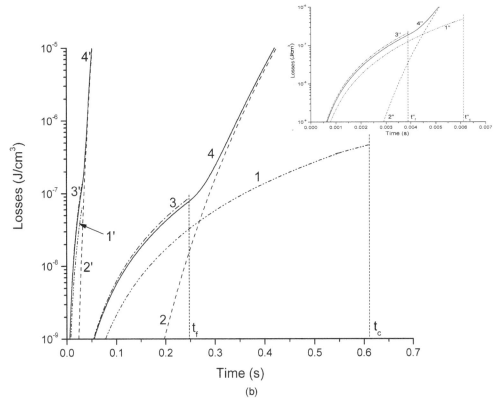

FIGURE 3.2 Change in time of the electric field on the surface (a) and self-field losses (b) at different current charging rates simulated in the framework of various approximations: 1, 1′, 1″ – CSM; 2, 2′, 2″ – zero-dimensional model; 3, 3′, 3″ – scaling approximation; 4, 4′, 4″ – numerical solution. Here, (1-4) - dI/dt = 1 A/s, (1′-4′) - dI/dt = 10 A/s, (1″-4″) - dI/dt = 100 A/s.

The presented results demonstrate the characteristic formation features of the operating modes in superconductors with continuously increasing V-I characteristics which must be taken into account when calculating or measuring losses.

First, there are three characteristic regimes of the current penetration modes: partial, full transient and full quasi-static. Their duration depends on the n-value, current charging rate as well as on the critical current density and cross-sectional area of the superconductor, as calculations show. In particular, the lower the n-value or the higher the current charging rate, the shorter the formation time of the partially penetrated states, the boundary of which is defined by the value E_f. It is important to note that under the existence of the full transient mode with a substantially heterogeneous electric field distribution at the initial stage (Figures 3.1 and 3.2), the measurement of the critical current density of a superconductor is incorrect near E_f value. The duration of the full transient mode depends on the current charging rate, n-value and other parameters. These dependencies lead to the fact that the electric field values of the practical static fully penetrated states, closed to zero-dimensional ones, are shifted toward the higher electric field values compared to the value of E_f. These features do not take into account the CSM, but they must be taken into consideration in the experimental determination of the critical current of superconductors.

Second, it turns out that the conventionally set value I_c may be noticeably higher than the full penetration current I_f of a superconductor with the constantly increasing V-I characteristic. For the power V-I characteristic, this feature is observed at any finite n-value. In other words, the full penetration modes can exist at currents significantly lower a priori defined critical current of the superconductor. For these subcritical states, the critical state model does not allow describing the electrodynamic states of superconductors because they are the fully penetrated modes. The value I_f

decreases with a decrease in n-value and the current charging rate as well as with an increase in the thickness of the superconductor and its critical current density according to (3.7). If the quantity I_f decreases, the range of the subcritical currents, which are in the fully penetrated mode, increases. Accordingly, the use of the CSM will lead to a noticeable deviation of the obtained results from the results that were obtained while taking into account the constantly increasing V-I characteristic. Thus, the CSM does not have its physical meaning in a wide range of the subcritical currents, namely, at $I_f < I < I_c$.

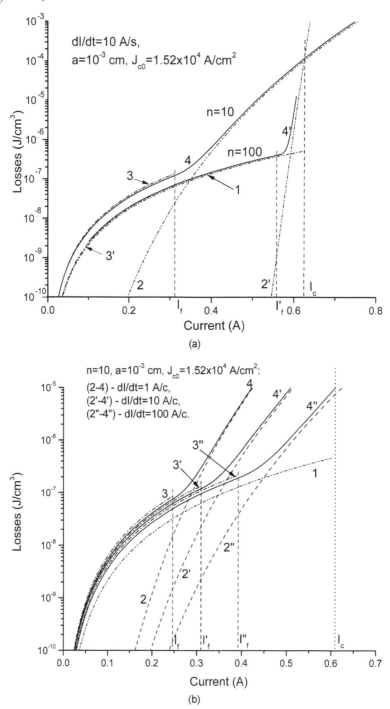

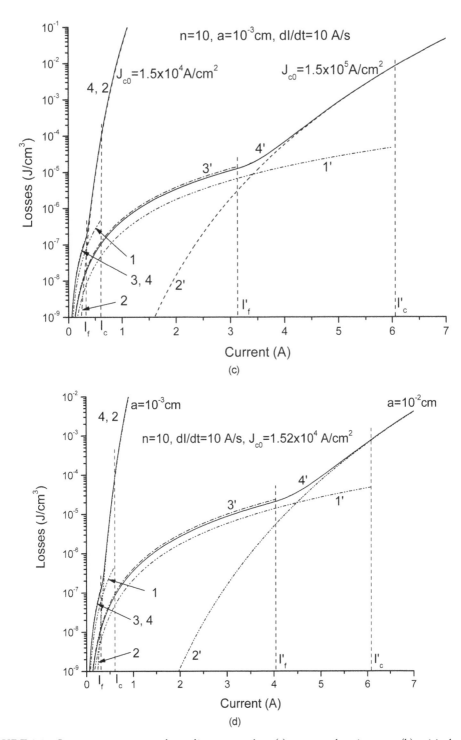

FIGURE 3.3 Losses versus current depending on n-values (a), current charging rates (b), critical current densities (c) and thicknesses of superconductor (d): 1, 1′ – CSM (3.3); 2, 2′, 2″ – zero-dimensional model (3.9); 3, 3′, 3″ – scaling approximation (3.6); 4, 4′, 4″ – numerical solution the system of equations (3.10).

To illustrate the validity of the proposed models, the detailed calculating results of the self-field losses as a function of the charging transport current are presented in Figure 3.3.

It is seen that the self-field losses are described with satisfactory accuracy by a scaling approximation during partial penetration modes, and the losses analysis can be carried out based on the simplest zero-dimensional model both at the subcritical and supercritical currents during full penetration modes ($E > E_f$). Figures 3.3a and 3.3b show that the self-field losses do not change noticeably during partial current penetration modes as in the full current penetration modes when the n-value and the current charging rate change. At the same time, variation in the critical current density of a superconductor or its thicknesses (Figures 3.3c and 3.3d) have a significant effect on the self-field losses, both in the partial and full current penetration modes. Accordingly, the difference in the loss calculations made within the framework of the CSM and scaling approximation becomes more noticeable with increasing the critical current density or thickness of a superconductor.

For all modes depicted in Figure 3.3, the calculated values of the losses obtained within the framework of the CSM are always lower than the corresponding values obtained for a superconductor with continuously increasing V-I characteristics. Moreover, the CSM leads to the results that have no physical meaning in the subcritical currents range from I_f to I_c, as discussed above.

To reduce the error of the CSM, the critical current of the superconductor I_c may be replaced with value I_f in the Formula (3.1). Then, the results of the calculations will turn out to be more correct and will show the dependence on the n-value or the current charging rate.

The discussed results lead to the conclusion that the volume density of the self-field losses may be calculated by the Formulae

$$G(I) = \frac{\mu_0 I_c^2}{4ab} \frac{n(n+1)}{(2n+1)(3n+2)} \left(\frac{n+1}{n}\right)^{\frac{2n+1}{n+1}} \left(\frac{a}{4b}\right)^{\frac{n}{n+1}} \left(\frac{I}{I_c}\right)^{\frac{3n+2}{n+1}} \left[\frac{E_c}{\mu_0 \, dI/dt}\right]^{\frac{1}{n+1}} \tag{3.11}$$

during partial penetration mode ($I < I_f$) and

$$G(I) = \frac{\mu_0 I_c^2}{4ab} \frac{n(n+1)}{(2n+1)(3n+2)} \left(\frac{n+1}{n}\right)^{\frac{2n+1}{n+1}} \left(\frac{a}{4b}\right)^{\frac{n}{n+1}} \left(\frac{I_f}{I_c}\right)^{\frac{3n+2}{n+1}} \left[\frac{E_c}{\mu_0 \, dI/dt}\right]^{\frac{1}{n+1}}$$

$$+ \frac{I_c^2}{4ab(n+2)} \left[\left(\frac{I}{I_c}\right)^{n+2} - \left(\frac{I_f}{I_c}\right)^{n+2}\right] \frac{E_c}{dI/dt} \tag{3.12}$$

during full penetration mode ($I_f < I$) when the current increases with a constant rate.

The simulated results of the self-field losses as a function of the transport current magnitude are presented in Figure 3.4, both for the partial and full penetration states during current charging with the break

$$I(t) = \begin{cases} \dfrac{dI}{dt} t, \, t < t_m, \, dI/dt = \text{const} \\ I_m = \text{const}, \, t \geq t_m \end{cases}$$

They were obtained using the Formula (3.3) (Curve 1), the Expressions (3.11) and (3.12) (Curves 2 and 2′ depicted by markers) and the numerical solution of the problem described by the system of Equation (3.10) (Curves 3 and 3′). Besides, the Formula (3.3) was used to simulate the self-field losses during partial penetration mode in the framework of the CSM replacing I_c with I_f (Curves 4 and 4′).

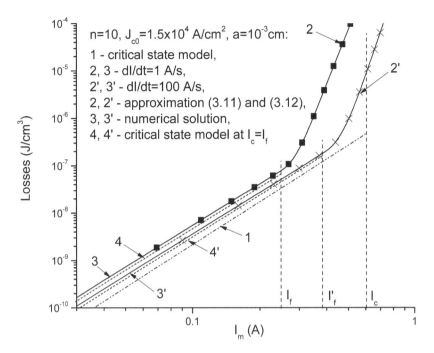

FIGURE 3.4 Self-field losses versus current magnitude I_m at different current charging rates. Values I_m are depicted by markers.

The presented curves, which confirm the previously formulated conclusions, show the use of legitimacy of the approximation based on the scaling and zero-dimensional approximations during self-field losses simulations. Besides, it is seen that the simplest Bean approximation can also be used when the full penetration current I_f calculated for the continuously increasing V-I characteristic is used instead of the a priori defined critical current I_c. In this case, the self-field losses are described by the formula

$$G_{c,f}(I) = \frac{\mu_0 I^3}{384 J_{c0} b^3 a^{(n+3)/n}} \left(\frac{n+1}{n} \right)^3 \left(\frac{4b}{\mu_0} \frac{E_c}{dI/dt} \right)^{3/n}, \quad I(t) = dI/dt \times t$$

in the partial penetration modes. This approximation clearly demonstrates the above-discussed dependence of the self-field losses on the n-value, current charging rate and the strong influence of the transverse dimensions of superconductor and its critical current density on the energy dissipation phenomena.

The Formulae (3.11) and (3.12) result in the following estimates

$$G \sim \left(\frac{I}{I_c} \right)^{\frac{3n+2}{n+1}} \left(\frac{dI}{dt} \right)^{-1/(n+1)} \tag{3.13}$$

during partial penetration mode and

$$G \sim \left(\frac{I}{I_c} \right)^{n+2} \left(\frac{dI}{dt} \right)^{-1} \tag{3.14}$$

during full penetration mode. They show that the losses increase in the partial penetration modes is less intense than in the full penetration ones. In the full penetration states, the self-field losses are strongly dependent on the current charging rate, transverse dimensions of the superconductor and its critical current density. Besides, according to the estimate (3.14), the losses in superconductors with sharply rising V-I characteristics (at high n-values) will rapidly increase after the current mode becomes fully penetrated (Figure 3.4). In the limiting case $G_f \to \infty$ at $n \to \infty$, since a jump-like transition from the superconducting state to the normal one occurs in superconductors with an ideal V-I characteristic.

Thus, direct use of the critical state model in which the voltage in a superconductor is missing at $I < I_c$, first, leads to an underestimate of the self-field heat release and, second, does not correctly describe the losses in superconductors with the real V-I characteristic in a wide range of the subcritical currents $I_f < I < I_c$. The inaccuracy depends on the character of the V-I characteristic rise, current charging rate, transverse dimensions of the superconductor and density of its critical current. At the same time, the scaling and zero-dimensional models make it possible to find the self-field losses with satisfactory accuracy, both at partial and full penetration states in superconductors with continuously increasing V-I characteristics.

3.2   SELF-FIELD LOSSES IN THE SERRATED (TRIANGULAR-WAVE) CURRENT CHARGING MODE

The formulae presented above allow one to investigate the self-field losses in more general cases, in particular during serrated modes (Figure 3.5a). In these modes, the current as a function of time is calculated as follows

$$I(t) = \begin{cases} \dfrac{dI}{dt}t, & t \le t_1 \\[2mm] \pm I_m \mp \dfrac{dI}{dt}t, & t_{2k-1} < t < t_{2k-1}, k = 1, 2, 3, \ldots \end{cases} \tag{3.15}$$

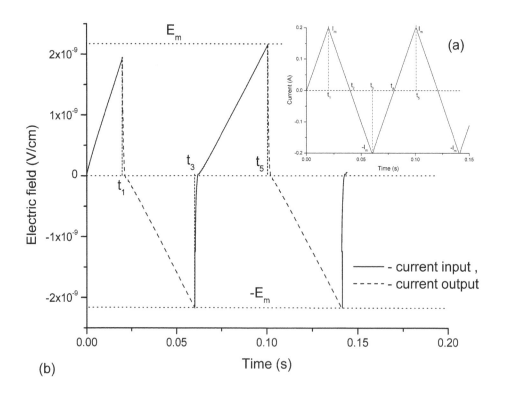

(a)

(b)

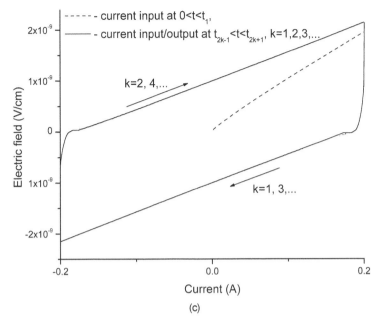

FIGURE 3.5 Time windows (a) and electric field on the surface of superconductor as a function of time (b) and current (c) during serrated mode at $I_m = 0.2$ A and $|dI/dt| = 10$ A/s.

Figure 3.5b shows the typical evolution of the electric field on the surface of superconducting slab $E_a = E(a, t)$ during partially penetrated mode. The corresponding $E_a(I)$ dependence is depicted in Figure 3.5c. The numerical simulation was made by means of the 1D model (3.10) for the current mode (3.15) using the following parameters: $a = 10^{-3}$ cm, $b = 10^{-2}$ cm, $n = 10$, $E_c = 10^{-6}$ V/cm, $J_{c0} = 1.52 \times 10^4$ A/cm^2, $T_{cB} = 26.1$ K.

As follows from Figures 3.5b and 3.5c, there exist the initial and repeated modes during which the variation of the electric field occurs. The initial stage takes place when the transport current increases from initial zero value to I_m ($0 < t < t_1 = t_m$, $t_m = I_m/|dI/dt|$). Therefore, the Formula (3.11) can be directly used to calculate the self-field losses in this partially penetrated mode existing in the time window $0 < t < t_1$. Accordingly, the following formula

$$G_{1,p}(t) = \frac{\mu_0 I_c^2}{S} \frac{n(n+1)}{(2n+1)(3n+2)} \left(\frac{n+1}{n}\right)^{\frac{2n+1}{n+1}} \left(\frac{a}{4b}\right)^{\frac{n}{n+1}} \left(\frac{1}{I_c}\frac{dI}{dt}t\right)^{\frac{3n+2}{n+1}} \left[\frac{E_c}{\mu_0 \frac{dI}{dt}}\right]^{\frac{1}{n+1}} \tag{3.16}$$

maybe utilized. In this initial stage, the maximum value of the electric field on the surface of the slab less than the subsequent maximum values $|E_m|$ which takes place in the repeated modes during times windows $t_{2k-1} < t < t_{2k+1}$, $k = 1, 2, 3, \ldots$ when the electric field changes in the range from E_m to $-E_m$ (Figures. 3.5b and 3.5c).

In the repeated modes, the electric field variation on the surface of a superconductor has two stages (Figure 3.5b) which are characterized by the following features. Initially, when the electric field decreases from $|E_m|$ to zero, the electric field changes very quickly. So, this time window is very small. However, its variation from zero to $|E_m|$ is taking place at an almost constant rate in the periods $t_{2k-1} < t < t_{2k+1}$, $k = 1, 2, 3\ldots$. Such a change in the electric field allows one to use Formulae (3.11) and (3.12) to calculate the self-field losses during serrated mode by introducing into them the effective values of the current change with time and its charging rate.

First, the value of the transport current I constantly increases with increasing time in the formula (3.11). Therefore, to use it and calculate the self-field losses during time windows $t_{2k-1} < t < t_{2k+1}$, $k = 1, 2, 3\ldots$, the following equivalent value of the current change in time

$$I^*(t) = \frac{1}{2}\frac{dI}{dt}(t - t_{2k-1}), \quad t_{2k-1} = (2k-1)t_m < t < (2k+1)t_m, \quad k = 1, 2, 3, \ldots$$

maybe utilized in both output and input modes.

Second, assuming that the electric field on the surface of the superconductor must have the same value in an equivalent approximation as in a real mode, the effective value of dI^*/dt in the Formula (3.11) may be defined as follows

$$\frac{dI^*}{dt} = \frac{1}{\sqrt{2}}\frac{dI}{dt}$$

Then the self-field losses may be described by the formula

$$G_{k+1,p}(t) = 2\frac{\mu_0 I_c^2}{S}\frac{n(n+1)}{(2n+1)(3n+2)}\left(\frac{n+1}{n}\right)^{\frac{2n+1}{n+1}}\left(\frac{a}{4b}\right)^{\frac{n}{n+1}}\left[\frac{dI}{dt}\frac{t-t_{2k-1}}{2I_c}\right]^{\frac{3n+2}{n+1}}\left[\frac{\sqrt{2}E_c}{\mu_0\,dI/_{dt}}\right]^{\frac{1}{n+1}} \quad (3.17)$$

during partially penetrated modes in the times windows $t_{2k-1} < t < t_{2k+1}$, $k = 1, 2, 3\ldots$. As a result, the total value of the self-field losses is equal to $G_p(t) = G_{1,p}(t) + G_{2,p}(t) + G_{3,p}(t) + \ldots$ in the framework of the scaling approximation.

The results of the losses simulations during partial penetration mode when the transport current is charged with different rates are depicted in Figure 3.6. The following models were used: the modified scaling approximation (3.16) and (3.17) and the 1D approximation based on the system of Equations (3.10) during the current mode described by the Formula (3.15). For comparison, the losses, which give the Bean model during serrated current charging mode (3.15) are also shown.

It is seen that the numerical and scaling simulations coincide with a good degree of accuracy, and the error between them decreases with increasing dI/dt. At the same time, the Bean model leads to the essential error even during partially penetrated mode because the value of I_c is defined by a priory defined critical value of the electric field E_c. The total increase of the error in the Bean approximation is explained by the fact that its error in a single cycle accumulates with time.

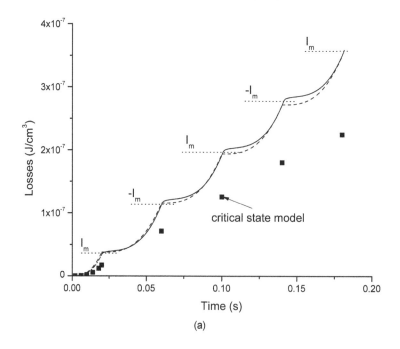

(a)

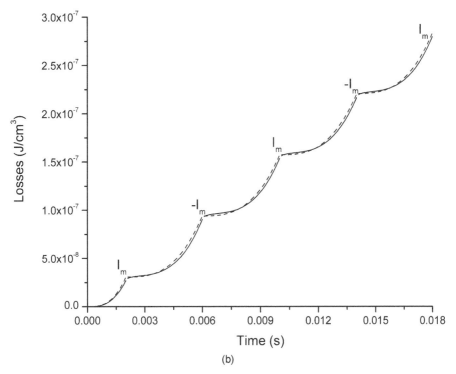

(b)

FIGURE 3.6 Results of the numerical and scaling computations of the self-field losses during partially penetrated cycle in the serrated mode ($I_m = 0.2$ A) with different current charging rates (a - $|dI/dt| = 10$ A/s, b - $|dI/dt| = 100$ A/s) simulated in the framework of various approximations: ———— - numerical simulation, – – – – – scaling approximation , ■ - Bean approximation. Here, $I_f = 0.305$ A, $I_c = 0.61$ A.

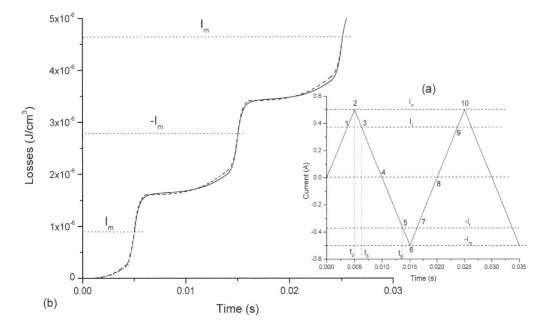

(b)

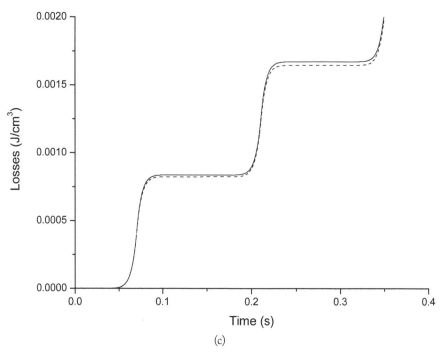

(c)

FIGURE 3.7 Results of numerical simulations (——) and approximate calculations (– – – –) of the self-field losses during fully penetrated cycle: (a, b) - $I_m = 0.5$ A, $|dI/dt| = 100$ A/s; (c) - $I_m = 0.7$ A, $|dI/dt| = 10$ A/s. Here, $I_f = 0.305$ A, $I_c = 0.61$ A.

Using the approximation given by the relations (3.16) and (3.17) and the Formula (3.9), it is easy to formulate the approximation allowing the analytical calculating of the self-field losses during fully penetrated states. The possibility of usage of such approximations depicts Figure 3.7 for the superconductor under consideration at different magnitudes of the charging currents and current charging rates. Numerical simulations were based on the 1D-approximation under the model described by Equations (3.10). The following formulae were used during approximate calculations ('scaling + full penetration' models) in the corresponding stages of the $[(I_m) \div (-I_m) \div (I_m) \div \ldots]$-cycles:

$$\text{mode } (0-1): G_{0,1}(t) = \frac{\mu_0 I_c^2}{S} \frac{n(n+1)}{(2n+1)(3n+2)} \left(\frac{n+1}{n}\right)^{\frac{2n+1}{n+1}} \left(\frac{a}{4b}\right)^{\frac{n}{n+1}} \left(\frac{1}{I_c}\frac{dI}{dt}t\right)^{\frac{3n+2}{n+1}} \left[\frac{E_c}{\mu_0 \, dI/dt}\right]^{\frac{1}{n+1}},$$

$$\text{mode } (1-2): G_{1,2}(t) = \frac{\mu_0 I_c^2}{S} \frac{n(n+1)}{(2n+1)(3n+2)} \left(\frac{n+1}{n}\right)^{\frac{2n+1}{n+1}} \left(\frac{a}{4b}\right)^{\frac{n}{n+1}} \left(\frac{I_f}{I_c}\right)^{\frac{3n+2}{n+1}} \left[\frac{E_c}{\mu_0 \, dI/dt}\right]^{\frac{1}{n+1}}$$
$$+ \frac{E_c I_c^2}{(n+2)S \, dI/dt}\left[\left(\frac{1}{I_c}\frac{dI}{dt}t\right)^{n+2} - \left(\frac{I_f}{I_c}\right)^{n+2}\right],$$

$$\text{mode } (2-3): G_{2,3}(t) = G_{1,2}(t_2) + \frac{E_c I_c^2}{(n+2)S \, dI/dt}\left\{\left(\frac{I_m}{I_c}\right)^{n+2} - \left[\frac{1}{I_c}\left(I_m - \frac{dI}{dt}(t-t_m)\right)\right]^{n+2}\right\},$$

mode $(3-4-5): G_{3,5}(t) = G_{2,3}(t_{23}) +$

$$2\frac{\mu_0 I_c^2}{S} \frac{n(n+1)}{(2n+1)(3n+2)}\left(\frac{n+1}{n}\right)^{\frac{2n+1}{n+1}}\left(\frac{a}{4b}\right)^{\frac{n}{n+1}}\left(\frac{1}{2I_c}\frac{dI}{dt}(t-t_{23})\right)^{\frac{3n+2}{n+1}}\left[\frac{\sqrt{2}E_c}{\mu_0 \frac{dI}{dt}}\right]^{\frac{1}{n+1}}, \; t_{23} = t_m + \frac{I_m - I_f}{dI/dt}$$

mode $(5-6): G_{5,6}(t) = G_{3,5}(t_{35}) +$

$$\frac{E_c I_c^2}{(n+2)S\frac{dI}{dt}}\left\{\left[\frac{1}{I_c}\left(I_f - \frac{dI}{dt}(t-t_{35})\right)\right]^{n+2} - \left(\frac{I_f}{I_c}\right)^{n+2}\right\}, \; t_{35} = t_m + \frac{I_m + I_f}{dI/dt}$$

and so on. The boundaries of the calculating ranges written are shown in Figure 3.7a.

It is seen that the proposed formulae are reasonable during both partially and fully penetrated modes. As a result, they allow one to find the maximum value of the self-field losses in the $[(I_m) \div (-I_m)]$-cycles. It is equal to

$$G_{max}\left(I_m, \frac{dI}{dt}\right) = 2\frac{\mu_0 I_c^2}{S}\left\{\begin{array}{l} \dfrac{n(n+1)}{(2n+1)(3n+2)}\left(\dfrac{n+1}{n}\right)^{\frac{2n+1}{n+1}}\left(\dfrac{a}{4b}\right)^{\frac{n}{n+1}}\left(\dfrac{I_m}{I_c}\right)^{\frac{3n+2}{n+1}}\left[\dfrac{\sqrt{2}E_c}{\mu_0 \frac{dI}{dt}}\right]^{\frac{1}{n+1}}, \quad I_m \leq I_f \\[4ex] \dfrac{n(n+1)}{(2n+1)(3n+2)}\left(\dfrac{n+1}{n}\right)^{\frac{2n+1}{n+1}}\left(\dfrac{a}{4b}\right)^{\frac{n}{n+1}}\left(\dfrac{I_f}{I_c}\right)^{\frac{3n+2}{n+1}}\left[\dfrac{\sqrt{2}E_c}{\mu_0 \frac{dI}{dt}}\right]^{\frac{1}{n+1}} + \\[4ex] \dfrac{E_c}{\mu_0(n+2)S\frac{dI}{dt}}\left[\left(\dfrac{I_m}{I_c}\right)^{n+2} - \left(\dfrac{I_f}{I_c}\right)^{n+2}\right], \quad I_m > I_f \end{array}\right.$$

Correspondingly, the maximum losses during the full cycle $[(I_m) \div (-I_m) \div (I_m)]$ are equal to $2G_{max}$. The results of the G_{max}-simulation are shown in Figure 3.8. It presents the dependence of the self-field losses on the charged current during half-cycle from I_m to $-I_m$ in the wide range of I_m-variation at $dI/dt = 10$ A/s for different thicknesses of the superconducting slab under consideration. The corresponding numerical and CSM calculations were also performed. In the latter case, the modified Bean approximation was also used to simulate the self-field losses during partial penetration mode replacing I_c with I_f. The total self-field losses as a function of the current charging rate are presented in Figure 3.9. In general, the results presented that the scaling and fully penetrated approximations may describe the electrodynamic states of the technical superconductors with satisfactory accuracy even at cycle operating modes.

Using the above written analytical formulae, it is easy to compare the losses that take place during partially and fully penetrated modes. Figure 3.10 shows the 'structure' of the self-field losses during different stages of the serrated cycles. It is seen that the higher dI/dt, the more important the role of the losses during partially penetrated mode according to the estimates (3.13) and (3.14).

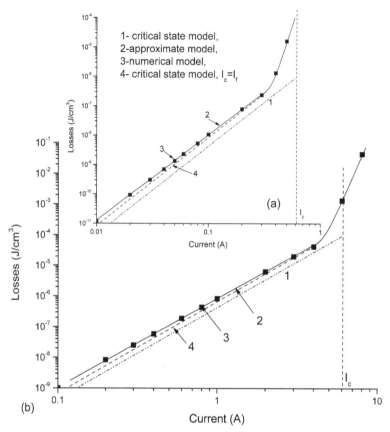

FIGURE 3.8 Maximum self-field losses versus current during serrated half-cycle $(I_m) \div (-I_m)$ in the superconductor with various thicknesses which were simulated under different approximations: (a) - $a = 10^{-3}$ cm, (b) - $a = 10^{-2}$ cm.

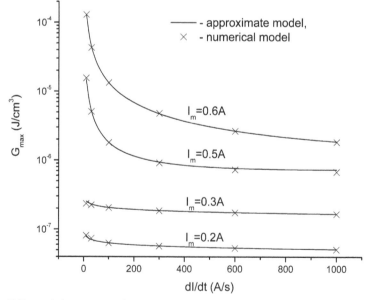

FIGURE 3.9 Effect of the current charging rate on the self-field losses during serrated half-cycles $(I_m) \div (-I_m)$.

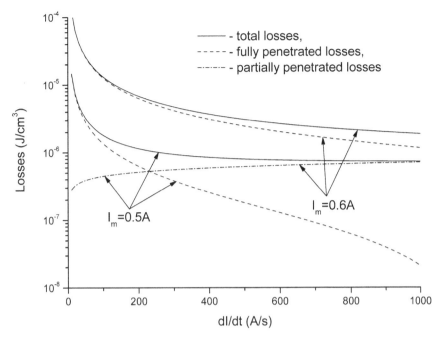

FIGURE 3.10   Influence of the partially and fully penetrated states on the total self-field losses during fully penetrated modes.

3.3   CONCLUSIONS

The proposed above-mentioned description of the self-field losses in superconductors allow one to analyze the dissipative states in a superconductor during different operating modes. They are useful to calculate the self-field losses at different current charging conditions without the fitting of the effective value of the critical current. The written solutions can explain the basic features of the dissipative phenomena in superconductors, in particular the frequency dependence of the losses. It is shown that there exists the boundary value of the current charging rate or frequency after which the losses during partially penetrated states may exceed the losses in the fully penetrated modes in the transport current cycles. This quantity depends on the properties of the superconductor and decreases with decreasing the magnitude of the charged current.

Direct use of the CSM cannot lead to correct dependencies of the self-field losses in the subcritical current range $I_f < I < I_c$. At the same time, the 'scaling + fully penetrated approximations' can be used to determine the losses in many practically important operating modes. First, it may be utilized for the analysis of the hysteresis losses, as it was done in section 2.4. Second, they are possible to calculate the losses in superconducting composites during the cycle-changing of the charging currents. Third, these approximations are realized for superconductors with arbitrary types of the nonlinearity of their voltage-current characteristics.

4

Thermo-Magnetic Phenomena in Low-T_c and High-T_c Superconductors

Many experimental and theoretical investigations of the electrodynamic phenomena both in low-T_c and high-T_c superconductors are devoted to the study of the magnetic instability conditions of their superconducting states. The magnetic instabilities lead to a spontaneous increase in the temperature of a superconductor with the consequence that the latter may undergo a transition to the normal state. The main conclusions of the existing magnetic instability theory were formulated on the basis of the initial stage study of the instability development but regardless of the formation features of their initial thermal states. As noted above, one of the main features of technical superconductors is the dissipative phenomena occurring in them which are due to many reasons. As a result, the thermal state change of a superconductor before the instability onset may change its operating mode. To understand the influence of the nonisothermal states on the electrodynamic phenomena occurring in low-T_c and high-T_c superconductors during the initial formation of their operating stages, the temporal and spatial evolution of the temperature and electromagnetic field induced by the applied magnetic field is discussed below in detail. The analysis focuses on the investigation of the fundamental physical mechanisms underlying the formation of the thermo-electrodynamic states of superconductors. It is based on the macroscopic description of diffusion phenomena in low-T_c and high-T_c superconductors by the system of Fourier and Maxwell equations for the different types of their voltage-current characteristics. The influence of the cooling conditions, sweep rate and parameters of a superconductor on the thermo-electrodynamic phenomena in superconductors are investigated in detail.

4.1 EVOLUTION OF INTERRELATED THERMAL AND ELECTRODYNAMIC STATES OF SUPERCONDUCTORS IN THE VISCOUS FLUX-FLOW MODES

In the general case, the dynamics of the macroscopic thermal and electrodynamic phenomena in superconductors should be described by the multidimensional nonstationary Fourier and Maxwell equations. These equations allow one to take into account the spatial and temporal features of the nonisothermal penetration of the electromagnetic field into a superconductor. However, this approach is associated with a significant amount of computations. Therefore, it significantly limits the formulation of the general physical regularities underlying the formation of the macroscopic states of superconductors. In this regard, let us investigate a simplified problem. Consider the cooled superconducting slab ($-a < x < a$, $-b < y < b$, $a \ll b$) located in the external magnetic field which is parallel to its surface and grows with a constant rate $dB_a/dt = \dot{B}_a$ from zero. The screening current with density J flowing in the superconductor leads to the corresponding heat release with a power EJ. Let us describe according to the system of equations 1.14 and 1.15 the thermal and electrodynamic states of a superconductor in the framework of a viscous flux-flow model based on the solution of one-dimensional Fourier and Maxwell system of equations of the form

$$C_s(T)\frac{\partial T}{\partial t} = \frac{\partial}{\partial x}\left(\lambda_s(T)\frac{\partial T}{\partial x}\right) + \begin{cases} 0, & 0 < x < x_p \\ EJ, & x_p \le x \le a \end{cases},$$

$$\mu_0 \frac{\partial J}{\partial t} = \frac{\partial^2 E}{\partial x^2}, x_p \le x \le a \qquad (4.1)$$

$$J = J_c(T, B) + E/\rho_f, x_p \le x \le a$$

setting the following thermal and electrodynamic initial-boundary conditions

$$T(x, 0) = T_0, \ \left.\frac{\partial T}{\partial x}\right|_{x=0} = 0, \ \left[\lambda_s\frac{\partial T}{\partial x} + h(T - T_0)\right]_{x=a} = 0,$$

$$E(x, 0) = 0, \quad E(x_p, t) = 0, \quad \left.\frac{\partial E}{\partial x}\right|_{x=a} = \dot{B}_a \qquad (4.2)$$

Here, C_s is the volumetric heat capacity of the superconductor, λ_s is its thermal conductivity coefficient, a is the half-thickness of the slab, h is the heat transfer coefficient, ρ_f is the differential resistivity of a superconductor in the viscous flow regime, T_0 is the temperature of the coolant, $J_c(T, B)$ is the critical current density and x_p is the penetration coordinate of the magnetic flux, which is the solution of equation

$$\mu_0 \int_{x_p}^{a} J(x, t)\,dx = B_a = \dot{B}_a t \qquad (4.3)$$

according to the results discussed in Chapter 2.

The problem (4.1)-(4.3) describes the dissipative penetration of the magnetic flux in a superconductor with finite velocity. Its dynamics depend not only on the sweep rate of the external magnetic field but also on the corresponding increase in temperature. The initial conditions describe the initial unperturbed thermal and electrodynamic states of the superconductor. Thermal boundary conditions take into account the symmetric temperature distribution in the superconductor and convective heat transfer on its surface. Electrodynamic boundary conditions take into account the finite penetration velocity of the magnetization front and the variation of the electric field on the surface of the slab, induced by an increasing external magnetic field. The results discussed below were obtained within the framework of various approximations describing the thermo-physical properties of superconductors.

The implicit form of the penetration law of magnetization front $x_p(t)$ described by equation (4.3) considerably complicates the use of the known methods for solving systems of the parabolic equations describing diffusion phenomena in solid with an unknown phase boundary. To solve this problem, a numerical method has been developed (Romanovskii 1997e), which allows one to investigate the nonlinear diffusion processes in solid with an implicit equation of the phase front. The main idea of the method is to define $x_p(t)$ as the root of the nonlinear equation (4.3). The made calculations are in the separation of the root, and then refining it. In the separation of the root, the sign of the following expression

$$r^{(s)} = \mu_0 \int_{x_p^{(s)}}^{a} J(x, t)\,dx - \dot{B}_a t$$

is found for each subsequent step of iterations $s = 1, 2\ldots$. According to the simple physical meaning of this equality, it is easy to understand that in the case when $x_p^{(s)}$ is greater than the true value x_p, then the sign of $r^{(s)}$ will be negative and vice versa if $x_p^{(s)} < x_p$, then $r^{(s)} > 0$. Therefore, the root separation stage is interrupted as soon as the sign of $r^{(s)}$ is reversed during two successive iterations. After that, it is easy to perform the refinement of the root with a given accuracy, the achievement of which was a condition for the termination of iterations. The described algorithm is reproduced without significant difficulties in analyzing the nonisothermal dynamics of the critical state in cylindrical superconductors as well as in superconducting composites.

To verify this solving method of the problem (4.1)-(4.3), the results of its numerical solution are compared with calculations based on the CSM in Figure 4.1. In the numerical solution, a linear dependence of the critical current density on the temperature

$$J_c(T, B) = J_{c0}(T_{cB} - T)/(T_{cB} - T_0)$$ (4.4)

was used. In the framework of the model under consideration, the critical parameters of a superconductor J_{c0} and T_{cB} may depend on the magnetic field induction. However, it was assumed that the values J_{c0} and T_{cB} are constant for comparison with the CSM approximation. Consequently, the change in the penetration front of the screening current x_c and the average volume density of the heat losses G_c were described by the expressions

$$x_c(t) = a - \dot{B}_a t / (\mu_0 J_{c0}) \quad \text{and} \quad G_c = (\dot{B}_a t)^3 / (6\mu_0^2 a J_{c0})$$

according to (2.4). In the numerical solution, the averaged volume density of the heat release arising from a varying external magnetic field was determined as follows:

$$G = \frac{1}{a} \int_0^t \int_0^{x_p(t)} EJ \, dx \, dt.$$

The following parameters of the superconductor were used $C_s = 10^3$ J/(m^3 × K), $\lambda_s = 0.1$ W/(m × K), $\rho_f = 5 \times 10^{-7}$ Ω × m, $J_{c0} = 4 \times 10^9$ A/m^2, $T_{cB} = 9$ K, $T_0 = 4.2$ K. They describe the averaged parameters of Nb-Ti cooled by the liquid helium. Simulations were performed at different values of the heat transfer coefficient and sweep rates of the external magnetic field.

Figure 4.1a demonstrates the existence of two phases of the magnetic flux penetration: first, the stable phase and second, the unstable one when the moving front of the magnetic flux has not yet reached the center of the slab. In this case, the onset and development of the instability at the partial penetration of the magnetic flux are called as the magnetic flux-jump. A more general definition of these phenomena, caused by a varying external magnetic field, is their definition as magnetic instability since they can also occur when the superconductor cross-section is filled with magnetic flux.

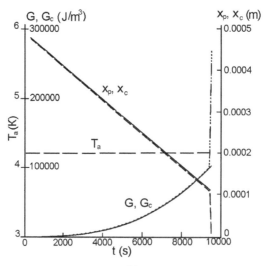

(a)

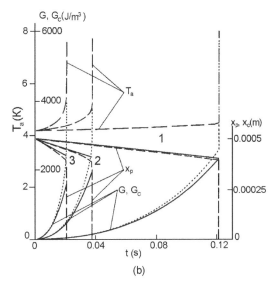

(b)

FIGURE 4.1 Time dependencies of the temperature on the surface of superconductor $T_a = T(a, t)$, coordinates of the magnetic front penetration (x_c – in the framework of the CSM, x_p – according to the numerical simulation) and heat release (G_c, G): (a) at $a = 5 \times 10^{-4}$ m, $h = 10$ W/(m² × K) and $dB_a/dt = 10^{-3}$ T/s; (b) – the same dependences at $h = 100$ W/(m² × K) and various sweep rates of an external magnetic field: 1 - $dB_a/dt = 2$ T/s, 2 - $dB_a/dt = 6$ T/s, 3 - $dB_a/dt = 10$ T/s.

It is seen that the diffusion of the magnetic flux occurs almost isothermally in the first stage due to the cooling of the superconductor and the very small sweep rate of the external magnetic field. Therefore, the numerical solution and the CSM-calculations coincide with a good degree of accuracy. However, the monotonic penetration of the magnetic flux is broken under certain conditions that are discussed below. As a result, the heat release sharply increases leading to a correspondingly sharp increase in its temperature and rapid motion of the magnetization front to the center of the slab. As follows from Figure 4.1b, the discrepancy between isothermal and nonisothermal states becomes more noticeable with an increase in the sweep rate even under the improved heat transfer conditions. Therefore, the temperature of the cooled superconductor before the onset of an unstable state differs from the temperature of the coolant. It is easy to understand that the isothermal character of the magnetic flux penetration will also be perturbed with the increasing of the superconductor transverse size. The most noticeable change in the thermal state of a superconductor will be observed under adiabatic conditions.

In Figure 4.2, the simulation results of the magnetization front dynamics and the average volume density of the magnetization losses, performed both in the nonisothermal approximation (the dashed lines) and the CSM-approximation (the solid lines) are shown for the most unfavorable heat transfer regime, namely during thermal insulation of the superconductor surface. The calculation was carried out for a superconductor of various thicknesses at $C_s = 40T^3$ J/(m³ × K), $\lambda_s = 0.0075T^{1.8}$ W/(m × K), $h = 0$, $T_0 = 4.2$ K, $J_{c0} = 4 \times 10^9$ A/m², $T_{cB} = 9$ K, $\rho_f = 5 \times 10^{-7}$ Ω × m. It can be seen that the nonisothermal dynamics of the critical state under adiabatic conditions are characterized by intense penetration of the magnetic flux into the interior of the superconductor compared with the CSM-calculations of $x_c(t)$. Therefore, the nonisothermal calculation of losses may lead to a deviation from the corresponding values calculated in the isothermal approximation and especially before the instability onset. As a result, the thermal prehistory of a thermally insulated superconductor may have a significant effect on the conditions of the occurrence and development of instability even with a nonintensive increase of the external magnetic field. In Figure 4.3, the results of a numerical calculation of the adiabatic states of a hard superconductor before and after the magnetic instability onset obtained for various dependences of the heat capacity on the temperature are shown. The initial parameters used in the simulations were written above. The solid lines show an increase in the surface temperature of the superconductor in cases when the screening currents do not fill the cross-section of the superconductor. The dashed lines show $T_a(t)$ after full penetration of the magnetic flux into the superconductor. The results of the performed numerical experiment show that taking into account the temperature dependence of the heat capacity of a thermally insulated superconductor significantly affects both the steady increase of its temperature and its final value as a result of the development of instability. This conclusion must be taken into account when analyzing the occurrence and development of instabilities in superconducting magnets.

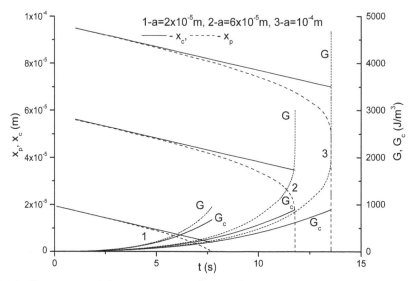

FIGURE 4.2 Penetration of the magnetization front and increase in the heat release at the diffusion of the magnetic flux in an uncooled superconductor at $dB_a/dt = 0.01$ T/s and various thicknesses of the superconductor.

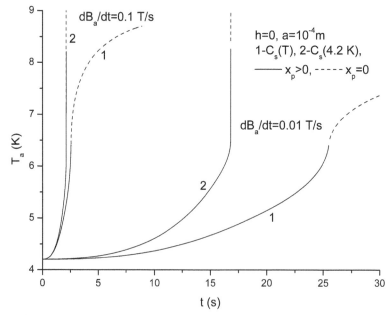

FIGURE 4.3 Change in the temperature of the surface of a thermally insulated superconductor with different dependencies of its heat capacity on temperature.

Thus, the characteristic feature of the nonisothermal electrodynamic states of superconductors is the more intense penetration of the magnetic flux into a superconductor. Therefore, the errors of the isothermal approximation increase when the thermal losses take into consideration. These regularities are most pronounced in the absence of cooling, high sweep rates and large transverse dimensions of superconductors. As a result, the unstable states (magnetic instability) may occur in superconductors located in the applied magnetic fields. Destruction of superconductivity is possible as a result of their development. Since the thermal prehistory influences even on the formation of the stable electrodynamic states, then it needs to correctly determine the allowable increase in temperature before the instability onset.

4.2   MECHANISMS OF THE MAGNETIC INSTABILITY ONSET AND DEVELOPMENT IN LOW-T_c SUPERCONDUCTORS

Let us investigate the formation mechanisms of the stable and unstable superconducting states in the framework of the problem formulated above. In Figures 4.4 and 4.5, the dynamics of the surface temperature of an Nb-Ti slab with a half-thickness $a = 10^{-4}$ m is shown when the external magnetic field was changed to values close to the so-called flux-jump field. Various cooling conditions were considered. The calculations were performed at

$$C_s = 0.812 \cdot 10^3 T \times B/B_{c0} + 42.73T^3 \ \text{J/(m}^3 \times \text{K)},$$
$$\lambda_s = 0.0075T^{1.8} \ \text{W/(m} \times \text{K)},$$
$$\rho_f = \rho_n B/B_{c2}(T) \quad \text{and} \quad \rho_n = 10^{-6} \ \Omega \times \text{m}.$$

In this case, the critical current density $J_c(T, B)$ of Nb-Ti was described by the expressions

$$J_c(T, B) = \frac{\alpha_0}{B + B_0}\left(1 - \frac{T}{T_{cB}}\right), \quad T_{cB} = T_{c0}\sqrt{1 - B/B_{c0}}, \quad B_{c2} = B_{c0}[1 - (T/T_{c0})^2] \tag{4.5}$$

where $\alpha_0 = 1.5 \cdot 10^{10}$ A·T/m^2, $T_{c0} = 9$ K, $B_{c0} = 14T$, $B_0 = 1.5T$.

In the general case, when the initial parameters of a superconductor are described by essentially nonlinear dependencies on the temperature and magnetic induction, to find the field of the flux jump B_m, it is apt to use the numerical methods. In particular, the method of the perturbation of the initial superconducting state was proposed in Keilin and Romanovsky (1982). It is based on the determination of the iterative sequence of the fixed inductions of external magnetic field $B_1 \rightarrow B_2 \rightarrow B_3 \rightarrow ...B_s \rightarrow B_{s+1}$, when the boundary values B_s and B_{s+1} may be found for a given accuracy of the calculation $|B_{s+1} - B_s| < \varepsilon$. At these boundary values, the temperature of the superconductor, determined from the solution of the problem (4.1)-(4.3) and (4.5), stabilizes at $B = B_s$ (there is no instability) or spontaneously increases at $B = B_{s+1}$, that is, the instability occurs. Accordingly, the electromagnetic boundary condition on the superconductor surface must be modified, namely $\dot{B}_a = $ const at $t < t_s$ and $\dot{B}_a = 0$ at $t > t_s$ in the (4.2). Here, t_s is the time of the continuous increase of the external magnetic field up to B_s.

The curves, which demonstrate the change in the temperature of the superconductor at stable and unstable states, calculated at $dB_a/dt = 0.01$ T/s and various cooling conditions, are shown in Figure 4.4a. The corresponding temperature distribution over the cross-section of the superconductor before the instability onset is presented in Figure 4.4b. The change in the electric field on the surface of a thermally insulated superconductor before and after the magnetic flux-jump and the avalanche-like penetration of the magnetization front into the superconductor, caused by the instability onset, are shown in Figure 4.5a and 4.5b, respectively. The states after which the sweep rate of the external magnetic field was set equal to zero (the external magnetic field B_a is fixed) are depicted. Correspondingly, the dashed lines demonstrate the stable states and the dashed-dotted lines correspond to the unstable states.

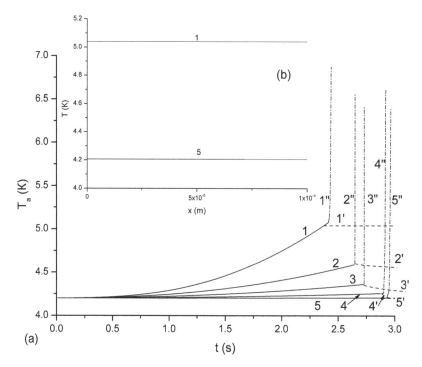

FIGURE 4.4 Temperature of the superconductor surface before and after the magnetic instability onset upon varying the heat transfer coefficient (a) and the temperature distribution in the superconductor ahead of the flux jump (b): 1, 1′, 1″ - h = 0; 2, 2′, 2″ - h = 1 W/(m² × K); 3, 3′, 3″ - h = 3 W/(m² × K); 4, 4′, 4″ - h = 10 W/(m² × K); 5, 5′, 5″ - h = 100 W/(m² × K). Details of the simulations are discussed in the main text.

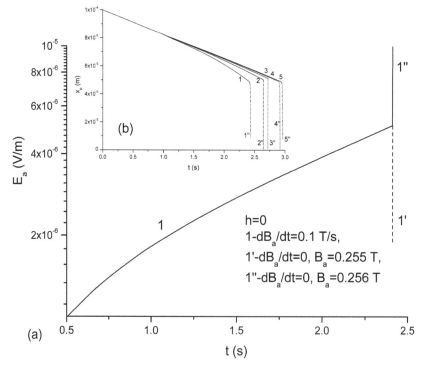

FIGURE 4.5 Dynamics of the electric field on the surface of the low-T_c superconductor (a) and the magnetization front (b) under the stable and unstable states (the same designations as in Figure 4.4).

In the calculations, the following modes were considered:

1. Curve 1 – continuous increase of the external magnetic field;
   Curve 1′ – stable state at $dB_a/dt = 0$, $B_a = 0.255$ T;
   Curve 1″ – unstable state at $dB_a/dt = 0$, $B_a = 0.256$ T.
2. Curve 2 – continuous increase of the external magnetic field;
   Curve 2′ – stable state at $dB_a/dt = 0$, $B_a = 0.279$ T;
   Curve 2″ – unstable state at $dB_a/dt = 0$, $B_a = 0.280$ T.
3. Curve 3 – continuous increase of the external magnetic field;
   Curve 3′ – stable state at $dB_a/dt = 0$, $B_a = 0.287$ T;
   Curve 3″ – unstable state at $dB_a/dt = 0$, $B_a = 0.288$ T.
4. Curve 4 – continuous increase of the external magnetic field;
   Curve 4′ – stable state at $dB_a/dt = 0$, $B_a = 0.303$ T;
   Curve 4″ – unstable state at $dB_a/dt = 0$, $B_a = 0.304$ T.
5. Curve 5 – continuous increase of the external magnetic field;
   Curve 5′ – stable state at $dB_a/dt = 0$, $B_a = 0.304$ T;
   Curve 5″ – unstable state at $dB_a/dt = 0$, $B_a = 0.305$ T.

The performed numerical experiment demonstrates the characteristic regularities of the variation in the thermo-electrodynamic states of LTS which occur during the increase of an external magnetic field up to values close to the field of the flux jump. It can be seen that the superconductor can either retain its superconducting properties despite the noticeable permissible overheating or transit to the normal state at a slight initial increase in its temperature, even after the termination of the external magnetic field rise. In particular, the temperature rise is noticeable in the thermally insulated superconductor (or during the rapid increase of the external magnetic field; see Figure 4.1b) before the magnetic instability onset when the time of the stable magnetic flux penetration t_s is relatively short. Note that the permissible overheating reaches 10% of the superconductor critical temperature in the considered numerical experiment. But with a decrease in dB_a/dt or improved heat transfer conditions, the permissible overheating decreases with a corresponding increase in the time of the stable diffusion of magnetic flux.

Qualitative analysis of the superconducting slab, permissible overheating in the dependence of the electrodynamic perturbation character, can be performed according to the estimate

$$\Delta T \sim \frac{1}{a} \int_0^{t_s} \int_0^{x_p(t)} EJ dx\, dt \left/ (C_s + ht_s/a), \quad t_s = B_a/\dot{B}_a \right.$$

which is easy to obtain passing from the heat equation to the heat balance equation. Here, it is assumed that $C_s \sim$ const for ease of the analysis. It follows from this estimate that if the instability is initiated in relatively short times ($t_s = B_a/\dot{B}_a \ll aC_s/h$), then the stable overheating of the superconductor depends on the total energy released during the diffusion of the magnetic field. In this case, the heat transfer conditions will have a weak effect on the value of ΔT, and the conditions of the instability onset will weakly depend on the nature of the changes of the external magnetic field. With an increase in the diffusion time of the magnetic flux t_s preceding the flux jump, the permissible overheating of the superconductor decreases and at $t_s \gg aC_s/h$, for example, at the small sweep rates or intensive cooling, the power of the dissipated energy will make the main contribution to the superconductor overheating. In this case, the nature of the change in time of the external magnetic field will have a noticeable effect on the stable formation of the superconducting states.

Summarizing the results of the numerical and qualitative analysis, let us formulate a general regularity that takes place before the magnetic instability onset: with an increase in the time of the stable diffusion of magnetic flux preceding its irreversible penetration into the superconductor, a stable overheating of the superconductor monotonously decreases and the amount of the dissipated energy

$$G_m = \frac{1}{a} \int_0^{t_s} \int_0^{x_p(t)} EJ dx dt$$

released during this time increases (Figure 4.6). This conclusion explains why, at first glance, there will be an unexpected decrease in permissible superconductor overheating before the magnetic instability onset as cooling conditions improve.

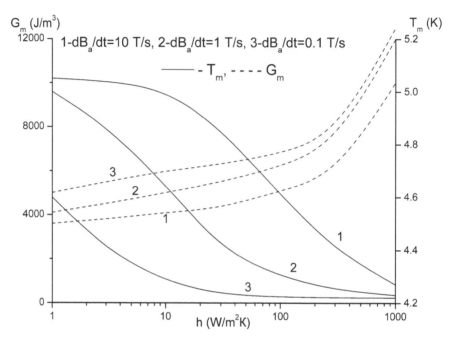

FIGURE 4.6 Dependence of the allowable temperature increase T_m of superconductor and the energy losses G_m before the flux jump on the heat transfer coefficient and sweep rate.

    Thus, the occurrence of a magnetic instability is always preceded by a finite allowable overheating of the superconductor, depending on the sweep rate of the external magnetic field, cooling conditions and thickness of the superconductor. Its most noticeable rise will be under the adiabatic cooling conditions or high sweep rates of the external magnetic field.

    Emphasize the importance of the conclusions formulated above, obtained on the basis of the use of a simplified viscous flux-flow model. Usually, as noted above, the stability of the critical state in this approximation is analyzed within the framework of models that assume that the temperature of the superconductor before the magnetic flux jump is equal to the temperature of the coolant. However, if the stability analysis of the critical state carries out from the point of view of the collective formation of thermo-electrodynamic states of a superconductor, then even in the case of using the V-I characteristic, described by the viscous flux-flow model, not only the finite stable overheating takes place before a jump of the magnetic flux but also its dependence on the conditions of an external magnetic field variation. That is why the magnetic field-jump even in the Bean model will depend on the sweep rate of the magnetic field and the cooling conditions. (It will be shown below that the characteristic thermal regularities of the electromagnetic field penetration formulated above also take place in the LTS with the real V-I characteristics). Consequently, if the stable overheating of the superconductor is not taken into account before the instability onset, the performed analysis of the conditions for its occurrence may misrepresent the results determining the boundary of the stable states. According to Figure 4.4 and 4.6, the error of such calculations increases if the sweep rate increases or the heat transfer coefficient decreases.

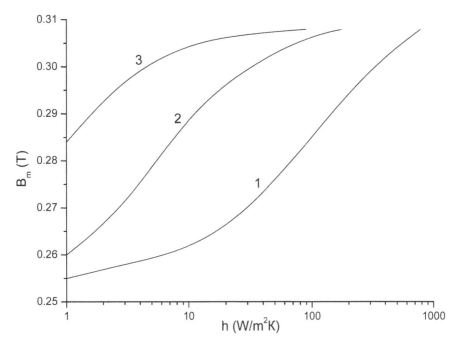

FIGURE 4.7 Effect of the heat transfer coefficient and sweep rate on the flux-jump field: 1 - $dB_a/dt = 10$ T/s, 2 - $dB_a/dt = 1$ T/s, 3 - $dB_a/dt = 0.1$ T/s.

The above-formulated relationship between the stable superconductor overheating and the stability conditions of its critical state initially underlie the dependence of the flux-jump field on the sweep rate. To illustrate this regularity, the curves presented in Figure 4.7 show the change in the flux-jump field as a function of the heat transfer coefficient for different values of the sweep rate. In the calculations performed on the basis of the proposed method for solving the system of the diffusion type equations with an unknown phase boundary, the full interrelated formation of thermo-electrodynamic states of a superconductor was taken into account.

The results obtained also lead to another important conclusion about the existence of a relationship between the permissible energy losses in a superconductor located in an applied magnetic field and a stable increase in its temperature before the instability onset. According to Figure 4.6, the allowable heat dissipation in a superconductor, which will not cause instability onset, increases against the background of its monotonically decreasing stable overheating at the improvement of the heat transfer conditions. Moreover, at an intense cooling of a superconductor, the insignificant thermal losses will lead to the instability onset, which will develop with a low permissible superconductor overheating. As a result, a stable increase in the temperature of the superconductor before the instability will be much lower than the critical temperature of the superconductor under an ideal cooling condition on the superconductor surface ($h \rightarrow \infty$). However, it will be higher than the coolant temperature. Therefore, the widespread assumption that the heat losses in a superconductor under the intense heat exchange conditions will lead to its transition to a normal state only after the temperature of the superconductor has exceeded its critical temperature is erroneous.

Thus, the nontrivial relationship between the stable heat release and overheating, existing in LTS, leads to a general formulation of the conditions of the superconducting state destruction under the magnetic perturbation: the magnetic instabilities occur due to the heat release exceeding the maximum allowable value (the critical value) that depends on the external conditions, for example the cooling conditions or the magnetic disturbance intensity. This conclusion and the regularities of the changes in stable overheating discussed above link together the theory of the magnetic instability, the theory of the losses and, as it will be shown below, the theory of the thermal stabilization (discussed in Chapters 7 and 8). It turns out that there is a common thermal mechanism for the occurrence of any instability, which is based on the existence of the critical energy that limits the heat release from any external disturbance. It allows one to determine the boundary of the stable states by means of a perturbation method of an initial superconductor state proposed in (Keilin, Romanovsky, 1982) taking into account stable formation of thermo-electrodynamic states of the superconductor before the beginning of instability.

4.3   ADIABATIC CRITERIA OF THE CRITICAL STATE STABILITY DURING PARTIAL PENETRATION OF THE SCREENING CURRENTS. RELATIONSHIP OF THE STABLE OVERHEATING WITH HEAT LOSSES

According to the results obtained above, a stable increase in the temperature of the superconductor can have a significant effect on the stability conditions of the superconducting states. Let us formulate the stability criteria of the critical state in the nonisothermal approximation in the case of the partial penetration of screening currents inside the superconductor, assuming that $C_s \sim$ const for ease of analysis. To do this, transform the system of equations (4.1), introducing the dimensionless variables

$$X = x/a, \tau = t/t_x \ (t_x = \mu_0 a^2/\rho_f), e = E/(J_{c0}\rho_f), j = J/J_{c0}, \theta = (T - T_0)/(T_{cB} - T_0).$$

Then, one can get

$$\frac{\partial \theta}{\partial \tau} = \Lambda \frac{\partial^2 \theta}{\partial X^2} + \beta \begin{cases} 0, & 0 < X < X_p \\ e[j_c(\theta) + e], & X_p \leq X \leq 1 \end{cases},$$

$$\frac{\partial e}{\partial \tau} = \frac{\partial^2 e}{\partial X^2} + \beta \left|\frac{dj_c}{d\theta}\right| e[j_c(\theta) + e] + \Lambda \left|\frac{dj_c}{d\theta}(\theta)\right| \frac{\partial^2 \theta}{\partial X^2}$$

(4.6)

after simple transformations by taking into account the decrease in the critical current density with temperature. Here,

$$\Lambda = \frac{\lambda_s \mu_0}{C_s \rho_f}, \quad \beta = \frac{\mu_0 J_{c0}^2 a^2}{C_s (T_{cB} - T_0)}, \quad j_c(\theta) = J_c/J_{c0}.$$

This system is reduced to the form

$$\frac{\partial \theta}{\partial \tau} = \Lambda \frac{\partial^2 \theta}{\partial X^2} + \beta \begin{cases} 0, & 0 < X < X_p \\ e(1 - \theta), & X_p \leq X \leq 1 \end{cases}$$

$$\frac{\partial e}{\partial \tau} = \frac{\partial^2 e}{\partial X^2} + \beta(1 - \theta)e + \Lambda \frac{\partial^2 \theta}{\partial X^2}$$

(4.7)

for a superconductor with a linear temperature dependence of the critical current density described by the relation (4.4) (in a dimensionless form $j_c = 1 - \theta$) and taking into account that before the instability onset, the resistive component E/ρ_f of the total current in the superconductor is much less than its critical current density J_c (in the dimensionless form $j_c \gg e$).

Within the framework of the performed transformations, the development of the thermal and electrodynamic phenomena in superconductors depends on two dimensionless quantities β and Λ. The quantity β (the magnetic instability quantity) can be rewritten as follows $\beta = (a/a_s)^2$ by entering the characteristic thicknesses of the superconductor $a_s = \sqrt{C_s(T_{cB} - T_0)/(\mu_0 J_{c0}^2)}$. Its physical meaning is discussed below. The formation of the thermo-electrodynamic states of superconductors will also be influenced by the nonlinear character of the decrease with temperature the values $dj_c/d\theta$. In the general case $dj_c/d\theta \neq$ const. This leads to the parameter of the magnetic instability decreasing with temperature $\beta_\theta = \beta \dfrac{dj_c}{d\theta}$. The quantity Λ is the ratio of the characteristic time of the magnetic flux diffusion t_m (in the viscous flux-flow model $t_m = \mu_0 a^2/\rho_f$) to the characteristic time of the heat flux diffusion t_λ ($t_\lambda = C_s a^2/\lambda_s$). Estimate the value of Λ using the averaged thermo-physical parameters of a hard superconductor specified in section 4.1. In this case, $\Lambda \sim 10^{-3}$. Usually, $\Lambda \ll 1$ for low-T_c and high-T_c superconductors. Physically this means that in superconductors the electrodynamic state changes first, and the temperature of the superconductor 'adjusts' to it in accordance with the first equation of the system (4.7). In this case, the temperature distribution in the superconductor will be almost uniform (Figure 1.4b) and can be determined from the solution of the simplified zero-dimensional heat equation

$$\frac{d\theta}{d\tau} = -\omega\theta + \beta(1-\theta)\int_{X_p}^{1} e\, dX \; .$$

Here,

$$\omega = \Lambda H = \frac{\mu_0}{\rho_f}\frac{ha}{C_s} \, ,$$

where $H = ha/\lambda_s$ is the dimensionless heat transfer coefficient (the so-called Bio parameter). The dimensionless quantity ω can also be rewritten as $\omega = t_m/t_h$ using the characteristic time of the magnetic flux diffusion t_m and the characteristic cooling time $t_h = aC_s/h$. This dimensionless parameter allows one to estimate the influence of the cooling conditions and properties of the superconductor on the diffusion character of the magnetic flux. In particular, it will be close to adiabatic at $\omega \ll 1$, that is at $\Lambda \ll 1$ or nonintensive heat transfer conditions $H \ll 1$. Conversely, any initial temperature change of a semi-infinite superconductor, which is often used to derive the stability criteria, will fade out because in this case $\omega \to \infty$. In other words, the superconducting state of the half-space is always stable. At the same time, the instability onset analysis, which is usually carried out in this case, is based on the utilization of the finite depth of the moving screening current front. In this approximation, the stability of the superconducting state of the half-space may be unstable contrary to the simple physical notion that it is impossible to irreversibly increase the temperature of the superconductor with unlimited geometric dimensions.

As noted in Chapter 1, the evolution of the electric field in superconductors refers to the type of diffusion phenomena with volumetric reproduction, for example, similar to the diffusion of neutrons. This feature is also retained in the viscous flux-flow regime, according to the second equation of the system (4.7). Correspondingly, the analog of the neutron multiplication factor equals $\gamma = \gamma_1 - \gamma_2$, where $\gamma_1 = \beta$ is the coefficient of the neutron's birth and $\gamma_2 = \beta\theta$ is the neutron absorption coefficient. A chain reaction occurs at $\gamma > 0$. According to this analogy and because $\beta(1 - \theta) > 0$, a spontaneous increase in the electric field when the critical state can be destroyed may occur in superconductors. However, an increase in the temperature of the superconductor before the magnetic instability onset, similar to an increase of the absorption coefficient, will lead, first of all, to an increase in the range of the stable states due to a corresponding decrease in the critical current density of superconductor.

In the general case, the analysis of the stability conditions carried out even on the basis of the simplified system (4.7) requires the use of the numerical methods. To write the criteria of the magnetic instability onset, let us investigate the initial stage of the electric field evolution. In this case, the dimensionless temperature of the superconductor θ_i (in the dimensional form of T_i) before the instability and the magnetization boundary X_p do not change significantly because $\Lambda \ll 1$. Besides, let us take into account that the temperature distribution in LTS is almost uniform at $\Lambda \ll 1$ (Figure 4.4b). Therefore, an analysis of the initial stage of the electric field redistribution in a superconductor, on the surface of which the external magnetic field is constant ($dB_a/dt = 0$), can be performed solving the simplified equation written in the form

$$\frac{\partial e}{\partial \tau} = \frac{\partial^2 e}{\partial X^2} + \beta^* e, \quad \beta^* = \beta(1-\theta_i) \sim \text{const} \tag{4.8}$$

with boundary conditions

$$e(X_p, \tau) = 0, \quad \partial e/\partial X(1, \tau) = 0 \tag{4.9}$$

The solution of the problem (4.8), (4.9) will be sought in the form

$$e(X, \tau) = \sum_{k=1}^{\infty} A_k \exp[(\beta^* - v_k)\tau]Q_k(X) \tag{4.10}$$

according to the method of the variable separation. Here, A_k are the integration constants, $Q_k(X)$ are the eigenfunctions and v_k are the eigenvalues that follow from the solution of the Sturm-Liouville problem. In the case under consideration, it has the form

$$\frac{d^2 Q_k}{dX^2} + v_k Q_k = 0, \quad Q_k(X_p) = 0, \quad dQ_k/dX(1, \tau) = 0$$

The increasing positive values of the eigenvalues of this spectral problem are equal to

$$\sqrt{v_k}(1 - X_p) = (2k-1)\pi/2, \quad 0 < v_1 < v_2 < \dots, k = 1, 2, 3\dots$$

Therefore, the electric field induced by a changing external magnetic field will spontaneously increase (Figure 4.5) even at a fixed value of the external magnetic field, if the condition $\beta^* - v_1 > 0$ is fulfilled for the first eigenvalue v_1. In this case, a multiplier appears which grows exponentially with time in the expression (4.10). Consequently, in the nonisothermal approximation, the superconducting state is stable if the condition

$$\beta < \frac{\pi^2}{4(1 - \theta_i)(1 - X_p)^2} \tag{4.11}$$

is fulfilled. It may be rewritten as follows

$$\frac{\mu_0 a^2 J_{c0}^2}{C_s} \frac{T_{cB} - T_i}{(T_{cB} - T_0)^2} \left(1 - \frac{x_p}{a}\right)^2 < \frac{\pi^2}{4} \tag{4.12}$$

relatively to dimensional variables.

Criteria (4.11) and (4.12), which comply with the limit transition to the isothermal stability criteria at $\theta_i \to 0$ and $X_p \to 0$ ($T_i \to T_0$ and $x_p \to 0$) (Wilson 1983), result in the need to limit the transverse size of the superconductor. They also lead to an unexpected result, namely they demonstrate a positive effect of a stable increase in the superconductor temperature θ_i on the stability conditions of the critical state, in particular on the permissible thickness of the superconductor and the value of the permissible increase in the magnetization area. According to (4.12), the first flux jump occurs if the coordinate of the magnetization front x_p satisfies the condition

$$a - x_p > \frac{\pi}{2} \sqrt{\frac{C_s(T_{cB} - T_0)^2}{\mu_0 J_{c0}^2(T_{cB} - T_i)}} = a_s \frac{\pi}{2} \sqrt{\frac{T_{cB} - T_0}{T_{cB} - T_i}} \tag{4.13}$$

and the external magnetic field (in general, the magnetic field drop in the superconductor) exceeds the value

$$B_m = \frac{\pi}{2} \sqrt{\mu_0 C_s(T_{cB} - T_i)} = \frac{\pi}{2} \mu_0 J_{c0} a_s \sqrt{(T_{cB} - T_i)/(T_{cB} - T_0)} \tag{4.14}$$

Here, a_s is the characteristic thickness of the superconductor introduced above. According to (4.13), its physical meaning can be defined as the minimum thickness of a superconductor at which magnetic instability does not occur.

Thus, the more stable increase in the superconductor temperature, the greater the stable value of the penetration depth $a - x_p$. Therefore, as the screening currents penetrate the superconductor, the conditions for the magnetic instability onset will transit beyond pure electrodynamic conditions and become more dependent on its thermal state. As a result, the greatest effect of the superconductor temperature on the instability conditions onset will be observed when the superconductor cross-section is filled with screening currents.

Since in deriving the criteria (4.13) and (4.14) the conditions of the heat transfer on the surface of the superconductor were not taken into account because $\Lambda \ll 1$, they are called adiabatic. As follows from Figure 4.7, the role of the heat transfer in the stability conditions of the hard superconductors is small and its consideration leads to insignificant increase of the flux-jump field, as it is shown in Wilson (1983) and Gurevich et al. (1997c). The formulated above criteria show that the probability of the magnetic instability will decrease with increasing the heat capacity of the superconductor which increases with increasing temperature. But the probability of the magnetic instability will increase with increasing of the critical current density. In general, it will increase with the increase of the steepness of the J_c drop with temperature. These regularities explain why LTS is more sensitive to the magnetic instabilities than HTS.

The simplified analysis of the temperature influence on the adiabatic stability conditions of the critical state was performed by studying only the initial stage of the redistribution of the electric field induced by the changing external magnetic field. Consider the other side of this problem, namely let us study the temperature dynamics of a superconductor assuming that the distribution of the electric field in the magnetization region satisfies the Bean approximation according to the results discussed in Chapter 2. Then, in terms of the above introduced dimensionless variables, the distribution of the electric field in the superconducting slab under consideration has the form

$$e(X, \tau) = \frac{a\dot{B}_a}{J_{c0}\rho_f}(X - X_p), \quad X_p \leq X \leq 1 \tag{4.15}$$

The heat balance equation that is written above under the adiabatic cooling conditions ($\omega = 0$) has the form

$$\frac{d\theta}{d\tau} = \beta(1-\theta)\int_{X_p}^{1} e(X, \tau)\,dX$$

Using this equation and the expression (4.15), let us write the relationship between the temperature rise rate and the depth of the magnetic flux penetration

$$(1 - X_p)^2 = \frac{2J_{c0}\rho_f}{\beta(1-\theta)a\dot{B}_a}\frac{d\theta}{d\tau}$$

Then according to criteria (4.11), the critical state is stable if the temperature rise rate of the superconductor does not exceed a certain characteristic value satisfying the condition

$$\frac{d\theta}{d\tau} < \frac{\pi^2}{8}\frac{a\dot{B}_a}{J_{c0}\rho_f}$$

In the dimensional form, it is written as follows

$$\frac{dT}{dt} < \frac{\pi^2}{8}\frac{(T_{cB} - T_0)\dot{B}_a}{\mu_0 J_{c0}a}$$

This condition determines the thermal boundary of the magnetic instability onset under the adiabatic cooling conditions. According to this estimate and the heat balance equation, written as

$$C_s \frac{dT}{dt} = \frac{1}{a} \int_{x_p}^{a} EJ dx$$

it is easy to find the corresponding constraint on the averaged value of the permissible heat release. Namely, when

$$\int_{x_p}^{a} EJ dx < \frac{\pi^2}{8} \frac{C_s(T_{cB} - T_0)\dot{B}_a}{\mu_0 J_{c0}},$$

and therefore, at

$$\int_{0}^{t}\int_{x_p}^{a} EJ dx dt < \frac{\pi^2}{8} \frac{C_s(T_{cB} - T_0)B_a(t)}{\mu_0 J_{c0}} = \frac{\pi^2}{8} J_{c0} a_s^2 B_a(t)$$

the critical state is stable. The physical meaning of the last inequality is obvious: the stable penetration of an external magnetic field in the superconductor is limited by the maximum allowable value of the dissipated energy, that is the critical value. After it is exceeded, the magnetic instability occurs. It is seen that the destruction of superconductivity does not depend on the sweep rate of the external magnetic field under adiabatic conditions and is determined only by the magnetic field drop.

Thus, the written down estimates strictly prove that there is a nontrivial relationship between the stable heat release and the corresponding overheating of a superconductor at violation of which magnetic instability occurs. Besides, its existence was already shown on the basis of a numerical simulation of the stability problem, considered in a more general formulation for a cooled superconductor with an arbitrary value of the heat transfer coefficient (Figure 4.6). Consequently, in addition to the electrodynamic stability conditions of the superconducting state, equivalent thermal stability conditions are existed, which also describes the appearance of the magnetic instability in superconductors. From the physical point of view, this conclusion is of more fundamental importance, as it leads to the unified method for determining the conditions of the instabilities of the superconducting state based on the determination of the critical energy at the action of any external perturbation. In the framework of this method, the limiting transitions to the previously obtained stability criteria for the superconducting state are observed regardless of the external perturbation nature (electromagnetic, thermal or any combination thereof), cooling conditions and properties of the superconductor.

4.4 NONISOTHERMAL ADIABATIC CRITERIA OF THE CRITICAL STATE STABILIZATION DURING FULL PENETRATION OF THE SCREENING CURRENTS: THE BEAN APPROXIMATION

As follows from the results presented above, the magnetic instability may occur when the magnetic flux does not completely penetrate the superconductors. For their practical use, the transverse size of the superconductor is reduced to such a thickness until they would retain superconductivity during the full penetration of the external magnetic field. In other words, to ensure an SMS operation, the thickness of the superconducting strands should not exceed the maximum permissible (critical) value. Let us estimate the possible values of the critical thickness of the superconductor and the corresponding value of the magnetic field drop.

The conditions of the magnetic instability onset in superconductors during the full penetration of the screening currents can be easily obtained from criteria (4.13) and (4.14). Let us use them, equating to zero the coordinate of the moving magnetization front and considering the extremely hard constraint on the stability conditions, setting $T_i = T_0$. Then, one may find that $a \sim 10^{-4}$ m and

$B_m \sim 0.1$ T at $T_{cB} - T_0 \sim 5$ K and $a \sim 10^{-3}$ m and $B_m \sim 0.5$ T at $T_{cB} - T_0 \sim 70$ K for the characteristic values $C_s \sim 10^3$ J/(m$^3 \times$ K) and $J_{c0} \sim 10^9$ A/m^2. Consequently, superconducting strands based on LTS should have a thickness of several tens of microns. The thickness of the HTS is much higher and therefore, the magnetic field drop is higher.

However, the parameters of a superconductor obtained in this way do not answer the question of what is the final state of a superconductor after instability occurs. Investigate this problem for a superconductor with the linear temperature dependence of the critical current density describing by the equality (4.4) in the case of the full magnetic flux penetration in the superconductor for the most unfavorable heat transfer condition, namely when it is thermally insulated. Suppose that the initial temperature of the superconductor differs from the temperature of the coolant as a result of the action of any external perturbation. To simplify the analysis, let us use the critical state model, neglecting the resistive component of the total current flowing through the superconductor. Then, the system of equations (4.1) leads to the solution of the following system written in a Cartesian coordinate system in the form

$$C_s(T)\frac{\partial T}{\partial t} = \frac{\partial}{\partial x}\left[\lambda_s(T)\frac{\partial T}{\partial x}\right] + EJ_c(T) \tag{4.16}$$

$$\mu_0 \frac{\partial J_c}{\partial t} = \frac{\partial^2 E}{\partial x^2} \tag{4.17}$$

with the following initial-boundary conditions

$$T(x, 0) = T_i, \quad \frac{\partial T}{\partial x}(0, t) = 0, \quad \frac{\partial T}{\partial x}(a, t) = 0, \quad E(0, t) = 0, \quad \frac{\partial E}{\partial x}(a, t) = 0 \tag{4.18}$$

The problem (4.16)-(4.18) describes the diffusion of the temperature and electric field in a superconducting heat-insulated slab placed in a constant external magnetic field in the case when the initial temperature of the slab T_i differs from the temperature of the coolant $T_0 (T_i > T_0)$ and the screening currents with density J_c completely penetrated the superconductor.

Integrate equation (4.16) from 0 to a and neglecting the spatial inhomogeneity of the temperature over the slab cross-section due to adiabatic cooling conditions, one can proceed to the heat balance equation of the form

$$C_s(T)\frac{dT}{dt} = \frac{J_c(T)}{a}\int_0^a E dx \tag{4.19}$$

According to (4.17), it is easy to find the distribution of the electric field over the cross-section of the slab

$$E(x, t) = \mu_0\left(\frac{x^2}{2} - ax\right)\frac{\partial J_c}{\partial t}$$

Then, equation (4.19) is converted to

$$C_s(T)\frac{dT}{dt} = -\frac{\mu_0 a^2}{6}\frac{dJ_c^2}{dt}, \quad T(0) = T_i \tag{4.20}$$

Since the temperature of the slab increases, then the critical current density decreases. Correspondingly, the solutions of equation (4.20) depend on the final thermal state of the slab, which will be as a result of the instability development.

In the case when the final temperature of the superconductor T_k will exceed its critical temperature, one may obtain the equation

$$\int_{T_i}^{T_k} C_s(T)\, dT = \frac{\mu_0 a^2 J_{c0}^2}{6} \left(\frac{T_{cB} - T_i}{T_{cB} - T_0} \right)^2 \tag{4.21}$$

to define T_k. The condition $T_k = T_{cB}$ determines the boundary values of the initial parameters at which the magnetic instability is always accompanied by a transition of the superconductor to the normal state. Therefore, in particular, for all

$$a > \frac{T_{cB} - T_0}{J_{c0}(T_{cB} - T_i)} \sqrt{\frac{6}{\mu_0} \int_{T_i}^{T_{cB}} C_s(T)\, dT} \tag{4.22}$$

the destruction of superconductivity will occur spontaneously as soon as the magnetic flux has filled the cross-section of the superconductor. Taking into account the relationship between the magnetic field induction on the slab surface and the critical parameters of the superconductor at the initial time, the following condition may be written

$$B_m > \sqrt{6\mu_0 \int_{T_i}^{T_{cB}} C_s(T)\, dT} \tag{4.23}$$

which determines the upper limit of the allowable magnetic field on the surface of a superconductor after which its spontaneous heating due to the development of the magnetic instability will be accompanied by the destruction of the superconducting state.

Equation (4.21) also leads to the existence of the characteristic value of the initial temperature T_i' of the superconductor, which is a solution of equation

$$\int_{T_i'}^{T_{cB}} C_s(T)\, dT = \frac{\mu_0 a^2 J_{c0}^2}{6} \left(\frac{T_{cB} - T_i'}{T_{cB} - T_0} \right)^2$$

Since the critical current density decreases with increasing of the initial perturbation temperature, which means that the energy of the screening currents decreases, then the physical meaning of T_i' corresponds to the limiting value of the initial perturbation temperature T_i after exceeding of which the superconductor will not transit to the normal state.

The written relations demonstrate the essential role of the superconductor's initial temperature in the adiabatic stability conditions of the superconducting state. Its difference from the coolant temperature leads not only to a quantitative change in the isothermal stability criteria but also directs to an important conclusion. Namely, there is a characteristic value of the initial perturbation temperature which separates the region with the inevitable transition of the superconductor to a normal state and the region with its heating to a temperature below the critical one after the instability development. At the same time, the formulated criteria answer the crucially important question: under what conditions are the critical state destroyed under infinitely small perturbations. Correspondingly, it will be only when the conditions (4.22) or (4.23) are satisfied for all $T_0 < T_i < T_i'$. Therefore, the superconductivity of the superconductor is retained in the temperature range of the perturbation $T_i' < T_i < T_{cB}$ despite the critical state that will change according to the temperature variation. According to the equality $T_i' = T_0$, for all parameters that satisfy the condition

$$\int_{T_0}^{T_{cB}} C_s(T)\, dT > \frac{\mu_0 a^2 J_{c0}^2}{6} \tag{4.24}$$

an arbitrary external change in the temperature of the superconductor up to the critical one will not lead to its subsequent transition to the normal state, despite the development of the instability in the superconductor. As above conclusion show, the partial thermal stabilization effect of the critical state of a superconductor exists when the resulting magnetic instability will not transit the superconductor to the normal state under any initial disturbance in its temperature in the range from T_0 to T_{cB}, when the condition (4.24) is satisfied. In this case, the higher the value T_i, the lower the value T_k. Accordingly, the conditions of stability of the critical state are improved with the increase of T_i.

Let us write the condition of the magnetic instability onset when the final temperature of a superconductor will be less than the critical one ($T_k < T_{cB}$) as a result of the instability development for all $T_i > T_i'$. In this case, it is not difficult to get the expression

$$\int_{T_i}^{T_k} C_s(T)dT = \frac{\mu_0 a^2 J_{c0}^2}{6(T_{cB} - T_0)^2}(T_k - T_i)(2T_{cB} - T_k - T_i) \tag{4.25}$$

from (4.20), which determines the final temperature of the superconductor heating. From simple physical considerations, it is clear that there are states when the left side of equation (4.25) can exceed its right side. Indeed, the energy of the screening currents decreases with an increase in the initial temperature of the superconductor. Therefore, starting with a certain value of T_i, any change in enthalpy will be greater than the energy released by the instability and therefore, the change in the initial thermal state will not occur. This means that the magnetic instability does not develop. Executing the limiting transition $T_i \rightarrow T_k$ in (4.25), the criterion for the absence of the instability is written as follows

$$\frac{\mu_0 a^2 J_{c0}^2}{C_s(T_i)(T_{cB} - T_0)} \frac{T_{cB} - T_i}{T_{cB} - T_0} < 3 \tag{4.26}$$

Note that this inequality can be deduced more strictly from (4.25) by analyzing the conditions for a change in the sign of the derivative dT_k/dT_i.

Let us find the restrictions imposed on the parameters of the superconductor following from (4.26), which provides the full stability of the critical state when its destruction does not occur.

First, for the stability of the critical state, the magnetic field drop in a superconducting slab must satisfy the condition

$$B_m < \sqrt{3\mu_0 C_s(T_i)(T_{cB} - T_i)} \tag{4.27}$$

This condition will take place when its half-thickness does not exceed the value

$$a < \sqrt{\frac{3C_s(T_i)(T_{cB} - T_0)^2}{\mu_0 J_{c0}^2(T_{cB} - T_i)}} = a_{s,0}\sqrt{3\frac{C_s(T_i)}{C_s(T_0)}\frac{T_{cB} - T_0}{T_{cB} - T_i}}$$

where $a_{s,0} = \sqrt{C_s(T_0)(T_{cB} - T_0)/(\mu_0 J_{c0}^2)}$ is the characteristic half-thickness at which the magnetic instability in a superconductor with a temperature-dependent heat capacity does not occur ($a_{s,0} = a_s$ at $C_s \sim$ const).

Second, if the initial temperature of the superconductor is in the range $T_i'' < T_i < T_{cB}$, where T_i'' is the solution of the equation

$$3C_s(T_i'')(T_{cB} - T_0)^2 - \mu_0 a^2 J_{c0}^2(T_{cB} - T_i'') = 0$$

then the spontaneous heating of the superconductor does not occur. Therefore, one can find the criterion

$$\mu_0 a^2 J_{c0}^2/[C_s(T_0)(T_{cB} - T_0)] < 3$$

from condition $T_i'' = T_0$, which means not only the electrodynamic stability of the critical state as it follows from the criterion (1.17) according to the Bean approximation but also its full stability to the temperature perturbations T_i varying in the range from T_0 to T_{cB}. In other words, if this condition is satisfied, then superconductivity is always saved at any initial change in the thermal state of the superconductor when it is not heated up to its critical temperature, that is at $T_0 < T_i < T_{cB}$.

Similar stability criteria of the critical state can be easily obtained for a superconductor with a circular cross-section placed in a longitudinal external magnetic field. In particular, its transition to the normal state will take place on the condition

$$\int_{T_i}^{T_{cB}} C_s(T)dT < \frac{\mu_0 a^2 J_{c0}^2}{4} \left(\frac{T_{cB}-T_i}{T_{cB}-T_0} \right)^2$$

and the magnetic instabilities will be absent at

$$\frac{\mu_0 a^2 J_{c0}^2}{C_s(T_i)(T_{cB}-T_0)} \frac{T_{cB}-T_i}{T_{cB}-T_0} < 2$$

Here, a is the radius of the superconductor.

Let us compare the formulated criteria with the criteria obtained in the framework of the CSM, assuming that $C_s \sim$ const. In the CSM approximation, which does not take into account the stable increase in the temperature of a superconductor before instability onset, the critical state of the superconducting slab is stable if $\beta < 3$ under the adiabatic conditions according to the condition (1.17). In the approximation of a viscous flux-flow model, the stability condition has the form $\beta < \pi^2/4$, which follows from (4.11) at $\theta_i \to 0$ and $X_p \to 0$. A slight difference in the numerical coefficient is explained by taking into account the finite value ρ_f in the second equation of the system (4.1) in contrast to equation (4.17), which is written in the approximation $\rho_f \to \infty$. However, consideration of the finite overheating of a superconductor, which is inevitable when the magnetic flux diffuses in a thermally insulated superconductor, leads to the appearance of a corresponding corrective temperature factor. In this case, the condition of the screening currents stability, filling the cross-section of the superconductor, takes the form

$$\beta < 3(T_{cB}-T_0)/(T_{cB}-T_i).$$

This criterion shows that the adiabatic stability condition of the critical state

$$\beta < 3$$

widely used in practical applications imposes excessively strong restrictions on the conditions of its stability, in particular the thickness of the superconducting strands. At the same time, this condition has a different physical interpretation: the magnetic instabilities are not only absent at $\beta < 3$, but the critical state is also stable with respect to any external temperature perturbation T_i in which the superconductor does not heat up to the critical temperature of the superconductor at the given magnetic field. In other words, the spontaneous increase in temperature of the superconductor will not occur in the range of the temperature perturbations $T_0 < T_i < T_{cB}$ at $\beta < 3$.

The noted feature of the stability conditions, demonstrating the stability of the critical state with respect to the temperature perturbations, takes place due to the existence of characteristic temperatures T_i' and T_i'' which lead to the thermal structure of the conditions of magnetic instability onset. As follows from the above formulae, these values are equal to $T_i' = T_{cB} - 6(T_{cB}-T_0)/\beta$ and $T_i'' = T_{cB} - 3(T_{cB}-T_0)/\beta$, respectively, in the approximation $C_s \sim$ const.

In Figure 4.8, the dependences of T_i' and T_i'' on the parameter β is shown schematically. It also depicts some initial temperature perturbation T_i which may occur in the temperature range $T_0 < T_i < T_{cB}$. Shaded Regions 1 and 2 correspond to the parameters at which the screening currents flowing in the entire volume of the superconductor are not stable and in area 3, the critical state in the entire volume of the superconductor is stable in spite of its high possible overheating. Accordingly, Region 1 is the region of the full magnetic instability when superconductor goes to the normal state while Region 2 is the region of the partial magnetic instability when it occurs but superconductor stays in the superconducting state and region 3 is the region of complete stabilization of the critical state when the magnetic instability is absent. In other words, Figure 4.8 demonstrates not only quantitative but also qualitative differences in the nonisothermal stability conditions from isothermal ones. Figure 4.8 depicts that the stability of the critical state can also be maintained at β > 3, if the initial temperature of the superconductor T_i exceeds the value T_i'' at given value β under any external thermal perturbation. In this case, the critical state is completely stable. At β > 6, the magnetic instability onset will not lead to the transition of the superconductor to the normal state if the initial temperature of the superconductor satisfies the condition $T_i' < T_i < T_i''$. These results prove the existence of the complete and partial thermal stabilization conditions of the critical state with respect to the external temperature perturbations.

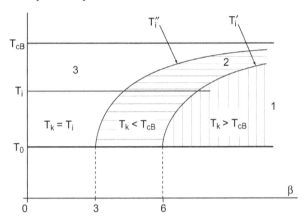

FIGURE 4.8 Thermal structure of the adiabatic conditions of magnetic instabilities under the temperature perturbation T_i.

To verify the formulated thermal stabilization criteria of the critical state, both stable and unstable time variation of the superconductor surface temperature are shown in Figures. 4.9 and 4.10 occurring under different temperature perturbations T_i. The calculations were performed for the thermally insulated Nb-Ti superconductor based on the numerical solution of the problem (4.1)-(4.4). In the calculations, the averaged parameters of the superconductor were assumed to be equal: $h = 0$, $C_s = 10^3$ J/(m^3 × K), $\lambda_s = 0.01$ W/(m × K), $T_0 = 4.2$ K, $T_{cB} = 9$ K, $J_{c0} = 4 \times 10^9$ A/m^2, $\rho_f = 5 \times 10^{-7}$ Ω × m and its thickness was varied. The two cases of the initial distribution of the screening current in the superconductor were considered.

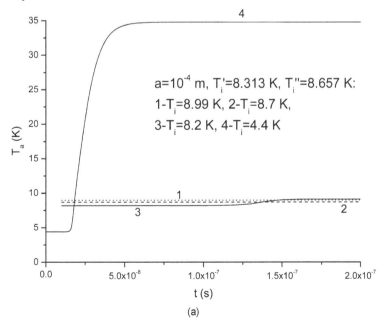

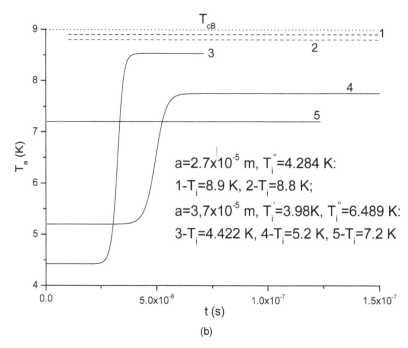

(b)

FIGURE 4.9 Thermal dynamics of the critical state of fully penetrated screening currents at various temperature perturbations and superconductor thicknesses at the adiabatic conditions: (a) - stable states of the superconductor (Curves 1 and 2) and its full transition to the normal state (Curves 3 and 4) during temperature disturbances closed to the critical temperature; (b) - stable states (Curves 1, 2 and 5) and states in which the transition to the normal state does not occur upon the magnetic instability onset (Curves 3 and 4).

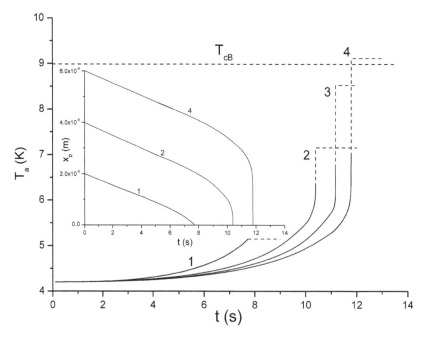

FIGURE 4.10 Effect of the self-heating temperature of the superconductor on the stability of its critical state depending on the thickness (1 - $a = 2 \times 10^{-5}$ m, 2 - $a = 4 \times 10^{-5}$ m, 3 - $a = 5 \times 10^{-5}$ m, 4 - $a = 6 \times 10^{-5}$ m) at different time dependences of the sweep rates: —— - $dB_a/dt = 0.01$ T/s, – – – – - $dB_a/dt = 0$.

First, the development of the magnetic instability was studied under the assumption that the screening currents already filled the cross-section of the slab at the initial time (Figure 4.9). In this case, the critical state may be stable at high initial temperatures of the superconductor T_i close to its critical temperature. Curves 1 and 2 in Figure 4.4a and Curves 1, 2 and 5 in Figure 4.9b show these stable states which exist depending on the values T_i' and T_i'', as discussed above. The condition of the complete thermal stabilization is satisfied for them since $T_i'' < T_i$. At the same time, there is a partial thermal stabilization of the critical state for all $T_i' < T_i < T_i''$. However, the instability develops without switching of the superconductor to the normal state (Curves 3 and 4 in Figure 4.9b). Finally, the irreversible destruction of superconductivity occurs when $T_i < T_i'$ (Curves 3 and 4 in Figure 4.9a).

Second, the change in the temperature of the superconductor was also determined when the external magnetic field was increased at the sweep rate $dB_a/dt = 0.01$ T/s from zero (Figure 4.10). In this case, its increase is ceased when the magnetization front reached the middle of the slab. The corresponding temperature dependencies are shown in Figure 4.10 by the solid and dashed lines. The presented results confirm the peculiarities of the formation of stable and unstable states in accordance with the above-discussed thermal structure of the adiabatic stability conditions. It can be seen that the inevitable heating of the superconductor has a noticeable effect on the stability of the critical state. As a result, the superconductor not only does not transit to the normal state (Curves 1-3) even at substantial overheating but can also save the stability of the superconducting state (Curve 1). In addition, the calculations performed at $a = 6 \times 10^{-5}$ m show that if one does not take into account the thermal prehistory of the electrodynamic state formation of superconductor, then its final temperature will be more than two times greater than the corresponding value as a result of the instability development which is determined by considering the self-heating of the superconductor.

The analysis performed allows one to give a more general explanation of some experimental results. First, it is well known that numerous experiments to determine the field of the magnetic flux jump have a significant scattering (Gurevich et al. 1997c). To explain this scattering along with the reasons discussed, one should consider the possibility of a change in the superconductor temperature preceding the instability onset which is usually not paid attention to. Second, a priori defined the initial temperature of the superconductor will misrepresent the final result of determining the flux-jump field as already noted. In particular, when the coolant temperature is close to absolute zero ($T_0 \to 0$), the field of the flux jump also tends to zero under adiabatic conditions, as follows from the formula $B_{m,0} = \sqrt{3\mu_0 C(T_0)(T_{cB} - T_0)}$, obtained in an isothermal approximation (Wilson 1983, Gurevich et al. 1997c). However, any temperature perturbation of the initial state of the thermally insulated superconductor, which is inevitable when the magnetic field is varied during adiabatic conditions, will not result in such limiting transition according to (4.27). In the nonisothermal analysis of the magnetic flux jump in the region of ultra-low coolant temperatures, the allowable values of the magnetic field induction will essentially differ from the corresponding isothermal values due to the corresponding increase in the heat capacity of the superconductor with temperature as proved early (Romanovskii 1998f). Besides, the isothermal approximation explains experimentally observed absence of the magnetic flux jumps at a coolant temperature close to the critical temperature of a superconductor is because in this temperature region the flux-jump field is greater than the corresponding magnetic induction value at which it completely penetrated the superconductor. However, the existence of characteristic temperatures T_i' and T_i'', which influence on thermal stabilization of the critical state, leads to another interpretation of this phenomenon: the magnetic instability is absent if the temperature perturbation T_i satisfies the condition $T_i'' < T_i < T_{cB}$, and the resulting instability will not lead to the destruction of superconductivity at $T_i' < T_i < T_i''$.

To illustrate these conclusions, the simulation results of the coolant temperature influence on the magnetic instability onset and development are presented in Figure 4.11. They were obtained during solving the system described by equations (4.1)-(4.3) in which a linear temperature dependence of the critical current density is defined as follows: $J_c(T, B) = J_k(1 - T/T_{cB})$. The following parameters $a = 2.5 \times 10^{-4}$ m, $dB_a/dt = 0.1$ T/s, $\rho_f = 5 \times 10^{-7}\,\Omega \times$m, $C_s = 30T^3$ J/(m$^3 \times$K), $\lambda_s = 0.0075T^{1.8}$ W/(m$\times$K) and $J_k = 7.5 \times 10^9$ A/m^2, $T_{cB} = 9$ K were used during simulations which corresponds to Nb-Ti.

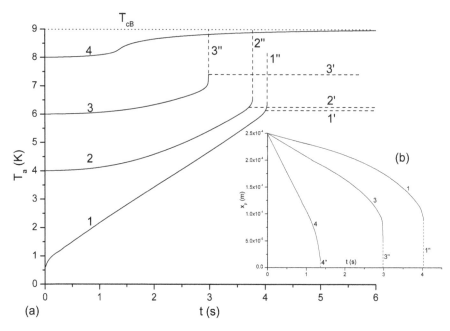

FIGURE 4.11 Time variations of the temperature of superconductor surface (a) and current penetration front (b) before and after the magnetic instability onset upon varying the coolant temperature: 1 - $T_0 = 0.5$ K, 2 - $T_0 = 4$ K, 3 - $T_0 = 6$ K, 4 - $T_0 = 8$ K. Here, the solid Curves 1-4 correspond to the constantly increasing external magnetic field, and the dashed curves describe the states when its increase is stopped: 1' - $B_a =$ 0.438 T, 1" - $B_a = 0.441$ T, 2' - $B_a = 0.408$ T, 2" - $B_a = 0.415$ T, 3' - $B_a = 0.328$ T, 3" - $B_a = 0.336$ T, 4" - B_a = 0.136 T.

As noted above, the stable temperature increases of the superconductor before the magnetic instability essential influence on the onset and development of the magnetic flux-jump. As a result, the high energy stored by the screening currents destroys very quickly the superconducting state of the slab at $T_0 = 0.5$ K during complete magnetic instability despite the high stable temperature rise before the instability onset (Figures 4.11a and 4.11b, Curve 1"). At the same time, an increase in the external magnetic field is not accompanied by the transition of the slab to the normal state at $T_0 = 8$ K although the flux jump occurs (Figure 4.11b, Curve 4").

Thus, the analysis made in the framework of the CSM and viscous flux-flow model shows that the interrelated electrodynamic and thermal phenomena in superconductors have the fission-chain-reaction nature (Equation (1.16) is written for superconductors with real V-I characteristics). They are characterized by the nontrivial connections between the stability conditions, allowable heat losses and corresponding overheating of a superconductor before the magnetic instability onset. Therefore, the stable heat release arising during flux penetration is responsible for the onset of the magnetic instability in superconductors when it exceeds the corresponding critical value. As mentioned above, this concept validates the use of a unified method which allows to find the conditions of the instability onset of different nature. Eventually, the instabilities are the direct consequence of the condition when the dissipated energy becomes larger than the corresponding critical energy, which will depend on the type of an external disturbance (electromagnetic, mechanical, thermal or a combination of them), cooling conditions and properties of the superconductor. As a consequence, this concept permits one to make an adequate analysis of the stability conditions. In particular, as discussed above, the criterion (1.21) loses the physical meaning during limiting transition J_δ, $T_\delta \to 0$. It leads to the conclusion that superconductors are not stable in an applied magnetic field in the critical state model. At the same time, the nonisothermal models formulated above yields an explanation of the critical state stability since a variation of the magnetic field changes the magnetic flux-jump field according to the permissible temperature increase versus the change of the sweep rate of external magnetic field or cooling conditions. Thereby, the nonisothermal stable states of the superconductors extend the class of their allowable states.

The analysis of the stability conditions performed by taking into account the collective nature of the thermal and electrodynamic phenomena shows that the characteristic temperatures T_i' and T_i'' exist, which influence on the mechanisms of the critical state destruction. Their existence proves that the criterion of the adiabatic stability $\beta < 3$ imposes strict conditions on the parameters of the superconductor and leads to a restriction of the range of allowable parameters at which superconductors retain their superconducting properties in relation to the magnetic perturbations. The dependence of the stability conditions of superconducting states on the thermal prehistory leads to the finite stable overheating of superconductors and provides them the stabilizing effect.

The nonisothermal conditions of the magnetic instability onset in the cooled superconductors, whose V-I characteristics are described even by the idealized critical state or viscous flux-flow models, demonstrate that the conditions of the critical state stability depend on the sweep rate of the external magnetic field due to the corresponding dependence of the stable overheating of the superconductor before the instability onset. The corresponding nonisothermal magnetic instability criteria prove that the generally accepted opinion according to which the thermal losses in a superconductor under intensive heat transfer conditions will lead to the transition of a superconductor to the normal state only when its temperature is higher than the critical temperature is erroneous. On the whole, the analysis of the violation of superconducting state stability made without accounting for the increase in temperature before instability may give underestimated values of the flux-jump field.

Since the macroscopic electrodynamics of superconductors is investigated, as a rule, in the isothermal approximation these results will be very important to describe the phenomena in nonintensively cooled superconducting devices with massive current-carrying elements placed in rapidly varying magnetic fields or alternating current.

4.5 MECHANISMS OF THE OCCURRENCE AND SUPPRESSION OF THE CRITICAL STATE OSCILLATIONS

The monotonous character of the formation of thermo-electrodynamic states of superconductors may be broken as a result of the occurrence of oscillations of the electric field, temperature and screening current in the superconductor that are observed in experiments by Zebouni et al. (1964), Del Gastillo and Oswald (1969) and Shimamoto (1974). Let us discuss the main physical causes leading to their occurrence and development. Let us consider this problem from the point of view of the collective development of the thermal and electromagnetic modes in hard superconductors that are located in a continuously increasing external magnetic field, as above.

In Figure 4.12, the curves demonstrating the effect of cooling conditions on the temperature oscillations on the surface of a superconducting slab in an external magnetic field are shown when the magnetic field increases at a constant rate $dB_a/dt = 1$ T/s. In the simulations performed on the basis of the problem (4.1)-(4.4), the initial parameters were assumed to be equal to $a = 5 \times 10^{-5}$ m, $C_s = 10^3$ J/(m^3 K), $\lambda_s = 0.01$ W/(m $\times$ K), $T_0 = 4.2$ K, $J_{c0} = 4 \times 10^9$ A/m^2 and $T_{cB} = 9$ K, $\rho_f = 5 \times 10^{-7} \Omega \times$ m.

The presented results show that oscillations are absent under the adiabatic condition. At the same time, the oscillation amplitude increases with an increase in the heat transfer coefficient at the corresponding decrease of the superconductor temperature on the background of which the oscillations occur. Consequently, the thermal state of a superconductor, preceding the oscillations occurrence, affects the conditions for their onset and development.

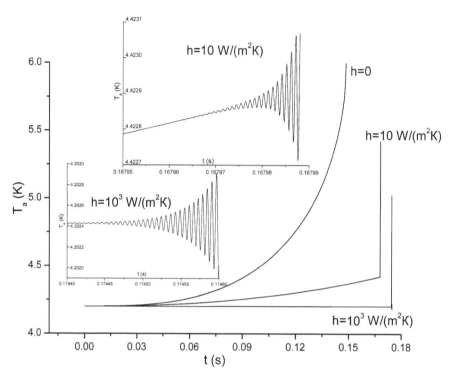

FIGURE 4.12 Initiation of the temperature oscillations in a superconductor during viscous flux-flow mode depending on the conditions of its cooling at a continuous increase of the external magnetic field. Their development is shown in more detail in the insets.

Let us discuss this regularity in more detail. To do this, integrate the second equation in the system (4.1) twice over spatial variable: first, from x to a and then from x_p to a. Then, considering the boundary conditions at $x = a$ and $x = x_p$, one may find the variation in the electric field on the surface of the superconductor

$$E(a, t) = (a - x_p)\frac{dB_a}{dt} - \mu_0 \int_{x_p}^{a} dy \int_{y}^{a} \frac{\partial J}{\partial t} dx$$

Determine its rate of rising by differentiating this equation over time. As a result, one will have

$$\frac{\partial E}{\partial t}\bigg|_{x=a} = -\frac{dx_p}{dt}\frac{dB_a}{dt} + \mu_0 \frac{dx_p}{dt} \int_{x_p}^{a} \frac{\partial J}{\partial t} dx - \mu_0 \int_{x_p}^{a} dy \int_{y}^{a} \frac{\partial^2 J}{\partial t^2} dx$$

Transform its right side by taking into account the second equation of the system (4.1) and the relationship between current and temperature, which may be described by an approximate formula

$$\frac{\partial^2 J}{\partial t^2} \approx -\left|\frac{dJ_c}{dT}\right| \frac{\partial^2 T}{\partial t^2}$$

The later equality is satisfied with a good degree of accuracy at $J_c \gg E/\rho_f$ and obtained under the simplifying assumption of linear dependence of the critical current density only on temperature. Performing the simple transformations and taking into account penetration of the magnetization front into the superconductor in the direction opposite to the direction of the spatial coordinate, finally, one gets

$$\left.\frac{\partial E}{\partial t}\right|_{x=a} = \left.\frac{\partial E}{\partial x}\right|_{x=x_p} \left|\frac{dx_p}{dt}\right| + \mu_0 \left|\frac{dJ_c}{dT}\right| \int_{x_p}^{a} dy \int_{y}^{a} \frac{\partial^2 T}{\partial t^2} dx \qquad (4.28)$$

Equation (4.28) demonstrates the effect of the intensity of the superconductor temperature rise in the entire magnetization region on the formation of its electrodynamic states. As a result, the dynamics of the electromagnetic field on the surface of a superconductor depends on the features of the variation of the magnitude and sign of $\partial^2 T/\partial t^2$. This regularity the more noticeable, the higher the value of $|dJ_c/dT|$.

The value of $\partial^2 T/\partial t^2$ is easy to find by differentiating the heat diffusion equation in the system (4.1). Assuming that C_s and λ_s do not depend on the temperature to simplify the analysis, equation (4.28) is reduced to the form

$$\left.\frac{\partial E}{\partial t}\right|_{x=a} = \left.\frac{\partial E}{\partial x}\right|_{x=x_p} \left|\frac{dx_p}{dt}\right| + \frac{\mu_0}{C_s} \left|\frac{dJ_c}{dT}\right| \int_{x_p}^{a} \left[\frac{\partial g}{\partial t} - \frac{\partial q}{\partial t}\right] dx \qquad (4.29)$$

after substitution of the corresponding value of $\partial^2 T/\partial t^2$. Here,

$$q(x,t) = \lambda_s \frac{\partial T}{\partial x}(x,t) + h(T - T_0)\big|_{x=a}$$

is the density of the conductive-convective heat flux in the cross-section of the superconductor,

$g(x,t) = \int_{x}^{a} EJ dx$ is the density of the Joule heat release per unit of the cross-sectional area of superconductor.

The value of the integral term on the right side of the expression (4.29) may be both positive and negative depending on the thermal state of the superconductor. Then, the value of $\partial E/\partial t\big|_{x=a}$ can also be alternating sign. Indeed, if $\int_{x_p}^{a} [\partial g/\partial t - \partial q/\partial t] dx > 0$, that is if the rate of rise of the heat release averaged over the entire magnetization region exceeds the rate of the averaged heat removal, then the electric field on the surface of the superconductor increases because within the framework of the flux-flow approximation $\partial E/\partial x\big|_{x=x_p} > 0$ throughout the diffusion process of the screening current. Conversely, the sign of $\partial E/\partial t\big|_{x=a}$ may become negative at a certain negative value $\int_{x_p}^{a} [\partial g/\partial t - \partial q/\partial t] dx$, and the electric field will begin to decrease even with the temperature increase of superconductor. According to the relation (4.29), the formation of states in which the value of $\partial E/\partial t\big|_{x=a}$ is negative depends on the features of the temperature distribution in the magnetization region, on the penetration velocity of the magnetization front, and the formation conditions for the electric field on the moving boundary of the magnetization region. Thus, the relation (4.29) proves the thermal cause of an occurrence of the oscillations in the hard superconductor: they begin when the rising rate of the density of the conductive-convective heat flux, which is averaged over the magnetization region, satisfies the condition

$$\int_{x_p}^{a} \frac{\partial q}{\partial t} dx > \int_{x_p}^{a} \frac{\partial g}{\partial t} dx + \frac{C_s}{\mu_0} \left.\frac{\partial E}{\partial x}\right|_{x=x_p} \left|\frac{dx_p}{dt}\right| \Big/ \left|\frac{dJ_c}{dT}\right| \qquad (4.30)$$

Based on this conclusion, let us discuss at first the qualitative regularities of the occurrence and development of the oscillations in the superconductor considered above by analyzing the features of their formation at $dB_a/dt = 1$ T/s and varying the heat transfer coefficient.

First of all, it should be emphasized that the inequality $\partial^2 T/\partial t^2 > 0$ is satisfied for the heat-insulated superconductor throughout the entire diffusion process of the magnetic field (Figure 4.10). Therefore, the condition (4.30) is not satisfied due to the constantly increasing the Joule heat release in the magnetization region. As a result, the oscillations under adiabatic cooling conditions are absent according to (4.28) since $\partial E/\partial t > 0$ during the entire process of the magnetic flux penetration. As will be shown below, condition (4.30) may not be also satisfied at a slow increase in the external magnetic field on the surface of the cooled superconductor.

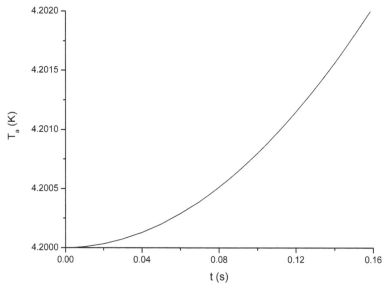

FIGURE 4.13 Temperature rise of the superconductor surface before the oscillations during intensive cooling ($h = 10^3$ W/(m^2 × K)).

The condition $\partial^2 T/\partial t^2 > 0$ is also satisfied at the initial stage of diffusion of the magnetic flux at any finite value of the heat transfer coefficient, including the cases of intensive cooling (Figure 4.13). However, a further increase in the surface temperature of the cooled superconductor, which occurs with an increase in the external magnetic field, will lead to the corresponding increase in the convective heat flux to the coolant. Besides, the cooling of the superconductor surface also changes the character of the temperature distribution inside the superconductor (Figure 4.14), and the cooling conditions also have the corresponding effect on the conductive heat flux. In other words, in the case of the cooled superconductors, there is a constant increase in the total heat flux into the coolant when the external magnetic field increases. Accordingly, after condition (4.30) is fulfilled, the intensity of the temperature rise in the superconductor, that is value $\partial^2 T/\partial t^2$ will change and the induced electric field will begin to decrease despite a further increase in the temperature of the superconductor ($\partial T/\partial t > 0$). As a result, the next stage of the oscillations will start at which the electric field on the superconductor surface will begin to decrease despite the increase in the external magnetic field. These states may be observed with the increasing temperature of the superconductor, that is at $\partial T/\partial t > 0, \partial^2 T/\partial t^2 < 0$ according to equation (4.28). In turn, the decrease in the electric field on the surface of the superconductor, which occurs when the external magnetic field increases, will be accompanied by the corresponding decrease in the Joule heat release in the region adjacent to the surface of the superconductor. Consequently, the temperature of the superconductor also begins to decrease after a certain time. To the greatest extent, it will occur on the surface of a superconductor as it follows from Figure 4.15 due to a decrease in the electric field in the superconductor. It also shows that the values of $\partial^2 T/\partial t^2$ change synchronously in the entire magnetization region when the oscillations occur in the case under consideration. Therefore, to analyze the oscillations, it is sufficient to consider the states of the superconductor that are formed on its surface.

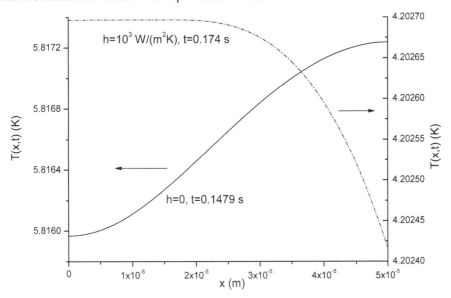

FIGURE 4.14 Temperature distribution in the superconductor at different values of the heat transfer coefficient.

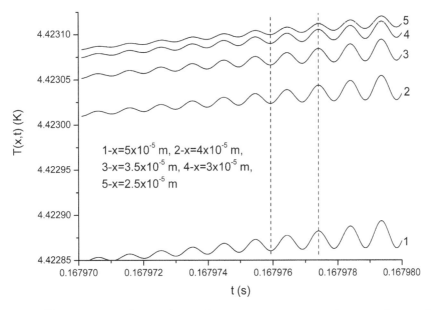

FIGURE 4.15 Temperature oscillations during the continuous increase in the external magnetic field at $h = 10$ W/(m^2 × K). The dashed lines correspond to individually selected points at which the temperature takes on either the minimum or maximum values.

Accordingly, as a result of lowering the temperature, the screening current will begin redistributing and flowing from the more heated region, located near the moving front of the magnetization region, to the less heated surface layer of the superconductor. An increase in the screening current density will lead to an increase in heat losses. That means that after some time it will lead to the occurrence of the states when the condition (4.30) will not be satisfied, that is to a subsequent increase in the electric field. In general, the continuous repetition of these states leads to oscillations, which can both fade out and increase, depending on the intensity of the heat release in a superconductor and the conditions of its cooling.

The interconnection of the main stages of the occurrence and development of the oscillations in the considered superconductor is presented in more detail in Figures 4.16-4.20 for the modes induced at $dB_a/dt = 1$ T/s.

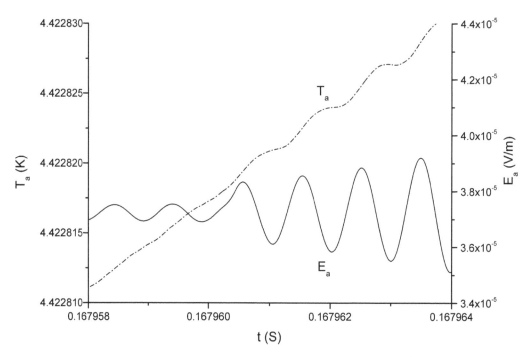

FIGURE 4.16 Initial oscillations phase of the temperature T_a and electric field E_a on the superconductor surface at the nonintensive cooling conditions.

The initial stage of the formation of the oscillating state on the surface of the nonintensively cooled superconductor ($h = 10$ W/(m² × K)) is shown in Figure 4.16. The change in the temperature of the superconductor, electric field and screening current at the stage of the increasing oscillations for which the value of $\left.\dfrac{\partial E}{\partial x}\right|_{x=x_p} \left|\dfrac{dx_p}{dt}\right|$ has a weak effect on the dynamics of the ongoing processes and the signs change of $\left.\partial E/\partial t\right|_{x=a}$ and $\left.\partial^2 T/\partial t^2\right|_{x=a}$ happen almost synchronously and are shown in more detail in Figure 4.17. It is seen that the values of $\left.\partial E/\partial t\right|_{x=a}$ are equal to zero at $t = t_1$ and $t = t_2$. Moreover, the electric field on the surface of the superconductor (the solid line) and its temperature (the dash-dotted line) increase simultaneously at $t < t_1$. At this stage, $\left.\partial^2 T/\partial t^2\right|_{x=a} > 0$ and the released heat prevails over the removed heat. However, the heat losses are effectively removed from the magnetization region at $t_1 < t < t_2$. Therefore, $\left.\partial^2 T/\partial t^2\right|_{x=a} < 0$ and the electric field on the surface of the superconductor monotonously decreases. Its value begins to increase again due to the corresponding redistribution of the screening current when $t > t_2$ (Figure 4.18). As a result, the temperature of the superconductor begins to increase after some time, namely when $t > t_3$ and the oscillations are repeated.

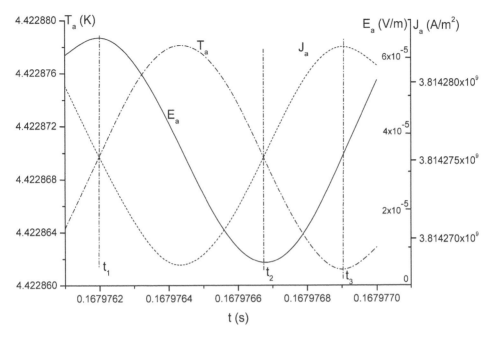

FIGURE 4.17 Interrelated oscillations of the temperature, electric field and current density on the superconductor surface.

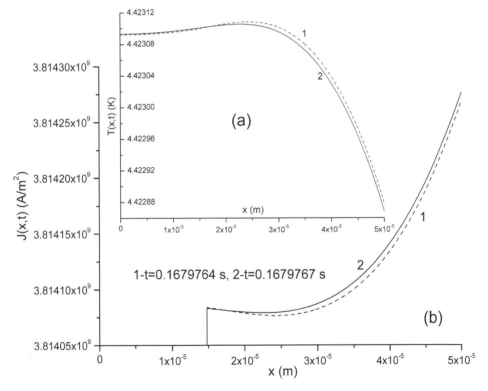

FIGURE 4.18 Redistribution of the temperature (a) and screening current (b) in the superconductor during oscillation.

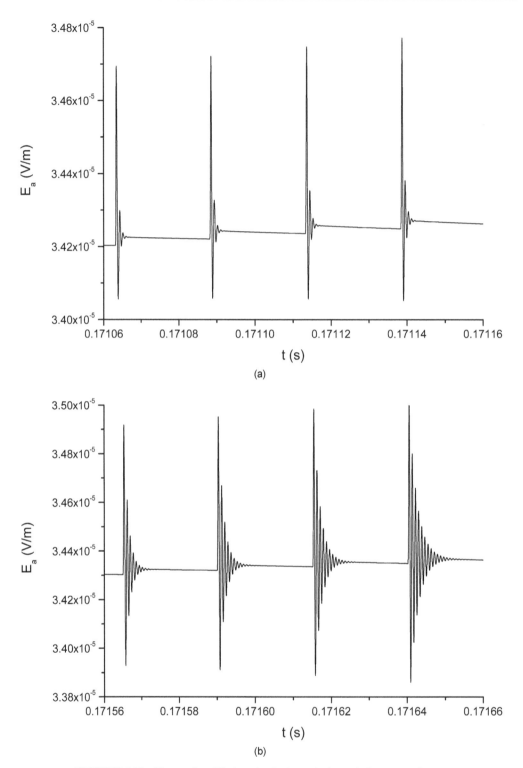

FIGURE 4.19   Damped oscillations in the intensively cooled superconductor.

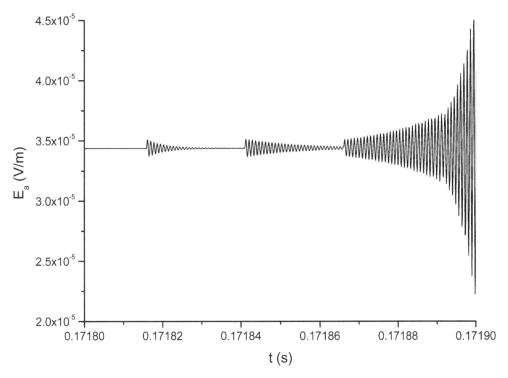

FIGURE 4.20 Transition from damped to increasing oscillations of the electric field in an intensively cooled superconductor.

The oscillations may be more complex. The results of a numerical experiment simulated the oscillations during intensive cooling ($h = 10^3$ W/(m$^2 \times$ K)) are shown for both the damping oscillations (Figure 4.19) and increasing ones (Figure 4.20). The damped oscillations are characterized by the existence of the oscillation trains. Their occurrence is necessarily preceded by a decrease in the electric field (Figure 4.19) which is the initial stage of any oscillations as shown above. It is not difficult to understand that the damped trains are due to the intense removal of the heat losses from the magnetization region of a superconductor. However, as the magnetic flux moves deeper into the superconductor, the heat release increases due to a corresponding increase in the magnetization region and the oscillations are becoming more and more intensive, eventually becoming irreversible (Figure 4.20).

Thus, the nonisothermal description of electrodynamic states in superconductors is necessary to understand the general laws describing the phenomena taking place in them, and to formulate the correct conditions of the magnetic instabilities onset. Using even simplified models of the Bean and viscous flux-flow, it allows one to understand the basic mechanisms for the stable penetration of the electromagnetic field into the superconductor at various cooling conditions, transverse dimensions and sweep rates of the external magnetic field.

In particular, it is shown that under certain conditions, the formation of the thermo-electrodynamic states of superconductors may be characterized by the oscillations occurrence of the electric field, temperature and screening current. Their occurrence is due to the existence of the critical value of the rate of the conductive-convective removal of the heat losses from the interior of the superconductor to the coolant. When it is exceeded, effective removal of the heat losses released in the magnetization region occurs, leading to the subsequent redistribution of the screening current over the cross-section of the superconductor and initiating a corresponding increase in the temperature of the superconductor, etc. Conversely, if the rate of the heat flux removal from the interior of the superconductor is small, then the oscillations are absent. In this case, an intensive monotonic increase in the temperature of the superconductor occurs throughout the entire mode of the critical state formation. Consequently, there is a thermal mechanism for the occurrence and suppression of oscillations, the development of which depends not only on the cooling conditions of the superconductor surface but also on other factors affecting the intensity of the temperature rise of the superconductor. In particular, it will be shown below that the oscillations disappear as the sweep rate of the external magnetic field increases or the current-carrying capacity of the superconductor deteriorates.

4.6   FEATURES OF THERMO-MAGNETIC PHENOMENA IN HIGH-T_c SUPERCONDUCTORS

At present, a lot of attention both experimental and theoretical researches of the electrodynamic phenomena occurring in HTS materials are paid. In this paragraph due to the limited volume of the results presented below, the basic thermo-electrodynamic formation features of the stable and unstable states are discussed only for $Bi_2Sr_2CaCu_2O_8$ superconductor without a stabilizing matrix placed in a changing external magnetic field at the temperature of the liquid helium. However, the formulated conclusions are general which should be taken into account when one investigates other high-T_c superconductors cooled by different coolants. As above, the analysis is carried out in a macroscopic approximation using the Fourier and Maxwell equations. The V-I characteristic of a superconductor is described by a power-law equation. A detailed study of the influence of the heat transfer conditions, the thickness of the superconductor, sweep rate of the external magnetic field and temperature dependences of the heat capacity and the critical current density on the development of the thermo-electrodynamic phenomena in HTS is performed.

As above, for ease of analysis, let us consider the cooled high-T_c superconductor in the form of infinitely long slab ($-a < x < a$, $-b < y < b$, $a \ll b$), which is placed instantly in the external magnetic field B_a, parallel to its surface which then begins to grow with a constant rate $dB_a/dt = \dot{B}_a$. Describe the macroscopic thermal and electrodynamic states of the HTS by one-dimensional Fourier and Maxwell equations of the form

$$C_s(T)\frac{\partial T}{\partial t} = \frac{\partial}{\partial x}\left(\lambda_s(T)\frac{\partial T}{\partial x}\right) + \begin{cases} 0, & 0 < x < x_p \\ EJ, & x_p \leq x \leq a \end{cases}, \tag{4.31}$$

$$\mu_0 \partial J/\partial t = \partial^2 E/\partial x^2, 0 \leq x_p \leq x \leq a$$

with the power V-I characteristic

$$E = E_c[J/J_c(T, B)]^n \tag{4.32}$$

setting the following initial-boundary conditions

$$T(x,0) = T_0, \quad \partial T/\partial x\big|_{x=0} = 0, \quad [\lambda_s \partial T/\partial x + h(T - T_0)]\big|_{x=a} = 0,$$

$$E(x,0) = 0, \quad E(x_p, t) = 0, \quad \partial E/\partial x\big|_{x=a} = \dot{B}_a \tag{4.33}$$

Here, as above, C_s is the volumetric heat capacity of the HTS, λ_s is its coefficient of the thermal conductivity, a is the half-thickness of the slab, h is the heat transfer coefficient, T_0 is the coolant temperature, $J_c(T, B)$ is the critical current density and x_p is the coordinate of the penetration front ($0 \leq x_p < a$) of magnetic flux determined by equation (4.3).

Thus, as in the analysis of the thermo-electrodynamic states of the LTS, study the dissipative diffusion of the magnetic flux in the HTS when it is in the unperturbed state at the initial time. Thermal boundary conditions include the symmetric temperature distribution in the superconductor and the convective heat transfer on its surface. Electrodynamic boundary conditions take into account the finite penetration velocity of the moving magnetization front and the variation of the electric field on the surface of the slab induced by an increasing external magnetic field.

The temperature dependences of the thermal properties of a superconductor define as follows

$$C_s\left[\frac{J}{m^3\times K}\right]=\begin{cases}58.5T+22T^3, & T\le 10\text{ K}\\ -10.54\times 10^4+1.28\times 10^4 T, & 10<T<80\text{ K}\end{cases}$$

$$\lambda_s\left[\frac{W}{m\times K}\right]=-1.234\times 10^{-3}+1.654\times 10^{-2}T+4.608\times 10^{-4}T^2-1.127\times 10^{-5}T^3+6.061\times 10^{-8}T^4$$

$$(4.34)$$

according to Junod et al. (1994) and Herrmann et al. (1993), respectively.

The critical current density in $Bi_2Sr_2CaCu_2O_8$ as a function of the temperature and magnetic induction describes by the Formula (1.8) using the parameters $J_0 = 8.655 \times 10^8$ A/m^2, $B_0 = 0.075$ T, $T_{c0} = 87.1$ K, $B_{c0} = 465.5$ T, $\alpha = 10.33$, $\beta = 6.76$, $\chi = 0.55$ and $\gamma = 1.73$ which describe the experimental data discussed in Bottura (2002b). The results of the investigations presented below were obtained at $T_0 = 4.2$ K, $E_c = 10^{-4}$ V/m, $n = 10$ and $B_a = 10$ T.

The general character of the thermo-electrodynamic phenomena in superconductors is described by equation (1.16). Accordingly, the nonisothermal electric field diffusion is described by equation

$$\frac{J_c(T,B)}{nE}\left(\frac{E}{E_c}\right)^{\frac{1}{n}}\frac{\partial E}{\partial t}=\frac{1}{\mu_0}\frac{\partial^2 E}{\partial x^2}-\frac{1}{C_s(T)}\frac{\partial J_c(T,B)}{\partial T}\left(\frac{E}{E_c}\right)^{\frac{1}{n}}\left[\frac{\partial}{\partial x}\left(\lambda_s\frac{\partial T}{\partial x}\right)+EJ_c(T,B)\left(\frac{E}{E_c}\right)^{\frac{1}{n}}\right]$$

$$-\frac{\partial J_c(T,B)}{\partial B}\left(\frac{E}{E_c}\right)^{\frac{1}{n}}\frac{\partial E}{\partial x}\quad(4.35)$$

in the magnetization region of a superconductor with the power V-I characteristic. Equation (4.35) directly shows that a stable increase in the temperature and magnetic induction in a superconductor can lead not only to more intense penetration of the magnetic flux into it but also to the transition of the superconductor to a normal state due to the fission-chain character of the magnetic flux diffusion under certain conditions. The latter happens as the spontaneous rise of the electric field and temperature. However, there exist the characteristic peculiarities that underlie the basis of the instability phenomena in HTS. To demonstrate this feature stricter, let us rewrite equation (4.35) in the form

$$\frac{\partial E}{\partial t}=D\frac{\partial^2 E}{\partial x^2}+\gamma E+q$$

assuming for the sake of simplicity that $n \gg 1$ and the critical current is only decreasing function of the temperature. Here,

$$D=\frac{nE}{\mu_0 J_c(T)},\quad \gamma=\frac{nE}{C_s(T)}\left|\frac{dJ_c(T)}{dT}\right|,\quad q=\frac{nE}{C_s(T)J_c(T)}\left|\frac{dJ_c(T)}{dT}\right|\frac{\partial}{\partial x}\left(\lambda_s\frac{\partial T}{\partial x}\right)$$

It is seen that the magnetic instability will develop less intensively with decreasing n-value because D, γ and q decrease with the decreasing n-value, and the V-I characteristic becomes smoother. As a result, instability has avalanche character in LTS with high n-value and will run smoother in the HTS with smaller n-value. Besides, the quantity γ is essentially temperature-dependent. First of all, it varies in the inverse proportion to the heat capacity of a superconductor. So, the fission-chain evolution of the induced electric field in HTS may begin to decay after an increase of their heat capacity with temperature as discussed in Chapter 1. Moreover, since the induced electric field may decrease in the full penetration mode (will be shown below) then the creation probability of the fission-like states will decrease with the corresponding temperature rise of superconductor in the fully penetrated mode.

Equation (4.35) demonstrates not only the stabilizing role of the heat capacity of the superconductor but also the stabilizing role of the small finite values of J_c and $\partial J_c/\partial T$ observed in the HTS due to the giant creep at the temperatures below the critical. To illustrate their existence, the temperature dependences of J_c and $\partial J_c/\partial T$ are shown in Figure 4.21 for the considered $Bi_2Sr_2CaCu_2O_8$. The linear approximations of the corresponding dependences $J_c(T, B)$ are also plotted by the dashed lines. One can see that the values of J_c and $\partial J_c/\partial T$ become very small in the subcritical temperature range after exceeding the temperatures which correspond to the effective values of the critical temperature of the superconductor at a given magnetic field in the linear temperature approximation. Accordingly, the electric field may have nonintensive development under certain operating temperatures when the quantities J_c and $\partial J_c/\partial T$ become small. So, the swept dynamics of the electric field may come close to the practically steady modes without the transition of the HTS to the normal state due to its large temperature margin, that is $T_{c0} - T_0$.

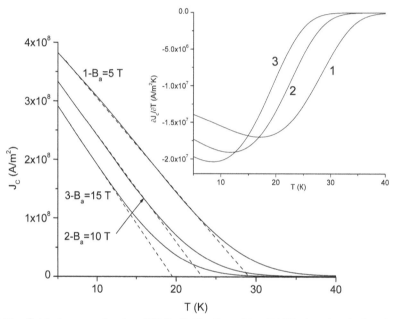

FIGURE 4.21   Critical current density of $Bi_2Sr_2CaCu_2O_8$ and its dJ_c/dT values (on the inset) as a function of the temperature at different magnetic fields. The critical temperature of the superconductor is $T_{c0} = 87.1$ K.

The temperature effect on the flux diffusion character depends also on the value and sign of $\partial \lambda_s(\partial T/\partial x)\partial x$-term. The latter may be both positive (generation mode) and negative (absorption mode) over the cross-section of a superconductor. In particular, this depends on the cooling conditions (see below Figure 4.27). Therefore, if this quantity is positive, then the diffusion of the flux is more intensive in this part of a superconductor. On the contrary, the temperature dynamics will be less intensive at $\partial \lambda_s(\partial T/\partial x)\partial x < 0$.

Thus, there exists the thermal self-stability effect according to which the electrodynamic states of HTS become more stable in the high intermediate temperature range, which is not close to the critical one. In other words, the temperature rise of HTS may lead to their stable states. For example, in the case of $Bi_2Sr_2CaCu_2O_8$ under consideration, similar states may exist in the temperature range close to about 30 K at high applied magnetic fields, as follows from Figure 4.21. As a whole, this conclusion is the result of the joint variation of $C_s(T)$, $J_c(T)$, $\partial J_c/\partial T$ and $\partial \lambda_s(\partial T/\partial x)\partial x$ quantities. The importance of this conclusion should be emphasized. To formulate the stability conditions of HTS, one usually considers only the heat capacity effect on the probability of the instability onset. In the meantime, the avalanche-like modes during electromagnetic field evolution in HTS may be gradually faded in the high intermediate temperature range due to the temperature dependencies of $J_c(T)$ and $\partial J_c/\partial T$ and the development of the electromagnetic field may become stable. As a result, unlike LTS, the magnetic instabilities in HTS may not lead to their transition to the normal state even in the high magnetic fields at low temperatures of the coolant. The discussed physical peculiarities of the nonisothermal electrodynamics behavior of HTS are pictured below in Figures. 4.22-4.28 where the simulation results of the characteristic thermo-electrodynamic states of the HTS are presented. They allow verifying the discussed above peculiarities of their development in HTS.

The typical electrodynamic states formation in the high-T_c superconducting slab under consideration is depicted in Figure 4.22 during both partial (Curves 1-4) and full (Curve 5) penetration modes, which were numerically simulated in the framework of the isothermal approximation ($T = T_0$) according to (1.8) and second equation (4.31) and relations (4.32)-(4.34). In this mode, the current density in the superconductor constantly increases. Therefore, the inequality $\partial^2 E/\partial x^2 > 0$ always takes place in accordance with the second equation of the system (4.31) when the term $\partial J_c/\partial B$ is negligible. Besides, as it has been shown in Chapter 2, the isothermal sweep flux dynamics of the electric field on the surface of the superconductor satisfies the estimator $E(a, t) \sim t^{n/(n+1)}$. Therefore, the next inequalities $\partial E(a, t)/\partial t > 0$ and $\partial^2 E(a, t)/\partial t^2 < 0$ will exist in the isothermal approximation. These conclusions confirm the curves depicted in Figure 4.23b discussed below.

The evolution of the temperature on the surface $T(a, t)$ and in center $T(0, t)$ of the slab, electric field $E(a, t)$ on its surface and magnetization front $x_p(t)$ at intensive cooling condition is shown in Figure 4.23. The time-dependent quantities $T(a, t)$, $T(0, t)$, $x_p(t)$, $E(a, t)$ and $J(a, t)$ are presented in Figure 4.24 for different values of the heat transfer coefficient. In both simulations, the computations were made at the continuously increasing external magnetic field with the sweep rate $dB_a/dt = 0.1$ T/s assuming that the slab half-thickness is equal to $a = 10^{-3}$ m. The heat capacity of the superconductor was calculated according to (4.34). It was also set to be constant and equal to $C_s(T_0)$ to estimate the effect of the thermal self-stabilization under the adiabatic conditions. The influence of the high sweep rate on the thermal and electric states of this superconductor is depicted in Figure 4.25. Figure 4.26 shows the simulation results of the nonisothermal flux penetration into the slab under consideration at its different thickness and the cooling condition close to the conduction-cooling. In this case, two kinds of the simulations were done: at continuous increase of the applied magnetic field and for the regime when the sweep rate of the applied field is equal to zero after a certain time t_s ($dB_a/dt = 0$ at $t > t_s$). The peculiarities of the spatial distributions of the swept electric field and screening currents, which are characteristic for the nonisothermal states, are depicted in Figures. 4.27 and 4.28.

The presented results not only confirm the conclusions already formulated for the LTS regarding the role of a stable increase of the superconductor temperature in the electrodynamic phenomena occurring in them but also expand them. The following characteristic formation features of the thermo-electrodynamics states of the HTS should be noted.

As discussed for the LTS with idealized V-I characteristics, the time character of the thermo-electrodynamic state development in the HTS with real V-I characteristic also depends on the sweep rate, heat transfer coefficient and thickness of a superconductor confirming the conclusions drawn for simplified V-I characteristics. These physical quantities influence on the temperature distribution in a superconductor, in particular, the difference between the temperatures on the surface of a superconductor and in its center. As follows from the estimator $ah/\lambda_s \ll 1$, defining the minor role of the transverse heat propagation and according to the present simulation results, this difference increases with the increase in the heat transfer coefficient, sweep rate and thickness of a superconductor. In the nonisothermal approximation, the hottest point may exist in an arbitrary part of the superconductor cross-section. In particular, it may be located both in the center of the slab and on its surface. For example, the sweep-rate states may be characterized by the nonuniform temperature distribution at intensive cooling conditions during which the temperature of a superconductor is maximal in its center both at partial and full penetration modes. In the meantime, the temperature of a superconductor may have the maximum on the slab surface or in the region, which is close to the surface, during partial penetration mode at nonintensive cooling conditions, but it becomes maximal in the center when the flux penetration mode is full.

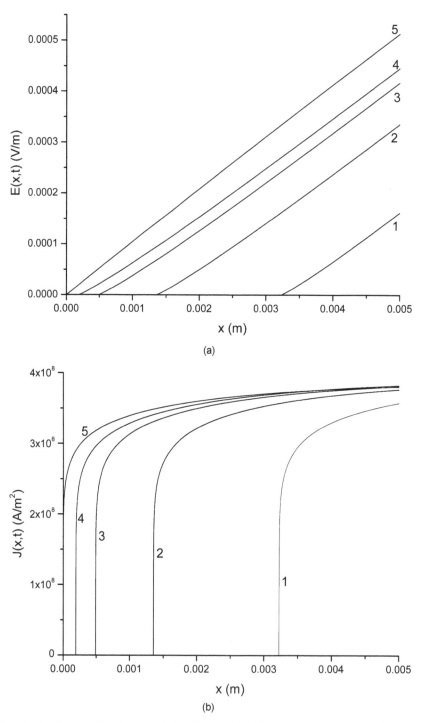

FIGURE 4.22 Isothermal distribution of the electric field (a) and screening current (b) in the high-T_c superconducting slab at $a = 5 \times 10^{-3}$ m and $dB_a/dt = 0.1$ T/s: 1 - $t = 7$ s, 2 - $t = 16$ s, 3 - $t = 20$ s, 4 - $t = 22$ s, 5 - $t = 30$ s.

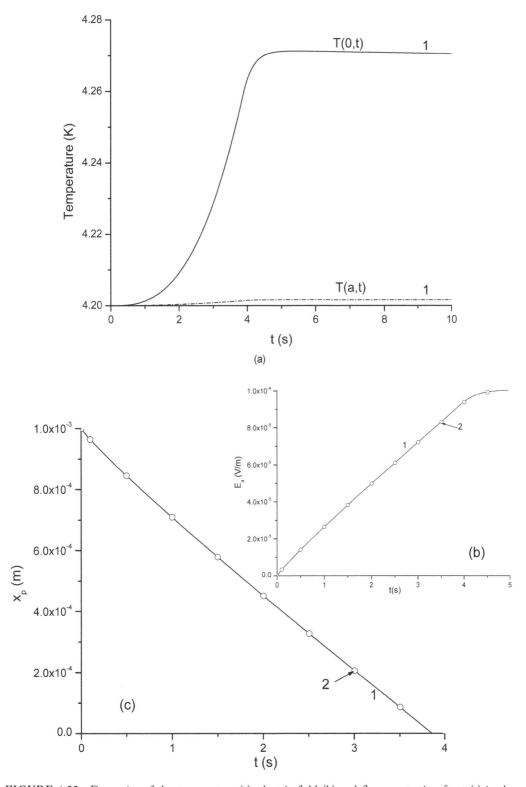

FIGURE 4.23 Dynamics of the temperature (a), electric field (b) and flux penetration front (c) in the intensive cooling ($h = 10^3$ W/(m^2 × K)) high-T_c superconducting slab at $a = 10^{-3}$ m and $dB_a/dt = 0.1$ T/s: 1 – nonisothermal model (1.8), (4.31)-(4.34); 2 – isothermal model ($T = T_0 = 4.2$ K), (1.8), (4.31)-(4.34).

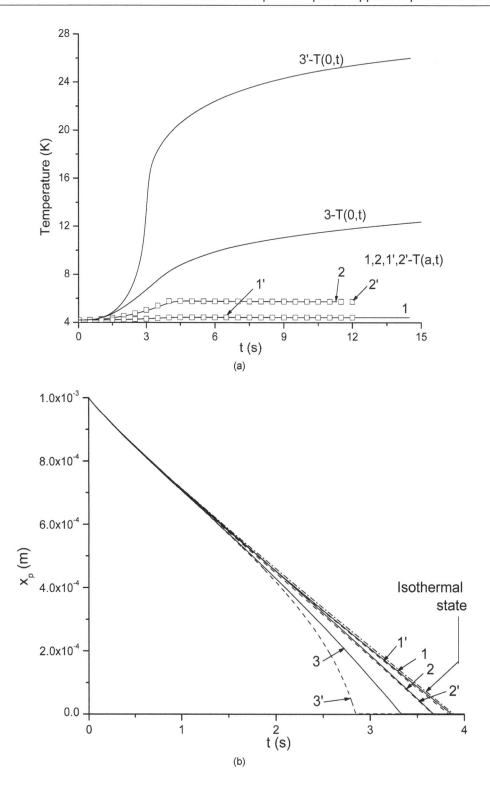

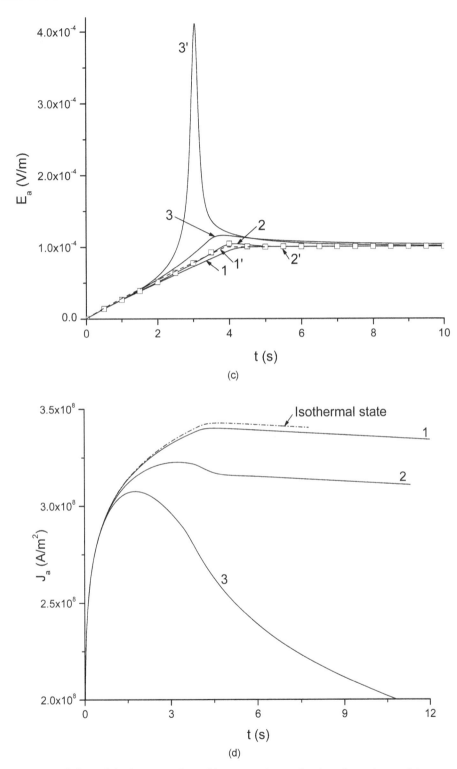

FIGURE 4.24 Effect of the heat transfer and heat capacity on the time dependence of the temperature (a), moving flux front (b), electric field (c) and current density (d) on the surface of high-T_c superconducting slab at $a = 10^{-3}$ m: 1, 1' - $h = 100$ W/(m² × K); 2, 2' - $h = 10$ W/(m² × K); 3, 3' - $h = 0$. Used models: 1, 2, 3 – nonisothermal model (1.8), (4.31)-(4.34); 1', 2', 3' – nonisothermal model (1.8), (4.31)-(4.34) at $C_s = C_s(T_0)$. Here, $E_s = 10^{-4}$ V/m.

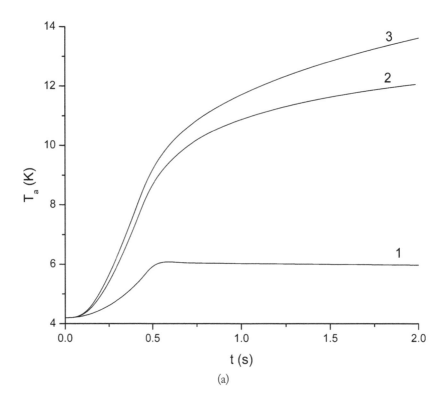

(a)

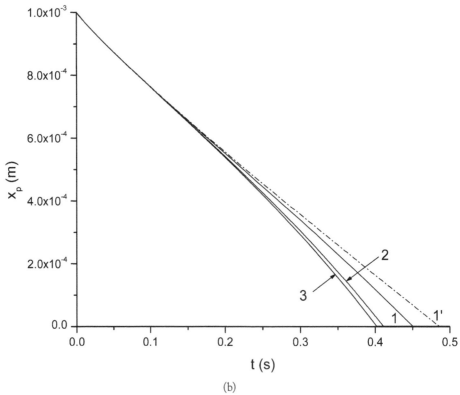

(b)

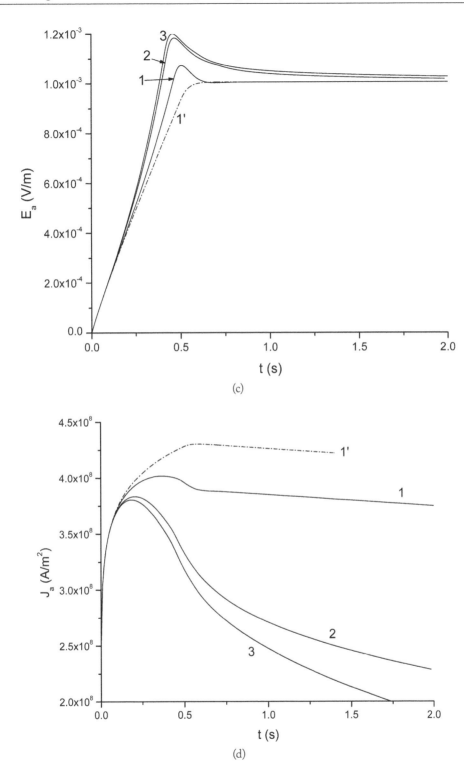

FIGURE 4.25 Time dependencies of the temperature (a), flux front (b), electric field (c) and current density (d) on the surface of high-T_c superconducting slab at $a = 10^{-3}$ m, $dB_a/dt = 1$ T/s at different values of the heat transfer coefficient: 1 - $h = 100$ W/(m² × K), 2 - $h = 10$ W/(m² × K), 3 - $h = 0$. Used models: 1, 2, 3 – nonisothermal diffusion model with nonlinear temperature dependences $C_s(T)$ and $J_c(T)$; 1′ – isothermal diffusion model ($T = T_0$). Here, $E_s = 10^{-3}$ V/m.

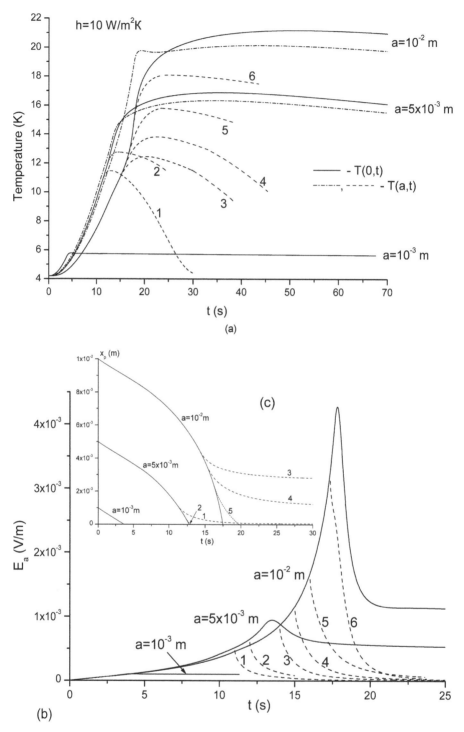

FIGURE 4.26 Formation of the operating modes of nonintensively cooled HTS slab (h = 10 W/(m²K)) with different thicknesses at dB_a/dt = 0.1 T/s and various increasing modes of the external magnetic field (——, —·— - continuous increase of the external magnetic field, — — — - relaxation of the thermo-electrodynamic states after different termination times of the increase of external magnetic field (1 - t_s = 11 s, 2 - t_s = 12 s, 3 - t_s = 14 s, 4 - t_s = 15 s, 5 - t_s = 16 s, 6 - t_s = 17.3 s). Here, (a) – the temperature change on the surface and in the center of the superconductor; (b) – the dynamics of the electric field on the surface of the superconductor; (c) – the magnetization front as function of the time.

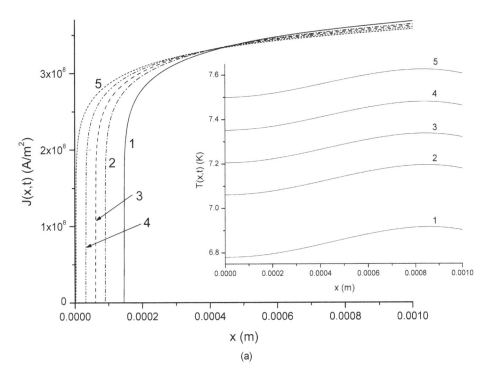

(a)

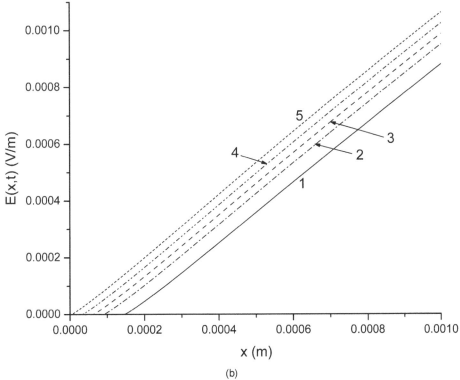

(b)

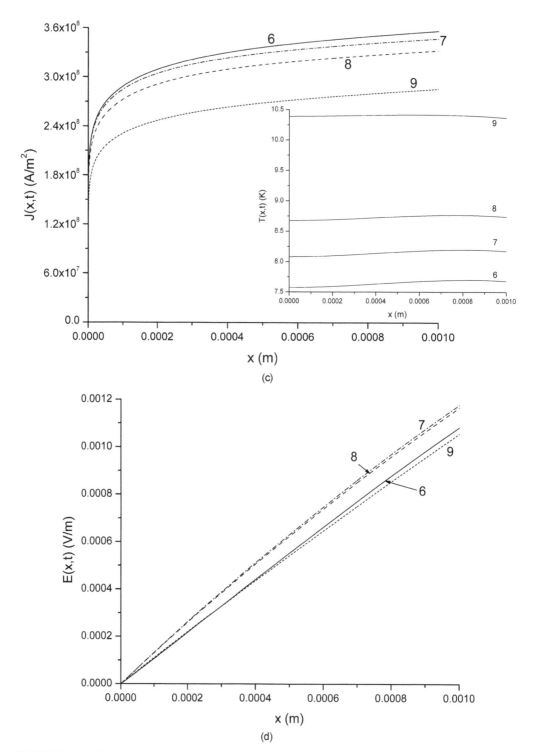

FIGURE 4.27 Formation of the partially (a, b) and fully (c, d) penetrated thermo-electrodynamic states of superconducting slab ($a = 10^{-3}$ m) at nonintensive cooling ($h = 10$ W/(m$^2 \times$ K)) and $dB_a/dt = 1$ T/s: 1 - $t =$ 0.36 s, 2 - $t = 0.38$ s, 3 - $t = 0.39$ s, 4 - $t = 0.40$ s, 5 - $t = 0.41$ s, 6 - $t = 0.415$ s, 7 - $t = 0.45$ s, 8 - $t = 0.5$ s, 9 - $t =$ 0.81 s. The insets to the Figures 4.27a and 4.27c show the corresponding temperature distributions.

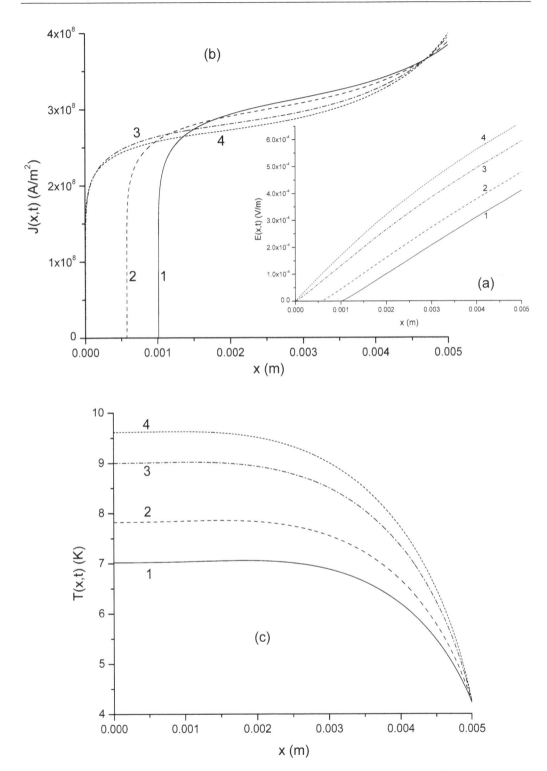

FIGURE 4.28 Partial penetration dynamics of the electric field (a), current density (b) and temperature (c) in the bulk superconductor ($a = 5 \times 10^{-3}$ m) at intensive cooling ($h = 10^3$ W/(m$^2 \times$ K)) and $dB_a/dt = 0.1$ T/s: 1 - t = 15.7 s, 2 - t = 17 s, 3 - t = 18.4 s, 4 - t = 19 s.

The magnetic flux penetration is nearly isothermal only at the intensive cooling of the thin superconductors located in slowly varying external magnetic fields. However, their temperatures noticeably differ from the coolant temperature when the heat transfer conditions deteriorate (it is most noticeable under adiabatic conditions, see Figure 4.24), at an increase in the sweep rate (Figure 4.25) or at an increase in the transverse size of the superconductor (Figure 4.26). The calculations show that the decrease of n-value will also lead to the nonisothermal character of the electrodynamic state formation.

In general, the temperature of the HTS is higher than the coolant temperature due to unavoidable Joule heat release, which exists in superconductors with continuously increasing V-I characteristics. The stable temperature rise of superconductor may lead to the self-stability effect as discussed above, first of all, during cooling regimes close to an adiabatic which may be observed at the nonintensive cooling conditions or at the high sweep rates. This effect illustrates the Curves 3 and 3′ in Figure 4.24 calculated for various dependencies of $C_s(T)$. They demonstrate the various characters of the temperature and electric field evolution in HTS at the adiabatic cooling condition. Therefore, the temperature dependence of the heat capacity must be taken into consideration to describe correctly the evolution of the adiabatic-like modes. As a consequence, the adiabatic stability criteria of the swept magnetic field penetration in HTS must be formulated allowing for the corresponding temperature dependence of the superconductor heat capacity. The importance of this conclusion should be emphasized because this peculiarity, as a rule, does not take into consideration during the investigation of the magnetic instabilities in HTS. Note that the intensity rise of curve 3′ decreases at $T > 24$ K. Since C_s was taken constant in this calculation, this effect is due to the corresponding change of the temperature dependencies of $J_c(T)$ and $\partial J_c/\partial T$ according to the results shown in Figure 4.21.

In accordance with the stable temperature increase of superconductor, the magnetic flux penetration begins to be nonisothermal. Let us discuss the characteristic features of these states.

As shown in Chapter 2, the velocity of the magnetic flux penetration in the partial penetration mode is equal to $|V_n| = \mathrm{const}\,(n, E_c, J_{c0}, T_0, dB_a/dt)/t^{1/(n+1)}$ during isothermal states, that is practically constant. The simulation results presented in Figure 4.23c and Curve 1 in Figure 4.24b prove this regularity. However, the absolute value of the nonisothermal velocity of magnetic flux penetration may have both the nonlinear time dependence and higher value then $|V_n|$ when the heat transfer coefficient decreases or the sweep rate or thickness of superconductor increase (see Figures 4.24b, 4.25b and 4.26c). This thermal acceleration of the moving flux boundary is the result of the corresponding interrelated increase of the temperature and electric field on the surface of a superconductor (see Figures 4.24c, 4.25c and 4.26b), according to which the value of dE_a/dt affects $|V_n|$ as proved in Chapter 2.

The monotonic electric field rise on the surface of the intensively cooled superconductor obeys the conditions $\partial E(a, t)/\partial t > 0$ and $\partial^2 E(a, t)/\partial t^2 < 0$ (Figure 4.23b and Figure 4.24c) as discussed above for isothermal states. At the same time, the nonisothermal increase of the electric field is characterized by the conditions $\partial E(a, t)/\partial t > 0$ and $\partial^2 E(a, t)/\partial t^2 > 0$. For example, they are observed at $t > 1$ s in Curves 2 and 3 in the partial penetration mode, as follows from Figure 4.24. These conditions exist due to the additional energy dissipation stored by the screening currents, which is accompanied by a more intense temperature increase and penetration of the magnetic flux as already discussed for LTS. That is why the nonisothermal states are characterized by the thermal acceleration of the magnetization front of both in LTS and HTS (Figures 4.2 and 4.24b). However, condition $\partial E(a, t)/\partial t > 0$ is violated in the full penetration mode. In this case, the distribution of the electric field in the superconductor, hence the temperature and electric field tend to their equilibrium states that clearly show the Figures. 4.24. In this case, the limiting value of the electric field on the surface will be equal to $E_s = a\dot{B}_a$, according to the solution of equation $d^2E(a, t)/dx^2 = 0$ under the conditions $E(0) = 0$, $dE/dx(a) = \dot{B}_a$.

Note that the transition from the regime of the partial penetration to the full penetration is characterized by several features. In a practically isothermal state, the value $E_a(t) = E(a, t)$ monotonously approach to E_s after the front of the magnetization reaches the center of the slab. Curves 1 in Figure 4.24 demonstrate this feature. At the same time, the nonisothermal values of $E_a(t)$ have a maximum, which interrelates with the increase in the temperature of the superconductor preceding the full penetration mode. The maximum increases with the sweep rate and becomes noticeable at the operating conditions close to adiabatic ones, in particular in the case of the bulk superconductors (Figure 4.24c, 4.25c and 4.26b). In these cases, the calculated maximum value of the electric field depends essentially on the used temperature dependence of the heat capacity of a superconductor (Figure 4.24c). The decrease of $E_a(t)$ during full penetration mode may have a rapidly decaying oscillatory character. Curve 1 depicts this mode in Figure 4.25c. The descending branch of $E_a(t)$ leads to a change in the intensity of the temperature rise when the condition $\partial^2 T(a, t)/\partial t^2 < 0$ is carried out in the fully penetrated modes unlike for the condition $\partial^2 T(a, t)/\partial t^2 > 0$ in the partially penetrated modes (Figures 4.24a, 4.25a, 4.26a). In other words, as expected, the intensity of the increase in the temperature of the superconductor correlates with the changes E_a during the transition from the partially penetrated mode to the fully penetrated mode.

The nonisothermal behavior of the induced electric field is accompanied by the specific evolution of the screening currents. Indeed, as mentioned above (Figure 4.22b), the screening current density increases monotonously over the cross-section of superconductors in the isothermal partially penetrated modes. Its subsequent decrease during full penetration mode is a result of the magnetic field dependence of the critical current density, which in the framework of the accomplished investigation is described by the relation (1.8). This tendency takes place at the intensive cooling conditions of the superconductor having a relatively small cross-section when its temperature is practically close to the coolant temperature, as depicted Figure 4.24d. At the same time, the intensive temperature rise of the superconductor on its surface leads to the nonmonotonous variation of the current density in the superconductor even during partial penetration modes. Namely, the time dependence of the partially penetrated current density on the surface has pronounced maximum at 'bad' cooling modes (Figures 4.24d and 4.25d). In these modes, the reduction of the current density in the region close to the surface and the simultaneous current penetration to the central part of the superconductor take plase despite the total electric field rise, as shown in Figures 4.27a and 4.27b. This feature may be proved taking into consideration the nonisothermal voltage-current characteristic. Indeed, differentiating equation (4.32) with respect to time in the case when the critical current density is only a function of the temperature, one can get the relation

$$\frac{\partial J}{\partial t} = \frac{J_c(T)}{nE}\left(\frac{E}{E_c}\right)^{1/n}\frac{\partial E}{\partial t} + \left(\frac{E}{E_c}\right)^{1/n}\frac{dJ_c}{dT}\frac{\partial T}{\partial t}$$

It is seen that the increase in current density directly follows the increasing electric field when the temperature change of a superconductor is small. Since the influence of the second term in this formula increases with increasing temperature and $dJ_c/dT < 0$, then the current density will decrease with increasing electric field after a certain temperature rise. Therefore, the time when the screening current density has the maximal value is smaller than that at which the relevant electric field is also maximal. (It should be noted that this peculiarity may lead to the existence the oscillation modes in a superconductor with the constantly increasing V-I characteristic, as it will be shown below). As a result, the current density in the nonisothermal approximation may monotonously decrease over the cross-section of the superconductor in the full penetration mode, as Figure 4.27c depicts. This regularity and the stabilization of the swept electric field distribution in the full penetration mode lead to such temperature change that its development may become more stable after a certain time even despite the nonintensive cooling conditions. As a whole, this thermal behavior of superconductors depends on the heat transfer coefficient, superconductor transverse size or sweep rate. However, the maximum temperature value of the HTS is not equal to the critical temperature even in the case of the bulk superconductor at the conduction-cooling conditions owing to the influence of the temperature-dependent properties of superconductors, as discussed above.

The discussed peculiarities of the nonisothermal electrodynamic state formation in the HTS allow one to predict the existence of various spatial profiles of the electric field and screening current induced in a superconductor by a swept magnetic field. They are important to understand the regularities under which the magnetic instability may occur in HTS.

Let us integrate the second equation of the system (4.31) with respect to x from x to a. Then one gets the following equality

$$\mu_0 \int_x^a \frac{\partial J}{\partial t} dx = \frac{dB_a}{dt} - \frac{\partial E}{\partial x} \tag{4.36}$$

This expression together with the second equation in the system (4.31) shows that $\partial E/\partial x < dB_a/dt$ and $\partial^2 E/\partial x^2 > 0$ (or $\partial E/\partial x > dB_a/dt$ and $\partial^2 E/\partial x^2 < 0$) at any point in the superconducting slab if the screening current monotonously increases (or it decreases) with time in the integration range. These conditions demonstrate the thermal effect on the slope of the electric field profile. As a result, the electric field profile in the superconductor, which has the high overheating, will satisfy the inequalities $\partial E/\partial x > dB_a/dt$ and $\partial^2 E/\partial x^2 < 0$. At the same time, the conditions $\partial E/\partial x < dB_a/dt$ and $\partial^2 E/\partial x^2 > 0$ will be observed during an insignificant temperature change of superconductor. These conditions typify the difference between the isothermal and nonisothermal character of the swept electric field and screening current penetration. Let discuss the reason leading to this peculiarity.

Integrating equality (4.36) from x_p to x, it is easy to write the next integral relationship

$$\mu_0 \int_{x_p}^x dy \int_y^a \frac{\partial J}{\partial t} dx = (x - x_p)\frac{dB_a}{dt} - E(x, t) \tag{4.37}$$

The first term in the right part of this equality corresponds to the Bean spatial-temporal electric field distribution in the superconducting slab for the mode under consideration. It is important than this formula demonstrates the possible existence of some nodal point x_n. This point will divide the magnetization region into two parts at which the character of the screening current redistribution is different. According to the (4.37), one can write the condition

$$\mu_0 \int_{x_p}^{x_n} dy \int_y^a \frac{\partial J}{\partial t} dx = 0 \tag{4.38}$$

which determines the nodal point location where the electric field values defined by Bean's and nonisothermal approximations are equal to each other. Consequently, the screening current will increase (or decrease) in the magnetization region when the value of the electric field is less (or higher) of the corresponding quantities defined by Bean's approximation. Since the total value of the screening current increases over the half cross-section of a superconductor ($0 < x < a$) during partial penetration mode, two different electric field distribution must simultaneously exist. They will have both the positive and negative values of $\partial^2 E/\partial x^2$ in a superconductor. So, the nonisothermal electric field distribution may have a profile consisting of two-piece satisfied by the conditions $\partial^2 E/\partial x^2 < 0$ and $\partial^2 E/\partial x^2 > 0$ where the screening current density will simultaneously both increase and decrease, respectively. Therefore, $\partial^2 E/\partial x^2 = 0$ at the nodal point x_n. As a result, the condition $\partial J/\partial t = 0$ must be fulfilled at this point according to the second equation of the system (4.31), that is, the value of the screening current density will be constant at this nodal point during the whole time of the current penetration in the partial penetrated mode. Such distributions of the swept electric field and screening current density are depicted in Figures 4.27a and 4.27b, which were calculated at the nonintensive cooling condition. In this case, the existing overheating of the surface layer is accompanied by the current density decrease and $\partial^2 E/\partial x^2 < 0$ in this region. Contrariwise, $\partial^2 E/\partial x^2 > 0$ in the central part of the superconductor, where the values of $J(x, t)$ increase in accordance with the relevant temperature change.

The influence of the nodal point on the redistribution of current density in the superconductor depends on the character of the temperature distribution in it.

The nodal point may also appear at intensive cooling conditions (Figure 4.28b). However, the high value of the heat transfer coefficient leads to the nonuniform temperature distribution. It becomes more visible in the bulk superconductors (Figure 4.28c). Such nonuniform character of the temperature distribution will result in the relevant current redistribution in a superconductor, which will be not similar to the corresponding current redistribution during isothermal approximations due to the existence of the nodal point and the noticeable nonuniform temperature distribution even at high values of the heat transfer coefficient. As a result, the partially penetrated screening current density induced by the swept magnetic field in the 'good' cooled superconductor will monotonously increase on its surface where the temperature is minimal, as depicted in Figure 4.28b. However, the current redistribution will have the nonmonotonous character in the internal part of the bulk superconductor ($0 < x < x_n$) since the overheating of a superconductor is high.

Formula (4.37) is reduced to

$$\mu_0 \int_0^x dy \int_y^a \frac{\partial J}{\partial t} dx = x \frac{dB_a}{dt} - E(x, t)$$

in the full penetration mode. It proves, in particular, that the fully penetrated stages are characterized by the high magnitudes of the induced electric field over the cross-section of a superconductor (Figures 4.27c and 4.27d), which exceeds Bean's values. Therefore, the electric field and screening current will monotonously decrease due to asymptotic approaches the swept electric field to Bean's distribution $x \frac{dB_a}{dt}$ at each point of the cross-section of superconductor. In this mode, the nodal point is absent. At the same time, the nonmonotonous redistribution of the fully penetrated screening current will happen at the intensive cooling conditions. This feature, which takes place in bulk superconductors, plays an essential role in the oscillation phenomena and stability conditions of the electromagnetic field evolution, as it will be proved below.

In accordance with the nonisothermal nature of the electromagnetic field penetration into the HTS, its magnetic moment $\mu_0 M = \dfrac{1}{a} \int_{x_p}^a B(x, t) dx - B(a, t)$ (Figure 4.29) and the magnetization energy losses $G = \dfrac{1}{a} \int_o^t \int_{x_p}^a EJ dx dt$ (Figure 4.30) change. Here, the calculation was made for the superconductor with a half-thickness $a = 10^{-3}$ m both at the partial ($x_p > 0$) and full ($x_p = 0$) magnetic flux penetration at a continuously increasing external magnetic field. Along with the numerical solution of the system of equations (4.31)-(4.33), the corresponding isothermal ($T = T_0$) dependencies $M(B_a)$ and $G(B_a)$ were determined, allowing one to understand the role of the temperature factor in their change.

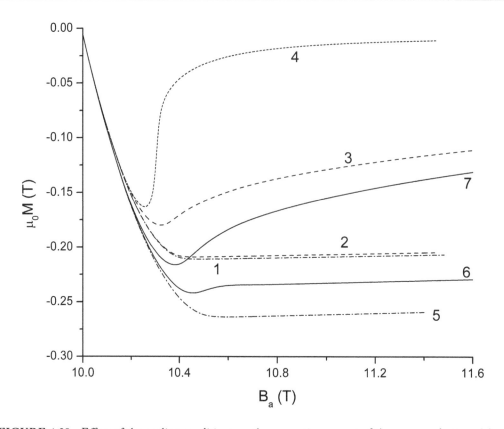

FIGURE 4.29 Effect of the cooling conditions on the magnetic moment of the superconducting slab at different operating modes. Here, according to the calculations at $C_s(T)$ and $dB_a/dt = 0.1$ T/s: 1 - isothermal state, 2 - $h = 100$ W/(m^2 × K), 3 - $h = 0$; 4 - according to the calculations at $h = 0$ and $C_s(T_0)$; 5 - isothermal state, 6 - $h = 100$ W/(m^2 × K), 7 - $h = 0$ according to the calculations at $C_s(T)$ and $dB_a/dt = 1$ T/s.

As follows from Figures 4.29 and 4.30, the nonisothermal character of the development of electrodynamic phenomena in superconductors can significantly affect the dependencies $M(B_a)$ and $G(B_a)$. Namely, the values $|M(B_a)|$ not only decrease during stable temperature change at full penetration of the magnetic flux but also lead them to a more peak-like form after the appearance of the full penetration mode. Moreover, the dependences $|M(B_a)|$ approach each other at an increase in the sweep rate of the external magnetic field even at high values of the heat transfer coefficient. In other words, the magnetic moment of superconductors during both partial and full penetration of the magnetic flux depends not only on the absolute value of the induction of the external magnetic field but also on its sweep rate due to the corresponding variation in the thermal state of superconductors. The values $|M(B_a)|$ are essentially reduced at a significant increase in the temperature of the superconductor. On other hand, if this regularity is observed in the experiment, then this will mean a significant deviation of the superconductor temperature from the temperature of the coolant. It is easy to understand that the noted thermal features of change in the magnetic moment of superconductors are retained if their electrodynamic states will be described using the critical state model. The 'smoothed' of the V-I characteristic will only lead to an increase in the role of the temperature factor.

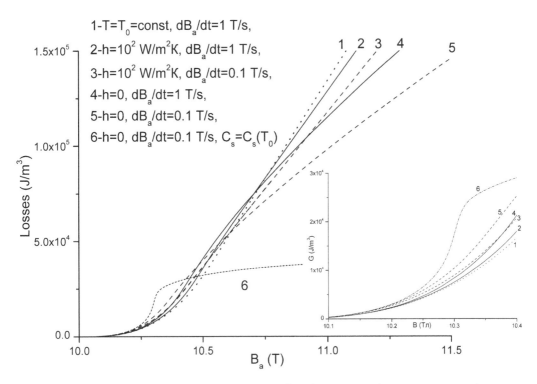

FIGURE 4.30 Energy losses in superconductor located in the continuously increasing external magnetic field at different operating modes. The losses in the partial penetration of the magnetic flux are shown in more detail in the inset.

Meanwhile, the nonisothermal character of superconducting states has an ambiguous effect on the energy losses during partial and full penetration of the magnetic flux. In the first case, an increase in the overheating of the superconductor increases them. Conversely, they decrease with increasing of a stable overheating of superconductor at the full penetration of the magnetic flux. Such a change in dissipative phenomena is explained in the following way. Energy losses primarily depend on the size of the magnetization region at the partial penetration of the magnetic flux. Accordingly, the higher the overheating, the higher the magnetization region, and hence, the higher are the losses. However, the energy losses decrease according to the decrease of the critical current density with the temperature at the full penetration of the magnetic flux when the losses depend primarily on the critical current density. Curve 6 calculated at constant value C_s also proves these peculiarities. In this mode, temperature rise of a superconductor is greatest. As a result, the calculated losses are not correct.

The nonisothermal nature of the electrodynamic phenomena in the HTS and their smooth character of V-I characteristics change the onset conditions and development of the flux jump in HTS, which differ from that taking place in LTS. Alongside the huge flux-creep, this is due to the high value of the temperature margin and the stabilizing role of the temperature-dependent properties of HTS. As discussed above, they lead to the smooth character of the magnetic flux penetration and, first of all, at the nonintensive cooling conditions. In Figures 4.26 and 4.31, the possible flux-jump modes under various operating conditions are presented. The solid curves correspond to the continuous increase of the applied magnetic field. The dashed curves were obtained under the condition when the applied magnetic field is fixed at $t \geq t_s$.

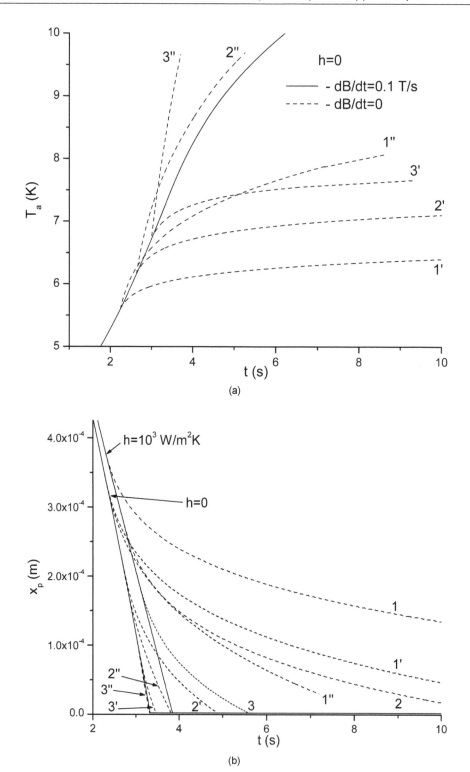

(a)

(b)

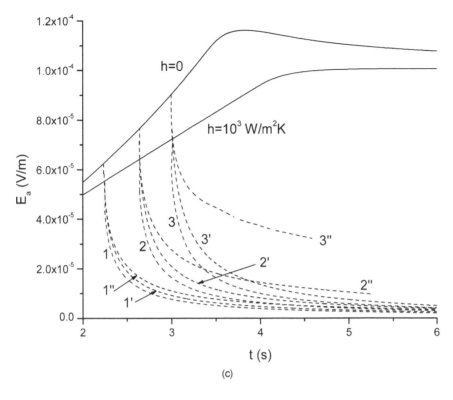

(c)

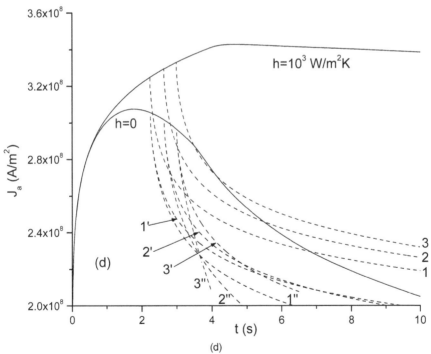

(d)

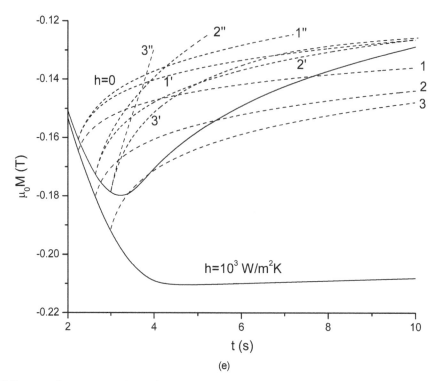

(e)

FIGURE 4.31 Operating modes and noncatastrophic flux jump in high-T_c superconductor ($a = 10^{-3}$ m) during continuous (——— - $dB_a/dt = 0.1$ T/s) and continuous-break (- - - - $dB_a/dt = 0$ at $t \geq t_s$) modes of the applied magnetic field: 1, 1', 1" - $t_s = 2.24$ s, 2, 2', 2" - $t_s = 2.63$ s, 3, 3', 3" - $t_s = 3$ s. Simulation conditions: 1, 2, 3 - $C_s = C_s(T)$, $h = 10^3$ W/(m$^2 \times$ K); 1', 2', 3' - $C_s = C_s(T)$, $h = 0$; 1", 2", 3" - $C_s = C_s(T_0)$, $h = 0$ at $t > t_s$.

Calculations given show the important physical peculiarities of the flux jump development in HTS. They are as follows. The flux jumps have no avalanche-like character even with a constant increase in the external magnetic field ($dB_a/dt > 0$ at $t > 0$). At these operating modes, even the temperature increase of the nonintensively cooled HTS does not reach the value of their critical temperature in spite of the high values of the induced electric field. However, the magnetization of HTS strongly depends on their thermal states. The development of the flux jump is becoming less intense when the external magnetic field is fixed even at the high induced electric field and the corresponding temperature rise of HTS. Moreover, the nonisothermal change of the electric field has only the decreasing character similar to the relaxation modes discussed in Chapter 2 after the termination of the increase in the external magnetic field ($dB_a/dt = 0$ at $t > t_s$). It occurs independently of the value of induced electric field in both the intensively cooled HTS and the case of the insulated bulk HTS. As a result, the bulk HTS may store its superconducting states at arbitrary value of t_s after the flux jump even in the conduction-cooling conditions (Figure 4.26).

These features lead to the crucial conclusion: the true boundary of the magnetic instability in HTS is defined 'bad'. Indeed, the boundary of the magnetic instability of a low-T_c superconductor is defined as the boundary between stable and unstable states that will be observed after the termination of the external magnetic field increase (Figures 4.4 and 4.5). In the latter case, the development of the magnetic instability is characterized by a spontaneous avalanche-like growth of the electric field and temperature in a superconductor. That is why the true boundary of the magnetic instability of LTS is 'well' defined. As it will be shown in Chapter 5, this definition of the magnetic instability boundary is also valid for superconducting composites based on the LTS. Besides, it allows one to find boundaries of the current or thermal instabilities both LTS and HTS as it will be proved in Chapters 6-8. However, any perturbation of the initial state of the high-T_c superconductor by the external magnetic field always fades after the termination of the perturbation action. The nature of these regularities is explained, first, by the thermal self-stability effect and second, by a smoother rise of the V-I characteristics of HTS as compared to the V-I characteristics of LTS. The first reason has been discussed above. The second reason is due to the fact that n-value of HTS is less than ones of LTS under other equal conditions. As follows from the results presented in Chapter 2, the lower the n-value, the smoother the decrease of the electric field induced in the superconductor by the external magnetic disturbance. In particular, the relaxation time of the disturbed states will be equal to zero at $n \to \infty$ during stable states. Figures 4.4 and 4.5 demonstrate this feature. As a consequence, the magnetic instabilities in HTS are not dangerous for superconducting magnets made of them. As a result, it is one of the least troubling instabilities for SMS with HTS current-carrying elements. It can be shown that this conclusion is true for the operating temperatures close to the temperature of liquid nitrogen. Meanwhile, the criteria formulated in section 4.4 can be used to assess the permissible parameters of the superconductor. They will impose the most severe restrictions on them when $T_i = T_0$.

Thus, the thermo-electrodynamic state formation of HTS is characterized by the following peculiarities:

- The thermal self-stability mechanism takes place according to which the probability of the fission-chain-reaction character development of the nonisothermal electromagnetic field evolution decreases.
- The temperature rise of high-T_c superconductor becomes noticeable at the cooling conditions close to adiabatic, for example, at a high sweep rate of magnetic field or in bulk HTS which, however, may occur without its transition to the normal state.
- The screening current redistribution depends on the temperature evolution in superconductor.
- The magnetic flux jumps in high-T_c superconductors have no catastrophic development.

These intrinsic electromagnetic properties of HTS are due to the joint temperature variation of their thermo-electric properties. That is why the cooling conditions, sweep rate and thickness of superconductor may have a pronounced effect on the thermal character of the induced electromagnetic field evolution. These peculiarities should be considered when the magnetization of the bulk superconductors and losses in nonintensively cooled current-carrying superconducting elements are analyzed even when their operating temperature is far from the critical temperature of a superconductor. In other words, the allowable temperature variation of HTS is required to give a complete description of the magnetic instability onset and development. In such theoretical studies, one has to solve the strongly nonlinear system of equation (4.31)-(4.33). Overall, namely the nonisothermal models permit one to understand the basic physical pecualiarities of magnetic phenomena in HTS and correctly describe their development in time.

4.7 OSCILLATING MAGNETIC INSTABILITIES IN HTS

The oscillations are experimentally observed in high-T_c superconductors (Tholence et al. 1988, Chen et al. 1991, Gerber et al. 1993, Muller and Andrikidis 1994, Legrand et al. 1996, Chabanenko et al. 1996, Khene and Barbara 1999). As shown in section 4.5, the conductive-convective mechanisms of the removal of magnetization energy losses may or may not result in oscillation operating modes. In particular, they do not occur in intensive cooling superconductor (Figure 4.23) or during its high overheating (Figure 4.26). However, if one decreases the overheating of a superconductor, for example, by improving the heat exchange conditions, then under certain conditions the oscillations may start. Conversely, they also may begin when the intensity of the heat release of an intensely cooled superconductor increases. Let us discuss in more detail the oscillation phenomena in the HTS slab considered above.

Figures 4.23 and 4.32 demonstrate the possible thermo-electrodynamic regimes that occur under slowly time-varying applied field (dB_a/dt = 0.1 T/s) at intensive cooling conditions (h = 10^3 W/(m^2 × K)) in the superconducting slab when its thickness is changed. The initial value of the external magnetic field is equal to B_a = 10 T. The numerical experiments based on the model (4.31)-(4.33) were performed for two types of the functional dependencies describing the variation of the critical current density on temperature and magnetic induction. First, the nonlinear approximation described by the Formula (1.8) was used, as above. The corresponding simulation results are plotted in Figure 4.32 by the solid curves. For a massive superconductor with a half-thickness $a = 5 \times 10^{-3}$ m, isothermal states were also calculated. In Figure 4.32, they are shown by the dash-dotted Curves to demonstrate the role of the temperature factor in the formation of the operating states of HTS. Second, the calculations were performed under the assumption of a linear temperature dependence $J_c(T)$ to estimate the effect of the temperature dependence of J_c on the thermo-electrodynamic states. In Figure 4.32, they are shown by the dashed curves. In this case, the expression

$$J_c(T) = J_k(1 - T/T_{cB}) \tag{4.39}$$

was used, assuming that $J_k = 4.21 \times 10^8$ A/m^2 and T_{cB} = 23.28 K, which approximate the critical parameters of the considered superconductor in the magnetic field of 10 T according to the results presented in Figure 4.21.

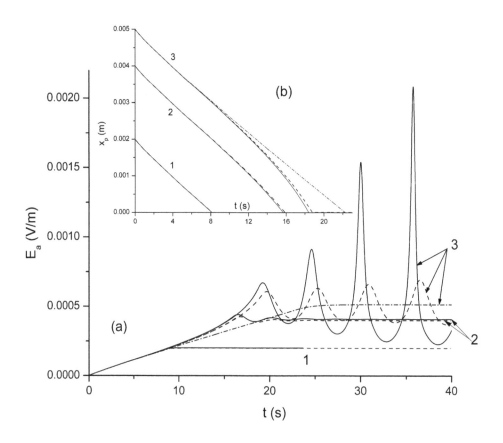

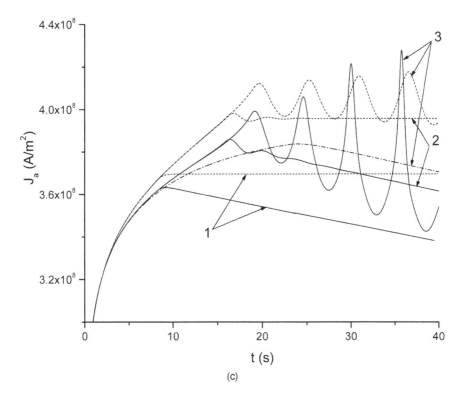

(c)

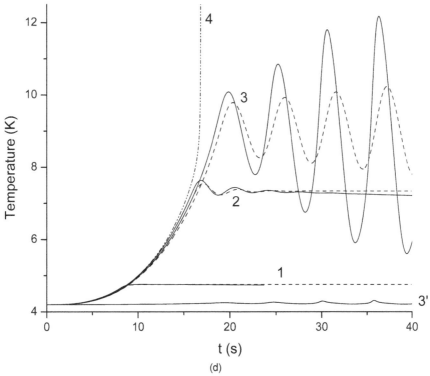

(d)

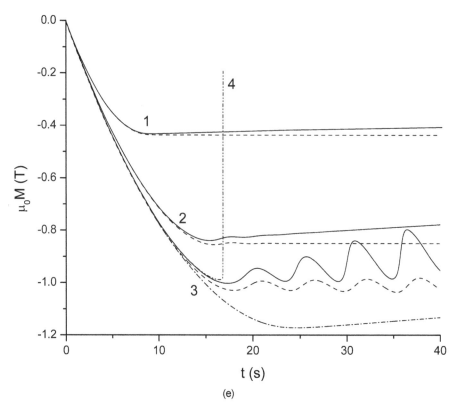

(e)

FIGURE 4.32 Dynamics of the electric field on the surface of the superconductor (a), boundary of the magnetization region (b), current density on the surface (c), temperature in the center (Curve 3) and on its surface (Curve 3′) (d), and magnetic moment (e) of the intensively cooled superconductor with different thickness: 1 - $a = 2 \times 10^{-3}$ m, 2 - $a = 4 \times 10^{-3}$ m, 3, 3′ - $a = 5 \times 10^{-3}$ m. Used J_c models: ——— – nonlinear approximation (1.8); - - - - – linear approximation (4.39, see below). Curves 1-3 correspond to the superconductor with $C_s = C_s(T)$, and Curves 4 are calculated at $a = 5 \times 10^{-3}$ m when in model (4.31)-(4.33) assumes $C_s = C_s(T_0)$. Curve (—·—·—) describes the calculations based on the isothermal model ($T = T_0$).

Figure 4.32 shows that the oscillation modes happen in the fully penetrated states even if the thickness of the superconductor changes in the wide range. As expected, the results presented in Figures 4.23 and 4.32 demonstrate the absence of the oscillations in an intensively cooled thin HTS (Curves 1) placed in the increasing external magnetic field since its temperature slightly differs from the coolant temperature and the oscillation is not observed in the isothermal state even in a massive superconductor. However, the oscillations of the thermo-electrodynamic states of HTS will appear when the transverse size of the superconductor increases. First, it is damped oscillation (Curves 2) and then they have an increasing character (Curves 3 and 3′). In these modes, the oscillations are characterized by a corresponding increase in the difference between the minimum and maximum values of the superconductor temperature and, thus, its inhomogeneous distribution across the cross-section.

As discussed above, the development of the magnetic instabilities in HTS is not an avalanche-like due to their high-temperature margin and the temperature-dependent of the superconductor properties. As a consequence, the heat capacity of the superconductor plays a significant stabilizing role in the development of the oscillations. To illustrate this peculiarity, Curves 4 in Figure 4.32d and 4.32e demonstrate the development of the magnetic instability in the considered superconductor if its heat capacity is constant and equal to $C_s(T_0)$. In this case, the magnetic instability leads to a sharp increase of the superconductor temperature and correspondingly, to a sharp change of the magnetic moment without the occurrence of the oscillations. In other words, the *oscillating magnetic instability* may occur in HTS that are in a continuously increasing external magnetic field due to the temperature dependence of the superconductor heat capacity.

It should be underlined that according to equation (4.35), the development of the oscillations will be affected by the dependence of the critical current density on the temperature and magnetic induction. As a consequence, the linear and nonlinear approximations of J_c will lead to a slight discrepancy between the results obtained, when the full penetration regime is achieved without a significant change in the initial temperature of the superconductor, for example, at its small thickness and intensive cooling (Curves 1 in Figure 4.32). However, the difference between approximations will increase, for example, at an increase in the transverse size of the superconductor even at its intensive cooling. (For LTS, a similar difference exists between approximations using the Bean and Kim models). Therefore, the occurrence and development of the oscillations in HTS with different temperature and magnetic field dependencies of the critical current density may differ.

To illustrate this peculiarity, the oscillations of the electric field $E_a = E(a, t)$ and current density $J_a = J(a, t)$ on the surface of a superconductor as a function of its temperature $T_a = T(a, t)$ are shown in Figure 4.33. They correspond to the above-mentioned linear and nonlinear approximations, which are depicted by Curves 2 in Figure 4.32. It is seen that the damped oscillations of $E_a(T_a)$ and $J_a(T_a)$ occurring in HTS in the linear approximation have the equilibrium values of the electric field $E_s = a \dfrac{dB_a}{dt}$ and temperature T_s during constantly increasing external magnetic field. However, if one takes into account the dependence of the critical current density on the magnetic induction, the equilibrium values of the dependencies $E_a(T_a)$ and $J_a(T_a)$ primarily will depend on the character of the variation in the external magnetic field.

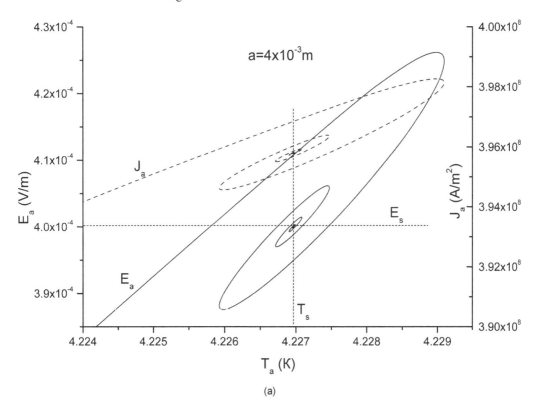

(a)

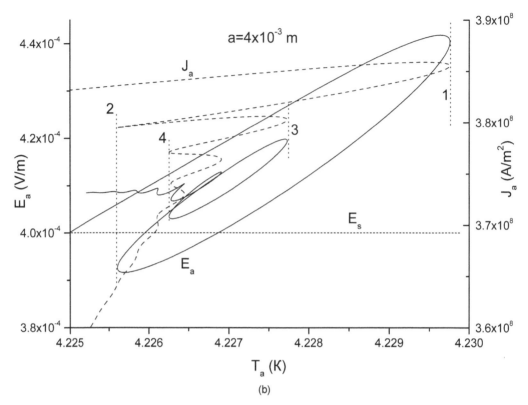

FIGURE 4.33 Influence the magnetic dependence of the critical current density on the damped oscillations at $h = 10^3$ W/(m² × K) and $dB_a/dt = 0.1$ T/s: (a) – oscillations of the electric field and current density at the linear approximation $J_c(T, B)$ (4.39); (b) – oscillations in superconductor with the nonlinear approximation $J_c(T, B)$ (1.8).

In Figure 4.34, the effect of the cooling conditions on the thermo-electromagnetic states, which take place in the superconducting slab having the half-thickness $a = 5 \times 10^{-3}$ m, are shown in detail. They were obtained at different temperature dependencies of the heat capacity of superconductor. It is seen that the various modes of the magnetic instability jump may be observed when the conditions, which influence on the thermal state of superconductors are varied. Correspondingly, the magnetization variation with time may have the following shape:

- The periodical onset of the magnetic instability at the intensive cooling condition (Curve 1).

- Aperiodic magnetic moment fluctuation at the conduction-cooling condition when its change occurs owing to the corresponding high stable temperature rise of a superconductor (Curve 2).

- The magnetic flux jump at a small value of the superconductor heat capacity even at high value of the heat transfer coefficient (Curve 3).

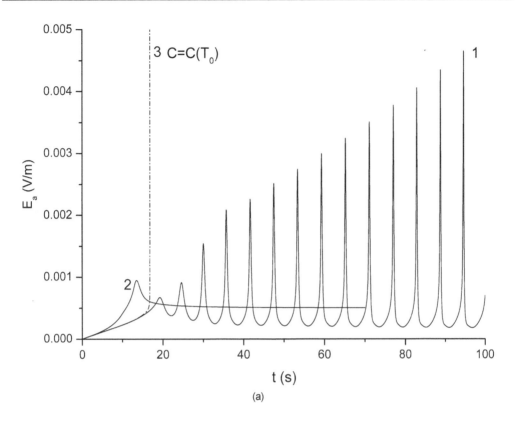

(a)

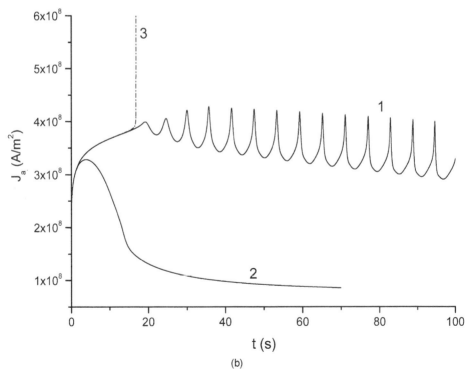

(b)

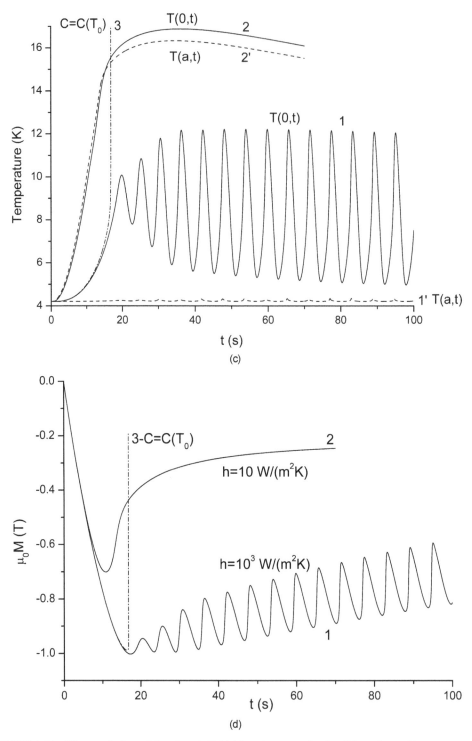

FIGURE 4.34 Time variation of the electric field (a) and current density (b) on the surface, temperature (c) and magnetic moment (d) of the bulk superconductor at dB_a/dt = 0.1 T/s and various heat transfer coefficient (1, 1′, 3 – h = 10^3 W/(m² × K); 2, 2′ – h = 10 W/(m² × K)). Used models: 1, 1′, 2, 2′ – nonisothermal diffusion model (4.31-4.33) with nonlinear $C_i(T)$-relation (4.34); 3 – nonisothermal diffusion model (4.31-4.33) at $C_i = C_i(T_0)$.

As discussed above, results presented in Figures 4.32 and 4.34 also prove the existence of the thermal reason underlying the emergence of the oscillations. It is seen that the size of the jumps and their duration depend appreciably on the character of the temperature rise occurring both before and during magnetic instability development. Besides, the electric field and temperature oscillations do not coincide. Their intensity in the central part of the superconductor and on its surface is different. Namely, the visible temperature fluctuation does not originate in the area close to the surface of a superconductor. At the same time, the noticeable electric field oscillations occur near the superconductor surface.

Huge oscillating thermo-electrodynamic states may be observed in the intensively cooled bulk superconductor. They are depicted in Figure 4.35 setting $h = 10^3$ W/(m$^2 \times$ K) and $a = 10^{-2}$ m. The solid curves demonstrate the existence of the periodic magnetic instability mode during the continuous increase of the applied field at $dB_a/dt = 0.1$ T/s from the initial value $B_a = 10$ T. In this case, the magnetic instability jumps have a larger amplitude in comparison with those pictured in Figures 4.33 and 4.34. They are also observed only during fully penetrated modes. It is interesting to note that the temperature of this huge oscillating mode of the massive superconductor does not exceed the critical one despite the large amplitude of the electric field fluctuation. As a result, the magnetic moment of the superconducting slab is not equal to zero after the huge jump development because the temperature of the HTS significantly lowers than its critical temperature. Such types of oscillations are observed in experiments. As Figure 4.35 shows, the huge oscillating the magnetic instability disappears when the applied magnetic field is fixed during both the partially and fully penetrated modes despite the cyclical repetition of jumps at a continuous increase in the external magnetic field. The dashed curves describe these relaxation stages when $dB_a/dt = 0$ at $t > t_s$.

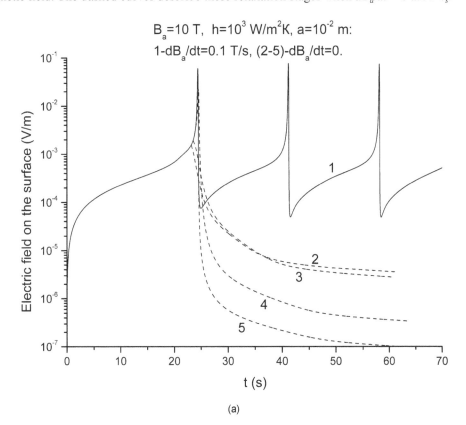

$B_a = 10$ T, $h = 10^3$ W/m^{2}K, $a = 10^{-2}$ m:
1-$dB_a/dt = 0.1$ T/s, (2-5)-$dB_a/dt = 0$.

(a)

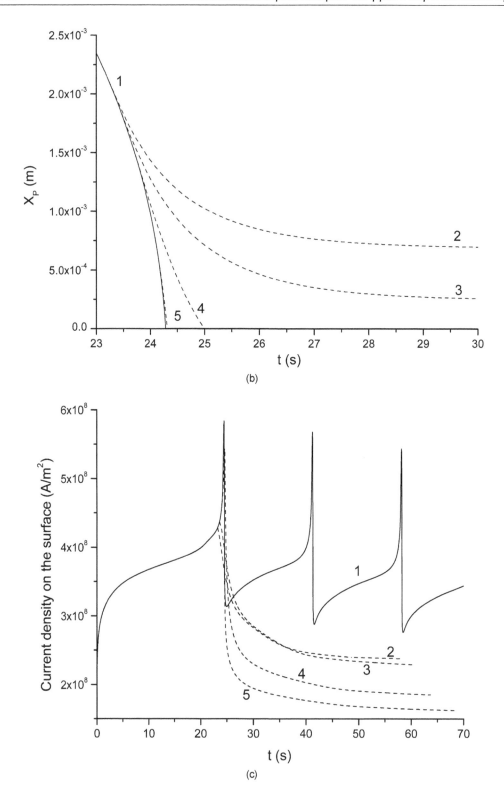

(b)

(c)

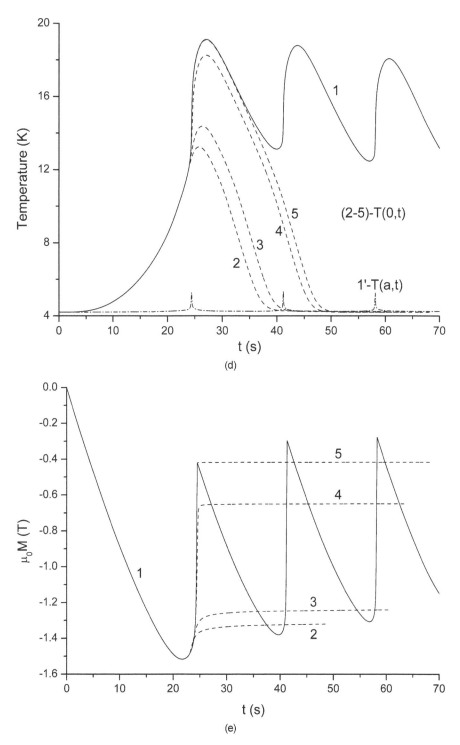

FIGURE 4.35 Evolution of the electric field (a), flux front (b), screening current (c), temperature (d) and magnetic moment (e) of the intensively cooled bulk superconductor during huge oscillations (Curve 1) and relaxations stages (the dashed curves: 2 - t_s = 23.2 s, 3 - t_s = 23.54 s, 4 - t_s = 24 s, 5 - t_s = 24.4 s). Used model: nonisothermal diffusion model (1.8), (4.31)-(4.33) with nonlinear $C_s(T)$-relation (4.34).

Thus, the curves plotted in Figures 4.32-4.35 show the physical peculiarities, which underlie the oscillation existence in HTS. In principle, the fluctuation behavior of the electric field, current density and temperature are determined by the thermal states of the superconductor. As a result, the cooling conditions, the thickness of the superconductor and temperature dependencies of the critical current density of superconductor and its heat capacity determine the magnitude and frequency of the oscillations. In particular, the oscillations are absent or have fading character, if the thermal state of superconductor close to the isothermal mode, for example, when the transverse size of the intensively cooled superconductor is not relatively large. However, they will have an increasing nature when the superconductor thickness exceeds a certain characteristic value. This peculiarity is observed in the experiments. So, the oscillation modes depend on the temperature of the HTS, which is influenced by the many operating parameters and is main of the reasons due to which the oscillations occur with different duration and amplitude.

The performed calculations also show that the oscillations start during full penetration modes, and they may not have avalanche-like nature leading to the heating of superconductors to their critical temperatures. It is due to the thermal self-stability effect. As a result, they may be accompanied by the huge oscillation modes unlike the possible critical state oscillations in low-T_c superconductors, discussed above.

The peculiarities of the thermo-electrodynamic phenomena in the HTS can be explained on the basis of the following relationship

$$\frac{\partial E}{\partial t}\bigg|_{x=a} = \mu_0 \left|\frac{dJ_c}{dT}\right| \int_{x_p}^a dy \int_y^a \frac{\partial^2 T}{\partial t^2} dx - \frac{\mu_0}{n} \int_{x_p}^a dy \int_y^a \left(\frac{\partial^2 E}{\partial t^2} - \frac{1}{E}\frac{\partial E}{\partial t}\right)\frac{J_c}{E} dx$$

existing during the formation of their thermal and electrodynamic states. The derivation of this equation is omitted due to the cumbersomeness of the mathematical transformations. Note that it may be obtained similarly to the above-written relations (4.28) and (4.29). This equation shows that the occurrence and development of the oscillations in HTS with the power V-I characteristic depend, first of all, on the intensity of the temperature variation of the term $\int_{x_p}^a dy \int_y^a \frac{\partial^2 T}{\partial t^2} dx = \frac{1}{C_s} \int_{x_p}^a \left[\frac{\partial g}{\partial t} - \frac{\partial q}{\partial t}\right] dx$,

as for the superconductor with the flux-flow V-I characteristic and also intensity of the electric field change throughout the volume of HTS in accordance with the variation of the alternating-sign term $(\partial^2 E/\partial t^2 - E^{-1}\partial E/\partial t)J_c/E$, which takes into account the continuous rise of the V-I characteristic. In particular, the lower n-value, the stronger the influence of the temporal variation of electric field induced in a superconductor, namely, the terms $\partial^2 E/\partial t^2$, $\partial E/\partial t$ and $J_c(T)$, as follows from Figures 4.34a and 4.35a. Their influence is negligible for $n \gg 1$. Therefore, in these superconductors, the main contribution to the development of the oscillations will be played by the mechanism of the conductive-convective removal of the heat losses from the volume of a superconductor, as discussed above.

The discussed peculiarities of the thermo-electrodynamic state formation in superconductors are observed in experiments, in which the magnetic moment of a superconducting sample is measured as a function of an external magnetic field. In particular, in the transition from the regime of the partial penetration of magnetic flux to the full penetration one, the dependence $M(t)$ and hence $M(B_a)$ changes, as already discussed and depicted, for example, in Figure 4.32e. In this case, as the temperature of the superconductor increases, the decrease in the values $|M(t)|$ may become more nonmonotonic due to the development of the oscillating magnetic instability, which also depends on the temperature-field dependence of the critical current density. As a result of the occurrence of increased overheating and the self-heating effect, the magnetic instabilities in superconductors may have a highly oscillating character (Figure 4.35).

4.8   CONCLUSIONS

The steady operating modes of both the LTS and HTS are characterized by their stable overheating induced by the magnetic flux diffusion. In particular, the lower the heat transfer coefficient or the higher the sweep rate or the superconductor thickness, the higher the stable overheating of superconductor. The steady temperature rise of the superconductor is nontrivially related to the conditions of the magnetic instability onset, the development of which has the fission-chain-reaction character.

In the stable formation of the superconducting state, an unambiguous link exists between the finite permissible overheating that does not lead to the transition of the superconductor to the normal state, and the energy losses that are inevitable in the superconductor before instability occurs. It allows one to formulate a general condition of the destruction of superconductivity under the action of perturbations of various natures. Namely, as a consequence of the existence of an allowable heat release, it is possible to find the magnetic stability conditions of the superconducting states using the general concept of the thermal destruction of superconductivity. Such concept may be based on the existence of the critical energy limiting the heat release from any external source of perturbation including the magnetic one. As a result, the nonisothermal approximation leads to the more general criteria of the magnetic instability onset. Therefore, the nonisothermal description of the electrodynamic states of superconductors is important to the adequate understanding of their electromagnetic properties, first of all, under the conditions close to an adiabatic.

The correct determination of the allowable temperature change of superconductors shows that the following regularities take place during the formation of their critical states:

- the destruction conditions of the critical state of cooled superconductor depend on the sweep rate of the external magnetic field, primarily due to the corresponding change in the stable increase in the temperature of the superconductor before the instability onset;
- there are characteristic temperatures of the superconductor that determine the thermal structure of the conditions of the magnetic instability onset, which leads to different conditions for the thermal stabilization of the critical state with respect to external initial temperature perturbations;
- there exists the effect of the full thermal stabilization of the critical state under the condition $\beta < 3$ when the critical state of a superconductor is stable with respect to the temperature perturbations, the amplitude of which can vary in the range from the coolant temperature to the critical temperature of superconductor.

An allowable temperature rise of the superconductor leads to the effect that may be called the thermal self-stabilization effect. This effect is due to its temperature-dependent properties of superconductors. It turns out that not only the absolute value of the heat capacity of superconductor influences on the character of the magnetic flux penetration. The peculiarities of the heat capacity and critical current variation with temperature also influence the intensity of the electromagnetic field penetration. As a result, the thermo-electrodynamic states of the superconductors maybe not accompanied by the irreversible superconducting-to-normal transition in the wide overheating ranges. According to these peculiarities, the nonisothermal behavior of the superconductors may influence their magnetization, the heat losses and also may lead to such magnetic instabilities that do not have the sharp avalanche-like development that lead to a normal state transition.

It shows that the basis for the occurrence of the oscillations in superconductors, first of all, consists of the conductive-convective mechanisms of the removal of the magnetization losses into the coolant. As a result, the thermal prehistory of a superconductor has an essential effect on the conditions of the oscillations occurrence. Namely, the effect of the thermal self-suppression of oscillations takes place: they will be absent when the high permissible overheating of superconductor arising before the instability onset. In particular, the probability of the oscillation occurrence decreases when the cooling conditions deteriorate, and they are completely absent under the adiabatic conditions. As a consequence, the development of the oscillations in HTS as compared with LTS has a more stable character due to their high value of the temperature margin leading to high allowable overheating.

5

Thermo-Magnetic Phenomena in Composites Based on LTS and their Magnetic Instabilities

As follows from the results obtained above, the superconducting wires for practical applications should be made of thin micron-thick superconducting filaments. A winding with a large number of ultra-thin turns is not only difficult to produce, but the filaments have very low critical currents due to small transverse size to ensure the achievement of high magnetic fields. For this reason filaments are placed in a non-superconducting matrix. Modern technologies for the production of multifilament composites make it possible to manufacture wires with any number of superconducting strands. The matrix for LTS-based composites is often copper. It is plastic and has high thermal and electrical conductivity. Bronze and nickel are also used. HTS wires of the first generation are produced with a silver matrix, and tapes of the second generation are additionally coated with copper which ensures their thermal stabilization.

The macroscopic phenomena occurring in them depend on the physical properties of the components of composite, their cross-section and the location of the filaments in it. The superconducting filaments in the composite are twisted to reduce the electromagnetic interaction between them. The twist pitch is small enough. (It is interesting to note that the first multifilament superconducting composites were not twisted). As a result, the description of the operating modes that take place in composites in the full formulation, that is to take into account the change in the thermo-electro-dynamic states of each superconducting filament and matrix involves considerable mathematical difficulties. Since composites usually contain a large number of thin superconducting filaments, the variation in temperature and electromagnetic field in the composite can be studied within the model of an anisotropic continuum, averaging its physical characteristics in the cross-sectional plane as it is usually done when analyzing phenomena in multicomponent conductors. (From a physical point of view, the use of an anisotropic continuum model is legitimate when the temperature and the electromagnetic field in the composite change slightly in the cross-section of superconducting filaments). As a result, the occurrence of unstable states is studied as a collective process in a superconducting composite with effective parameters. In the framework of this concept, the nonisothermal conditions of the stability of the superconducting composite state under the influence of an external magnetic field are investigated below.

5.1 FORMATION FEATURES OF THE CRITICAL STATE OF COMPOSITE SUPERCONDUCTORS: THE MODEL OF AN ANISOTROPIC CONTINUUM

Study the basic features of nonisothermal formation of the critical state in a superconducting composite with filaments uniformly distributed over its cross-section. To simplify the analysis, let us consider the composite in the form of the slab ($-a_k < x < a_k$, $-b_k < y < b_k$, $a_k \ll b_k$) placed in an external magnetic field parallel to its surface which increases with a constant sweep rate dB_a/dt at $0 < t < t_s$ and becomes constant at $t > t_s$ ($dB_a/dt = 0$). Also, assume that the twist pitch of superconducting filaments allows one to ignore the change in the magnetic field in the longitudinal direction. This approximation will lead to the lower boundary of the stability conditions for the critical state of the composite. In this case, the time-space distributions of the temperature and electric field in the composite in which the V-I characteristic of the superconducting filaments is described by the viscous flux-flow model and the critical current density decreases linearly with temperature are the solutions of the following system of equations

$$C_k(T)\frac{\partial T}{\partial t} = \frac{\partial}{\partial x}\left(\lambda_k(T)\frac{\partial T}{\partial x}\right) + \begin{cases} 0, & 0 < x < x_p \\ EJ, & x_p \le x \le a_k \end{cases}, \tag{5.1}$$

$$\mu_0 \frac{\partial J}{\partial t} = \frac{\partial^2 E}{\partial x^2}, \tag{5.2}$$

$$J = \eta J_c(T) + E/\rho_k, \tag{5.3}$$

$$J_c(T) = J_{c0}(T_{cB} - T)/(T_{cB} - T_0) \tag{5.4}$$

Let us set the following initial-boundary conditions

$$T(x,0) = T_0, \quad E(x,0) = 0, \quad \left.\frac{\partial T}{\partial x}\right|_{x=0} = 0, \quad \lambda_k \frac{\partial T}{\partial x} + h(T - T_0)\Big|_{x=a_k} = 0,$$

$$E(x_p, t) = 0, \quad x_p \ge 0, \quad \left.\frac{\partial E}{\partial x}\right|_{x=a_k} = \begin{cases} \dfrac{dB_a}{dt}, & 0 < t < t_s \\ 0, & t \ge t_s \end{cases} \tag{5.5}$$

in which the front of magnetization follows from the solution of equation

$$\mu_0 \int_{x_p}^{a_k} J(x,t)\,dx = \frac{dB_a}{dt}\,t.$$

Here, C_k is the volumetric heat capacity of a composite, λ_k is the coefficient of thermal conductivity of a composite, ρ_k is the specific electrical resistivity of a composite, a_k is its half-thickness, η is the filling factor of a composite by the superconductor and J_m is the current density in the matrix. The physical meaning of the other variables remained unchanged.

The system of Equation (5.1)-(5.5) describes the one-dimensional dissipative diffusion of magnetic flux in the superconducting composite. According to this formulation, the thermal and electrodynamic states of the composite, which are unperturbed at the initial time, will be considered. The thermal boundary conditions take into account the symmetric distribution of the temperature in composite and convective heat transfer on its surface, and the conditions written for the electric field describe the increase in the external magnetic field at a given sweep rate and the finite character of the penetration of magnetic flux. In this case, the electric field in the composite and the currents in the superconducting core of the composite with density ηJ_c and the matrix with density E/ρ_k follows from the Kirchhoff equations written for two conductors connected in parallel.

To determine the volumetric heat capacity of a composite, one may use the mixture (additivity) rule, according to which the heat capacity of a composite conductor with unidirectional filaments is equal to the sum of the values of the heat capacities of its parts. Then, the effective heat capacity of the composite ('superconductor + matrix') is equal to $C_k = \eta C_s + (1 - \eta) C_m$. Here, C_s and C_m are the heat capacities of the superconductor and the matrix, respectively. The thermal conductivity and electrical resistivity of a composite strongly depend on the direction in which the heat flux and the current propagate. This is because the composite has the superconducting filaments with very low values of the thermal conductivity coefficient and electrical conductivity in relation to the corresponding values of the matrix. To their definitions in the longitudinal direction relative to the direction of the filaments, the following formulae can be used $\lambda_k = \eta\lambda_s + (1-\eta)\lambda_m \approx (1-\eta)\,\lambda_m$ and $1/\rho_k = \eta/\rho_f + (1-\eta)/\rho_m \approx (1-\eta)/\rho_m$, accordingly, taking into account the low values of λ_s and $1/\rho_f$ for superconductors. These values in the transverse direction relative to the filaments substantially depend on their shape and structure of the arrangement in the composite. In its simplest form, the transverse thermal conductivity of a composite can be defined as follows

$$\lambda_k = \lambda_m \frac{(1+\eta)\lambda_s + (1-\eta)\lambda_m}{(1-\eta)\lambda_s + (1+\eta)\lambda_m} \approx \lambda_m \frac{(1-\eta)}{(1+\eta)}$$

and to calculate the effective electrical resistance of the multifilament composite in the transverse direction, one can use the formula

$$\frac{1}{\rho_k} = \frac{1}{\rho_m}\frac{(1+\eta)+(1-\eta)\chi}{(1-\eta)+(1+\eta)\chi}, \quad \chi = \frac{r_f\rho_b}{d_b\rho_m}$$

where r_f is the filament radius, ρ_b is the specific contact resistance between the superconducting filaments and the matrix and d_b is the thickness of the contact layer. In two limiting cases, this formula takes the form $\rho_k \approx \rho_m(1-\eta)/(1+\eta)$ and $\rho_k \approx \rho_m(1+\eta)/(1-\eta)$, if the contact resistance is absent or is very large, respectively.

The use of the anisotropic continuum model makes it easy to use the methods of analysis of thermo-electrodynamic states developed for a hard superconductor without a stabilizing matrix but takes into account the division of the current between its superconducting filaments and the matrix. The results obtained on the basis of the numerical solution of the system of Equations (5.1)-(5.5) are discussed below.

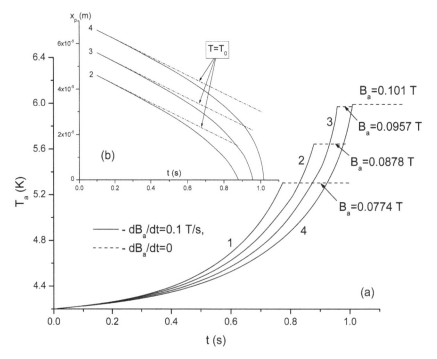

FIGURE 5.1 Stable overheating of a thermally insulated superconducting composite (a) and the dynamics of the magnetization front (b) 1 - $a_k = 4 \times 10^{-5}$ m, 2 - $a_k = 5 \times 10^{-5}$ m, 3 - $a_k = 6 \times 10^{-5}$ m, 4 - $a_k = 7 \times 10^{-5}$ m.

The curves describing a stable increase in the surface temperature $T_a = T(a_k, t)$ of a heat-insulated superconducting composite of various thicknesses are plotted in Figure 5.1a. In the calculations, the initial parameters of the composite were taken equal to: $h = 0$, $\eta = 0.5$, $C_k = 2 \times 10^3$ J/(m³ × K), $\lambda_k = 200$ W/(m × K), $\rho_k = 2 \times 10^{-10}$ Ω × m, $T_0 = 4.2$ K, $J_{c0} = 3.5 \times 10^9$ A/m², $T_{cB} = 9$ K, which correspond to the average values of the Nb-Ti in the copper matrix. The solid lines show the temperature change with a continuous increase in the external magnetic field up to its full penetration in the composite, and the dashed lines depict the temperature change after the growth termination of the external magnetic field as soon as the screening currents fill the cross-section of the composite. To illustrate the existence of stable penetration states, the corresponding dynamics of the magnetization front is presented in Figure 5.1b. To demonstrate the role of the temperature factor in the character of the magnetic flux penetration, the change in the screening boundary is shown here in the isothermal state of the composite ($T = T_0$).

The calculation results of the flux-jump field in the superconducting composite ($a_k = 10^{-4}$ m) are shown in Figure 5.2. These simulations were made according to the perturbation method of the initial state.

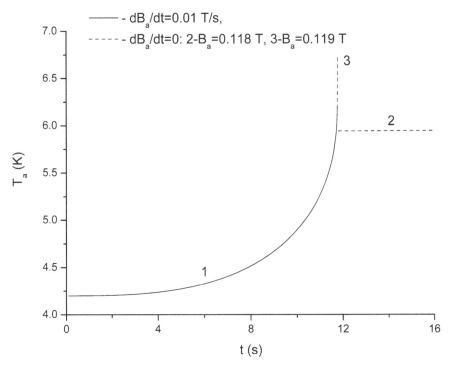

FIGURE 5.2 Change in the surface temperature of the thermally insulated composite at stable and unstable states.

These results demonstrate the existence of a noticeable stable overheating reaching about 20% of the superconductor critical temperature, its influence on the dynamics of the stable moving boundary penetration of the magnetization region and the conditions of the magnetic instability onset. Such a significant stable increase in the composite temperature compared to the overheating of superconductors without a stabilizing matrix occurs due to the damping effect of the matrix. As a consequence, the determination of the stability boundary of the critical state of the composite without taking into account the peculiarities of its temperature change will misrepresent the results of the analysis of stable states.

The peculiarities of the critical state formation in the superconducting composite can be explained on the basis of the following relationship

$$\left.\frac{\partial E}{\partial t}\right|_{x=a_k} = \left.\frac{\partial E}{\partial x}\right|_{x=x_p}\left|\frac{dx_p}{dt}\right| + \mu_0\eta\left|\frac{dJ_c}{dT}\right|\int_{x_p}^{a_k} dy \int_y^{a_k} \frac{\partial^2 T}{\partial t^2}\,dx - \mu_0(1-\eta)\int_{x_p}^{a_k} dy \int_y^{a_k} \frac{\partial^2 J_m}{\partial t^2}\,dx \qquad (5.6)$$

It may be obtained similarly to the above-written relation (4.28), taking into account the current sharing mechanism between the superconductor and the matrix which affects the thermo-electrodynamic phenomena in superconducting composites. This feature takes into account the last term on the right-hand side of the Formula (5.6).

Figures 5.1 and 5.2 demonstrate that the formation of thermo-electrodynamic states of the thermally insulated composite occurs without oscillations of the critical state. As shown above, they may occur if the cooling exists. Figures 5.3-5.8 depict the characteristic peculiarities of the critical state dynamics in the cooled composite superconductor placed in an external magnetic field, which changes at different sweep rates. The calculations were performed for the composite considered above at $a_k = 5 \times 10^{-4}$ m and $h = 100$ W/(m$^2 \times$ K).

The results depicting the time variation of the temperature and corresponding value of the time derivative on the surface of the cooled composite at a relatively low sweep rate are shown in Figure 5.3. It is seen that $\left.\partial T/\partial t\right|_{x=a_k} > 0$ and $\left.\partial^2 T/\partial t^2\right|_{x=a_k} > 0$ throughout the entire diffusion of the screening current for the regime under consideration. Therefore, the temperature of the composite increases without oscillations. However, the oscillations in the cooled composite may occur at increasing the sweep rate.

The time variation of the temperature, which occurs on the surface of the composite under consideration at $dB_a/dt = 0.1$ T/s at the stage of the transition from the phase of the damped oscillations to the phase of the increasing oscillations, is shown in Figure 5.4. In this case, the stabilizing role of the matrix is insignificant and almost the entire screening current flows through the superconducting core of the composite. Therefore, the oscillations in superconducting composites are primarily due to the interrelated change in the values $\partial E/\partial t$ and $\partial^2 T/\partial t^2$ according to the Formula (5.6). Figure 5.5 confirms the existence of this relationship. Indeed, the electric field increases ($\partial E/\partial t > 0$) with the temperature when $\partial^2 T/\partial t^2 > 0$. In this case, these conditions take place when $t < t^*$. (To demonstrate the sign change in the values $\partial^2 T/\partial t^2$, a tangent to a dashed curve describing the variation in temperature of the composite surface $T_a = T(a_k, t)$ is drawn in Figure 5.5 at $t = t^*$). However, $\partial^2 T/\partial t^2 < 0$ when $t > t^*$ and therefore, the electric field begins to decrease at $t > t^*$ despite the corresponding increase in temperature. This is explained by the fact that the removal rate of the dissipated energy from the interior of the composite into the coolant exceeds the rate of the increase of heat generation. This is the phase of the oscillations in which a decrease in the electric field and a subsequent decrease in the density of the screening current in the superconductor will lead to a decrease in the heat losses in the magnetization region. As a result, after a certain period, the temperature of the composite begins to decrease. This is the stage of the effective conductive-convective removal of the heat losses in the development of which the sign of $\partial J/\partial t$ on the surface of composite changes from negative to positive due to current redistribution inside composite. After that, the next phase of the oscillations takes place when an increase in the density of the screening current leads to an increase in the heat generation and hence, to a subsequent increase in the electric field, etc. The periodic repetition of these states underlies the oscillations.

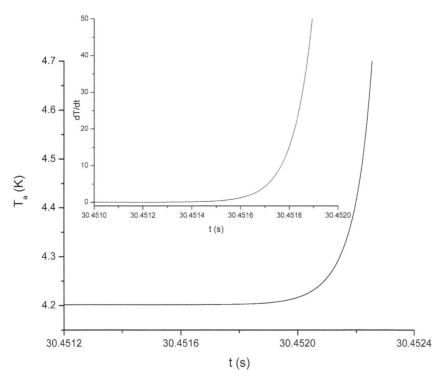

FIGURE 5.3 Monotonous temperature increase in the surface of the cooled composite at dB_a/dt = 0.01 T/s. The corresponding change of its ramp rate is presented on the inset.

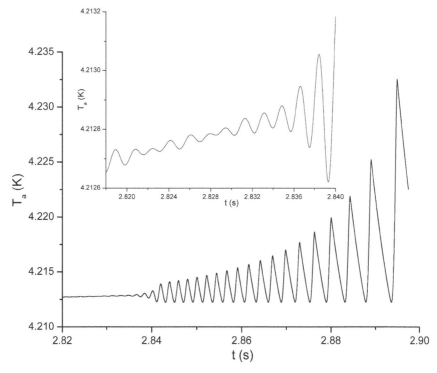

FIGURE 5.4 Temperature oscillation on the surface of composite near the boundary of the unstable states at dB_a/dt = 0.1 T/s. This oscillation is shown in more detail on the inset.

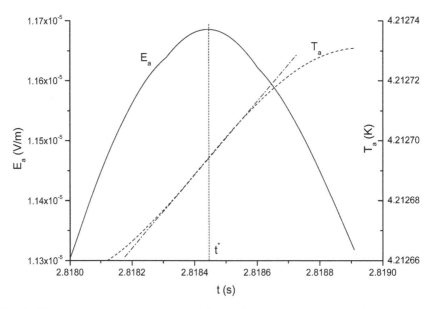

FIGURE 5.5 Thermo-electrodynamic nature of the oscillating electric field and the temperature on the surface of the superconducting composite with real V-I characteristic at dB_a/dt = 0.1 T/s.

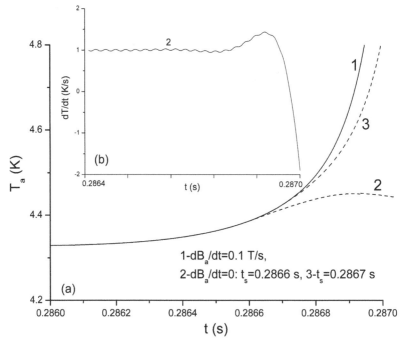

FIGURE 5.6 Oscillations of the temperature (a) and its time derivative (b) on the surface of superconducting composite near the boundary of the unstable state at dB_a/dt = 0.1 T/s.

The curves describing weak oscillations on the surface of a composite that occur before the magnetic instability onset at a higher sweep rate are shown in Figure 5.6. The simulation results of the corresponding stability boundary defined according to the perturbation method of the initial state are also shown in Figure 5.6. It can be seen that the oscillations before the instability onset are not as intense as in the previous case. First of all, this is associated with a corresponding increase in the temperature of the composite that occurs before the instability onset.

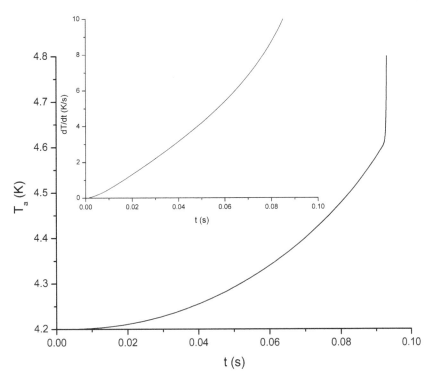

FIGURE 5.7 Monotonous formation of the temperature and its time derivative on the surface of composite during a continuous increase in the external magnetic field with the high sweep rate (dB_a/dt = 3 T/s).

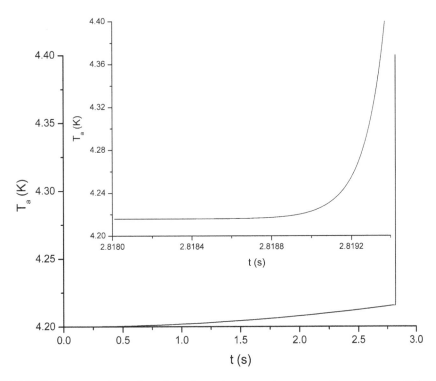

FIGURE 5.8 Monotonous formation of the thermal state on the surface of composite at dB_a/dt = 0.1 T/s and J_{c0} = 3.2 × 10⁹ A/m². This state is shown in more detail on the inset.

The oscillations disappear during a further increase in the sweep rate (Figure 5.7). As noted above, these regularities occur due to the effect of the thermal suppression of oscillation when the rate of the heat flux removal from the interior of the superconductor to the coolant is less than the growth rate of the heat release induced by a rapidly changing external magnetic field.

Since the oscillations onset depends on the intensity of a stable increase in the temperature of the composite, their occurrence or suppression also depends on the critical parameters of the superconductor, which influences the temperature rise of a composite. In particular, the probability of their onset should decrease with decreasing of $|dJ_c/dT|$ as noted above. Therefore, the smaller J_{c0}, the lesser the probability of the oscillation onset of all other conditions being equal due to the more intense increase in the composite temperature. To illustrate this peculiarity, the curves describing a monotonous increase in the time of the temperature surface of the composite are presented in Figure 5.8. The simulation was made at J_{c0}, which is less than the value of J_{c0}, resulting in the operating mode shown in Figures 5.1-5.7.

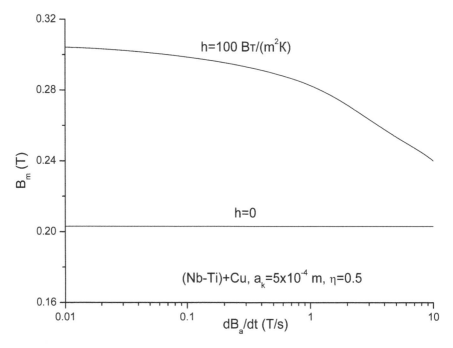

FIGURE 5.9 Dependence of the flux-jump fields of the critical state of a superconducting composite on the sweep rate.

In accordance with the noted features of the thermo-electrodynamic state formation of a composite superconductor, the values of the magnetic flux-jump were determined as a function of the sweep rate under various cooling conditions. The results of these calculations based on the perturbation method of the initial state are presented in Figure 5.9. It can be seen that the values B_m under the adiabatic cooling do not depend on the sweep rate of the external magnetic field, as in superconductors without a stabilizing matrix. At the same time, the falling dependence $B_m(dB_a/dt)$ for the cooled composite, whose V-I characteristic is described by the viscous flux-flow model, is primarily due to the corresponding change in its temperature preceding the instability onset which depends on the sweep rate. It increases with increasing of dB_a/dt, as it follows from Figures 5.3, 5.4 and 5.6, 5.7. As a result, the critical current density of the superconductor decreases, respectively, and therefore, there is a corresponding decrease in the magnetic field of the flux jump.

5.2 DYNAMIC CRITERIA OF THE CRITICAL STATE STABILITY OF COMPOSITE SUPERCONDUCTORS AT FULL PENETRATION OF THE MAGNETIC FLUX

As already noted, the development of thermo-electrodynamic states of superconductors depends on the ratio between the characteristic diffusion times of the electromagnetic field and heat flux. This is taken into account in the dimensionless parameter Λ, which is equal to the ratio of the time of magnetic diffusion to the time of thermal diffusion. For superconductors without a stabilizing matrix $\Lambda \ll 1$. The situation is different for the composite superconductors. In this case, the value of $\Lambda_k = \lambda_k \mu_0 / C_k \rho_k$ may vary from 10^3 to 10^4. Indeed, for Nb-Ti in the copper matrix, the average values of the thermo-electrical parameters may be set as follows $\lambda_k \sim 100$ W/(m × K), $C_k \sim 10^3$ J/m^3 × K, $\rho_k \sim 10^{-10}$ Ω × m. Therefore, $\Lambda_k \sim 10^4$. This means that the temperature of the multifilament composites determines the development of the electromagnetic states in the superconducting composite. Let us take advantage of this feature and write the criteria for the stability of the critical state of a superconducting composite during full penetration of the magnetic flux assuming the values λ_k, C_k and ρ_k temperature-independent and supposing the linear reduction of the critical current density with temperature.

Introduce the following dimensionless variables

$$X = x/a_k, \tau = t/t_x \, (t_x = C_k a_k^2 / \lambda_k), e = E/(J_{c0}\rho_k), j = J/J_{c0}, \theta = (T - T_0)/(T_{cB} - T_0)$$

using the time of the thermal diffusion as a characteristic time since $\Lambda_k \gg 1$. Then, the system of equations (5.1)-(5.5) will be reduced to the form

$$\frac{\partial \theta}{\partial \tau} = \frac{\partial^2 \theta}{\partial X^2} + \frac{\beta_k}{\Lambda_k} ej,$$

$$\frac{\partial j}{\partial \tau} = \frac{1}{\Lambda_k} \frac{\partial^2 e}{\partial X^2},$$

$$j = \eta j_c(\theta) + e$$

Here,

$$\beta_k = \frac{\mu_0 J_{c0}^2 a_k^2}{C_k(T_{cB} - T_0)}, \quad j_c(\theta) = J_c/J_{c0} = 1 - \theta.$$

Simplify this system taking into account that $\Lambda_k \gg 1$. In this case, $\partial j/\partial \tau \approx 0$ according to the second equation of the system written. Therefore, $j \sim$ const, that is the heat flux is redistributed in the composite in the 'frozen' magnetic flux regime. According to the third equation of the system, one get $e = j - \eta j_c$. Then, the heat equation takes a simpler form

$$\frac{\partial \theta}{\partial \tau} = \frac{\partial^2 \theta}{\partial X^2} + \frac{\beta_k}{\Lambda_k} j(j - \eta j_c) = \frac{\partial^2 \theta}{\partial X^2} + \beta^*\theta + q,$$

where $\beta^* = \eta j \beta_k / \Lambda_k, q = j^2 \beta_k / \Lambda_k - \beta^*$. This is a typical equation describing a diffusion process of a chain nature, which may lead to an avalanche-like increase in the temperature of the composite under certain conditions. In this case, β^* is a parameter on which the conditions of the magnetic instability onset in a composite are dependent on. Let us analyze this equation, taking into account the thermal boundary conditions. With respect to the introduced dimensionless variables, they are written as follows

$$\left.\frac{\partial \theta}{\partial X}\right|_{X=0} = 0, \quad \left.\frac{\partial \theta}{\partial X} + H_k \theta\right|_{X=1} = 0, \quad H_k = ha_k/\lambda_k$$

According to the method of the separation of variables, the solution will be found in the form

$$\theta(X, \tau) = \sum_{k=1}^{\infty} B_k \exp[(\beta^* - \mu_k)\tau]R_k(X) - q/\beta^*$$

Here, B_k are the integration constants, $R_k(X)$ are the eigenfunctions and μ_k are the eigenvalues that follow from the solution of the corresponding Sturm-Liouville problem. In this case, it has the form

$$\frac{d^2 R_k}{dX^2} + \mu_k R_k = 0, \quad \left.\frac{dR_k}{dX}\right|_{X=0} = 0, \quad \left.\frac{dR_k}{dX} + H_k R_k\right|_{X=1} = 0$$

Its increasing positive eigenvalues are a solution to the equation

$$\sqrt{\mu_k}\, tg\sqrt{\mu_k} = H_k, \quad 0 < \mu_1 < \mu_2 < ..., k = 1, 2, 3...$$

Therefore, the temperature of the composite induced by the changing external magnetic field will spontaneously increase even at its fixed value, if the inequality $\beta^* - \mu_1 > 0$ is satisfied for the first eigenvalue. Consequently, the superconducting state is stable when the condition $\eta j \beta_k / \Lambda_k < \mu_1$ is met which can be written with respect to dimensional variables as follows

$$\frac{\eta^2 J_{c0}^2 \rho_k a_k^2 (T_{cB} - T_i)}{\lambda_k (T_{cB} - T_0)^2} < \mu_1(h)$$

taking into account that before the magnetic instability, the temperature of the composite T_i is different from the temperature of the coolant, and the current density in the superconducting core of the composite exceeds the current in the matrix. This condition is satisfied if $J \approx \eta J_c(T_i) \gg E/\rho_k$.

The written criterion has a simplified form in two extreme cases. Firstly, at $H_k \gg 1$, that is at $\lambda_k \ll ha_k$, the estimate $\mu_1 \approx \pi^2/4$ is valid. Therefore, the stability criterion of the critical state of a composite at intensive cooling is written as follows

$$\frac{\eta^2 J_{c0}^2 \rho_k a_k^2 (T_{cB} - T_i)}{\lambda_k (T_{cB} - T_0)^2} < \frac{\pi^2}{4}$$

Secondly, the stability criterion at $0 < H_k \ll 1$, that is at $\lambda_k \gg ha_k$, $h \neq 0$, by virtue of the evaluation $\mu_1 \approx H_k$ is as follows

$$\frac{\eta^2 J_{c0}^2 \rho_k a_k (T_{cB} - T_i)}{h(T_{cB} - T_0)^2} < 1$$

These criteria are called dynamic stability criteria. They take into account the relationship between the power of the heat release in the composite and the conditions of its removal into the coolant. That leads to the existence of the corresponding characteristic thickness of the superconducting composite, which determines the full stability of its critical state. It is equal to

$$a_x = \sqrt{\frac{\lambda_k (T_{cB} - T_0)}{\eta^2 J_{c0}^2 \rho_k}}$$

Accordingly, the dynamic stability criterion is written as follows

$$a_k < a_x \sqrt{(T_{cB} - T_0)/(T_{cB} - T_i)\mu_1}$$

In the general case, the generalized stability curves $\beta_k(\Lambda_k)$ under various cooling conditions can be found on the basis of a numerical solution of the initial system of the dimensionless equations, which allows one to take into account the stable overheating of the composite placed in a varying external magnetic field. Let us discuss the possibility of using a more simplified method for determining the stability curves $\beta_k(\Lambda_k)$ based on the method of the infinitely small perturbations Gurevich et al, 1997c and analyzing the solution of the simplified system

$$\frac{\partial \theta}{\partial \tau} = \frac{\partial^2 \theta}{\partial X^2} + \frac{\beta_k}{\Lambda_k} e \eta (1 - \theta_i),$$

$$\frac{\partial e}{\partial \tau} = \frac{1}{\Lambda_k} \frac{\partial^2 e}{\partial X^2} + \eta \frac{\partial \theta}{\partial \tau}$$

In this case, the heat conduction equation is written taking into account that the dimensionless temperature of the composite before the magnetic instability is equal to θ_i, and the current density in the superconducting core of the composite exceeds the current in the matrix. In dimensionless form, the last assumption is satisfied if $j = \eta j_c + e \approx \eta (1 - \theta_i) \gg e$. The equation describing the change in the electric field is derived from equality $\partial j / \tau = \eta \partial j_c / \partial \tau + \partial e / \partial \tau = -\eta \partial \theta / \partial \tau + \partial e / \partial \tau$.

Let us differentiate the heat conduction equation with respect to time. As a result, one obtains the following equation

$$\frac{\partial^2 \theta}{\partial \tau^2} = \frac{\partial^3 \theta}{\partial \tau \partial X^2} + \frac{\eta \beta_k (1 - \theta_i)}{\Lambda_k} \frac{\partial e}{\partial \tau} = \frac{\partial^3 \theta}{\partial \tau \partial X^2} + \frac{\eta \beta_k (1 - \theta_i)}{\Lambda_k} \left(\frac{1}{\Lambda_k} \frac{\partial^2 e}{\partial X^2} + \eta \frac{\partial \theta}{\partial \tau} \right)$$

In this equation, the value $\partial^2 e / \partial X^2$ can be found by differentiating twice the heat conduction equation by X. Then, one will finally have

$$\frac{\partial^4 \theta}{\partial X^4} - (1 + \Lambda_k) \frac{\partial^3 \theta}{\partial \tau \partial X^2} + \Lambda_k \frac{\partial^2 \theta}{\partial \tau^2} - \eta^2 \beta_k (1 - \theta_i) \frac{\partial \theta}{\partial \tau} = 0$$

As usual, let us find the stable solutions of this equation in the form $\theta \propto \exp(\mu \tau) R(X)$. As a result, $R(X)$ is determined by the solving of the equation

$$\frac{\partial^4 R}{\partial X^4} - \mu (1 + \Lambda_k) \frac{\partial^2 R}{\partial X^2} - \mu [\eta^2 \beta_k (1 - \theta_i) - \mu \Lambda_k] R = 0$$

which requires four boundary conditions. They follow from the thermal and electrodynamic conditions (5.5). For the regime of the full penetration and a constant external magnetic field, they lead to the conditions

$$\left. \frac{dR}{dX} \right|_{X=0} = 0, \quad \left. \frac{dR}{dX} + H_k R \right|_{X=1} = 0,$$

$$\left. \frac{d^2 R}{dX^2} - \mu R \right|_{X=0} = 0, \quad \left. \frac{d^3 R}{dX^3} - \mu \frac{dR}{dX} \right|_{X=1} = 0$$

The solution of this equation is determined with an accuracy of up to four integration constants. It leads to a transcendental equation for μ, which arises from the condition for the existence of the nontrivial values of integration constants in accordance with the written boundary conditions. In this case, depending on $\eta^2 \beta_k (1 - \theta_i)$, there is a certain range of the values μ that can have both imaginary and real parts. Oscillations occur in the range of values of $\text{Im}(\mu)$. Unstable states correspond to the solution when the first positive value μ_1 appears in the spectrum of the values μ_k ($k = 1, 2, 3...$), as discussed above when determining the simplified stability criteria for both a hard superconductor and a superconducting composite. Since the value of $\eta^2 \beta_k (1 - \theta_i)$ decreases with an increase in the stable overheating of the composite, then the oscillations may not occur and the stability conditions improve. Thus, by defining the set of μ values, one can find the generalized stability curves $\beta_k(\Lambda_k)$.

5.3   ADIABATIC STABILITY OF THE CRITICAL STATE OF COMPOSITE SUPERCONDUCTORS AT FULL PENETRATION OF THE MAGNETIC FLUX: MULTILAYER MODEL 'SUPERCONDUCTOR + NORMAL METAL'

Let us formulate the nonisothermal criterion for the magnetic instability of the composite in the adiabatic approximation, taking into account the inevitable difference between the initial temperature of a composite and the coolant temperature.

To simplify the analysis, let us consider the thermally insulated composite slab, which consists of alternating layers of superconducting and nonsuperconducting materials (Figure 5.10), which is in a constant external magnetic field with induction B_a parallel to its surface. Suppose that the magnetic flux is fully penetrated in the composite, the initial temperature of the composite is T_i and it exceeds the temperature of the coolant T_0. In general, this may occur due to some perturbation. Let the critical current density depend only on temperature and changes according to the law (4.4). The temperature of the composite may rise to the value T_k as a result of the release of a certain amount of the magnetic energy stored by the screening currents. This changes the enthalpy of the composite. Let us consider the two characteristic cases, assuming that the temperature distribution in the composite is uniform.

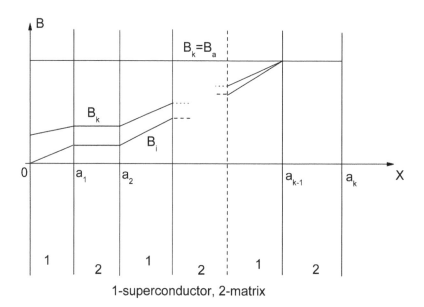

1-superconductor, 2-matrix

FIGURE 5.10 Distribution of the magnetic field in the right part of the composite at full penetration of the screening currents: B_i – the initial distribution, B_a – the magnetic field distribution outside and inside composite after the superconducting layer transition to the normal state, B_k – the magnetic field after partial release of the stored magnetic energy. Here, layers '1' are the superconductors, layers '2' are the normal metal.

Full magnetic instability during superconductor-to-normal transition ($T_k > T_{cB}$, $J_c(T_k) = 0$).

As a result of the development of the full magnetic instability, the composite will be in a normal state. Then, according to the law of conservation of energy, the change in its enthalpy is equal to

$$\int\limits_{0}^{a_k}\int\limits_{T_i}^{T_k} C_k(T)dTdx = \int\limits_{0}^{a_k} \frac{B_i^2(x)}{2\mu_0}dx$$

due to the complete release of the magnetic energy of the screening currents in the superconducting layers. When calculating the heat capacity of a composite, taking into account both its components and considering that the distribution of the magnetic field in each superconducting layers of the composite in the initial state is described by the relation $dB_i/dx = \mu_0 J_c(T_i)$ according to the critical state model, it is not difficult to rewrite the last equality in the form

$$\int_{T_i}^{T_k} [\Delta_s C_s(T) + \Delta_m C_m(T)] dT = \frac{B_a^3}{6\mu_0^2 J_c(T_i)}$$

Here, Δ_s and C_s are the total thickness of superconducting layers and their volumetric heat capacity, Δ_m and C_m are the total thickness of the matrix and its volumetric heat capacity. Since the induction of the magnetic field on the surface of the composite before the instability is equal to $B_a = \mu_0 J_c(T_i)\Delta_s$, then by taking into account the linear dependence of J_c on temperature one can get

$$\int_{T_i}^{T_k} [\eta C_s(T) + (1-\eta)C_m(T)] dT = \frac{\mu_0 \eta^3 J_{c0}^2}{6} \left(\frac{T_{cB} - T_i}{T_{cB} - T_0} \right)^2 a_k^2 \tag{5.7}$$

Here, $\eta = \Delta_s/a_k$ is the filling factor for the considered multilayer slab. This expression allows one to determine the final temperature of the composite. Besides, it is easy to find the criterion for the irreversible transition of the superconducting composite to the normal state from (5.7). Namely, its final temperature due to the development of the full magnetic instability will exceed the critical temperature of the superconductor when

$$\int_{T_i}^{T_{cB}} [\eta C_s(T) + (1-\eta)C_m(T)] dT < \frac{\mu_0 \eta^3 J_{c0}^2}{6} \left(\frac{T_{cB} - T_i}{T_{cB} - T_0} \right)^2 a_k^2 \tag{5.8}$$

or when the induction of an external magnetic field exceeds the value

$$B_a > B_m' = \sqrt{6\frac{\mu_0}{\eta} \int_{T_i}^{T_{cB}} [\eta C_s(T) + (1-\eta)C_m(T)] dT} \tag{5.9}$$

The Criteria (5.8) and (5.9), first of all, impose limitation on the transverse size of the composite, and also makes it possible to determine the characteristic value of its initial temperature T_i after exceeding which the energy stored by the screening currents is not enough to completely destroy the superconducting state of the composite. (The role of the quantity T_i was discussed in section 4.4 in detail). Let us consider the peculiarities of the formation of these states.

Partial magnetic instability after the incomplete release of magnetic energy stored by screening currents $(T_k < T_{cB}, J_c(T_k) > 0)$.

When a partial magnetic instability occurs, the energy conservation law is written in the form

$$\int_0^{a_k} \int_{T_i}^{T_k} C_k(T) dT dx = \int_0^{a_k} \frac{B_i^2}{2\mu_0} dx - \int_0^{a_k} \frac{B_k^2}{2\mu_0} dx$$

In the framework of the model under consideration, it leads to the equality

$$\int_{T_i}^{T_k} [\eta C_s(T) + (1-\eta)C_m(T)] dT = \frac{\mu_0 \eta^3 J_{c0}^2 a_k^2}{6(T_{cB} - T_0)^2} (T_k - T_i)(2T_{cB} - T_i - T_k) \tag{5.10}$$

Since its right-hand side decreases with increasing of T_i, the corresponding value of the final temperature of the superconducting composite, as a result of the magnetic instability development, also decreases with increasing of T_i. Physically, this is explained by the fact that with an increase of T_i the energy of the screening currents decreases. As a result of the energy release of screening currents, a spontaneous increase in the temperature of the composite may occur. However, it will not be enough to change the temperature of the composite under certain conditions. This means that the magnetic instability is absent and $T_k = T_i$. Making the limiting transition $T_k \to T_i$ in (5.10) and considering that the enthalpy of the composite does not change, one can find the criterion

$$\frac{\mu_0 a_k^2 \eta^2 J_{c0}^2}{[\eta C_s(T) + (1-\eta)C_m(T)]|_{T=T_i}(T_{cB} - T_0)} < \frac{3}{\eta}\frac{T_{cB} - T_0}{T_{cB} - T_i} \qquad (5.11)$$

According to this criterion, the critical state is stable despite the difference in the temperature of composite and the coolant temperature after the full penetration of the magnetic flux. In this case, the induction of an external magnetic field must satisfy the condition

$$B_a < B_m'' = \sqrt{\frac{3\mu_0}{\eta}[\eta C_s(T) + (1-\eta)C_m(T)]|_{T=T_i}(T_{cB} - T_i)}$$

Let us analyze the criteria presented, which were obtained in the CSM-approximation. First of all, it should be noted that they fully comply with the limiting transition at $\eta \to 1$ to the nonisothermal stability conditions formulated above, which describe the stability criteria of the critical state of a hard superconductor without a stabilizing matrix. At the same time, they demonstrate the influence of the composite structure on the stability conditions, which cannot be formulated within the framework of the anisotropic continuum model. To evaluate the influence of the heterogeneity of the composite physical properties on the conditions of the magnetic instability onset, let us use the well-known criterion obtained under the assumption C_s, $C_m \sim$ const and $T_i = T_0$. Then, the critical state of the thermally insulated composite is stable if the magnetic instability parameter β satisfies the condition (Wilson 1983)

$$\beta = \frac{\mu_0 a_k^2 \eta^2 J_{c0}^2}{[\eta C_s + (1-\eta)C_m](T_{cB} - T_0)} < 3 \qquad (5.12)$$

However, in this approximation, the criterion (5.11) leads to the condition

$$\beta < 3/\eta$$

For the composite $\eta < 1$ then the stability condition (5.12) turns out to be more stringent. It is easy to understand that this peculiarity is due to the averaging of the screening current over the entire cross-section of the composite, which is used in the anisotropic continuum model. When deriving conditions (5.8) and (5.11), this assumption was not used and they were obtained for the case when the screening currents flow only in the superconducting layers of the composite. Thus, the model of the anisotropic continuum underestimates the stability range of the critical state with an accuracy factor of $1/\eta$. However, the model of an anisotropic continuum is widely used in the simulations due to its simplicity.

5.4 THERMAL FEATURES OF THE SCREENING CURRENTS DIFFUSION IN LTS COMPOSITES WITH AN EXPONENTIAL V-I CHARACTERISTIC

The above discussed features of the critical state formation of low-temperature superconductors and composites based on them show that the analysis of the conditions of the magnetic instability onset made above even within the framework of a simplified model of the viscous flux-flow but taking into account the thermal prehistory of the magnetic flux penetration expands the boundaries of the stable states of superconducting composites. However, the real V-I characteristic of both LTS and HTS monotonically increases with increasing current. Taking into account the thermal prehistory of the electrodynamic state formation of composite, the peculiarities of the conditions of the magnetic instability onset in a superconducting composite are discussed below, whose V-I characteristic is described by the exponential equation (1.4).

Let us investigate the nonisothermal diffusion of the electric field and screening current in the cooled composite slab $(-a_k < x < a_k, -b_k < y < b_k, a_k \ll b_k)$ within the framework of the anisotropic continuum model during the increase of the magnetic field on its outer surface. To analyze the penetration of the electromagnetic field, let us use the system of equations (5.1), (5.2) with initial-boundary conditions (5.5) and linear dependence of the critical current density (5.4). At the same time, the relationship between the electric field E and the currents in the superconductor J_s with the exponential V-I characteristic and nonsuperconducting matrix J_m, let us write according to the Kirchhoff equations as follows

$$E = E_c \exp\left(\frac{J_s - J_c}{J_\delta}\right) = J_m \rho_m, \quad J = \eta J_s + (1-\eta) J_m \tag{5.13}$$

During numerical calculations, the averaged values of the composite parameters were assumed to be equal to $\eta = 0.5$, $C_k = 2 \times 10^3$ J/(m^3 × K), $\lambda_k = 200$ W/(m × K), $T_0 = 4.2$ K, $E_c = 10^{-4}$ V/m, $J_{c0} = 4 \times 10^9$ A/m^2, $J_\delta = 4 \times 10^7$ A/m^2, $T_{cB} = 9$ K, $\rho_m = 2 \times 10^{-10}$ Ω × m.

The results of a numerical experiment, which determine the stability boundary of the superconducting state of the slab at $h = 10$ W/(m^2 × K) are presented in Figures 5.11 and 5.12. It was performed in accordance with the perturbation method of the initial state. Accordingly, the temperature dynamics both during the continuous increase of the external magnetic field (Curves 1) and at its fixed values $(dB_a/dt = 0)$ close to the field of the flux jump (Curves 2, 3) is shown. In the inset to Figure 5.11, the stable and unstable temperature variation near the boundary of the instability is shown in more detail.

The presented results show that the initial stage of the temperature change of the composite at the stable relaxation of the induced electric field, that is at $B(a, t) = \text{const}$ is characterized by its increase, and then it begins to decrease as in the case of a composite superconductor whose V-I characteristic is described by the viscous flux-flow model, that is the aperiodic oscillation exists. Figures 5.11 and 5.12 also demonstrate that their existence depends on the sweep rate of the external magnetic field. This dependence follows from equation

$$\left.\frac{\partial E}{\partial t}\right|_{x=a_k} = \mu_0 \eta \left|\frac{dJ_c}{dT}\right|_{x_p}^{a_k} \int dy \int_y^{a_k} \frac{\partial^2 T}{\partial t^2} dx - \mu_0(1-\eta)\int_{x_p}^{a_k} dy \int_y^{a_k} \frac{\partial^2 J_m}{\partial t^2} dx - \mu_0 \eta J_\delta \int_{x_p}^{a_k} dy \int_y^{a_k} \left(\frac{\partial^2 E}{\partial t^2} - \frac{1}{E}\frac{\partial E}{\partial t}\right)\frac{dx}{E}$$

which can be obtained similarly to Equations (4.28) and (5.6). This equation shows that the oscillations in composite superconductors with the real V-I characteristic will also depend on the intensity of the electric field increase throughout the volume in accordance with the variation of the term $\eta J_\delta(\partial^2 E/\partial t^2 - E^{-1}\partial E/\partial t)$, that is smooth character of the real V-I characteristic, which is the intensity of the composite temperature rise $(\partial^2 T/\partial t^2)$ and current in the matrix $(\partial^2 J_m/\partial t^2)$ in addition to the already discussed mechanisms affecting the occurrence and development of the oscillations. As can be seen from Figure 5.11, the smooth character of V-I characteristic will have the most effect, immediately before the instability onset.

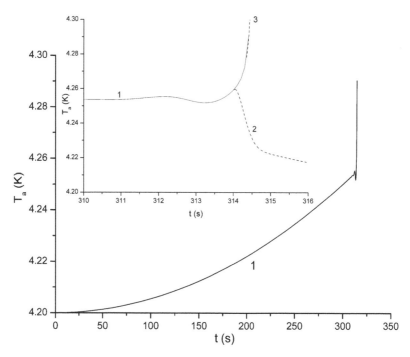

FIGURE 5.11 Temperature change of the composite surface under stable and unstable states: 1 - continuous increase of the external magnetic field at $dB_a/dt = 0.003$ T/s; $dB_a/dt = 0$, 2 - $B_a = 0.942$ T, 3 - $B_a = 0.943$ T.

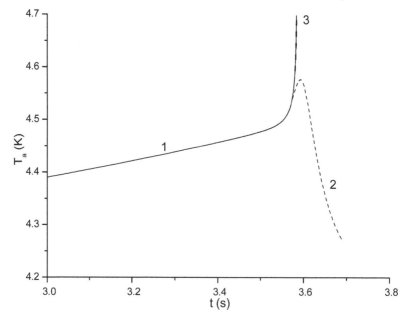

FIGURE 5.12 Temperature change in the surface of composite at stable and unstable states. Here, 1 - $dB_a/dt = 0.1$ T/s, 2 - $dB_a/dt = 0$, $B_a = 0.357$ T, 3 - $dB_a/dt = 0$, $B_a = 0.358$ T.

Figures 5.13-5.15 demonstrate the characteristic features of the occurrence and development of the oscillations, which may be observed in the composite superconductors with a real V-I characteristic. Numerical experiments were performed for the superconducting composite, the thermal and electrical parameters of which were written above. At the same time, the variable parameters were the thickness of the composite (Figure 5.13), the heat transfer coefficient (Figure 5.14) and the sweep rate of the external magnetic field (Figure 5.15).

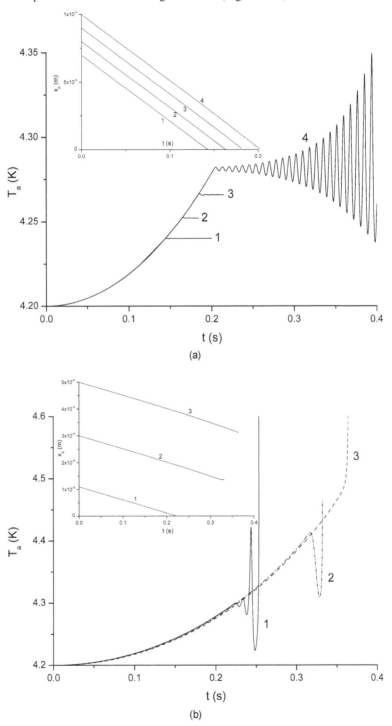

(a)

(b)

FIGURE 5.13 Occurrence and suppression of the temperature oscillations on the surface of the cooled composite at an increase in its thickness: (a): 1 - $a_k = 7 \times 10^{-5}$ m, 2 - $a_k = 8 \times 10^{-5}$ m, 3 - $a_k = 9 \times 10^{-5}$ m, 4 - $a_k = 10^{-4}$ m; (b): 1 - $a_k = 1.1 \times 10^{-4}$ m, 2 - $a_k = 3 \times 10^{-4}$ m, 3 - $a_k = 5 \times 10^{-4}$ m. The insets show the dynamics of the magnetization front.

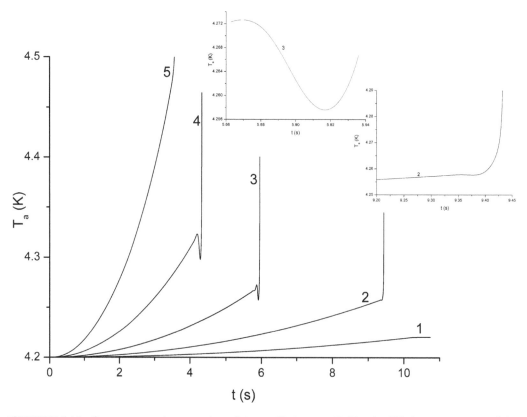

FIGURE 5.14 Occurrence and suppression of the oscillations at $dB_a/dt = 0.1$ T/s during variation of the heat transfer coefficient: 1 - $h = 1000$ W/(m² × K), 2 - $h = 300$ W/(m² × K), 3 - $h = 100$ (m² × K), 4 - $h = 30$ (m² × K), 5 - $h = 10$ (m² × K). On the inset, Curves 2 and 3 are shown in more detail.

The temperature oscillations on the surface of the cooled superconducting composite ($h = 100$ W/(m²K)) with different thickness occurring at a continuous increase in the external magnetic field at the sweep rate $dB_a/d = 1$ T/s are shown in Figure 5.13. In the case under consideration, an increase in the thickness of the composite is accompanied by a transition from the damping to the increasing oscillations during full penetrated mode (Figure 5.13a). This is due to a corresponding increase in the magnetization region in which the energy of the screening currents dissipates due to the corresponding temperature change. Correspondingly, damping oscillations occur when the screening current is fully penetrated in composites of small thickness. The increasing oscillations occur both during full and partial penetration of the screening current into more massive composites (Figure 5.13b). However, oscillations, first, become anharmonic and second, cease during a further increase in the thickness of the composite. As discussed above, this change in the character of the oscillations is associated with the effect of their thermal suppression when the conductive and convective mechanisms of the heat transfer cannot effectively dissipate the released energy at the stable penetration stage of the screening current. As a result, by virtue of a corresponding increase in the temperature of the composite, the rate of the removal of heat flux from the interior of the composite to the coolant is less than the rate of the increase of heat generation induced by a varying external magnetic field. This peculiarity is demonstrated by Curve 3 in Figure 5.13b when the corresponding stable increase of the composite temperature before the instability onset is the greatest.

Similar thermal features of the formation of oscillating states are observed with a decrease in the heat transfer coefficient (Figure 5.14) and with an increase in the sweep rate of the external magnetic field (Figure 5.15). The numerical experiments presented in these figures were obtained at $a_k = 5 \times 10^{-4}$ m.

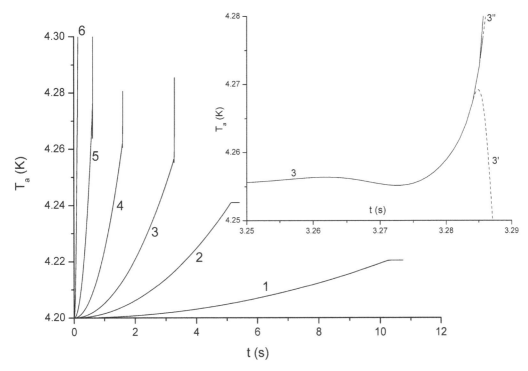

FIGURE 5.15 Oscillations and their suppression at $h = 10^3$ W/(m^2 × K) under different sweep rate: 1 - $dB_a/dt = 0.1$ T/s, 2 - $dB_a/dt = 0.2$ T/s, 3 - $dB_a/dt = 0.3$ T/s, 4 - $dB_a/dt = 0.5$ T/s, 5 - $dB_a/dt = 1$ T/s, 6 - dB_a/dt = 3 T/s. The inset shows the definition of the flux-jump field at $dB_a/dt = 0.3$ T/s. Here, 3' - $dB_a/dt = 0$, B_a = 0.985 T, 3'' - $dB_a/dt = 0$, B_a = 0.986 T.

To explain the causes of the onset and development of oscillations that take place in superconducting composite with a real V-I characteristic, the occurrence of the oscillations is presented in Figure 5.16 in more detail. The onset of the damping oscillations during full penetration of the screening current is shown in Figure 5.16a, anharmonic oscillations during partial penetration of the screening current are shown in Figure 5.16b. Their general formation was also shown above in Figure 5.13: Curve 3 in Figure 5.13a and Curve 2 in Figure 5.13b, respectively. The following characteristic times are also given in Figure 5.16a. Here, t_f is the time when the cross-section of a composite is fully penetrated with a screening current; t_1 is the time of the beginning of a decrease in the electric field; t_2 is the time of the beginning of a decrease in temperature; t_3 is the time of the beginning of a rise of current density and t_4 is the time of the beginning of a rise of electric field.

Figure 5.16 shows the existence of three stages that underlie the existence of the oscillations. It consists of the following time intervals. The first stage occurs when $t_1 < t < t_2$, the development of which is characterized by an initial decrease in the electric field, occurring with the increasing temperature of the composite and appropriate decreasing of the current density. In this case, $\partial E/\partial t < 0$, $\partial T/\partial t > 0$, $\partial^2 T/\partial t^2 < 0$ and $\partial J/\partial t < 0$. The second stage occurs at $t_2 < t < t_3$ when the temperature of the composite begins to decrease due to the decreasing values of the electric field and current density. In this case, the following conditions $\partial T/\partial t < 0$, $\partial^2 T/\partial t^2 < 0$, $\partial E/\partial t < 0$, $\partial J/\partial t < 0$ exist. The third stage begins at $t > t_3$ and it is characterized by an increase in the current density and conditions $\partial J/\partial t > 0$, $\partial T/\partial t < 0$, $\partial^2 T/\partial t^2 < 0$, $\partial E/\partial t < 0$. When $t > t_4$ (not exist in the states shown in Figure 5.16b), a new oscillation cycle begins.

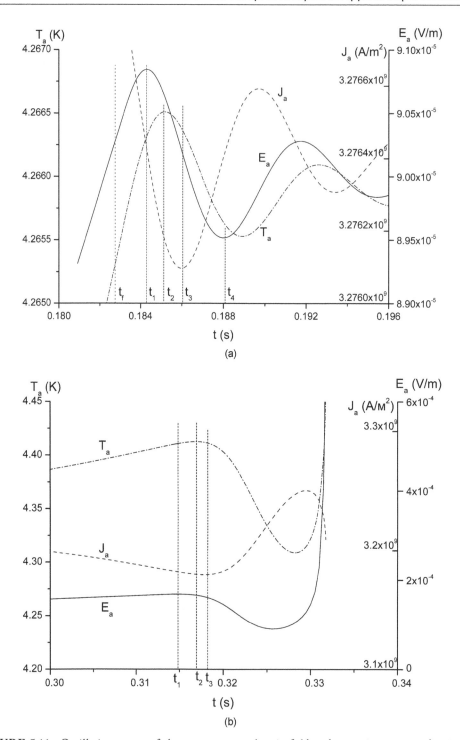

FIGURE 5.16 Oscillation stages of the temperature, electric field and screening current density on a composite surface with different transverse dimensions at $dB_a/dt = 1$ T/s and $h = 100$ W/(m² × K): (a) - $a_k = 9 \times 10^{-5}$ m, (b) - $a_k = 3 \times 10^{-4}$ m.

The discussed peculiarities of the thermo-electrodynamic state formation of a superconducting composite show that the general features of the oscillations onset and development are existed for the superconducting composites both with the ideal and real V-I characteristic. They are based on a corresponding variation in the electric field and screening current with temperature. To formulate the thermal conditions of their occurrence, let us rebuild the dependencies shown in Figure 5.16 as a function of temperature. The corresponding curves are shown in Figures 5.17 and 5.18. They make it possible to formulate the thermal conditions of the oscillations onset. The beginning of both damping and increasing oscillations is described by the condition $\partial E/\partial T = 0$ at $t = t_1$. The boundary of the next stage of oscillations at $t = t_2$ follows from the condition $\partial E/\partial T \to \infty$. And finally, the following stage develops when the condition $\partial J/\partial T = 0$ is met at $t = t_3$. The damping oscillations occur around the corresponding stationary state. As it follows from equation (5.2), the stationary value of the electric field on the surface of the composite E_s, shown in Figure 5.17, is equal to $E_s = a_k \dfrac{dB_a}{dt}$.

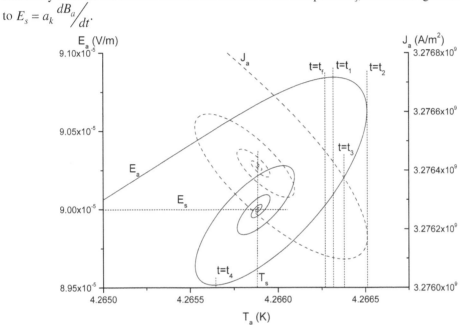

FIGURE 5.17 Dependence of the electric field and current density on the surface of the composite as a function of its temperature during damping oscillations ($a_k = 9 \times 10^{-5}$ m).

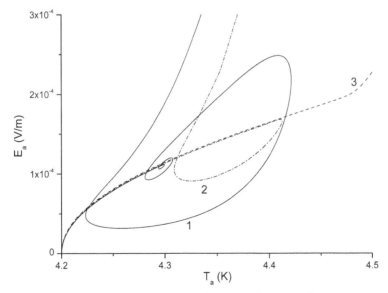

FIGURE 5.18 Electric field on the composite surface as a function of its temperature at unstable oscillations: 1 - $a_k = 1.1 \times 10^{-4}$ m, 2 - $a_k = 3 \times 10^{-4}$ m, 3 - $a_k = 5 \times 10^{-4}$ m.

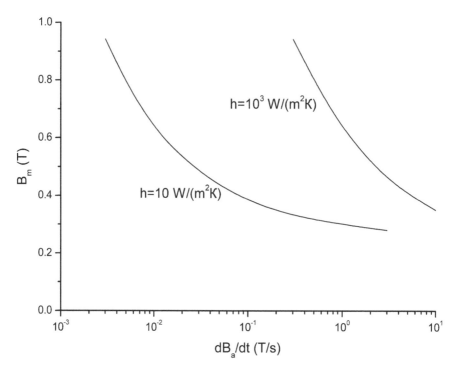

FIGURE 5.19 Dependence of the flux-jump field on the sweep rate at different values of the heat transfer coefficient at $a_k = 5 \times 10^{-4}$ m.

These features of the superconducting state formation were taken into account when determining the boundary of the magnetic instability onset in a composite superconductor with an exponential V-I characteristic. The values of the field of the magnetic flux jump as a function of the sweep rate are presented in Figure 5.19. They were determined by the perturbation method of the initial state in accordance with the noted features of the boundary formation of stable thermo-electrodynamic states of the superconducting composite considered above. The calculation was carried out on the basis of the numerical solution of the problem (5.1), (5.2), (5.4), (5.5) and (5.13).

Curves presented show that in the area of high values of dB_a/dt the values of the flux-jump field tend to the corresponding limiting value (adiabatic limit), which is independent of the heat transfer coefficient. This conclusion once again underlines the importance of the correct consideration of the effect of a stable temperature variation on the conditions of the magnetic instability onset in superconducting composites.

5.5   DYNAMIC STABILITY CRITERIA FOR SUPERCONDUCTING COMPOSITES WITH AN EXPONENTIAL V-I CHARACTERISTIC

Let us formulate the criteria for the stability of the superconducting state of the cooled composite with a slab geometry $(-a_k < x < a_k, -b_k < y < b_k, a_k \ll b_k)$ in the framework of the anisotropic continuum model for the case when the growth of the magnetic field on its outer surface stops after the magnetic flux has filled the cross-section of the composite. To analyze the variation of the temperature and electric field in the composite over the time and space, the V-I characteristic of which has an exponential form, thermal and electrical parameters do not depend on temperature and the critical current density decreases linearly with temperature; the following system of equations may be used

$$C_k \frac{\partial T}{\partial t} = \lambda_k \frac{\partial^2 T}{\partial x^2} + EJ, \quad \mu_0 \frac{\partial J}{\partial t} = \frac{\partial^2 E}{\partial x^2},$$

$$E = E_c \exp\left(\frac{J_s - J_c}{J_\delta}\right) = J_m \rho_m, \quad J = \eta J_s + (1 - \eta)J_m, \tag{5.14}$$

$$J_c(T) = J_{c0}(T_{cB} - T)/(T_{cB} - T_0)$$

under the following boundary conditions

$$\left.\frac{\partial T}{\partial x}\right|_{x=0} = 0, \quad \lambda_k \frac{\partial T}{\partial x} + h(T - T_0)\big|_{x=a_k} = 0, \quad E(0, t) = 0, \quad \partial E/\partial x\big|_{x=a_k} = 0 \tag{5.15}$$

Introduce the following dimensionless variables

$$X = x/a_k, \tau = t/(C_k a_k^2/\lambda_k), \theta = (T - T_0)/(T_{cB} - T_0),$$
$$e = E/(J_{c0}\rho_m), j = J/J_{c0}, \quad j_s = J_s/J_{c0}, \quad j_m = J_m/J_{c0}, \quad j_c = J_c/J_{c0}$$

Then, the system of Equations (5.14) with the Conditions (5.15) will be reduced to the form

$$\frac{\partial \theta}{\partial \tau} = \frac{\partial^2 \theta}{\partial X^2} + \frac{\beta_k}{\Lambda_k} ej, \quad \frac{\partial j}{\partial \tau} = \frac{1}{\Lambda_k} \frac{\partial^2 e}{\partial X^2},$$

$$e = e_c \exp\left(\frac{j_s - j_c}{\delta}\right) = j_m, \quad j = \eta j_s + (1 - \eta)j_m,$$

$$\left.\frac{\partial \theta}{\partial X}\right|_{X=0} = 0, \quad \frac{\partial \theta}{\partial X} + H_k \theta\big|_{X=1} = 0, \quad e(0, \tau) = 0, \quad \partial e/\partial X(1, \tau) = 0$$

Here,

$$\beta_k = \frac{\mu_0 J_{c0}^2 a_k^2}{C_k(T_{cB} - T_0)}, \quad \Lambda_k = \frac{\mu_0 \lambda_k}{C_k \rho_m}, \quad H_k = ha_k/\lambda_k, \quad e_c = \frac{E_c}{J_{c0}\rho_m}, \quad j_c(\theta) = 1 - \theta, \quad \delta = \frac{J_\delta}{J_{c0}}.$$

Simplifying this system and taking into account that $\partial j/\partial \tau \approx 0$ at $\Lambda_k \gg 1$, that means $j \sim$ const, that is, as discussed above, the heat flux is redistributed in the composite in the 'frozen' mode of the magnetic flux. From the dimensionless V-I characteristic equation, one may find the current density in the superconducting core of the composite: $j_s = j_c + \delta \ln(e/e_c)$. Then, the total current density in the composite may be rewritten as follows $j = \eta[j_c + \delta \ln(e/e_c)] + (1 - \eta)e$. Introduce a new variable $e = 1 - \varepsilon$. Using the expanding into the power series $\ln(1 - \varepsilon) \approx -\varepsilon - \ldots$, one may obtain the following expression for the current density

$$j \approx \eta j_c - \eta \delta \ln e_c + (1 - \eta) - \varepsilon[\eta \delta + (1 - \eta)]$$

Note that the accuracy of the formulae written below is improved if the members of a higher order of the smallness are taken into account in the expansion of $\ln(1 - \varepsilon)$. However, this leads to cumbersome final expressions. Therefore, only linear approximation was used below.

Substituting this expression for j in the heat conduction equation and considering that $e = 1 - \varepsilon$, the more simplified form may be obtained

$$\frac{\partial \theta}{\partial \tau} = \frac{\partial^2 \theta}{\partial X^2} + \beta^* \theta + q$$

where

$$\beta^* = \frac{\eta j \beta_k}{\Lambda_k(1 - \eta + \eta\delta)}, \quad q = \frac{j\beta_k}{\Lambda_k} \frac{j - \eta(1 - \delta - \delta \ln e_c)}{1 - \eta + \eta\delta}.$$

In Section 5.2, a similar equation has already been solved with the same thermal boundary conditions. According to the results obtained, the increasing spectrum of positive eigenvalues μ_k is a solution of equation $\sqrt{\mu_k} tg\sqrt{\mu_k} = H_k, 0 < \mu_1 < \mu_2 < \dots, k = 1, 2, 3\dots$. Accordingly, the temperature of the superconductor may be presented in the form

$$\theta(X, \tau) = \sum_{k=1}^{\infty} B_k \exp[(\beta^* - \mu_k)\tau] R_k(X) - q/\beta^*$$

It will begin to grow exponentially when the inequality $\beta^* - \mu_1(H_k) > 0$ is satisfied for the first eigenvalue. (Remind that the estimate $\mu_1 \approx \pi^2/4$ takes place at $H_k \gg 1$ and $\mu_1 \approx H_k$ at $H_k \ll 1$ is valid.) Consequently, the superconducting state of the composite will be unstable at the condition

$$\frac{\eta j \beta_k}{\Lambda_k} \frac{1}{1 - \eta + \eta\delta} > \mu_1(H_k)$$

Taking into account the increase in temperature of the composite arises to θ_i before the instability and simplifying the expression for j, as was done above, bringing it to the form $j \approx \eta(1 - \theta_i - \delta \ln e_c) + 1 - \eta$, this condition can be written as follows

$$\frac{\eta \beta_k}{\Lambda_k} \frac{1 - \eta(\theta_i + \delta \ln e_c)}{1 - \eta + \eta\delta} > \mu_1(H_k)$$

It allows one to estimate the influence of the smooth character of the V-I characteristic of superconductors on the conditions of the magnetic instability onset in superconducting composites. In particular, the second factor on the left side of this condition decreases with increasing δ. Therefore, the magnetic instability onset in the composites with more smoothed V-I characteristic is less likely at all other things being equal.

5.6 COMPARISON OF THEORETICAL RESULTS WITH EXPERIMENTS

Let us use the analysis methods of the thermo-electrodynamic phenomena discussed for superconducting composites with real V-I characteristic and determine the boundaries of the stable states of the current-carrying element of the so-called helicoidal-type superconducting magnet[1], comparing the results of calculations with experiments.

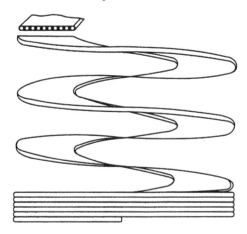

FIGURE 5.20   Superconducting helicoid.

[1] Keilin VE, Kovalev IA, Kruglov SL. et al. Superconducting helicoid – an alternative to conventional superconducting windings. Doklady of Russian Academy of Science. 1988. Vol. 303. No 6. p. 1366-1370.

A superconducting helicoid is a set of separate parallel superconducting composites that are connected in a single unit along a helical plane (Figure 5.20). It has several advantages compared to the traditional SMS: the increased engineering density and high mechanical rigidity. However, their operability is limited by some circumstances. Since the helicoid is a massive composite, it is sensitive to electromagnetic disturbances like a superconducting composite. In particular, it is necessary to study the stability conditions of the superconducting state depending on the sweep rate of the external magnetic field.

Let us consider a cooled helicoid, which may be approximated in the simplest case by a hollow infinitely long cylinder. The nonisothermal diffusion of the electric and magnetic fields when the external magnetic field changes on the inner surface of the helicoid will be determined from the numerical solution of the following system of equations

$$C_k(T)\frac{\partial T}{\partial t} = \text{div}(\lambda_k(T)\,\text{grad}\,T) + EJ,$$

$$\mu_0\frac{\partial J}{\partial t} = \Delta E, \quad B = \frac{dB}{dt}t - \mu_0\int_{r_1}^{r} J(r,t)r\,dr$$

with the initial conditions

$$T(r,0) = T_0, \quad E(r,0) = 0, \quad B(r,0) = 0$$

and conditions on the inner and outer surfaces

$$-\lambda_k\partial T/\partial r + h(T - T_0)\big|_{r=r_1} = 0, \quad \text{rot}\,E\big|_{r=r_1} = -dB_a/dt,$$

$$\lambda_k\partial T/\partial r + h(T - T_0)\big|_{r=r_2} = 0, \quad \text{rot}\,E\big|_{r=r_2} = 0$$

To describe the relationship between the electric field and the currents in the superconducting core and matrix, the exponential V-I equation and the Kirchhoff equations may be used

$$E = E_c\exp\left(\frac{J_s - J_c}{J_\delta}\right) = J_m\rho_m, \quad J = \eta J_s + (1-\eta)J_m, \quad J_c(T,B) = J_{c0}(T_{cB} - T)/(T_{cB} - T_0)$$

During numerical calculations, the initial parameters were assumed to be equal to

$$r_1 = 1.2\cdot10^{-2}\text{m}, \quad r_2 = 2.7\cdot10^{-2}\text{m}, \quad \eta = 0.2, \quad T_0 = 4.2\text{ K}, \quad h = 2500\text{ W/(m}^2\times\text{K)},$$

$$J_\delta = 8\cdot10^7\text{ A/m}^2, \quad T_{cB} = 9.5 - 0.643\times B, \quad E_c = 10^{-4}\text{ V/m},$$

$$\rho_m = (2.13 + 0.605B)10^{-10}\,\Omega\times\text{m}, \quad \lambda_k = 2.45\cdot10^{-8}\frac{T}{\rho_m}\frac{1-\eta}{1+\eta}\frac{\text{W}}{\text{m}\times\text{K}},$$

$$C_k = \eta C_s + (1-\eta)C_m, \quad C_s = 13T^3\frac{\text{J}}{\text{m}^3\times\text{K}}, \quad C_m = 8T^3\frac{\text{J}}{\text{m}^3\times\text{K}},$$

They describe the thermal and electrical properties of the helicoid containing 71 planar turns. Each turn consisted of 15 niobium-titanium composite wires in a copper matrix with a diameter of 0.5 mm and 15 copper wires of the same diameter that were soldered to each other. The value of the heat transfer coefficient was chosen in such a way that the theoretical and experimental values of the magnetic field of flux jump insignificantly differed from each other. This is the effective value of the heat transfer coefficient. It allows one to take into account a significant number of the experimental factors that affect the value of the heat transfer coefficient. This determination is often used when comparing experimental data with the results of calculations. Note that the effective value of the heat transfer coefficient introduced in this way does not go beyond the values commonly used in evaluating the conditions of the magnetic instability onset in composite superconductors.

The results of the numerical experiment are presented in Figure 5.21 during which the stability conditions of the superconducting state of the helicoid were determined. It was performed in accordance with the perturbation method of the initial state. The general dynamics of its temperature variation throughout the entire diffusion of the magnetic flux, both with a continuous increase in the external magnetic field (the solid lines) and with its fixed values close to the field of the flux jump (the dashed lines) are shown in Figure 5.21a. The stable and unstable temperature variation near the boundary of the instability is shown in Figure 5.21b in more detail at $dB_a/dt = 0$. It demonstrates the same thermo-electrodynamic features that were discussed above and is the basis for determining the boundaries of the magnetic instability onset in a composite superconductor with a real V-I characteristic.

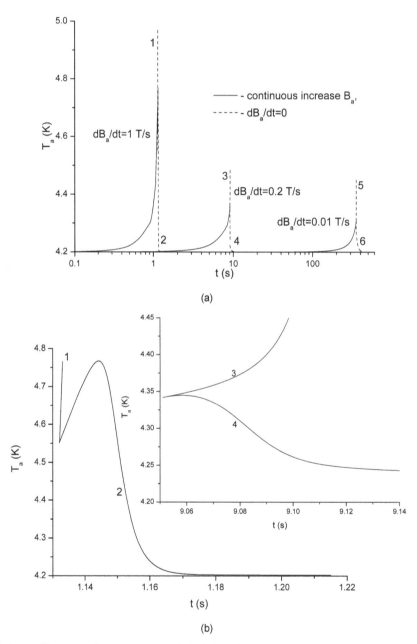

FIGURE 5.21 Change in the temperature of the helicoid inner surface at the stable and unstable states ((a): 1 - B_a = 1.14 T, 2 - B_a = 1.13 T, 3 - B_a = 1.811 T, 4 - B_a = 1.810 T, 5 - B_a = 3.527 T, 6 - B_a = 3.526 T) and its dynamics for the external magnetic field close to the field of the flux jump (b).

The corresponding values of the field of the flux jump as a function of the sweep rate are presented in Figure 5.22. It also shows the results of direct measurements $B_m(dB_a/dt)$. The satisfactory coincidence of the theoretical and experimental results verifies the used nonisothermal approximation, confirming the existence of a dependence of the permissible overheating on the conditions for variation of the external magnetic field (Figure 5.21).

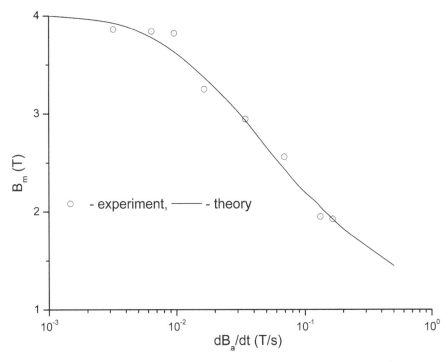

FIGURE 5.22 Effect of the sweep rate on the stability of the helicoid superconducting state.

5.7 CONCLUSIONS

The stable temperature fluctuation of the superconducting composites may be noticeable. As a result, it may affect the electrodynamic properties of them. Therefore, if the temperature of a superconducting composite is chosen a priori before the magnetic instability, then the incorrect conditions for its performance will be formulated. Especially, the magnetic instability criteria, which are written in the isothermal approximation, lead to the essential restriction of the stability range. Hence, a correct definition of the conditions of magnetic instability onset should be based on an analysis involving the interrelated penetration of the magnetic flux and the corresponding change in temperature. This approximation, first, shows that the thermo-electrodynamic phenomena in superconducting composites also have the nature of the fission chain reaction. However, the presence of the matrix and the smoothing nature of the real voltage-current characteristic of superconductor smooth the temporal development of the transition of the superconducting composite from the superconducting state to the normal one. Second, it extends the magnetic instability boundaries of superconducting composites even it is made in terms of the simplified models, especially, in the framework of the critical state model but accounting for the thermal prehistory of the occurring phenomena. These features must be taken into account during development of the large-scale SMS with massive current-carrying elements placed in rapidly changing magnetic fields.

The written down stability criteria of the superconducting state of the composite show that the model of the anisotropic continuum leads to stricter requirements imposed on the parameters of the composite, ensuring its superconducting properties.

The performed numerical experiments demonstrate that the laws underlying the occurrence and development of the oscillations in superconductors without a stabilizing matrix with the V-I characteristic, described by the viscous flux-flow model, are also fulfilled when the oscillations occur in composite superconductors with the real V-I characteristic. They are based on the existence of the critical rate of conductive-convective heat removal from the interior of the superconducting composite to the coolant. At the same time, with an increase in the sweep rate of the external magnetic field or deterioration of heat transfer conditions and, as calculations show, with an increase in the values of the smoothing parameters of the V-I characteristic, the effect of thermal self-suppression of oscillations may occur at higher stable overheating.

6

Current-Carrying Capacity of Superconducting Composites and Tapes: Formation and Destruction of Stable Current Modes

The study of macroscopic phenomena in superconducting composites and tapes that carry maximally high transport currents underlies the solution of important problems of the stability theory of the superconducting state, namely, the problem of the stable current charging in SMS. It allows determining the stable current load of SMS without its transition to the normal state studying the mechanisms of the current instability. The performed analysis is important both from the physical and applied point of view. In this context, this chapter examines the current-carrying capacity of superconducting composites based on both LTS and HTS, taking into account the temperature rise during current charging. Zero-dimensional and one-dimensional models are used to formulate the stability conditions of the transport current charged into the superconducting composites and tapes as a function of their thermophysical properties, transverse dimensions, cooling conditions and current charging rate.

6.1 FEATURES OF THE TRANSPORT CURRENT DIFFUSION INTO A COMPOSITE SUPERCONDUCTORS

One of the main properties of superconductors is the current that can be charged into them without transition to the normal state. Its existence is a consequence, first of all, of the magnetic flux-creep when the finite electrical voltage appears in the superconductor long before the transition to the normal state during current charging that leads to the heat release. Under certain conditions, the thermal balance between the heat released and heat removed into the coolant may be disturbed. As a result, the charged current may become unstable. Therefore, the problem of determining stable current loads should be solved already at the design stage of SMS.

As already noted, the operating modes of the superconducting current-carrying elements should be described by the nonlinear multi-dimensional unsteady Fourier and Maxwell equations in the general case. However, to simplify the analysis, the one-dimensional nonisothermal model of the electromagnetic field diffusion in the superconducting composite under the current charging will be used below to discuss the general physical laws of the transport current penetration into the composite superconductors and tapes.

Let us consider an infinitely long composite of a circular cross-section with a radius r_0 when the current is charged into this composite in the direction of the longitudinal axis (z-axis) with a rate dI/dt. Suppose that there is no current at the initial time and the composite has the temperature of the coolant T_0. Let us also assume that

- The transport current is charged by a third-party source after an external magnetic field, which is parallel to the surface of the composite, has completely penetrated it and its magnetic field is small compared to the external field.
- Thin superconducting filaments are evenly distributed over the cross-section of the composite with a filling factor of η, and they do not lead to the magnetic instability.
- The variation of the magnetic field may be neglected in the longitudinal direction.
- The V-I characteristic of the superconductor is described by the exponential Equation (1.4).
- The critical current density of the superconductor decreases linearly with temperature and depends on the magnetic field.
- On the surface of the composite, the convective heat exchange takes place with a heat transfer coefficient h.
- External thermal perturbations are absent.

To describe the modes arising in the superconducting composite under the current charging, let us use the model of an anisotropic continuum. Then, in a cylindrical coordinate system, one-dimensional nonstationary distribution of the temperature, electric field and current are the solution of the following system of equation

$$C_k(T)\frac{\partial T}{\partial t} = \frac{1}{r}\frac{\partial}{\partial r}\left(\lambda_k(T)r\frac{\partial T}{\partial r}\right) + \begin{cases} 0, & 0 < x < r_p \\ EJ, r_p \leq x \leq r_0 \end{cases}, \tag{6.1}$$

$$\mu_0\frac{\partial J}{\partial t} = \frac{1}{r}\frac{\partial}{\partial r}\left(r\frac{\partial E}{\partial r}\right) \tag{6.2}$$

with initial conditions

$$T(r,0) = T_0, \quad E(r,0) = 0 \tag{6.3}$$

and with conditions on the surface

$$\lambda_k\frac{\partial T}{\partial r} + h(T-T_0)\bigg|_{r=r_0} = 0, \quad \frac{\partial E}{\partial r}\bigg|_{r=r_0} = \frac{\mu_0}{2\pi r_0}\frac{dI}{dt} \tag{6.4}$$

Here, $r_p(t)$ is the current penetration coordinate, which is described by equation

$$2\pi\int_{r_p}^{r_0} J(r,t)rdr = I(t) = \frac{dI}{dt}t, \quad E(r_p,t) = 0 \tag{6.5}$$

at $r_p > 0$ according to the results discussed in Chapter 2.

The electric field and currents in the superconducting core of the composite J_s and matrix J_m follow from the Kirchhoff equations written for a circuit formed by the parallel-connected superconducting filaments and nonsuperconducting matrix. Then, the system of Equations (6.1)-(6.5) must be supplemented with the following relations

$$E = E_c\exp\left(\frac{J_s - J_c}{J_\delta}\right) = J_m\rho_m, \quad J = \eta J_s + (1-\eta)J_m \tag{6.6}$$

To approximate the critical current density, the dependence of the form $J_c(T, B) = J_{c0}(T_{cB} - T)/(T_{cB} - T_0)$ will be used as above.

Equations (6.1) and (6.2) describe the dissipative modes during current charging into the composite superconductor with the 'smoothed' V-I characteristic. The initial conditions (6.3) describe the initial undisturbed thermal and electrodynamic states of the composite. The boundary conditions (6.4) take into account the convective heat transfer and variation in the electric field on its surface. The physical meaning of the parameters used was discussed above.

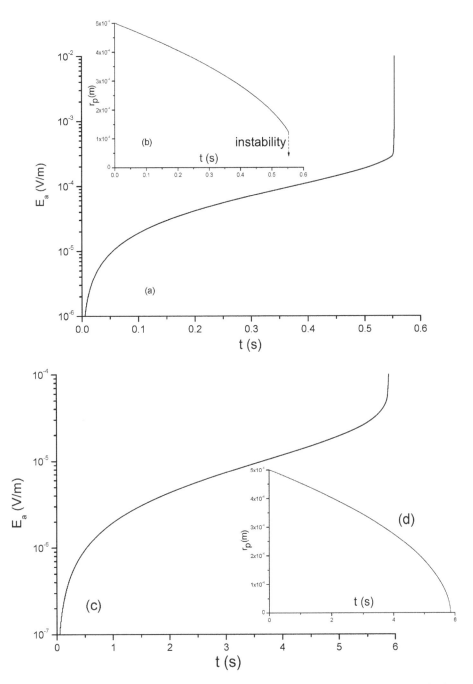

FIGURE 6.1 Stable and unstable stages of the electric field increase on the composite surface (*a*, *c*) and the dynamics of the current penetration front (*b*, *d*) during continuous current charging at the partial (*a*, *b*) and full (*c*, *d*) current penetration modes: a, b - *dI/dt* = 10^3 A/s; c, d - *dI/dt* = 10^2 A/s.

In Figure 6.1, the simulation results of the electric field evolution caused by the continuous current charging with different rates are presented. They describe the characteristic regularities of the variation in the current penetration coordinate and electric field on the surface of the intensively cooled (*h* = 10^3 W/(m² × K)) superconducting composite based on Nb-Ti in the copper matrix. Calculations were performed at

$$r_0 = 5\times10^{-4}\,\text{m}, T_0 = 4.2\,\text{K}, E_c = 10^{-4}\,\text{V/m}, T_{cB} = 6.5\,\text{K}, \eta = 0.5, J_{c0} = 4\times10^{9}\,\text{A/m}^2$$

$$J_\delta = 2\times10^{7}\,\text{A/m}^2, \quad \rho_m = 2\times10^{-10}\,\Omega\times\text{m}, \quad \lambda_k = 200\frac{\text{W}}{\text{m}\times\text{K}}, \quad C_k = 2\times10^{3}\frac{\text{J}}{\text{m}^3\times\text{K}} \quad (6.7)$$

The presented curves show the existence of both stable and unstable phases of the current penetration into the composite. The current instability is accompanied by a sharp increase in the electric field and hence, an increase in its temperature. In the case under consideration, the onset of the current instability depends on the rate of the current charging and occurs either during partial (Figures 6.1a and 6.1b) or full (Figures 6.1c and 6.1d) penetration modes.

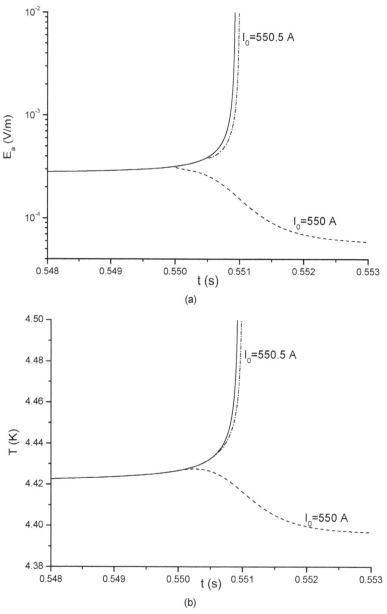

FIGURE 6.2 Determining the boundary of the stable current mode during partial penetration mode (—— - continuous current charging at $dI/dt = 10^3$ A/s; ————, — · — · — · — - $dI/dt = 0$): (a) – the current instability boundary defined according to the change in the electric field; (b) – the current instability boundary defined according to the change in the temperature.

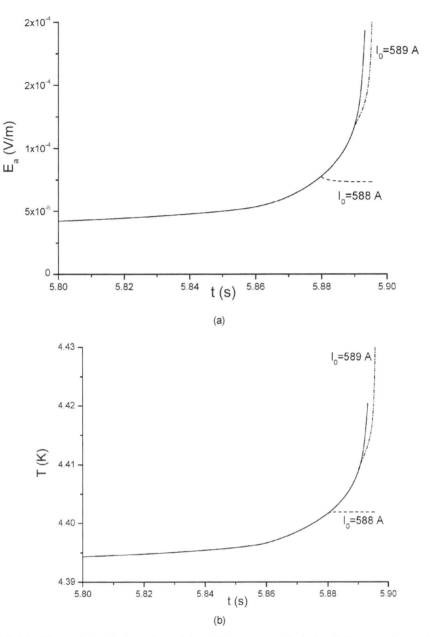

FIGURE 6.3 Determining the boundary of the stable current mode during full penetration mode (———⋅ continuous current charging at $dI/dt = 100\,\text{A/s}$; ————, —⋅—⋅—⋅- $dI/dt = 0$): (a) – the current instability boundary defined according to the change in the electric field; (b) – the current instability boundary defined according to the change in the temperature.

To find the boundary of the stable states, let us use the finite perturbation method of the initial state proposed in Chapter 4. The corresponding results of the numerical experiments are shown in Figures 6.2 and 6.3. They show the time variations of the electric field and temperature that occur on the surface of the composite under consideration, both at continuous current charging (the solid curves) and when current is charged with a stop (the dotted and dash-dotted curves). In the calculation model, dI/dt was assumed to be equal to zero when the currents I_0 specified in the figures are reached in the composite.

In Figure 6.4, the results determining the current instability boundary are shown for the composite with higher J_δ values when the instability arose at the partially penetrated mode. Comparing them with the results shown in Figure 6.2, it is seen that the range of the stable currents decreases with increasing J_δ. As calculations show, this effect will be observed to the greatest extent in cases of violation of the stable current distribution when it completely fills the cross-section of the composite. Thus, the 'smoothness' of the V-I characteristic of a superconductor plays a destabilizing role during current charging.

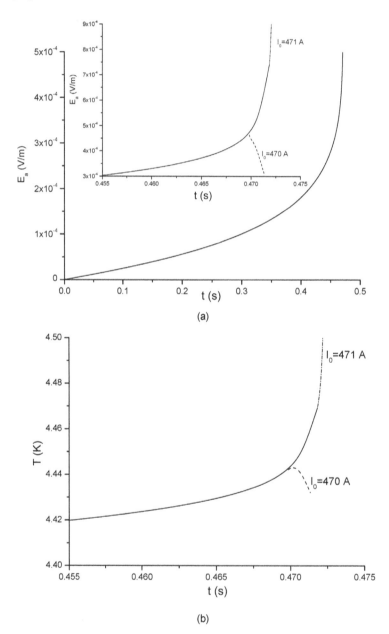

(a)

(b)

FIGURE 6.4 Determining the stable boundary of the charged current at $dI/dt = 10^3$ A/s and $J_\delta = 4 \times 10^7$ A/m^2 (⎯⎯⎯ - continuous current charging; ⎯ ⎯ ⎯ ⎯, ⎯ · ⎯ · ⎯ · ⎯ - $dI/dt = 0$): (a) – the current instability boundary defined according to change in the electric field, (b) – the current instability boundary defined according to change in the temperature.

In general, the results presented in Figures 6.2-6.4 demonstrate that the variation in the thermo-electrodynamic states of the superconducting composite before the current instability depends on the character of the current diffusion. For the considered modes, it changes in accordance with the change in the rate of current charging and the 'smoothness' parameter of the V-I characteristic. As a result, the electric field and temperature of the composite increase markedly in comparison with the initial unperturbed state. Moreover, the value of the electric field on the surface may be either lower or higher than a conventionally defined critical value E_c. As a consequence, a stable increase in the temperature of the composite before the current instability increases with an increase of the current charging rate or the 'smoothness' parameter of the V-I characteristic. Therefore, a priori setting of the permissible overheating of the composite will lead to incorrect results of determining the current instability conditions. At the same time, Figures 6.2-6.4 as well as the results of Chapters 4 and 5 show that the finite perturbation method of the initial state allows one to determine the boundary of the stable states regardless of the nonlinearity type of the V-I characteristic, the character of the variation in the external magnetic field or the transport current. Therefore, the general definition method of both magnetic and current instabilities exists. Thus, when the external magnetic field or the transport current changes, the current of the instability onset (screening or transport) is the current whose value lies between the maximum fixed current—when the spontaneous transition of the superconducting composite to the normal state does not yet occur—and the minimum stop current after which this transition occurs spontaneously. The generality of this definition makes it possible to determine the boundary of the stable states in more general cases, for example, when there is a simultaneous variation in the external magnetic field and the transport current and as it will be shown below, under the action of external thermal perturbations.

For the first time, this method of the instability current definition (the so-called quenching current) was given in (Keilin and Romanovsky 1982) when analyzing the conditions for a current instability into superconducting composites based on LTS. At the moment, it is widely used in the experimental determination of the stable currents charged into HTS.

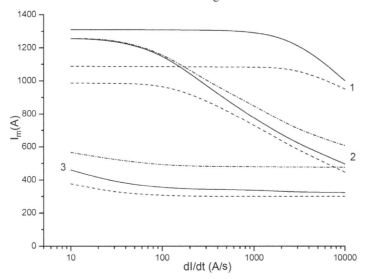

FIGURE 6.5 Characteristic dependencies of the instability current on the current charging rate at different values of the 'smoothness' parameter of the V-I characteristic and cooling conditions: 1- h = 3,000 W/(m² × K), 2 - h = 100 W/(m² × K), 3 - h = 0.1 W/(m² × K); (——, — · — · — · — - J_δ = 4 × 10⁷ A/m², — — — - J_δ = 8 × 10⁷ A/m²).

Let us use the formulated definition of the instability current and analyze the influence of the current charging rate, heat transfer coefficient and the 'smoothness' parameter of the V-I characteristic on the boundary of the stable currents of the above-considered superconducting composite. The corresponding results are presented in Figure 6.5 (the solid and dashed curves). It also shows the instability currents, calculated under the assumption that the thermal properties of the composite depend on temperature (the dash-dotted curves). In this case, the following dependencies were used

$$\lambda_k = 2.45 \cdot 10^{-8} \frac{T}{\rho_m} \frac{1-\eta}{1+\eta} \frac{W}{m \times K}, \quad C_k = \eta C_s + (1-\eta)C_m, \quad C_s = 50T^3 \frac{J}{m^3 \times K}, \quad C_m = 8T^3 \frac{J}{m^3 \times K}$$

It can be seen that the nonlinearity of the thermal properties of the composite leads only to a quantitative change in the dependence $I_m(dI/dt)$ in areas where the operating modes are characterized by significant stable overheating, in particular at high value of the charging rate or nonintensive cooling conditions. As a whole, the presented results allow one to formulate the following conclusions.

First, the dependence $I_m(dI/dt)$ has an area where the instability currents are maximum and weakly dependent on the current charging rate. This is the limiting current-carrying capacity of the superconducting composites. Its existence was first noted by Polak et al. (1973). It is characterized by the complete filling of the cross-section by the transport current and a finite permissible overheating on the background of which a transition to a normal state develops. The mode, which is similar to this, is shown in Figure 6.3. The limiting quenching current I_q for these modes given in (Keilin, Romanovsky, 1982) is the root of equation that may be written as follows

$$i_q = 1 - \delta - \delta \ln \frac{\alpha i_q - 1}{\delta \varepsilon_1} \tag{6.8}$$

Here,

$$i_q = \frac{I_q}{\eta J_{c0} S}, \quad \alpha = \frac{J_{c0}^2 \eta^2 \rho_m S}{(1-\eta)hp(T_{cB} - T_0)}, \quad \delta = \frac{J_\delta}{J_{c0}}, \quad \varepsilon_1 = \frac{\eta J_{c0} \rho_m}{(1-\eta)E_c}$$

Formally, this quantity I_q corresponds to the current when instability occurs during infinitely slow current charging. The physical meaning of the parameters α and ε_1 will be discussed in Section 6.2 where the corresponding limiting quenching parameters will be derived in a universal form for superconducting composites with different nonlinearity types of V-I characteristics.

Second, there is a region of the smallest current-carrying capacity. It is also weakly dependent on the current charging rate. In this case, the thermo-electrodynamic states of the composite asymptotically approach adiabatic ones. Their formation is characterized by the high overheating and a small depth of the current penetration.

Third, in the transition region of $I_m(dI/dt)$ variation, the values of the quenching currents monotonously decrease with increasing the current charging rate. In these modes, as follows from Figures 6.2 and 6.3, the permissible overheating of the composite increases respectively. This result is not difficult to explain by analogy with the conclusions formulated in Chapter 4 when the permissible heat dissipation induced by the magnetic perturbation decreases and the temperature on the background of which the magnetic instability develops, increases with a decrease in the time of the stable penetration of the screening current. Therefore, the higher the current charging rate, the smaller the depth of the stable penetration of the transport current and the higher the temperature of the composite preceding the beginning of the current instability.

To illustrate these regularities, the corresponding values of the permissible composite overheating

$(\Delta T = T(r_0, t_s) - T_0)$, the Joule heat density per unit length $(G_i = \int_0^{t_s} \int_S EJdSdt)$ and depth of current

penetration $(\Delta_r = r_0 - r_p)$ as a function of the current charging rate just before instability are presented in Figure 6.6. Here, t_s is the current charging time to the quenching value I_q, S is the cross-sectional area of the composite. It can be seen that there are the same qualitative regularities that were formulated in Chapter 4 when determining the flux-jump field in a superconductor with an ideal V-I characteristic, that is at $J_\delta \to 0$. Thus, the noted regularities strictly prove that the occurrence of the unstable states is preceded by the general thermal regularities that affect the stable formation of the electrodynamic states of superconductors regardless of the nonlinearity type of V-I characteristic, which are characteristic of any external disturbances. They must be considered when determining the conditions of stability. Otherwise, the results obtained will not only incorrectly describe the stability boundary of the superconductivity states but also will not comply with the corresponding limiting transitions from superconductors with a 'smoothness' V-I characteristic to superconductors with an ideal V-I characteristic.

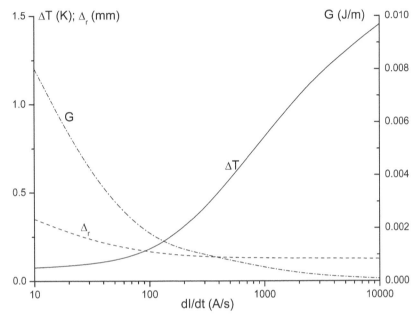

FIGURE 6.6 Characteristic dependencies of the permissible overheating of the composite ΔT, dissipated energy G allocated just before the current instability and the depth of the current penetration Δ_r on the current charging rate at $h = 10$ W/(m²× K), $J_\delta = 4 \times 10^7$ A/m².

Thus, the features of the occurrence of the unstable states are associated with the corresponding variations in the temperature of the composite during current diffusion (screening or transport). Its stable increase depends on the properties of the superconductor and the matrix, conditions of external disturbance, etc. The finite perturbation method of the initial state allows one to determine the alloable overheating for any type of the V-I characteristics of a superconductor (including ideal) avoiding the mistakes that are inevitable when the permissible overheating of the composite is a priori defined. The errors will be at maximum assuming that the temperature just before the current instability is equal to the coolant temperature especially in the operating modes close to the adiabatic.

The discussed results were based on the numerical solution of the heat propagation and Maxwell equations. At the same time, for practical applications, the solution of the stability problem in analytical form is of undoubted interest. Let us write the corresponding stability criterion taking into account the physical features noted above, that is, will consider the existence of a stable overheating, which depends on the character of the current penetration, assuming that the heat capacity of the composite does not depend on temperature to ease the analysis.

Let us integrate equation (6.1) in the cross-sectional of the composite. Considering the thermal boundary condition (6.4), it is easy to obtain the heat balance equation of the form

$$C_k \int_S \frac{\partial T}{\partial t} ds = -\int_p (T - T_0) dp + \int_S EJds$$

Here, p is the cooled perimeter and S is the cross-sectional of the composite. Let us formulate on the basis of this equation the general condition for preserving the superconducting properties of a composite. Namely, the energy released during diffusion of the transport current, which causes a change in its thermal state and removed to the coolant, should not lead to a heating of the composite above the permissible overheating. Accordingly, in the approximation of uniform temperature distribution over the cross-section of the composite ($hp/S\lambda_k \ll 1$), the stable states are satisfied with the condition

$$\left(\frac{C_k}{\Delta t} + \frac{hp}{S} \right) \Delta T \le \frac{1}{S} \int_S EJ\, ds$$

where Δt is the time of the stable current penetration and ΔT is the corresponding overheating. To calculate the Joule losses, the approximate formula may be applied

$$\int_S EJ ds \approx \eta J_c (T_0)(1 - \Delta_j) \int_S E ds$$

in which the distribution of the electric field in the composite of the circular cross-section determines as follows

$$E = \frac{\mu_0}{2\pi} \frac{dI}{dt} \ln \frac{r}{r_p}$$

according to the Formula (2.42). Here, Δ_j is some constant to be determined and takes into account a change in the current-carrying capacity of superconductor due to the V-I nonlinearity, r_p is the coordinate of the current penetration front, determined from the equality

$$I_m = 2\pi\eta \int_{r_p}^{r_0} J_c(T_0)(1 - \Delta_j)r\, dr$$

according to the current conservation law. Here, the current I_m is the quenching current at which the destruction of the superconductivity occurs during partial penetration mode ($r_p > 0$) and finite overheating ΔT. Since $\Delta t = I_m/(dI/dt)$, the following system of equations may be obtained

$$\frac{4\pi\Delta T}{\mu_0 \eta J_{c0} S(1 - \Delta_j) dI/dt} \left(hp + \frac{C_k S}{\eta J_{c0} I_m} \frac{dI}{dt} \right) + 2\ln \frac{r_p}{r_0} + 1 - \left(\frac{r_p}{r_0} \right)^2 = 0, \quad I_m = (1 - \Delta_j)\left[1 - \left(\frac{r_p}{r_0} \right)^2 \right] \eta J_{c0} S$$

Complementing it with the limiting transition $\lim I_m \to I_q = i_q \eta J_{c0} S$ at $r_p \to 0$ and performing simple transformations, one may obtain the following equation defining the quenching current I_m

$$I_m \left[\ln\left(1 - \frac{I_m}{I_q} \right) + \frac{I_m}{I_q} \right] + \frac{4\pi C_k S \Delta T}{\mu_0 I_q} + \frac{4\pi hp \Delta T}{\mu_0 I_q} \frac{I_m}{dI/dt} = 0$$

The results of the many numerical experiments show that to evaluate the value of ΔT, one can use the following equality

$$\Delta T = \frac{J_\delta}{J_{c0}} (T_{cB} - T_0) \left[1 + \left(\frac{I_q}{I_m} \right)^3 - \frac{32}{(1 + I_m/I_q)^5} \right]$$

that is observed with a satisfactory degree of accuracy at $0.01 \le \delta \le 0.05$.

Note, that equation for I_m allows the estimation of the overheating of a composite if the experimental values of I_m will be used for a given value dI/dt.

Let us introduce the dimensionless variables

$$i_m = I_m/(\eta J_{c0}S) \text{ and } \theta_m = \frac{J_{c0}\Delta T}{J_\delta(T_{cB} - T_0)}.$$

Then, the dimensionless value of the quenching current i_m is the solution of the equation

$$i_m + i_q \ln\left(1 - \frac{i_m}{i_q}\right) + \frac{\theta_m}{\alpha_{\text{eff}}}\left(1 + \frac{C_k}{hp\eta J_{c0}i_m}\right) = 0 \qquad (6.9)$$

where i_q follows from Equation (6.8)

$$\theta_m = 1 + \left(\frac{i_q}{i_m}\right)^3 - \frac{32}{(1 + i_m/i_q)^5}$$

and the value α_{eff} is defined as follows

$$\alpha_{\text{eff}} = \frac{\mu_0 S J_{c0}\eta}{4\pi hp J_\delta}\frac{dI}{dt}\left|\frac{dJ_c}{dT}\right|$$

according to Mints and Rakhmanov (1975).

If in Equation (6.9) put $C_k \to 0$, $i_q \to 1$, $\theta_m \to 1$, that is not to take into account the stabilizing role of the heat capacity and the temperature effects taking place during current charging, which is discussed above, then the limiting transition to the simplified equation (1.23) is observed. In the framework of the introduced variables, it is written as follows

$$i_m + \ln(1 - i_m) + \frac{1}{\alpha_{\text{eff}}} = 0 \qquad (6.10)$$

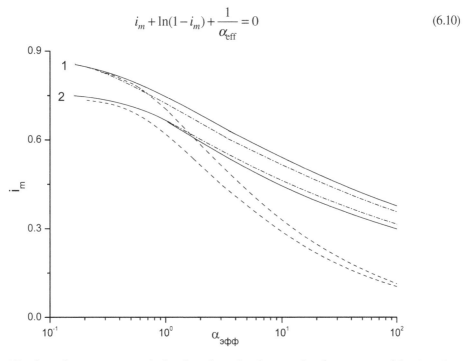

FIGURE 6.7　Quenching currents calculated within the framework of various models: 1 - h = 3,000 W/(m² × K), 2 - h = 0.1 W/(m² × K); (——— - numerical calculation according to (6.1)-(6.6), — · — · — · — - calculation according to (6.9), — — — - calculation according to (6.10)).

The calculations of the instability currents made under different models are compared in Figure 6.7. The solid curves show the dependencies $i_m(\alpha_{\text{eff}})$, which were obtained based on the numerical solution of the problem (6.1)-(6.6). The dash-dotted curves were calculated using Equation (6.9), and the dashed curves were calculated using Equation (6.10). It is easy to see that the temperature effects discussed above are small if $\alpha_{eff} \ll 1$. Therefore, the parameter α_{eff} can be used to estimate the effect of the permissible overheating of the composite on the values of the instability currents.

Thus, there is a general interconnection between the current penetration depth, the finite value of overheating and the amount of energy released before the instability. This connection takes place both when determining the conditions of the magnetic instability of the critical state and when determining the stability conditions of a current charged into a superconducting composite with a 'smoothness' V-I characteristics. It is a consequence of the dissipative phenomena occurring in a composite superconductor during the diffusion of the screening or transport currents. Therefore, the models based on a priori set the permissible composite overheating can lead even to the physically incorrect conclusions.

The analysis performed shows that the currents of the instability monotonously decrease when increasing the charging rate, deterioration of the heat transfer conditions and asymptotically approaching the minimum value other than zero. The 'smoothness' parameters of the V-I characteristics will have the most noticeable effect on the current-carrying properties of superconducting composites at low current charging rates (in the limiting case at $dI/dt \to 0$) degrading them. Such operating modes are discussed in detail in next Section.

6.2   NONISOTHERMAL V-I CHARACTERISTICS AND LIMITING QUENCHING CURRENTS OF SUPERCONDUCTING COMPOSITES

So, the current-carrying capacity of the superconducting composites is characterized by the dependence on the current charging rate. The highest values of the quenching current I_m will be at a low charging rate and the maximum value of the quenching current exists at $dI/dt \to 0$. It is the limiting quenching current I_q ($I_m < I_q$). Let us analyze it in the framework of the so-called zero-dimensional static approximation in the fully penetrated mode. Accordingly, the temperature and electric field distributions over the cross-section of a thin composite are uniform due to the infinitely low current charging rate ($dI/dt \to 0$). This approximation may allow one to obtain the corresponding analytical formulae to calculate I_q.

Consider an infinitely long composite superconductor, whose V-I characteristic is described by the power or exponential equations. Under this approximation, the temperature T and the electric field E in the composite with cross-section S and the cooling perimeter p with heat transfer coefficient h obey the heat balance equation of the form

$$EJ = q(T)p/S, \quad q(T) = h(T - T_0) \tag{6.11}$$

for a given density of a transport current J. It is equal to the sum of the currents density in the superconducting core J_{sc} and matrix J_m according to the Kirchhoff equations

$$J = \eta J_{sc} + (1 - \eta)J_m \tag{6.12}$$

Correspondingly, the static electric field induced in the parallel circuit on the superconducting core and matrix satisfies the relations

$$E = E_c[J_{sc}/J_c(T, B)]^n = J_m \rho_m(T, B) \tag{6.13}$$

for the superconductor with the power V-I characteristic and

$$E = E_c \exp\{[J_{sc} - J_c(T, B)]/J_\delta\} = J_m \rho_m(T, B) \tag{6.14}$$

for the superconductor with the exponential V-I characteristic. Here, ρ_m is the matrix resistivity.

As above-mentioned, assume that

$$J_c(T) = J_{c0}(T_{cB} - T)/(T_{cB} - T_0) \tag{6.15}$$

Let us rewrite Equations (6.12)-(6.14) in the form

$$E = E_c\{[J - E(1-\eta)/\rho_m(T, B)]/(\eta J_c(T, B))\}^n$$

for the superconductor with the power V-I characteristic and

$$E = E_c \exp\{[J - E(1-\eta)/\rho_m(T, B) - \eta J_c(T, B)]/\eta J_\delta\}$$

for the superconductor with the exponential V-I characteristic.

Let us exclude the temperature in the written equations of the V-I characteristic using Equation (6.11) and the linear dependence of the critical current density on temperature (6.15) by assuming that $\rho_m(T, B) \approx \rho_m(T_0, B) = $ const to simplify the analysis. Then, the nonisothermal V-I characteristic of the composite can be written as follows

$$J = \left[\eta J_{c0}\left(\frac{E}{E_c}\right)^{1/n} + \frac{1-\eta}{\rho_m}E\right] / \left[1 + \frac{\eta J_{c0}SE}{hp(T_{cB} - T_0)}\left(\frac{E}{E_c}\right)^{1/n}\right] \qquad (6.16)$$

for the superconductor with the power V-I characteristic and

$$J = \left[\eta J_{c0} + \eta J_\delta \ln\frac{E}{E_c} + \frac{1-\eta}{\rho_m}E\right] / \left[1 + \frac{\eta J_{c0}SE}{hp(T_{cB} - T_0)}\right] \qquad (6.17)$$

for the superconductor with the exponential V-I characteristic. These relationships may be rewritten as follows

$$J = \eta J_{c0}\frac{(E/E_c)^{1/n} + E/E_\rho}{1 + (E/E_h)(E/E_c)^{1/n}}, \quad J = \eta J_{c0}\frac{1 + (J_\delta/J_{c0})\ln(E/E_c) + E/E_\rho}{1 + E/E_h} \qquad (6.18)$$

for the superconductor with the power or exponential V-I characteristics, respectively. Here,

$$E_\rho = \frac{\eta J_{c0}\rho_m}{1-\eta}, \quad E_h = \frac{hp(T_{cB} - T_0)}{\eta J_{c0}S} \qquad (6.19)$$

These quantities determine the characteristic values of the electric field, which affect the character of the thermo-electrodynamic state's formation of the superconducting composite in zero-dimensional approximation. They depend on the properties of the superconductor and matrix, cooling conditions and may satisfy the condition $E_\rho \ll E_h$ or $E_\rho \gg E_h$. In the first case, the current sharing between the superconducting core and matrix will be practically isothermal, and the current sharing becomes nonisothermal in the second case. (The physical meaning of E_ρ and E_h is discussed below in detail). Let us evaluate their possible values assuming, in particular $\rho_m \sim 10^{-9}\ \Omega \times$ m, $J_{c0} \sim 10^9$ A/m^2, $p \sim 10^{-3}$ m, $S \sim 10^{-6}$ m^2, $\eta \sim 0.5$, $T_{cB} - T_0 \sim 20$ K. Then, $E_\rho \sim 1$ V/m and $E_h \sim 4 \times 10^{-4}$ V/m under nonintensive cooling conditions ($h \sim 10$ W/(m$^2 \times$ K)) when cryorefrigerators are used. Therefore, in these cases, it is necessary to take into account the corresponding difference between the temperatures of the composite and cryorefrigerator, even when the main part of the current flows in the superconducting core of the composite.

According to (6.11) and (6.18), the temperature of the composite as the function of the electric field is described as follows

$$T = T_0 + (T_{cB} - T_0)\frac{(E/E_c)^{1/n} + E/E_\rho}{(E/E_c)^{1/n} + E_h/E} \qquad (6.20)$$

for the superconductor with the power V-I characteristic and

$$T = T_0 + (T_{cB} - T_0)\left(1 + \frac{J_\delta}{J_{c0}}\ln\frac{E}{E_c} + \frac{E}{E_\rho}\right) / (1 + E_h/E) \qquad (6.21)$$

for the superconductor with the exponential V-I characteristic.

Relations (6.18)-(6.21) show that the main part of the charged current will flow in the superconducting core of the composite, when

$$E \ll E_c (E_\rho/E_c)^{n/(n-1)}$$

for the superconductor with the power V-I characteristic and

$$\frac{E}{1 + \dfrac{J_\delta}{J_{c0}} \ln \dfrac{E}{E_c}} \ll E_\rho$$

for the superconductor with the exponential V-I characteristic. In this case, the V-I characteristic is close to isothermal $(T \sim T_0)$ if

$$E \ll E_c (E_h/E_c)^{n/(n+1)} \quad \text{or} \quad E \ll E_h$$

for the superconductors with the power or exponential V-I characteristics, respectively.

Using the relationship (6.20) or (6.21), one can formulate the so-called condition of the cryogenic thermal stabilization in the framework of the linear approximation (6.15) for a superconducting composite with the real V-I characteristic. (This condition is discussed in detail in Chapter 7.) In this mode, the stable temperature of the superconductor is equal to the critical temperature and the transport current stably flows through the matrix. Such modes satisfy the conditions $T = T_{cB}$ and $\partial E/\partial J > 0$, which lead to the following criteria of the complete thermal stabilization

$$\left(\frac{\eta J_{c0}}{E_c^{1/n}}\right)^2 \left[\frac{S}{hp(T_{cB}-T_0)}\right]^{1-\frac{1}{n}} \left(\frac{\rho_m}{1-\eta}\right)^{1+\frac{1}{n}} < 1, \quad 1 - \frac{J_\delta}{J_{c0}} < \frac{E_0}{E_\rho}$$

for the superconductors with the power or exponential V-I characteristic, respectively. Here, E_0 is a solution to the equation

$$\frac{E_h}{E_0} - \frac{E_0}{E_\rho} = \frac{J_\delta}{J_{c0}} \ln \frac{E_0}{E_c}$$

The written criteria satisfy the limiting transition to the known Stekly criterion (Stekly and Zar 1965)

$$\alpha = \frac{\eta^2 J_{c0}^2 \rho_m S}{hp(1-\eta)(T_{cB}-T_0)} < 1$$

at $n \to \infty$ and $J_\delta \to 0$, respectively.

According to Stekly, the dimensionless parameter α (the so-called thermal stabilization parameter) is equal to the ratio of the Joule heat release in the matrix to the heat flux into the coolant. At the same time, it is not difficult to find that

$$\alpha = E_\rho/E_h$$

In other words, the above characteristic values E_ρ and E_h allow one to give another definition of the thermal stabilization parameter. It is equal to the ratio of the characteristic value of the electric field, above which the current between the superconductor and the matrix is shared, to the characteristic value of the electric field separating the isothermal and nonisothermal modes. Because usually $\alpha \gg 1$ (Wilson 1983), it is necessary to take into account the nonisothermal character of the real operating modes.

The above-written relations allow one to find the conditions for the violation of the stable current distribution in the composite superconductors for the considered nonlinearity types of V-I characteristics. To do this, let us find the corresponding values of the composite differential resistivity. According to (6.18) one may get

$$\frac{\partial E}{\partial J} = \frac{\dfrac{E_\eta}{\eta J_{c0}}\left[1+\dfrac{E}{E_h}\left(\dfrac{E}{E_c}\right)^{1/n}\right]^2}{1+\dfrac{E_\eta}{nE}\left(\dfrac{E}{E_c}\right)^{\frac{1}{n}}-\dfrac{E_\eta}{E_2}\left(\dfrac{E}{E_c}\right)^{\frac{2}{n}}-\dfrac{E}{nE_h}\left(\dfrac{E}{E_c}\right)^{\frac{1}{n}}} \tag{6.22}$$

for the superconductor with the power V-I characteristic and

$$\frac{\partial E}{\partial J} = \frac{\dfrac{E_\eta}{\eta J_{c0}}\left(1+\dfrac{E}{E_h}\right)^2}{1+\dfrac{J_\delta}{J_{c0}}\dfrac{E_\eta}{E}-\dfrac{E_\eta}{E_h}\left(1-\dfrac{J_\delta}{J_{c0}}+\dfrac{J_\delta}{J_{c0}}\ln\dfrac{E}{E_c}\right)} \tag{6.23}$$

for the superconductor with the exponential V-I characteristic.

Formulae (6.22) and (6.23) show that the differential resistivity of a composite depending on the sign of the denominator can be either positive or negative. Therefore, in the considered static zero-dimensional approximation, the boundary between stable and unstable states follows from the condition

$$\partial E/\partial J \to \infty \tag{6.24}$$

both for the superconductors with the power and exponential V-I characteristics. Note that it is not difficult to obtain an equivalent thermal condition of the current instability in the form $\partial T/\partial J \to \infty$, as it follows from (6.20) and (6.21). The condition (6.24) was formulated for the first time by Polak et al. (1973) for LTS with the exponential V-I characteristic.

According to these criteria, the limiting quenching values of the electric field E_q, current I_q and temperature T_q during infinitely low current charging are described by the following relations

$$\frac{\eta}{n}\frac{J_{c0}}{E_c^{1/n}}E_q^{(1-n)/n}-(\eta J_{c0}/E_c^{1/n})^2\frac{S}{hp(T_{cB}-T_0)}E_q^{2/n}=\frac{1-\eta}{\rho_m}\left[\frac{\eta}{n}\frac{J_{c0}}{E_c^{1/n}}\frac{S}{hp(T_{cB}-T_0)}E_q^{(n+1)/n}-1\right]$$

$$I_q=\frac{\eta\dfrac{J_{c0}}{E_c^{1/n}}E_q^{1/n}+\dfrac{1-\eta}{\rho_m}E_q}{1+\dfrac{J_{c0}}{E_c^{1/n}}\dfrac{\eta S}{hp(T_{cB}-T_0)}E_q^{(n+1)/n}}S,\quad T_q=T_0+(T_{cB}-T_0)\frac{\eta+\dfrac{1-\eta}{\rho_m}\dfrac{E_c^{1/n}}{J_{c0}}E_q^{(n-1)/n}}{\eta+\dfrac{hp(T_{cB}-T_0)}{S}\dfrac{E_c^{1/n}}{J_{c0}}E_q^{-(n+1)/n}} \tag{6.25}$$

for the superconductor with the power V-I characteristic and

$$1+\frac{1-\eta}{\eta J_\delta\rho_m}E_q=\frac{\eta J_{c0}SE_q}{hp(T_{cB}-T_0)}\left(\frac{J_{c0}}{J_\delta}-1+\ln\frac{E_q}{E_c}\right)$$

$$I_q=\frac{J_\delta}{J_{c0}}\left(1+\frac{1-\eta}{\eta J_\delta\rho_m}E_q\right)\frac{hp(T_{cB}-T_0)}{SE_q},\quad T_q=T_0+\frac{J_\delta}{J_{c0}}\left(1+\frac{1-\eta}{\eta J_\delta\rho_m}E_q\right)(T_{cB}-T_0) \tag{6.26}$$

for the superconductor with the exponential V-I characteristic.

Let us introduce the dimensionless variables $\varepsilon_q=E_q/E_c$, $i_q=J_q/(\eta J_{c0})$, $\theta_q=(T_q-T_0)/(T_{cB}-T_0)$. In this case, the Expressions (6.25) and (6.26) can be converted to the form

$$\frac{1}{n\varepsilon_2}\varepsilon_q^{1+\frac{1}{n}}+\frac{\varepsilon_1}{\varepsilon_2}\varepsilon_q^{\frac{2}{n}}-\frac{\varepsilon_1}{n}\varepsilon_q^{\frac{1}{n}-1}=1,\quad i_q=\frac{\varepsilon_q^{1/n}+\varepsilon_q/\varepsilon_1}{1+\varepsilon_q^{1+\frac{1}{n}}/\varepsilon_2},\quad \theta_q=\frac{\varepsilon_q^{1+\frac{1}{n}}+\varepsilon_q^2/\varepsilon_1}{\varepsilon_q^{1+\frac{1}{n}}+\varepsilon_2} \tag{6.27}$$

$$\frac{\varepsilon_q}{\varepsilon_1} + \delta = \frac{\varepsilon_q}{\varepsilon_2}(1 - \delta + \delta \ln \varepsilon_q), \quad i_q = \delta \frac{\varepsilon_2}{\varepsilon_1}\left(\frac{\varepsilon_1}{\varepsilon_q} + 1\right), \quad \theta_q = \delta + \frac{\varepsilon_q}{\varepsilon_1} \tag{6.28}$$

for the superconductors with the power and exponential V-I characteristics, respectively. Here,

$$\delta = J_\delta/J_{c0}, \quad \varepsilon_1 = E_p/E_c, \quad \varepsilon_2 = E_h/E_c$$

As discussed above, the stability conditions of a charged current may be both subcritical and supercritical. Let us determine the boundaries of their existence in the approximation under consideration. Putting $\varepsilon_q = 1$, it is easy to find the boundary between the subcritical and supercritical stable values of the electric field. Then the initial parameters must satisfy the equality

$$\varepsilon_2 = \begin{cases} (1+n\varepsilon_1)/(\varepsilon_1+n) \\ \varepsilon_1(1-\delta)/(1+\delta\varepsilon_1) \end{cases}$$

for the superconductor with the power or exponential V-I characteristics, respectively. For a given value of ε_2, the corresponding value of the limiting quenching current is equal to

$$i_q = \begin{cases} \dfrac{n}{n+1} + \dfrac{1}{(n+1)\varepsilon_1} \\ 1-\delta \end{cases}$$

that is it is obviously less than the critical current. Then the current instability will occur in the subcritical region ($E_q < E_c$, $I_q < I_c$) when the conditions are carried out

$$\frac{\eta J_{c0}E_cS}{hp(T_{cB}-T_0)} > \frac{\eta\rho_m J_{c0} + n(1-\eta)E_c}{n\eta\rho_m J_{c0} + (1-\eta)E_c}, \quad \frac{\eta^2 J_{c0}^2 E_c S\rho_m}{hp(T_{cB}-T_0)} > \frac{\eta\rho_m J_\delta + (1-\eta)E_c}{1-J_\delta/J_{c0}}$$

for the superconductors with the power or exponential V-I characteristics, respectively. In terms of the thermal stabilization parameter, the stable current charging regimes are subcritical if

$$\alpha > 1 + \frac{\varepsilon_1^2 - 1}{\varepsilon_1 n + 1}, \quad \alpha > 1 + \frac{1+\varepsilon_1}{1-\delta}\delta$$

for the superconductor with the power or exponential V-I characteristics, respectively. When these conditions are violated, then the value of the electric field, after which the current instability occurs, will exceed E_c. However, in this case, the limiting quenching currents may be either subcritical or supercritical. The latter ($i_q > 1$) exist if $\varepsilon_q > \varepsilon_{q,v}$, where the value $\varepsilon_{q,v}$ follows from the solution of the equation

$$\varepsilon_1 = \varepsilon_{q,v}^{1/n} + \varepsilon_{q,v} - \alpha\varepsilon_{q,v}^{1+(1/n)}$$

for the superconductor with the power V-I characteristic and is equal to

$$\varepsilon_{q,v} = \frac{\delta\varepsilon_1}{\alpha-1}$$

for the superconductor with the exponential V-I characteristic. Then the possible stable states of the superconducting composite will be characterized by the supercritical values of the electric field and subcritical currents at

$$\frac{\eta J_{c0}E_cS}{hp(T_{cB}-T_0)} < \frac{\eta\rho_m J_{c0} + n(1-\eta)E_c}{n\eta\rho_m J_{c0} + (1-\eta)E_c}, \quad \frac{\eta^2 J_{c0}^2 E_c S\rho_m}{hp(T_{cB}-T_0)} < \frac{\eta\rho_m J_\delta + (1-\eta)E_c}{1-J_\delta/J_{c0}}$$

for the superconductor with the power or exponential V-I characteristics, respectively.

The results determining the subcritical and supercritical parameters of the limiting quenching current are presented in Figures 6.8 and 6.9. They were obtained at $\varepsilon_1 = 10^4$, which is typical of superconducting composites as it may be estimated. In this case, the values n and δ were set in such a way that the condition $n = 1/\delta$ was met. As mentioned above, under this condition the power and exponential V-I characteristics touch each other in a point $\{E_c, J_{c0}\}$. It allows one to compare the calculations carried out for superconductors with these nonlinearity types of V-I characteristics.

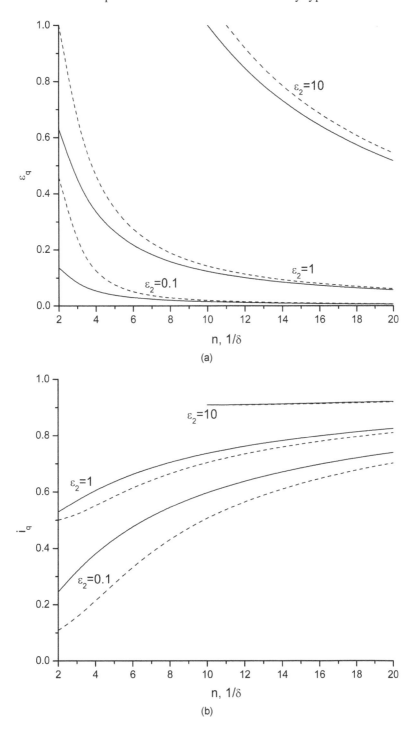

(a)

(b)

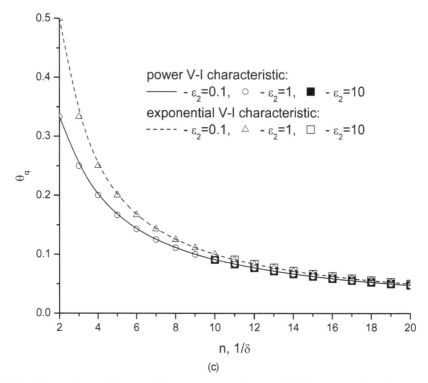

FIGURE 6.8 Dependence of the quenching parameters (electric field (a), current (b) and temperature (c)) on the 'smoothness' parameters of the V-I characteristics: (———) - the power V-I characteristic, (- - - -) - the exponential V-I characteristic.

Figure 6.8 shows the effect of the "smoothness" parameters of V-I characteristic on the subcritical limiting quenching values of the electric field, current and temperature determined for different values of ε_2 (that is for different cooling conditions according to the dimensioning used). In accordance with (6.27) and (6.28) for a given range of permissible parameters, it is easy to obtain the following estimates

$$\varepsilon_q \sim \left(\frac{\varepsilon_2}{n}\right)^{\frac{n}{n+1}}, \quad i_q \sim \frac{n}{n+1}\left(\frac{\varepsilon_2}{n}\right)^{\frac{1}{n+1}}, \quad \varepsilon_q \sim \frac{\delta\varepsilon_2}{1-\delta}, \quad i_q \sim 1-\delta+\delta\ln\frac{\delta\varepsilon_2}{1-\delta}$$

for the superconductor with the power or exponential V-I characteristics, respectively. They show that with a deterioration in the quality of a superconductor, that is with a decrease of n or an increase of δ, the conditions for the current instabilities that will occur in superconducting composites with the power or exponential V-I characteristics are characterized not only by increased values of permissible overheating but also by a noticeable difference in the limiting quenching currents all other conditions being equal. At the same time, as follows from Figure 6.8, the most significant difference in the quenching parameters will be observed when $n = 1/\delta < 10$. Therefore, by taking into account the results of Chapter 2, in both isothermal and nonisothermal states, the values $n = 1/\delta = 10$ describe the boundary between states that can be defined as states with the strong or weak creep.

In general, the results in Figure 6.8 show that the quenching current and electric field cannot only be significantly lower than the conditionally set critical values $\{E_c, J_{c0}\}$ but can also lead to a finite overheating depending on the thermal stabilization parameter. This thermal formation feature of the electrodynamic states of superconducting composites should be taken into account in the experimental determination of the critical parameters of a superconductor, since it is generally assumed that the temperature of the superconductor is equal to the coolant temperature in the subcritical range.

The influence of parameter α on the subcritical and supercritical values of the electric field and currents is shown in Figure 6.9. Note that within the used dimensionless analysis the variation of the values α at a corresponding variation of ε_2 is due to a change in the factor hp/S, in particular due to the change in cooling conditions. As follows from Figure 6.9, the subcritical quenching currents i_q increasingly depend on the V-I characteristic nonlinearity type with an increase of α. Correspondingly, superconducting composites with the power V-I characteristic are more stable than composites with the exponential V-I characteristic. First of all, these regimes will be observed under the nonintensive cooling conditions or when current is charged into the massive composites. The numerical experiments show that this range of stability also depends on the properties of the matrix: the smaller the resistance of the matrix, the larger the range of the stability. At the same time, in the region of the supercritical stability parameters, which exists for small values of α, the influence of the V-I characteristic nonlinearity type on the quenching current becomes negligible.

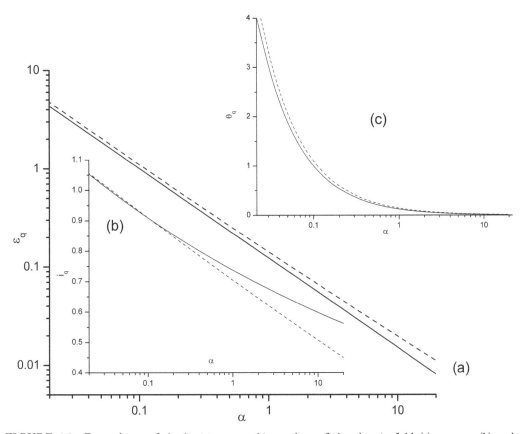

FIGURE 6.9 Dependence of the limiting quenching values of the electric field (a), current (b) and temperature (c) on the thermal stabilization parameter at $n = 1/\delta = 10$: (——) - the power V-I characteristic, (- - - -) - the exponential V-I characteristic.

Using the obtained criteria, let us estimate the influence of the matrix properties, the critical current of the superconductor and the filling factor on the conditions for the occurrence of the current instabilities. To do this, let us consider the problem of determining the current instability boundary of the superconducting composite based on $Bi_2Sr_2CaCu_2O_8$ in a silver matrix (Ag/Bi2212) placed in an external magnetic field $B = 10$ T and cooled at the initial time-up to the temperature of the liquid helium. Varying the values of RRR = $\rho_m(273 \text{ K})/\rho_m(4.2 \text{ K})$, J_{c0} and η, let us define the limiting quenching parameters for the composite with the parameters

$S = 1.3 \times 10^{-4} \, \text{m}^2, \, p = 5 \times 10^{-3} \, \text{m}, \, T_0 = 4.2 \, \text{K}, \, E_c = 10^{-4} \, \text{V/m}, \, T_{cB} = 26.1 \, \text{K}, \, n = 1/\delta = 10$ (6.29)

In this case, the electrical resistivity of the silver as a function of temperature and magnetic induction will be approximated, as above, using the results presented in Seeber (1998) assuming that $\rho_m(273 \, \text{K}) = 1.48 \times 10^{-8} \, \Omega \times \text{m}$.

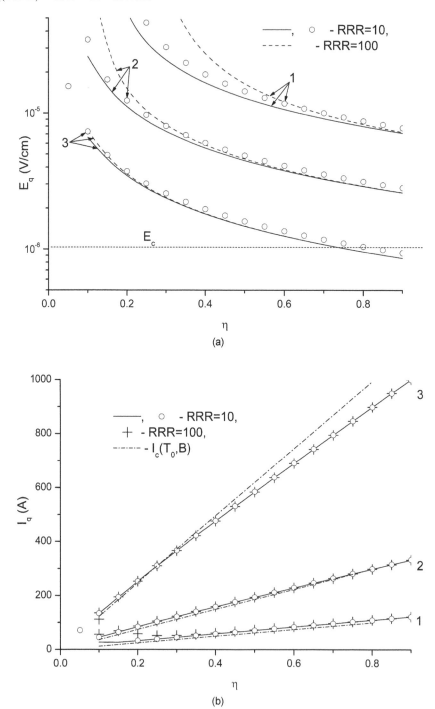

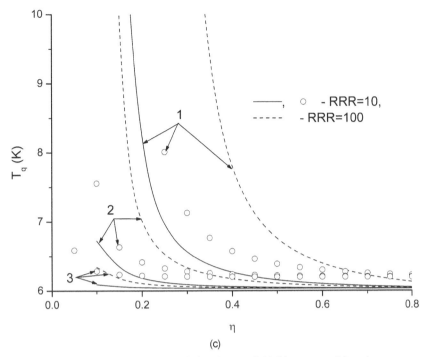

(c)

FIGURE 6.10 Limiting quenching values of the electric field (a), current (b) and temperature (c) as a function of the filling factor: (——, +) - the power V-I characteristic, o - the exponential V-I characteristic, 1 - J_{c0} = 10^8 A/m²; 2 - J_{c0} = 3×10^8 A/m²; 3 - J_{c0} = 10^9 A/m².

In Figure 6.10, the results of calculating the boundary stable values of the electric field, current and temperature depending on the filling factor of the composite with a superconductor at h = 10 W/(m² × K), various values of RRR and J_{c0} are presented. They show the characteristic features of the existence of a stable mode of composite superconductors which were mentioned above. First, the subcritical stable modes ($E_q < E_c$, $I_q < I_c$) may occur, which primarily exist at increased values of the filling factor after exceeding the corresponding value of the critical current density of the superconductor. Second, in a wide range of η variation, the supercritical stable modes may be observed ($E_q > E_c$, $I_q > I_c$), resulting in the noticeable allowable overheating of the composite at relatively low values of J_{c0}. At high values of J_{c0}, the supercritical modes will exist in the region of the small values of the filling factor. Thirdly, at relatively high values of η, intermediate stable modes are possible, the permissible parameters of which are characterized by supercritical values of the electric field and subcritical currents. In particular, at J_{c0} = 10^9 A/m² similar modes occur when the value of η is varied from 0.4 to 0.7, as it follows from Figure 6.10.

The existence of the discussed stable modes also depends on the induction of the external magnetic field. Numerical experiments show that there is a tendency of the occurrence of subcritical stable modes in a wide range of variation η when the external magnetic field increases.

From Figure 6.10 follows an important conclusion according to which the current-carrying capacity of the superconducting composite degrades with an increase in the filling factor or the critical current density. In other words, the currents of instability will not increase in proportion to the increase in the critical current of the composite. This regularity takes place both in the supercritical and subcritical modes.

The degradation effect is associated with the nonisothermal character of the electrodynamic state formation in the composite, that is with the inevitable difference in its temperature from the temperature of the coolant before the instability onset. As follows from (6.16) and (6.17), the temperature effect on the electrodynamic state formation is due to the presence of the following term

$$\frac{\eta J_{c0} S E}{h p (T_{cB} - T_0)} \left(\frac{E}{E_c} \right)^{1/n}$$

for the superconductor with the power V-I characteristic and

$$\frac{\eta J_{c0} S E}{h p (T_{cB} - T_0)}$$

for the superconductor with the exponential V-I characteristic. Their increase leads to the corresponding deviation operating modes from the isothermal ones. As a result, with an increase in the filling factor or the critical current density, the thermal degradation of the current-carrying capacity of the composite becomes inevitable. In addition, the degradation effect will increase with an increase in the cross-section of the composite or with a decrease in the temperature margin of the superconductor (to a greater extent in LTS) or the heat transfer coefficient. The effect of the cross-section of the composite on its current-carrying capacity will be discussed below by taking into account the size effect.

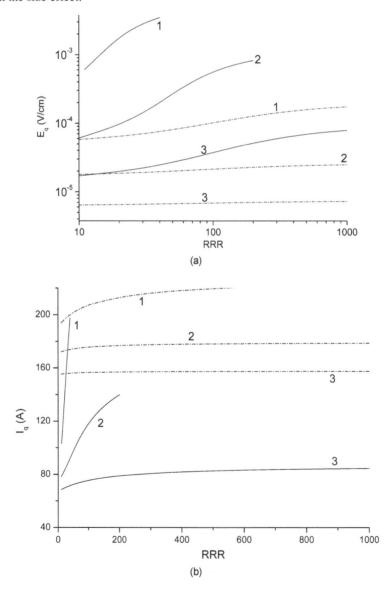

(a)

(b)

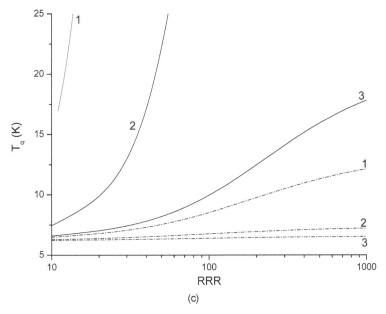

(c)

FIGURE 6.11 Limiting quenching values of the electric field (a), current (b) and temperature (c) as a function of the matrix resistivity: 1 - h = 100 W/(m² × K); 2 - h = 30 W/(m² × K); 3 - h = 10 W/(m² × K); - - - - η = 0.2; —— - η = 0.5.

The influence of the matrix resistivity on the maximum permissible values of the electric field, current and temperature in the superconducting composite with the power V-I characteristic is shown in Figure 6.11. The calculations were carried out at J_{c0} = 1.52 × 10⁸ A/m², different values of a heat transfer coefficient and filling factor. It can be seen that the supercritical stable modes will exist under intensive cooling conditions or in composites with a low-resistive matrix. Besides, depending on the values η and h, there are areas of the stability where the variation of the RRR can significantly affect the stability conditions of the charged current. These modes are typical for the composites with a relatively low filling factor. The probability of their occurrence increases with the improvement of the heat transfer conditions. In these cases, the allowable increase in the electric field and temperature is quite significant. Namely, the permissible overheating can exceed 10 K due to the supercritical values of the induced electric field. Therefore, to correctly determine the boundary of the stable states, it is necessary to take into account the corresponding properties variation of the superconductor and matrix with the temperature. This conclusion should also be taken into account when measuring the critical properties of a superconductor, when it is usually a priori assumed that the temperature in the composite is equal to the temperature of the coolant.

Thus, the stability conditions of the currents charged into composite superconductors, whose V-I characteristics are described by power or exponential equations, are not equivalent in general. The difference increases with the deterioration of the superconducting material quality and the conditions of the thermal stabilization of the composite. In this case, the most noticeable difference will be observed in the superconductors with a strong creep ($n < 10$, $J_\delta/J_{c0} > 0.1$).

The limiting quenching values of the electric field and current may be either subcritical or supercritical. Their existence depends on the properties of the superconductor and matrix and the conditions of the heat exchange with the coolant. In particular, the subcritical modes will be observed for composites with a high value of the filling factor or in the nonintensive cooling conditions. Stable supercritical current modes will take place at small values of the filling factor or thermal stabilization parameter ($\alpha < 1$). In this case, the supercritical stable modes are characterized by the high permissible overheating of the composite. Thus, a priori set critical values of the electric field and current do not define the boundary of the stable states of the composite superconductors.

This means that these values do not correspond to their physical meaning. To determine the stability boundary of the charged current, it is necessary to find the quenching parameters. Only they define the current-carrying capacity of technical superconductors correctly.

The inevitable overheating of the composite before the occurrence of the current instability leads to the thermal degradation of its current-carrying capacity. Correspondingly, the currents of instability do not increase proportionally to the increase of its critical current. The degradation effect significantly affects the stability conditions of the charged currents.

6.3   MULTISTABLE STATIC CURRENT STATES OF SUPERCONDUCTING COMPOSITES BASED ON HTS AND CURRENT INSTABILITIES AT VARIOUS OPERATING TEMPERATURES

The regularities of the current instability in superconducting composites discussed in the previous section were formulated under the assumption of linear dependence of the critical current density on temperature and the resistivity of the matrix that does not vary with temperature. These assumptions are practically true for LTS. However, the dependence of the critical current density on the temperature of HTS ceases to be linear at high magnetic fields and operating temperatures, as already noted. In particular, it can be described by a substantially nonlinear dependence of the form (1.8) for Bi-based superconductors. In addition, it is necessary to take into account that the silver resistivity, which is used in the manufacture of HTS tapes, depends on the temperature at the operating temperatures higher than 20 K. Figure 6.12 depicts the dependencies on the temperature the values ρ_m and $\rho_m^{-2}\partial\rho_m/\partial T$ for the silver at $B = 10$ T and various values of RRR = $\rho_m(273$ K$)/\rho_m(4.2$ K$)$. These dependencies are calculated at $\rho_m(273$ K$) = 1.48 \times 10^{-8} \Omega \times$ m according to Seeber (1998). In this connection, let us study the influence of nonlinear temperature dependencies on the critical current density and resistivity of the matrix on the current instability conditions of Ag/Bi2212.

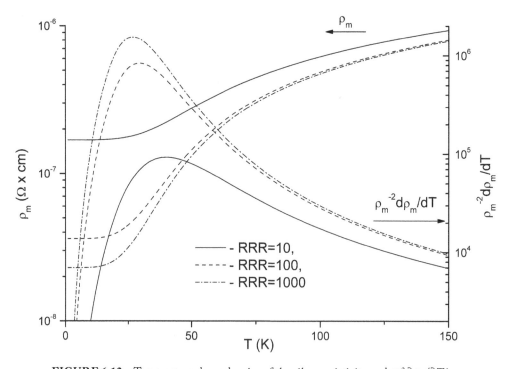

FIGURE 6.12   Temperature dependencies of the silver resistivity and $\rho_m^{-2}\partial\rho_m/\partial T$ in.

Let find the differential resistivity of the superconducting composite with the power V-I characteristic but by taking into account the nonlinear variation with the temperature of the quantities J_c, ρ_m, h and n using zero-dimensional approximation. For the problem described by equations (6.11)-(6.13), it can be written in the form

$$\frac{\partial E}{\partial J}(T) = \frac{1 + \dfrac{ES}{hp}\xi}{\dfrac{1-\eta}{\rho_m} + \dfrac{\eta J_c}{nE}\left(\dfrac{E}{E_c}\right)^{1/n} - \dfrac{JS}{hp}\xi} \tag{6.30}$$

where

$$\xi = \frac{\eta\left|\dfrac{\partial J_c}{\partial T}\right|\left(\dfrac{E}{E_c}\right)^{1/n} + E\dfrac{1-\eta}{\rho_m^2}\dfrac{\partial \rho_m}{\partial T} + \eta J_c\dfrac{\partial n}{\partial T}\left(\dfrac{E}{E_c}\right)^{1/n}\ln\left(\dfrac{E}{E_c}\right)^{1/n^2}}{1 + \dfrac{EJS}{ph^2}\dfrac{dh}{dT}}$$

This expression demonstrates the main regularities that affect the growth of the nonisothermal V-I characteristic of superconducting composites and the causes that lead to the nonstandard (multistable) conditions for the occurrence of the current instabilities. Indeed, as follows from Formula (6.30), the shape of the superconducting composite V-I characteristic depends on the peculiarities of the joint variation of the values $\eta|\partial J_c/\partial T|$, $(1-\eta)\rho_m^{-2}\partial\rho_m/\partial T$, $h^{-2}dh/dT$ and $\partial n/\partial T$ $\ln(E/E_c)$. As a result, its differential resistivity may be positive at high values of the electric field and temperature of the composite. In other words, unlike the conventional shape of the static V-I characteristic, when there are only one stable ($\partial E/\partial J > 0$) and unstable ($\partial E/\partial J < 0$) branches, as discussed above, the formation of the V-I characteristic of HTS composite can be characterized by the occurrence of the additional stable branches at high values of the electric field and permissible overheating. Accordingly, their existence will depend on the variation with temperature or external magnetic field of the critical current and the resistivity of the matrix, cooling conditions and filling factor. Let us discuss the possible mechanisms leading to the formation of the multistable V-I characteristics in more detail.

First, the stability of the charged current will be improved with a decrease of $\eta|\partial J_c/\partial T|$. Accordingly, it is necessary to take into account the corresponding variation of the values $\partial J_c/\partial T$ with the temperature over the entire range of the superconductor temperature variation up to its critical temperature when determining the stable current modes. As follows from Figure 4.21, the values $\partial J_c/\partial T$ of Bi2212 have nonmonotonic temperature dependencies at high magnetic fields. Moreover, there is a characteristic temperature when the values $\partial J_c/\partial T$ begin to decrease sharply and then becomes very small. Such a variation in $\partial J_c/\partial T$ can lead to an increase in the stability range of the transport current in the high operating temperature and additional branches with a positive slope $\partial E/\partial J$ may appear on the V-I characteristic of the superconducting composite.

The influence of the stabilizing matrix on the V-I characteristic of the composite is described by the factor $(1-\eta)\rho_m^{-2}\partial\rho_m/\partial T$. This means that the shape of the V-I characteristic depends on both the value of the matrix resistivity and the temperature dependence of the ρ_m. As follows from Figure 6.12, this effect will be most noticeable in the silver matrix when $T > 20$ K and at high values of RRR. Accordingly, it is necessary to take into account the temperature dependence ρ_m to correctly describe the V-I characteristic in these cases.

The role of the coolant in the formation of stable modes at high electric fields depends on the value of $h^{-2}dh/dT$. Taking into account the existence of a boiling crisis when the superconducting composite is cooled by a liquid refrigerant, it is easy to understand that the transition from nucleate boiling to film one may be accompanied by a sharp increase in its differential resistivity. Hence, the stability conditions of the transport current deteriorate accordingly, which leads to the occurrence of current instability.

The effect of temperature dependence $n(T)$ depends on the value of $\partial n/\partial T \ln(E/E_c)$. The condition $\partial n/\partial T < 0$ occurs in superconductors due to the magnetic flux-creep. Therefore, the dependence $n(T)$ will have a destabilizing effect on the charged current at $E < E_c$, and conversely, cause a stabilizing effect at $E > E_c$. In other words, $E(J)$-branch with a positive slope may appear on the V-I characteristic of the superconducting composite at the supercritical electric fields.

Thus, the generally accepted shape of the static V-I characteristic of the superconducting composite may be noticeably changed if one takes into account the nonlinear character of the temperature dependencies J_c, ρ_m, h and n.

Let us write the corresponding temperature dependence of the differential resistivity of composite by taking into account only the temperature dependence of J_c and ρ_m for the simplicity of the performed analysis. According to (6.30), it can be written in the form

$$\frac{\partial E}{\partial J}(T) = \frac{1+\left[\eta\left|\frac{\partial J_c}{\partial T}\right|\left(\frac{E}{E_c}\right)^{1/n} + E\frac{1-\eta}{\rho_m^2}\frac{\partial \rho_m}{\partial T}\right]\frac{ES}{hp}}{\frac{1-\eta}{\rho_m}+\frac{\eta}{n}\frac{J_c}{E}\left(\frac{E}{E_c}\right)^{1/n} - \left[\eta\left|\frac{\partial J_c}{\partial T}\right|\left(\frac{E}{E_c}\right)^{1/n} + E\frac{1-\eta}{\rho_m^2}\frac{\partial \rho_m}{\partial T}\right]\frac{JS}{hp}} \tag{6.31}$$

This formula allows determining the range of the stable values of the electric field for a given density of the transport current J. Then, the boundaries of the stable modes are a consequence of the existence of multi-valued solutions of the following nonlinear equation

$$\frac{1-\eta}{\rho_m(T,B)}+\frac{\eta}{n}\frac{J_c(T,B)}{E_q}\left(\frac{E_q}{E_c}\right)^{1/n} - \left[\eta\left|\frac{\partial J_c(T,B)}{\partial T}\right|\left(\frac{E_q}{E_c}\right)^{1/n} + E_q\frac{1-\eta}{\rho_m^2(T,B)}\frac{\partial \rho_m(T,B)}{\partial T}\right]\frac{JS}{hp} = 0$$

In this case, the density of the transport current in the composite and its temperature as a function of the electric field are described by equalities

$$J = \eta J_c(T,B)(E/E_c)^{1/n} + (1-\eta)E/\rho_m(T,B),$$

$$T = T_0 + ES[\eta J_c(T,B)(E/E_c)^{1/n} + (1-\eta)E/\rho_m(T,B)]/hp .$$

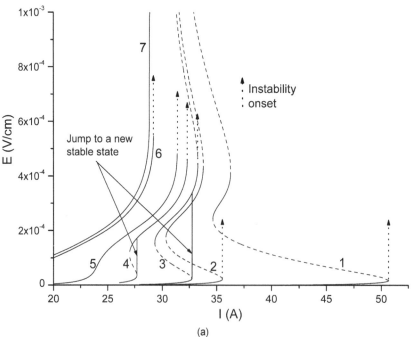

(a)

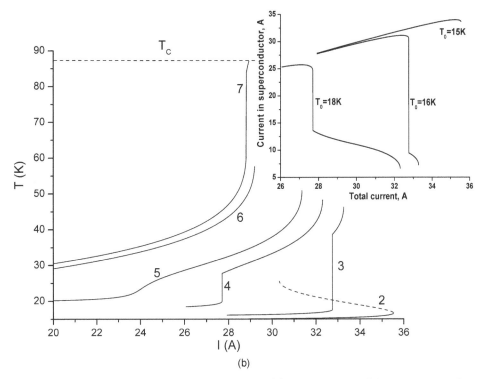

(b)

FIGURE 6.13 Voltage-current (a), temperature-current (b) characteristics of the composite and current sharing (in the inset) during multistable current modes: 1 - T_0 = 10 K; 2 - T_0 = 15 K; 3 - T_0 = 16 K; 4 - T_0 = 18 K; 5 - T_0 = 20 K; 6 - T_0 = 25 K; 7 - T_0 = 26 K, (———) - stable states, (- - - -) - unstable states.

The voltage-current and temperature-current characteristics of the nonintensively cooled composite Ag/Bi2212 are plotted in Figure 6.13. Here, the stable and unstable branches are shown by the solid and dashed curves, respectively. The calculations were carried out according to (6.29) at h = 10 W/(m^2 × K) and various temperatures of the coolant exceeding the temperature of liquid helium. In this case, the critical current density was calculated by the Formula (1.8) at

$$J_0 = 5.9 \times 10^8 \, \text{A/m}^2, \, T_c = 87.1 \, \text{K}, \, B_{c0} = 465 \, \text{T}, \, B_0 = 1 \, \text{T}, \quad (6.32)$$
$$\alpha = 10.3, \, \beta = 6.76, \, \gamma = 1.73, \, \chi = 0.27.$$

These parameters J_c describe the experimentally measured dependencies $J_c(T, B)$ discussed in Seto et al. (2001).

The presented results show the possible existence of the nontrivial shapes of V-I characteristics. Namely, the following characteristics may be observed:

1. The multistable V-I characteristic with two stable branches in which the currents in the second stable branch do not exceed the instability current (Curves 1 and 2).
2. The multistable V-I characteristic with two stable branches, when the stable currents in the second branch exceed the allowable currents in the first branch (Curves 3 and 4).
3. The monostable V-I characteristic. At its increasing, a significant stable temperature increase of the composite will be observed, but its maximum allowable value is below the critical temperature of the superconductor (Curves 5 and 6).
4. The monostable V-I characteristic. The stable shape of it takes place over the entire range of temperature variation, that is from the temperature of the coolant to the critical temperature of the superconductor (Curve 7).

The discussed types of multistable V-I characteristics are the result of two effects: a stable increase in the superconductor temperature affecting the values of $|\partial J_c/\partial T|$ and $\rho_m^{-2}\partial\rho_m/\partial T$ and the current sharing between the superconductor and matrix. Indeed, for the modes presented in Figure 6.13 by Curves 3 and 4, there is an abrupt redistribution of the current between the superconductor and matrix (Figure 6.13c) during the transition from the first stable branch to the second one. But at the same time, the composite does transit to a normal state even despite a significant increase in its temperature. In this case, the matrix plays a stabilizing role in the formation of a new stable state. As a result, the stable thermo-electrodynamic states of HTS composites can be observed at the current loads significantly exceeding the conventionally set values of E_c and J_c. Let us emphasize that the jump-like redistribution of the current is a consequence of using the static approximation ($dI/dt \to 0$). Below, we will discuss the change in the multistable V-I characteristic when current is charged at a finite rate.

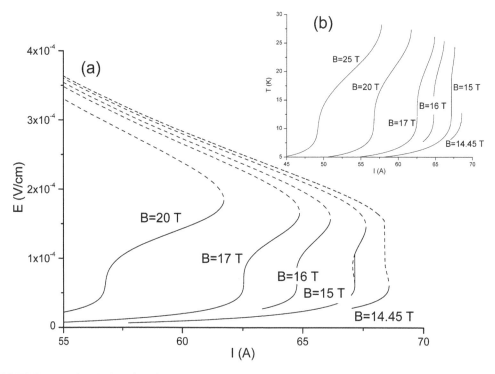

FIGURE 6.14 Stable (——) and unstable (- - - -) branches of the voltage-current (a) and temperature-current (b) characteristics of the superconducting composite at varying the external magnetic field.

Nontrivial E-I and T-I characteristics of the superconducting composites, which have additional areas of the stability, can also be observed by varying the external magnetic field due to the corresponding variation in the properties of the superconductor and matrix. Figure 6.14 demonstrates the possible modification of the voltage-current and temperature-current characteristics of the above-considered composite with a low-resistance coating (RRR = 1,000). The calculations were performed at $T_0 = 4.2$ K and $h = 10$ W/(m$^2 \times$ K). They show the existence of the characteristic values of the external magnetic field induction. The growth of E-I and T-I characteristics depend on them. Namely, at $B < 14.45$ T, the latter has one stable and one unstable branch. When $B > 14.45$ T, there is an additional area of the stability, which merges with the main area of the stability at $B > 16$ T. As a result, the current modes of the considered nonintensively cooled composite are stable at high magnetic fields and temperatures exceeding 25 K when the allowable increase in the electric field is more than two orders of magnitude higher than a priori defined value E_c.

As already noted, the variation of the cooling conditions may also lead to the occurrence of the multistable E-I and T-I characteristics. The corresponding dependencies are presented in Figure 6.15, which were calculated for the composite under consideration at $T_0 = 4.2$ K, $B = 10$ T, RRR = 10 and various values of the heat transfer coefficient. As above-mentioned, the stable branches E-I and T-I characteristics are shown by the solid curves and unstable characteristics are depicted by the dashed curves. It is seen that the E-I and T-I characteristics have a standard form at the nonintensive cooling and the permissible overheating does not exceed 10 K. However, the stable operating temperature range of the composite expands at its more intensive cooling. As a consequence, an additional stable branch takes place at $h = 30$ W/(m^2 × K). For these states, its stable temperature increase exceeds by 20 K, when, first of all, the values $\partial J_c/\partial T$ decrease significantly, as follows from (1.8). As a consequence, the additional branch merges with the main one at $h = 100$ W/(m^2 × K), and a significant stable overheating of the composite is observed. Therefore, E-I and T-I characteristics are monostable at intensive cooling ($h > 100$ W/(m^2 × K)) and they are characterized by stable supercritical states.

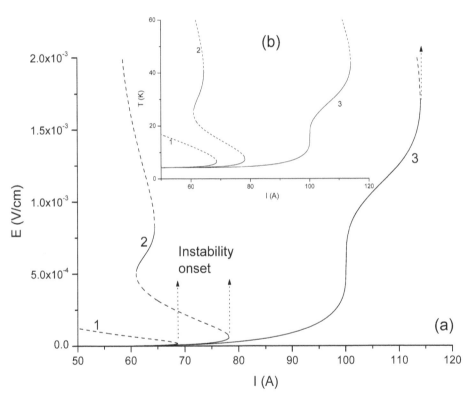

FIGURE 6.15 Influence of the heat transfer coefficient on the voltage-current (a) and temperature-current (b) characteristics of Ag/Bi2212 composite: 1 - $h = 10$ W/(m^2 × K), 2 - $h = 30$ W/(m^2 × K), 3 - $h = 100$ W/(m^2 × K).

The existence of the additional branch of the stability can also be explained by a corresponding variation of the Joule heat release in the composite. The possible temperature dependencies are shown in Figure 6.16. They were obtained for the composite considered above at $I = 32$ A, $T_0 = 4.2$ K, $B = 10$ T, RRR = 10 for two filling-factor values. The solid curves show the temperature variation of the total amount of the heat released in the composite ($G = EJS$). The dashed curves show the heat release in the superconductor ($G_s = \eta J_s ES$) and the matrix ($G_m = (1 - \eta)J_m ES$). The dash-dotted straight lines describe the heat transfer to the coolant ($W = hp(T - T_0)/S$) for some values of h and T_0.

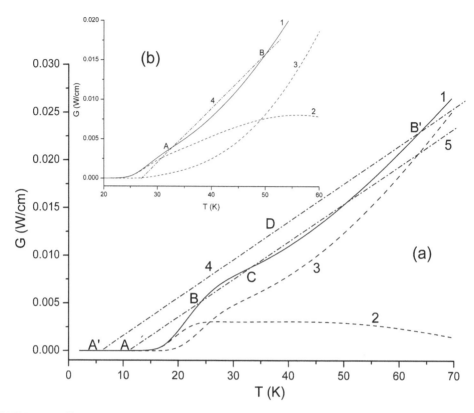

FIGURE 6.16 Temperature dependencies of the heat flux into the coolant (Curves 4 and 5) and the Joule heat release in the composite (Curve 1), superconductor (Curve 2) and matrix (Curve 3) under different cooling conditions of the composite and values of the filling factor: (a) - $\eta = 0.2$, (b) - $\eta = 0.5$.

It can be seen that the dependencies of G and W may have several intersection points due to the nonmonotonic character $G(T)$. Accordingly, a new stable state may arise in the intermediate temperature range. In the considered cases, it corresponds to the point C along with the points A and A', which standardly determine the equilibrium states. Here, the points B, B' and D determine the nonequilibrium ones. The nonmonotonous character of $G(T)$ is due to the corresponding variation of the heat release in the superconductor and matrix, which in turn depends on the value of the filling factor. At the same time, it follows from Figure 6.16 that the lower η, the more nonmonotonic the dependence is $G(T)$. Therefore, the occurrence of multistable states is most likely in the composites with a low superconductor amount, when the current sharing effect underlying the existence of multistable states plays a significant role. Figure 6.16 also proves the existence of the boundary values for the coolant temperature when the multistable states do not arise below or above of these values, even despite the nonmonotonic character of $G(T)$. This explains the existence of additional stable branches depending on the initial parameters of the composite and conditions for its cooling.

The existence of the multistable states will also be reflected in the character of the occurrence and development of instabilities. First, for the modes described by Curves 1 and 2 in Figure 6.13, the current instability in the first region means the common border of the stable states of these modes. After exceeding this branch, all charged currents become unstable. Second, the stable current modes with maximum currents in the second stable branches of the V-I characteristics that are depicted by Curves 3 and 4 in Figure 6.13 will be accompanied by the current redistribution between the superconductor and matrix with a corresponding stable increase in its temperature after the charge of currents that exceed the maximum current in the first branches of the V-I characteristics. In this case, the general boundary of stable states is determined by the instability current of the second region. Third, it is possible to combine both areas of stability. Then, the stability conditions the monostable V-I characteristic will be characterized by a noticeable permissible overheating and electric field (Curves 5 and 6 in Figure 6.13). Finally, the current instability may be absent during the stable increase in the temperature of the superconductor from the coolant temperature up to the critical temperature (Curve 7). Let us emphasize that the stabilizing role of the heat capacity of the superconductor and matrix do not underlie the existence of such stable regimes. They arise due to the stable current sharing between the superconductor and matrix.

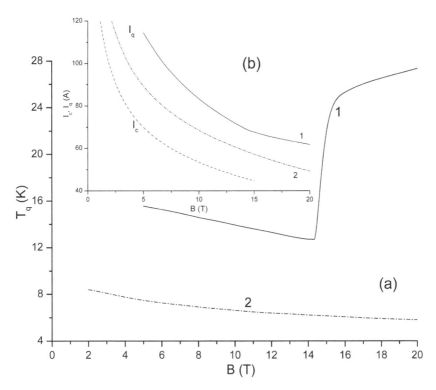

FIGURE 6.17 Influence of an external magnetic field on the quenching values of temperature (a) and current (b) in multistable (1 - RRR = 1,000) and monostable (2 - RRR = 10) states.

In Figure 6.17 the results of determining the boundary of the stable current states depending on the induction of an external magnetic field are presented. They were calculated at T_0 = 4.2 K and two characteristics RRR values corresponding to the matrix with low and high resistance. The corresponding dependence of the critical current on the magnetic field is also shown here.

The results show that the multistability that exists in a composite with a well-conducting matrix (Curve 1) leads to a nonmonotonic variation in the quenching temperature and, as the calculations show, in the quenching electric field. This is due to the current sharing between the superconductor and matrix during the transition from the first stable region to the second. At the same time, the permissible temperature of the composite decreases with an increase in the magnetic field induction before the jump in the $T_q(B)$ dependence and increases after a jump. Let us emphasize the importance of this feature. It shows that not only the temperature rise of the composite is accompanied by a corresponding decrease in the values $|\partial J_c/\partial T|$ but also a variation in the induction of an external magnetic field, interrelatedly changing the critical current of the superconductor and the matrix resistance, can also lead to multistable states, even if the permissible overheating decreases. Consequently, there is a thermomagnetic mechanism for the occurrence of multistable states in general, when their appearance is caused by a change in the properties of the superconductor and matrix as the temperature or magnetic field induction is changed.

The importance of the noted reasons leading to an increase in the range of stable current states of HTSC composites due to the occurrence of multistable states should be emphasized. The discussed nontrivial modes show that the multistable modes can lead to the states whose stability is observed throughout the entire range of temperature variations of the composite, namely from the coolant temperature to the critical temperature of the superconductor. Their appearance occurs due to the thermal change in the properties of the superconductor and the matrix. Since in the region of the high electric fields the significant stable overheating of the composite can be observed, then this has to be taken into account in experiments to correctly determining the properties of superconductors.

6.4 MECHANISMS OF THE CURRENT INSTABILITIES IN HTS COMPOSITES COOLED BY A LIQUID COOLANT

The results discussed above showed that noticeable permissible overheating may take place at the full penetration of the transport current into the composite. However, the formulated above conditions for the occurrence of the current instabilities allow to find the stable modes if the heat transfer coefficient is constant. Such modes correspond to the cooling conditions when cryocoolers or gaseous refrigerants are used or an indirect cooling of the current-carrying elements due to the conductive mechanism of the heat redistribution between them. However, as follows from (6.30), if the current-carrying element of a superconducting magnet will be cooled by a liquid coolant, then it is necessary to take into account the peculiarities of the change in the heat-transfer capacity of the refrigerant, in particular the existence of a boiling crisis. Besides, the peculiarities of the current sharing between the superconductor and matrix can also affect the stability conditions. From the results discussed above, it follows that the consideration of this mechanism will be noticeable when determining the stability conditions of HTS composites cooled by refrigerants with high temperatures at which the transition from nucleate boiling to film one takes place. Let us discuss the characteristic regularities of the current instabilities onset in a composite superconductor cooled, in particular by the liquid hydrogen.

Consider the problem of determining the boundary of the stable current modes of the superconducting composite Ag/Bi2212 in the framework of zero-dimensional model (6.11)-(6.13). The critical current density of a superconductor is described by Formula (1.8) as a function of temperature and magnetic field using the parameters (1.9). The results presented below were obtained at $B = 10$ T, $S = 10^{-6}$ m^2, $p = 5 \times 10^{-3}$ m, $n = 15$ and various values of η and RRR by taking into account the dependence of the matrix resistance on temperature and magnetic field in the same way as assumed above. The heat flux into the liquid hydrogen is described by the expression (Brentari and Smith 1965)

$$q(T)\,[\text{W/m}^2] = \begin{cases} 0.66 \times 10^4 (T - T_0)^{\nu_1}, & T \le T_0 + \Delta T_{cr} \\ 0.024 \times 10^4 (T - T_0)^{\nu_2}, & T > T_0 + \Delta T_{cr} \end{cases} \tag{6.33}$$

at $T_0 = 20$ K, $\Delta T_{cr} = 3$ K, $\nu_1 = 2.6$, $\nu_2 = 1.1$.

The temperature-current and voltage-current characteristics of the composite calculated at RRR = 10 for nucleate (Curves 1, 2, 3) and film (Curves 1′, 1″, 1‴, 2′, 3′) boiling regimes are shown in Figure 6.18. Stable and unstable states in the film boiling regime are limited by the corresponding currents I'_f, I_{f1}, I_{f2}, I''_f and I'''_f (Figure 6.18a). The temperature rise in the mode 1 is depicted in Figure 6.18b. The results presented show that the current modes are more unstable during film boiling regime. Therefore, the instability of the charged current is a consequence of the transition from the nucleate boiling regime to the film in the considered operating mode. As a result, the stable increase in temperature with the current is monotonic from T_0 to $T_0 + T_{cr}$ (Figure 6.18b). The corresponding quenching current is presented on the inset to Figure 6.18b. It should be noted that the intensive cooling of the composite in the nucleate boiling regime leads to the stable supercritical states.

Figure 6.18 also shows that the stable states of a composite cooled by a liquid refrigerant may be multistable. An additional area of stability occurs in the film boiling regime. The stable jumps of the electric field, current and temperature without the transition of the superconducting composite to the normal state may occur in the current range that is not close to the quenching current. The transition from one stable state to another one may be caused by the action of some external thermal perturbation. These states are characterized by high stable values of the electric field and temperature. Moreover, the temperature may stably increase up to the critical temperature of the superconductor. Accordingly, an additional stable heat-generating region may appear in the magnetic system. It exists in the range of currents from $I_{f,1}$ to $I_{f,2}$ (Curve $1'''$ in Figure 6.18b) and will result in substantial magnet heating.

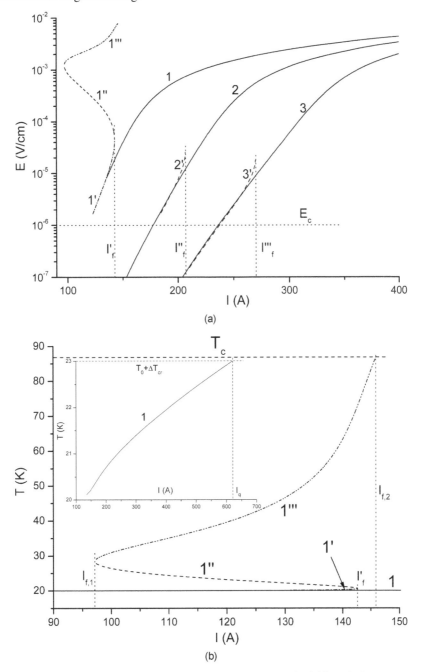

FIGURE 6.18 Monostable and multistable variation of the electric field (a) with the current at different values of the filling factor (1 - η = 0.2, 2 - η = 0.3, 3 - η = 0.4) and the temperature of the composite at η = 0.2 (b) during various cooling regimes.

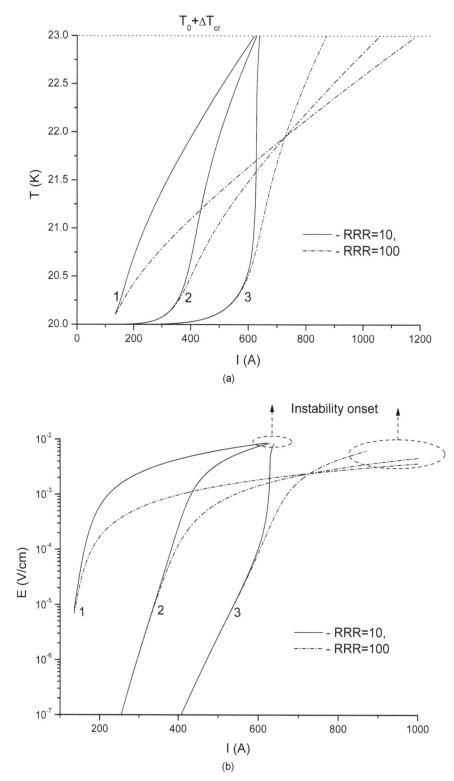

FIGURE 6.19 Dependancies of the composite temperature (a) and the electric field (b) on the current during nucleate boiling of hydrogen: 1 - $\eta = 0.2$, 2 - $\eta = 0.5$, 3 - $\eta = 0.8$.

Figure 6.19 shows the formation of the stable branches of the voltage-current and temperature-current characteristics of the composite in the nucleate boiling regime calculated for different values of RRR and η. It is seen that the occurrence of new stable states is possible, which will be characterized by the existence of a common nodal point. Its location does not depend on the filling factor. But it affects the character of the increase in static dependencies $E(I)$ and $T(I)$. Indeed, if the nodal point is absent, then the stable values of the temperature and the electric field decrease with an increase in the filling factor before the occurrence of the instability. However, when the nodal point exists, other regularities will be observed before the instability. Namely, the higher η, the higher the stable values of the temperature and electric field. This feature affects the conditions for the occurrence of the current instability. The corresponding results of determining the quenching values of the electric field E_q and current I_q are presented in Figure 6.20. They demonstrate the existence of the following stable modes.

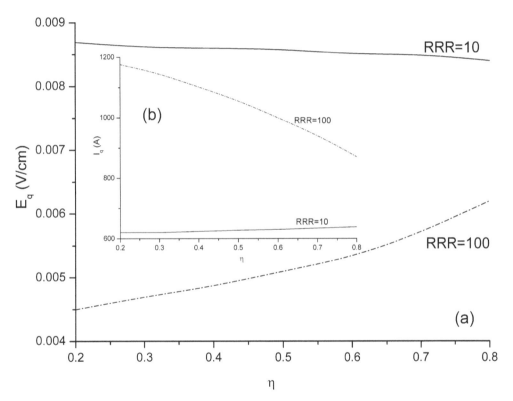

FIGURE 6.20 Dependencies of the maximum allowable values of the electric field (a) and current (b) on the filling factor.

First, if the nodal point is absent (RRR = 10), then the values E_q slightly but monotonously decrease with an increase in the filling factor and I_q increase accordingly. Second, if there is the nodal point, these dependencies have the opposite tendency: the higher η, the more noticeable the increase of E_q and the decrease of I_q.

To explain the reason for the existence of these regimes, let us consider the variation of the matrix resistance and the superconductor with the current. According to the model used, it is easy to find

$$J = \eta \frac{E}{\rho_s} + (1-\eta) \frac{E}{\rho_m}$$

where ρ_s is the resistivity of the superconductor, which is equal to

$$\rho_s = \frac{E_c}{J_s}\left(\frac{J_s}{J_c}\right)^n$$

These relationships show that the values of the density of the charged current and the induced electric field will not depend on the filling factor if $\rho_s = \rho_m$. In other words, the nodal point occurs when the current density in the superconductor is equal to the current density in the matrix, that is $J_s = J_m = J$. In this case, the equality will take place

$$\rho_m(T, B) = \frac{E_c}{J}\left(\frac{J}{J_c(T, B)}\right)^n$$

Then, it is easy to find the temperature at the nodal point T_N solving equation

$$\left[\frac{h_2(T_N - T_0)^{\upsilon_2} p}{4S\rho_m(T_N, B)}\right]^{\frac{n-1}{2}} = \frac{\rho_m(T_N, B)J_c^n(T_N, B)}{E_c}$$

Therefore, the nodal point exists if

$$T_N > T_0 + \Delta T_{cr}$$

The corresponding values of the current density and electric field are equal to

$$J\big|_{T=T_N} = \sqrt{h_2(T_N - T_0)^{\upsilon_2} \frac{p}{S\rho_m(T_N, B)}}, \quad E\big|_{T=T_N} = \sqrt{h_2(T_N - T_0)^{\upsilon_2} \frac{p\rho_m(T_N, B)}{S}}$$

The first of these quantities allows one to formulate another reason underlying the occurrence of a nodal point. Let us rewrite it as

$$J^2\big|_{T=T_N} \rho_m(T_N, B) = h_2(T_N - T_0)^{\upsilon_2} p/S$$

Consequently, the nodal point exists at current at which the heat release in the matrix (or in the superconductor) is equal to the heat flux into the refrigerant. This definition shows that the mechanism of the current sharing plays a significant role in the existence of such modes. Therefore, they will be absent when cooling HTS composites with liquid helium (these modes are discussed below for YBCO-based tape) but will necessarily be observed when using liquid nitrogen. The occurrence of these states must be taken into account when determining the critical properties of superconducting materials by taking into account the corresponding influence of the current sharing mechanism.

The discussed features of the current instabilities in HTS composites cooled by a liquid coolant are due to the transition from nucleate boiling regime to film one. However, the solid Curve 3 in Figure 6.19 shows that another mechanism is possible for the occurrence of the current instability in the nucleate cooling regime when the condition (6.24) may be compliant. Figure 6.21 demonstrates the existence of such states. Here, the solid curves correspond to stable states and the dashed curves describe the unstable states defining according to the condition (6.24). It is seen that there is a range of magnetic fields, when the current instability occurs before the boiling crisis (Figure 6.21a). In this case, even multistable states may occur (Figure 6.21b).

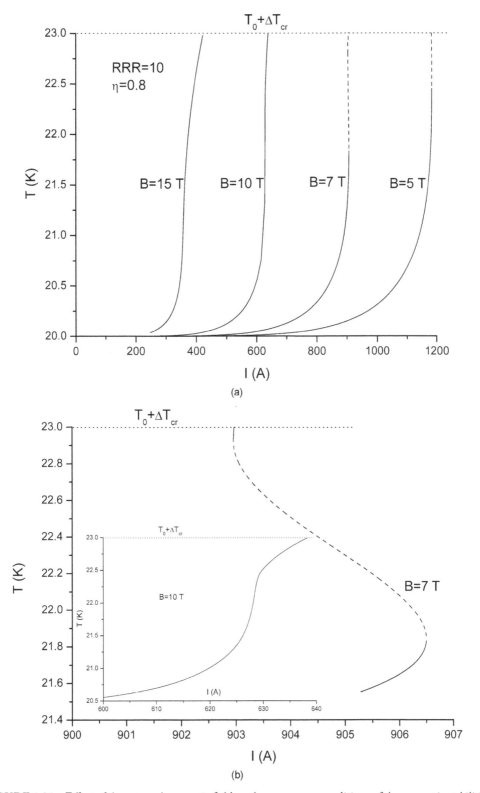

FIGURE 6.21 Effect of the external magnetic field on the occurrence conditions of the current instabilities in superconducting composites cooled by liquid coolant.

Let us find a condition separating these mechanisms from each other for the superconducting composite with arbitrary temperature dependencies $J_c(T)$, $\rho_m(T)$ and $q(T)$. Excluding the electric field in the system of equations (6.11)-(6.13) and (6.33), one can obtain

$$J = \eta J_c(T)\left[\frac{pq(T)}{JSE_c}\right]^{1/n} + \frac{1-\eta}{\rho_m(T)}\frac{pq(T)}{JS}$$

Then, according to the condition (6.24), one may find the temperature T_q, which follows from the solution of equation

$$\frac{q(T_q)p}{S}\left(\frac{1-\eta}{\eta}\frac{E_c}{J_c(T_q)\rho_m(T_q)}\right)^{\frac{n+1}{n-1}} = \eta J_c(T_q)E_c[\xi(T_q)]^{\frac{n+1}{n-1}}(1+\xi(T_q)).$$

Here,

$$\xi(T) = \left[\frac{1}{J_c(T)}\frac{dJ_c}{dT} + \frac{1}{nq(T)}\frac{dq}{dT}\right]\bigg/\left[\frac{1}{\rho_m(T)}\frac{d\rho_m}{dT} - \frac{1}{q(T)}\frac{dq}{dT}\right].$$

If this temperature is exceeded, unstable states occur. Consequently, when cooling HTS composite with a liquid refrigerant, current instability occurs before the boiling crisis if $T_q < T_0 + T_{cr}$.

Thus, two mechanisms are possibly causing the transition to the normal state when cooling superconducting composites by liquid coolant. First, it is due to the trivial transition from nucleate boiling regime to film one. Second, the current instability may occur as a result of the unlimited increase of the differential resistivity of the composite according to the condition $\partial E/\partial J \to \infty$ or $\partial T/\partial J \to \infty$. As shown in the numerical experiments, the probability of the latter is higher when there is higher ΔT_{cr}. Therefore, when the composites are cooled with liquid nitrogen, the probability of such transition of the charged currents increases.

6.5　MAXIMUM ALLOWABLE CURRENTS IN MULTILAYER SUPERCONDUCTING TAPES BASED ON $YBa_2Cu_3O_7$

It is important to conduct measurements in a strong constant magnetic field for many physical studies. The discovery of $YBa_2Cu_3O_7$ allows one to develop a new generation of the current-carrying elements for many practical applications since their unique critical properties are preserved under very high magnetic fields. At present, the manufacture of long YBCO tapes coated with stabilizing layers of silver and copper satisfies the real practical requirements. Therefore, superconducting sections based on them are now successfully used to create magnetic systems with a magnetic field above 20 T.

To ensure the operating temperature of such magnets, both liquid refrigerants, for example liquid helium, and indirect cooling methods that can be provided by a cryocooler may be used. Such SMS free of a liquid cryoagent significantly expands the range of practical use of superconductivity, making it possible to create compact devices.

Let us estimate the limiting current-carrying capacity of a multilayer superconducting tape based on YBCO with stabilizing coatings of silver and copper under various conditions of its cooling. Focus on studying the effect of coating thickness on the conditions for the occurrence of current instabilities.

Consider an infinitely long cooled superconducting tape of width b, consisting of a superconductor of thickness a_s with stabilizing silver coating of thickness a_{ag} and copper coating of thickness a_{cu} ($b \gg a = a_s + a_{ag} + a_{cu}$). Assume that the current charging time is sufficiently long, and there are no external thermal perturbations. The nonuniform temperature distribution over the cross-section of the tape is negligible. Let us also suppose that the V-I characteristic of a superconductor is described by a power equation; the critical current density of the superconductor and n-value of the V-I characteristic depend on the temperature and the magnetic field; the permanent external magnetic field completely penetrated in the tape; the transport current flows over the entire cross-section of the tape with a cross-sectional area $S = ab$ and the magnetic field of the charged current compared with the external magnetic field is small; the heat exchange with a coolant whose temperature is T_0 occurs on a part of the tape surface with a cooling perimeter $p = \gamma p_0$, where $0 < \gamma < 1$, $p_0 = 2(a + b)$. Formally, this approximation corresponds to the current charging into a tape with high-conductive coatings at an infinitely small velocity ($dI/dt \to \infty$).

According to the assumptions made, the thermal and electrodynamic states of the superconducting tape may be investigated using zero-dimensional model of the anisotropic continuum. To do this, the heat balance equation and the Kirchhoff equations will be used. Therefore, let us determine the temperature T and the electric field E for the given current density J redistributed between the superconducting tape layer, silver and copper coatings, based on the numerical solution of the following system of equations

$$EJ = q(T)p/S \tag{6.34}$$

$$E = E_c \left(\frac{J_s}{J_c(T, B)} \right)^{n(T,B)} = J_{ag}\rho_{ag}(T, B) = J_{cu}\rho_{cu}(T, B) \tag{6.35}$$

$$J = \eta_s J_s + \eta_{ag} J_{ag} + \eta_{cu} J_{cu} \tag{6.36}$$

Here, as above, E_c is a priori set value of the electric field used in determining the critical current density of a superconductor $J_c(T, B)$; ρ_{ag} and ρ_{cu} are the resistivity of silver and copper, respectively; $\eta_s = a_s/a$, $\eta_{ag} = a_{ag}/a$ and $\eta_{cu} = a_{cu}/a$ are the relative amount of silver and copper in the tape, respectively; $q(T)$ is the heat flow to the refrigerant. Two regimes of cooling the tape are considered below: cooling with a cryocooler, which is characterized by a constant heat transfer coefficient h, and cooling with liquid helium when nucleate and film boiling regimes take place. The corresponding dependencies $q(T)$ are written below.

The dependencies of the resistivity of silver and copper on the temperature and magnetic field for the given value of RRR $= \rho(273 \text{ K})/\rho(4.2 \text{ K})$ will be calculated in accordance with the results given in (Seeber B. 1998), namely at $\rho_{ag}(273 \text{ K}) = 1.48 \times 10^{-8} \Omega \times \text{m}$ and $\rho_{cu}(273 \text{ K}) = 1.55 \times 10^{-8} \Omega \times \text{m}$. To calculate the nonlinear dependencies $J_c(T, B)$ and $n(T, B)$, let us apply the theory developed in (Inoue et al. 2007).

The typical voltage-current and temperature-current characteristics of the tape initially cooled to the coolant temperature $T_0 = 4.2$ K are presented in Figure 6.22. The calculations were carried out at $E_c = 10^{-3}$ V/m, $a_s = 2.25 \times 10^{-6}$ m, $a_{ag} = 5 \times 10^{-6}$ m, $b = 4 \times 10^{-3}$ m and RRR $= 100$ (for both silver and copper coatings) for different values of the copper thickness and cooling perimeter. The plotted curves correspond to the states preceding the occurrence of the current instability and were calculated under the nonintensive cooling conditions. Therefore, the heat flux $q(T)$ was calculated as follows $q(T) = h(T - T_0)$ at $h = 10$ W/(m$^2 \times$ K).

As one would expect, the boundary of the stable current modes for the considered cooling conditions follows from the equivalent current instability conditions $\partial E/\partial J \to \infty$ or $\partial T/\partial J \to \infty$, which lead to the limiting quenching current I_q, electric field E_q and temperature T_q. After exceeding these values, the unstable states occur as discussed above. Their respective values are given in Tables 1 and 2 for different values of the magnetic induction of the external magnetic field. The calculations performed at $b = 2 \times 10^{-3}$ m, $a_s = 10^{-6}$ m, RRR $= 100$ and various copper thickness are presented in Tables 3 and 4. The temperature of the coolant was varied. The presented results allow one to draw the following conclusions.

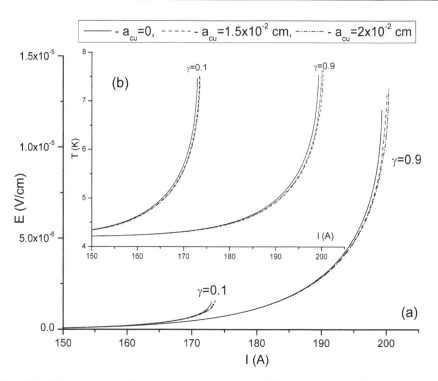

FIGURE 6.22 Voltage-current (a) and temperature-current (b) characteristics of the indirectly cooled YBCO tape at $B = 17$ T during variation cooling conditions.

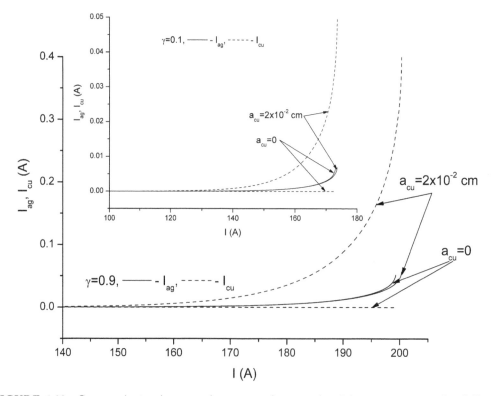

FIGURE 6.23 Current sharing between the superconductor and stabilizing coatings under different cooling conditions.

Firstly, the stabilizing effect of the thickness of the copper coating on the instability current at nonintensive cooling is almost absent. As Figure 6.23 shows, this feature is a direct consequence of the fact that the main part of the transport current flows in its superconducting core up until before the occurrence of the current instability in the case under consideration. Therefore, the engineering density of the permissible currents $J = I/S$ will decrease in inverse proportion to the increase of the copper coating thickness.

TABLE 1 Limiting quenching parameters at $B = 17$ T, $T_0 = 4.2$ K, $a_{ag} = 5 \times 10^{-6}$ m, $I_c(T, B) = 233$ A

a_{cu}, m	I_q, A	E_q, V/m	T_q, K
$\gamma = 0.1$			
0	172.99	0.15143×10^{-3}	7.4685
10^{-4}	173.29	0.15835×10^{-3}	7.5404
2×10^{-4}	173.58	0.15912×10^{-3}	7.4823
$\gamma = 0.5$			
0	191.88	0.71447×10^{-3}	7.6212
10^{-4}	192.29	0.72894×10^{-3}	7.6128
2×10^{-4}	192.70	0.73996×10^{-3}	7.5892
$\gamma = 0.9$			
0	199.30	0.12490×10^{-2}	7.6512
10^{-4}	199.80	0.12811×10^{-2}	7.6621
2×10^{-4}	200.31	0.13255×10^{-2}	7.7006

TABLE 2 Limiting quenching parameters at $B = 23$ T, $T_0 = 4.2$ K, $a_{ag} = 5 \times 10^{-6}$ m, $I_c(T_0, B) = 194$ A

a_{cu}, m	I_q, A	E_q, V/m	T_q, K
$\gamma = 0.1$			
0	143.41	0.17844×10^{-3}	7.3931
10^{-4}	143.67	0.17996×10^{-3}	7.3474
2×10^{-4}	143.93	0.18379×10^{-3}	7.3438
$\gamma = 0.5$			
0	160.16	0.81857×10^{-3}	7.4717
10^{-4}	160.53	0.82864×10^{-3}	7.4387
2×10^{-4}	160.90	0.85151×10^{-3}	7.4565
$\gamma = 0.9$			
0	166.77	0.14122×10^{-2}	7.4651
10^{-4}	167.22	0.14495×10^{-2}	7.4786
2×10^{-4}	167.68	0.14924×10^{-2}	7.5046

TABLE 3　Limiting quenching parameters at $B = 23$ T, $T_0 = 4.2$ K, $a_{ag} = 5 \times 10^{-6}$ m, $I_c(T_0, B) = 39$ A

a_{cu}, m	I_q, A	E_q, V/m	T_q, K
$\gamma = 0.1$			
0	34.999	0.10099×10^{-2}	13.010
0.2×10^{-4}	35.048	0.10123×10^{-2}	12.956
10^{-4}	35.243	0.10382×10^{-2}	12.887
$\gamma = 0.5$			
0	41.747	0.42422×10^{-2}	13.029
0.2×10^{-4}	41.842	0.43612×10^{-2}	13.207
10^{-4}	42.228	0.45963×10^{-2}	13.461
$\gamma = 0.9$			
0	44.549	0.72487×10^{-2}	13.143
0.2×10^{-4}	44.683	0.74903×10^{-2}	13.377
10^{-4}	45.234	0.80864×10^{-2}	13.849

TABLE 4　Limiting quenching parameters at $B = 23$ T, $T_0 = 20$ K, $a_{ag} = 5 \times 10^{-6}$ m, $I_c(T_0, B) = 32$ A

a_{cu}, m	I_q, A	E_q, V/m	T_q, K
$\gamma = 0.1$			
0	27.978	0.95332×10^{-3}	26.648
0.2×10^{-4}	28.020	0.97682×10^{-3}	26.755
10^{-4}	28.183	0.10187×10^{-3}	26.816
$\gamma = 0.5$			
0	33.455	0.41200×10^{-2}	26.871
0.2×10^{-4}	33.536	0.41895×10^{-2}	26.935
10^{-4}	33.867	0.44293×10^{-2}	27.123
$\gamma = 0.9$			
0	35.731	0.70312×10^{-2}	26.958
0.2×10^{-4}	35.846	0.71259×10^{-2}	27.004
10^{-4}	36.322	0.77140×10^{-2}	27.391

It should also be noted that the currents of unstable states occurrence during nonintensive cooling decrease by approximately 20% with an increase in the coolant temperature from 4.2 K to 20 K according to the results presented in Tables 3 and 4. This result demonstrates the possibility of varying the coolant temperature without a noticeable decrease in the boundaries of the stable states, which is undoubtedly important when choosing the operating temperature of the winding.

Secondly, the currents and electric field may take both subcritical and supercritical values before the current instability. The importance of the existence of subcritical values should be emphasized in particular. This result demonstrates that a priori defined critical parameters of the superconductor will lead to distorted estimates of the stability range of the current regimes under nonintensive cooling conditions. Therefore, the nonisothermal formation of the operating modes of a YBCO tape should be taken into account to correctly assess the boundary of the stable states. This feature takes place at the stage of the stable states despite the low currents flowing in stabilizing coatings. Note that this conclusion is also important to consider when measuring its critical parameters.

Thirdly, when cooling conditions are improved, there is a tendency for the limiting quenching current to approach the critical current of the tape. Furthermore, as the numerical experiments show, the stable current modes are possible if the permissible values of the current and the electric field are supercritical even with indirect cooling.

Fourthly, the limiting quenching currents decrease at increasing of the magnetic induction, and the corresponding quenching values of the electric field increase. Thus, the probability of the current instability increases at high magnetic fields which will be characterized by a deliberately subcritical quenching current. In this case, the induced electric fields may be either subcritical or supercritical. In practical applications, this fact should be taken into account when diagnosing the occurrence of unstable current states if the diagnosing is based on the permissible level measurement of electrical voltages, determined from a priori-defined the critical value of the electric field.

Let us discuss the current-carrying properties of the considered tape when it is cooled by liquid helium. Analysis of the ultimate current-carrying ability is carried out under parameters $E_c = 10^{-3}$ V/m, $a_s = 10^{-6}$ m, $a_{ag} = 5 \times 10^{-6}$ m, $b = 10^{-2}$ m, $T_0 = 4.2$ K and RRR = 100, varying the thickness of the copper coating and induction of the external magnetic field. In this case, assume that the cooled perimeter is $p = a + 2(a_s + a_{ag} + a_{cu})$. As above-mentioned, the critical current density and the n-value of the V-I characteristic will be calculated by taking into account their nonlinear character. The heat flux $q(T)$ in the assumption of a jump-like transition from nucleate to film boiling is defined as follows

$$q(T)[\text{W/m}^2] = \begin{cases} 2.15 \times 10^4 (T - T_0)^{1.5}, & T - T_0 < \Delta T_{cr} = 0,6 \text{ K} \\ 0.06 \times 10^4 (T - T_0)^{0.82}, & T - T_0 \geq \Delta T_{cr} = 0,6 \text{ K} \end{cases} \tag{6.37}$$

The temperature-current and voltage-current characteristics of the tape calculated for the nucleate boiling regime, which exists in the temperature range from T_0 to $T_0 + \Delta T_{cr}$ are shown in Figure 6.24. Within the framework of the approximation used, the overheating of the tape above ΔT_{cr} will lead to its transition to film boiling which, however, does not ensure the stability of the current and this will cause an irreversible temperature increase of the tape as discussed above. Therefore, the stability of the current modes of the tape will be violated after its trivial overheating above ΔT_{cr}. The corresponding stability boundary is shown in Figure 6.24 by the dashed line. The results of the analysis for the occurrence conditions of the current instability in a tape with different thickness of the copper coating are shown in Figure 6.25 and 6.26. They lead to the following conclusions.

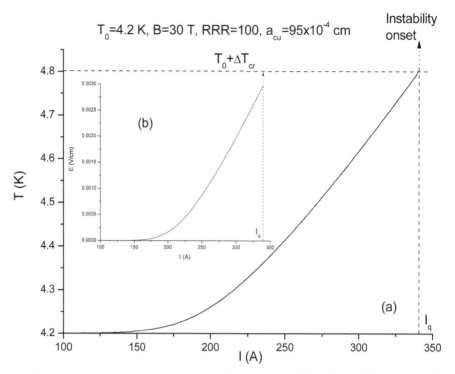

FIGURE 6.24 Permissible increase in temperature (a) and electric field (b) in YBCO tape cooled by liquid helium.

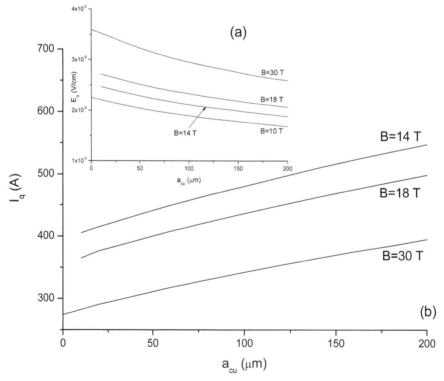

FIGURE 6.25 Limiting quenching values of electric field (a) and currents (b) depending on the thickness of the copper coating.

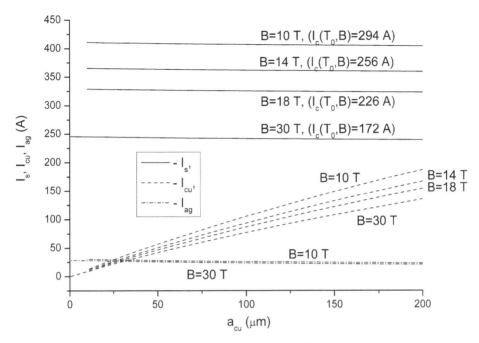

FIGURE 6.26 Influence of the copper coating thickness on the limiting quenching currents in the superconductor and stabilizing coatings.

The current instability in YBCO tape cooled by liquid helium occurs at supercritical values of the electric field and current. As a result, the limiting quenching electric field is more than two orders of magnitude greater than a priori defined critical electric field E_c. Accordingly, the limiting quenching currents increase significantly when the tape is cooled with liquid helium. So, for example, the critical current of the tape is equal to 226 A at $B = 18$ T. As follows from Figure 6.25, this value is even less than the limiting quenching currents of the tape, which is in an ultra-high magnetic field of $B = 30$ T. An increase in the thickness of its copper coating leads to an almost linear increase in limiting quenching currents in the entire considered range of changes in the magnetic field. Despite this, the corresponding values of the electric field are reduced. This regularity is a direct consequence of the current sharing mechanism between the superconductor and stabilizing coatings. Figure 6.26 demonstrates these features. Note, first, that a noticeable current sharing occurs in the supercritical voltage region. Therefore, in the subcritical region of the electric field, the unwanted heat release is small. Second, in the operating modes under consideration, the main role in the mechanism of current sharing plays the copper coating. Namely, the currents in the superconductor and in the silver layer with an increase in copper thickness practically do not change, and the current in the copper layer monotonously increases. In particular, at $B = 30$ T, the increase of current in copper is approximately 0.7 A for every 1 µm of the copper layer. As a result, the instability current is approximately 140% of the tape critical current in a given magnetic field and $a_{cu} = 200$ µm. The growth in the limiting quenching current increases with decreasing induction of an external magnetic field. However, the main part of the current flows in the superconductor. Therefore, if with an increase in the thickness of the copper layer, the total amount of heat released in the tape increases but only slightly. The calculations confirm it.

It is also important to emphasize that the presented results demonstrate the existence of the characteristic value of the copper coating thickness, after which its role in the stability of the transport currents is most noticeable. For the considered tape, this stabilizing effect is performed approximately at $a_{cu} > 50$ µm.

The results discussed above were obtained within the framework of the model that assumes a uniform distribution of the electric field and temperature over the cross-section of the tape. Let us consider a more general case, taking into account the nonuniform temperature distribution across the thickness of the tape ($0 < x < a = a_s + a_{ag} + a_{cu}$) when its geometric parameters satisfy the condition $b >> a$. Suppose that current is charged into the above-discussed tape at infinitely low rate ($dI/dt \to 0$); a perfect thermal contact is maintained between its layers; conditions for cooling the tape are so that the surface $x = 0$ is thermally insulated due to the design features of 2G tapes in which the superconductor is deposited on a poorly conducting substrate separating it from the superconductor by buffer layers and on the surface $x = a$ the coolant, either at a constant heat transfer coefficient during nonintensive cooling or the tape is cooled with liquid helium, as discussed above; the transition from the nucleate to film boiling regime is described by the Formula (6.37).

Under this formulation, let us numerically solve a system of one-dimensional stationary equations of the form

$$\frac{d}{dx}\left[\lambda_s(T)\frac{dT_s}{dx}\right] + EJ_s = 0,$$

$$\frac{d}{dx}\left[\lambda_{ag}(T)\frac{dT_{ag}}{dx}\right] + EJ_{ag} = 0,$$

$$\frac{d}{dx}\left[\lambda_{cu}(T)\frac{dT_{cu}}{dx}\right] + EJ_{cu} = 0$$

under the following thermal conditions

$$\frac{dT_s}{dx} = 0, \quad x = 0,$$

$$T_s = T_{ag}, \quad \lambda_s(T)\frac{dT_s}{dx} = \lambda_{ag}(T)\frac{dT_{ag}}{dx}, \quad x = a_s,$$

$$T_{ag} = T_{cu}, \quad \lambda_{ag}(T)\frac{dT_{ag}}{dx} = \lambda_{cu}(T)\frac{dT_{cu}}{dx}, \quad x = a_s + a_{ag}$$

$$\lambda_{cu}(T)\frac{dT_{cu}}{dx} + q(T) = 0, \quad x = a_s + a_{ag} + a_{cu}$$

and the Kirchhoff Equations (6.35) and (6.36). Here T_s, T_{ag} and T_{cu} are the temperatures in the superconductor, silver and copper layers, respectively; λ_s, λ_{ag} and λ_{cu} are the corresponding thermal conductivity coefficients; $q(T)$ is the heat flux into the coolant.

The calculations were also carried out in the framework of zero-dimensional model (6.34)-(6.36) formulated above. In this case, the value of S/p (the geometric factor of the tape, which has the shape of a slab with a cross-sectional area $S = a \times b$ and cooled perimeter $p = b$), is $S/p = a$.

The analysis of the thermo-electrodynamic states of the tape was carried out at $T_0 = 4.2$ K. As above, the nonlinear character of dependencies was taken into account $J_c(T, B)$, $n(T, B)$, $\rho_{ag}(T, B)$ and $\rho_{cu}(T, B)$. The values of the thermal conductivity of the silver and copper layers were determined under the Wiedemann-Franz law at RRR = 100 while for YBCO it was approximated at $T < 90$ K with the expression

$$\lambda_s = -1.6906 + 10.9355T - 9.47339 \times 10^{-3}T^2 - 1.2475 \times 10^{-3}T^3 + 6.00586 \times 10^{-6}T^4 \ [\text{W/(m} \times \text{K)}]$$

under the results of (Uher 1990).

On Figure 6.27 the dependencies $E(I)$ and $T(I)$, which take place on the surface of the tape ($a_s = 2 \times 10^{-6}$ m, $a_{ag} = 5 \times 10^{-6}$ m, $a_{cu} = 2 \times 10^{-4}$ m, $b = 4 \times 10^{-3}$ m), calculated in the framework of zero-dimensional and one-dimensional models at $B = 23$ T and nonintensive cooling $h = 10$ W/(m$^2 \times$ K). The corresponding dependencies for the tape with the parameters $a_s = 10^{-6}$ m, $a_{ag} = 5 \times 10^{-6}$ m, $a_{cu} = 2 \times 10^{-4}$ m, $b = 10^{-2}$ m when it is cooled by liquid helium, which is in a magnetic field of 30 T, are shown in Figure 6.28.

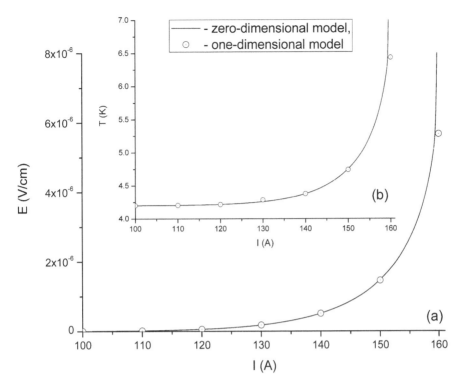

FIGURE 6.27 Dependencies of the electric field (a) and temperature (b) of the YBCO-surface tape on the current before instability during nonintensive cooling.

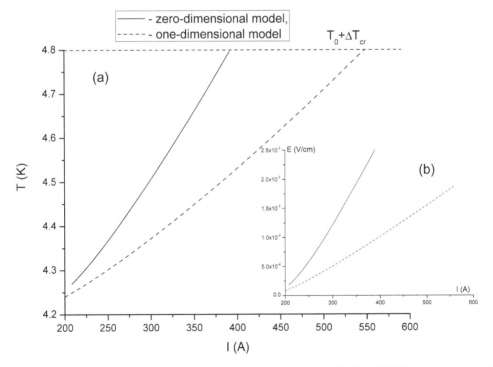

FIGURE 6.28 Dependencies of the temperature (a) and electric field (b) of the YBCO-surface tape on the current before instability at intensive cooling by liquid helium.

The results show that, first, both computational models adequately describe the mechanisms for the occurrence of the current instabilities depending on the cooling conditions. (The influence of the size effect on the conditions of the current instabilities is discussed below in more detail). Accordingly, in the case of the indirect cooling, the instabilities occur against the background of noticeable overheating which, as it was shown above, depend on the properties of the superconductor and stabilizing coatings, the cooled perimeter and other factors. If liquid helium is used, the destruction of superconductivity occurs as a result of the transition from the nucleate to the film boiling regime.

Second, the result of the analysis of conditions for the occurrence of the current instability practically does not depend on the simulation models used in the case of the nonintensive cooling although there is a slight heterogeneity in the temperature distribution in this cooling mode. In this case, the most heated is the superconducting layer where flows practically all the charged current (Figure 6.23) and the instability occurs in it. During intense cooling of the tape by liquid helium, a noticeable nonuniform temperature distribution observes in it for obvious reasons, with a maximum at $x = 0$, that is in the superconducting layer. This feature leads to an additional conductive heat flow, which has a stabilizing role. As a result, the one-dimensional model leads to more optimistic results when calculating the parameters of the limiting currents.

Thus, analysis of the conditions for the occurrence of the current instability in a multilayer tape based on YBCO shows that:

1. The thickness of the coatings practically does not affect the limiting quenching currents at indirect cooling of the YBCO tape when using a cryocooler as a coolant. In other words, they do not have a stabilizing effect on the stability conditions of the charged currents under these cooling regimes.
2. The instability currents almost linearly increase with increasing thickness of the copper layer when the YBCO tape is cooled by liquid helium. This regularity is due to the stable current sharing between the superconductor the copper coating.
3. The destruction condition of the stable current states of YBCO tapes can be either subcritical or supercritical. Therefore, when designing superconducting magnets based on YBCO, it is necessary to take into account that their transition to the normal state may occur at subcritical operating currents because the quenching currents may be lower than a priori defined critical current of the tape used.

6.6   THE HEAT CAPACITY EFFECT ON THE NONSTATIONARY FORMATION OF THE CURRENT MODES OF SUPERCONDUCTING COMPOSITES

The criteria for the stability of the current modes of superconducting composites formulated above were obtained in the stationary approximation. That is, the possible change in the temperature of the composite over time, which will inevitably occur when the transport current is charged into it at a finite rate, was not taken into account. Let us discuss the physical regularities that must be considered in the experimental study of the current modes of superconducting composites when a current is charged into them at a finite rate.

To simplify the analysis, the nonstationary zero-dimensional model will be used. This model will allow one to evaluate the effect of the heat capacity of the composite on the formation of its thermo-electrodynamic states during continuous current charging within the assumptions formulated in section 6.2, omitting their spatial heterogeneity from attention. Note that below the problem of the current instability will be investigated in a more general formulation, taking into account the spatially inhomogeneous distribution of the temperature and electric field.

Let us consider a superconductor with the power V-I characteristic and a linear dependence of the critical current density on temperature. Let us describe the spatially homogeneous dynamics of the thermo-electrodynamic states of the composite superconductor, unperturbed at the initial time $(T(0) = T_0, E(0) = 0, I(0) = 0)$, by the system of equations

$$C_k(T)S\frac{dT}{dt} = -hp(T - T_0) + EI$$

$$I(t) = \frac{dI}{dt}t = \eta J_s S + (1 - \eta)J_m S \qquad (6.38)$$

$$E = E_c[J_s/J_c(T)]^n = J_m\rho_m(T), \quad J_c(T) = J_{c0}(T_{cB} - T)/(T_{cB} - T_0)$$

where the physical meaning of the variables used was discussed before.

The results of the analysis of the characteristic formation features of voltage-current and temperature-current characteristics of the composite Ag/Bi2212 ($S = 2 \times 10^{-6}\,\mathrm{m}^2$, $p = 5 \times 10^{-3}\,\mathrm{m}$, $\eta = 0.2$, $n = 10$, $h = 10\,\mathrm{W/(m^2 \times K)}$, $T_0 = 4.2\,\mathrm{K}$, $E_c = 10^{-4}\,\mathrm{V/m}$, $T_{cB} = 26.1\,\mathrm{K}$, $J_{c0} = 1.5 \times 10^8\,\mathrm{A/m^2}$) originally placed in the external magnetic field $B = 10\,\mathrm{T}$ are presented below. The temperature dependence of the heat capacity of the composite, taking into account the heat capacity of the superconductor C_s and matrix C_m was defined as follows $C(T) = \eta C_s(T) + (1 - \eta)C_m(T)$. The temperature dependence of the heat capacity $\mathrm{Bi_2Sr_2CaCu_2O_8}$ was calculated by formula $C_s = 58.5T + 22T^3\mathrm{J/(m^3 \times K)}$, $T \leq 10\,\mathrm{K}$. The heat capacity of the silver matrix C_m was calculated according to the results of Dresner (1993). The value of RRR $= \rho_m(273\,\mathrm{K})/\rho_m(4.2\,\mathrm{K})$ is varied, taking into account the dependence of the matrix resistivity on temperature and magnetic field induction in the same way as described above. The results discussed below were obtained based on the numerical solution of the problem (6.38).

In Figure 6.29, the voltage-current and temperature-current characteristics calculated for various values of dI/dt and RRR $= 10$ are depicted. Here, the dashed curves show the stationary dependencies $E(I)$ and $T(I)$. They are based obtained on the numerical solution of the problem (6.11)-(6.13) and (6.15). It is easy to see that the slope of the $E(I)$ and $T(I)$ curves in the steady-state approximation differs from the corresponding curves calculated for a finite value of the current charging rate, namely, the greater dI/dt, the greater the difference. It should be emphasized that the nonstationary dependencies $E(I)$ and $T(I)$ do not satisfy the condition for the occurrence of the current instability (6.24) since during continuous current charging the slope $T(I)$ is always positive and decreases with increasing of the heat capacity as follows

$$\frac{dT}{dI} = \frac{EI - hp(T - T_0)}{C_k(T)S\dfrac{dI}{dt}} > 0$$

Therefore, appropriately changing the slope of $E(I)$ characteristics takes place, which may be written in the form

$$\frac{dE}{dI} = \frac{1 + \left[\eta\left|\dfrac{dJ_c}{dT}\right|\left(\dfrac{E}{E_c}\right)^{\frac{1}{n}} + E\dfrac{1 - \eta}{\rho_m^2}\dfrac{d\rho_m}{dT}\right]\dfrac{dT}{dI}S}{\dfrac{\eta SJ_c(T)}{nE}\left(\dfrac{E}{E_c}\right)^{\frac{1}{n}} + S\dfrac{1 - \eta}{\rho_m}} > 0$$

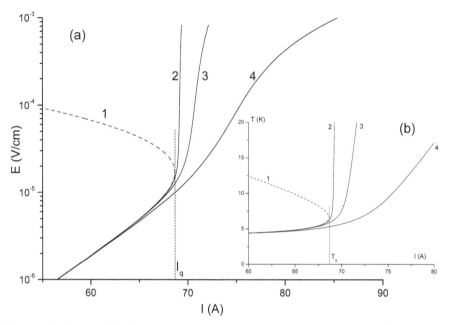

FIGURE 6.29 Influence of the charging rate on the formation of the voltage-current (a) and temperature-current (b) characteristics of the superconducting composite: 1 - $dI/dt \to 0$, 2 - dI/dt = 0.1 A/s, 3 - dI/dt = 1 A/s, 4 - dI/dt = 10 A/s.

As follows from this relationship, the higher the resistance of the matrix, the sharper the V-I characteristic of the composite will increase. The corresponding thermal effect of the heat capacity and the resistivity of the matrix on the character of the dependencies $T(I)$ and $E(I)$ is illustrated in Figure 6.30.

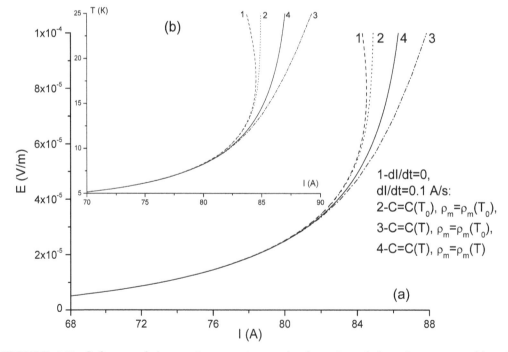

FIGURE 6.30 Influence of the matrix properties on the formation of the voltage-current (a) and temperature-current (b) characteristics of the composite at RRR = 1,000.

The existence of the positive slope of $T(I)$ and $E(I)$ curves explains the need to use the method of the finite perturbation of the initial state in the experimental determination of the instability current in HTS composites.

In section 6.3, it was shown that in the range of the operating temperatures at which a significant decrease in the slope of the dependence $J_c(T)$ is observed, the multistable states can occur. Their formation, investigated in the stationary approximation, is accompanied by a significant increase in the temperature of the composite. Obviously, in the experimental observation of these states, the influence of the heat capacity cannot be neglected. In this regard, let us investigate the formation of the corresponding multistable dependencies $E(I)$ and $T(I)$, which may be observed in real experiments when the current charging rate is finite. For this, model (6.38) will be used in which the critical current density of the considered composite superconductor Ag/Bi2212, which is in a constant external magnetic field $B = 10$ T, will be determined according to (1.8) and (6.32). In this case, the other parameters, except for the temperature of the coolant, remain are unchanged.

Figure 6.31 demonstrates the effect of the current charging rate on the voltage-current and temperature-current characteristics at $T_0 = 18$ K, when multi-stable states are observed, as shown above (Figure 6.13). Here, Curve 1 corresponds to the static states and Curves 2-4 describe states at continuous current charging. The corresponding limiting quenching values, determined in the stationary approximation, after which the charged current are unstable, are also presented here. The simulation results demonstrate the stabilizing role of the heat capacity. The latter essentially influences the formation of $E(J)$ and $T(J)$ characteristics when their jump rise transfers into the continuously increasing dependencies, even at very low current charging rates. It can be seen that the dependencies $E(I)$ and $T(I)$ not only shift to higher currents but also significantly change their shape during continuous current charging. As a result, the transition from the first stable branch to the second one becomes smoother. In particular, the transient V-I characteristic (Curve 4) increases so slowly at $dI/dt = 1$ A/s that at currents exceeding the limiting quenching current I_q, the values of the induced electric field are much less than the conventionally defined value E_c. At the same time, at a very current charging ($dI/dt = 0.01$ A/s, Curve 2), the nonstationary voltage-current characteristic begins to increase sharply even at currents noticeably smaller I_q and the electric field begins to exceed the value E_c at a certain point. Such a change in V-I characteristics can be perceived as the beginning of instability if one does not take into account the additional stable branch that exists in the supercritical region. These features must be taken into account in experiments to measure the critical current of a superconductor and during investigation of the existence of multistable regions. To do this, the current in the composite must be charged at extremely low rates. Another alternative method for the restoration of a multistable V-I characteristic of the composite can be a quick current charging up to a specified value and its subsequent relaxation to this value. But in this case, the relaxation time of the charged current will increase significantly with increasing current charging rate as follows from Figure 6.31 due to the corresponding influence of the heat capacity of the composite.

As shown above (Figure 6.13 and Curve 7), the thermo-electrodynamic states of the superconducting composites can maintain their stability over the entire range of the temperature variations at the high coolant temperatures, namely from the coolant temperature to the critical temperature of the superconductor. In Figure 6.32, the corresponding stationary and nonstationary voltage-current and temperature-current characteristics of the composite under consideration are compared which will be observed at $T_0 = 30$ K. The corresponding current I_{TC} is also depicted at which the critical temperature of the superconductor is reached in the stationary approximation. It can be seen that the stationary and nonstationary voltage-current and temperature-current characteristics not only can differ significantly from each other but also the latter are characterized by the nonintensive increase in the region of the unstable states. In experiments, this unstable current range, in which the electric field and temperature gradually increase while current increases, can be taken as a stable one. This feature, which is due to the stabilizing effect of the heat capacity, must be taken into account when conducting experiments in the region of the high coolant temperatures.

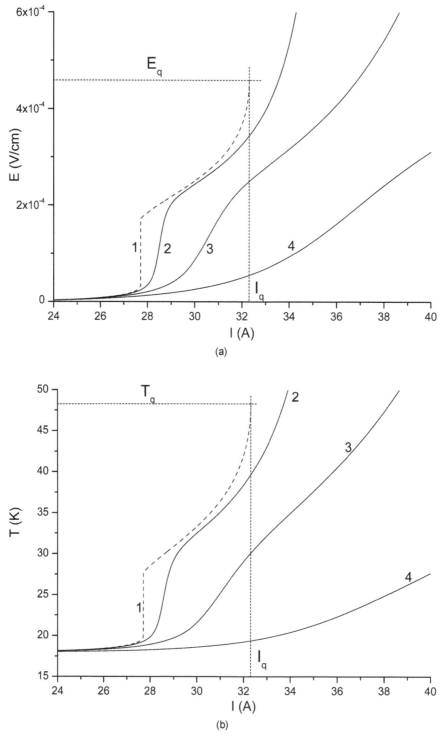

FIGURE 6.31 Effect of the current charging rate on the formation of the multistable voltage-current (a) and temperature-current (b) characteristics of the Ag/Bi2212 composite at $h = 10$ W/(m^2 × K) and RRR = 10: 1 - $dI/dt \rightarrow 0$, 2 - $dI/dt = 0.01$ A/s, 3 - $dI/dt = 0.1$ A/s, 4 - $dI/dt = 1$ A/s.

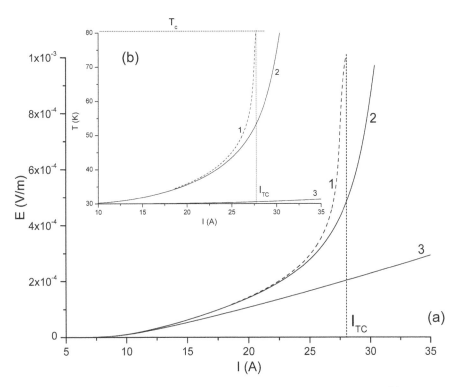

FIGURE 6.32 Effect of the current charging rate on the formation of voltage-current (a) and temperature-current (b) characteristics of the HTS composite at a high coolant temperature: B_a = 10 T, RRR = 10, T_0 = 30 K, h = 10 W/(m² × K); 1 - dI/dt = 0, 2 - dI/dt = 0.01 A/s, 3 - dI/dt = 10 A/s.

6.7   SIZE EFFECTS DURING FORMATION OF THE V-I CHARACTERISTIC OF SUPERCONDUCTING COMPOSITES AND THE MECHANISMS OF THE CURRENT INSTABILITIES

The above investigation of the reasons for the transition of the composite superconductors to the normal state led to the simplest criteria for the occurrence of the current instabilities using zero-dimensional approximations. However, the nonuniform distribution of both temperature and electric field in superconducting composites should be taken into account in the general case as was noted above. In this regard, let us discuss the general mechanisms underlying the changes in the steepness of their V-I characteristic, which must be taken into account in experiments to determine both the critical current and the instability currents taking into account the spatial heterogeneity of their thermo-electrodynamic states.

As above, consider an infinitely long composite of a rectangular cross-section ($-a < x < a, -b < y < b, -\infty < z < \infty, b \gg a$), placed in a constant external magnetic field B_a, which has fully penetrated the composite. Suppose that the tape has the temperature of the coolant T_0 at the initial time, there is no current and then it is charged in the direction of the longitudinal axis with a rate dI/dt. Let us also assume that a change in the external magnetic field and the critical current of the superconductor in the longitudinal direction may be neglected, and thin superconducting filaments are uniformly distributed over the cross-section of the tape with a filling factor η, which do not lead to the magnetic instabilities. Let there is a convective heat exchange with a constant heat transfer coefficient h on the surfaces of the tape. Suppose that the V-I characteristic of the superconductor is described by the power equation, and its critical current density J_c decreases linearly with temperature.

According to these assumptions, largely due to the condition $b >> a$, the time variation of the temperature, current and electric field in the cross-section of the tape will be determined by solving the following system of one-dimensional nonstationary equations

$$C_k(T)\frac{\partial T}{\partial t} = \frac{\partial}{\partial x}\left(\lambda_k(T)\frac{\partial T}{\partial x}\right) + \begin{cases} 0, & 0 < x < x_p \\ EJ, & x_p < x < a \end{cases}$$

$$\mu_0 \frac{\partial J}{\partial t} = \frac{\partial^2 E}{\partial x^2}, \quad t > 0, \quad 0 \le x_p < x < a$$

$$J = \eta J_s + (1-\eta)J_m \tag{6.39}$$

$$E = E_c[J_s/J_c(T, B)]^n = J_m\rho_m(T, B),$$

$$J_c(T, B) = J_{c0}(B)[T_{cB}(B) - T]/[T_{cB}(B) - T_0]$$

under the following initial-boundary conditions

$$T(x, 0) = T_0, \quad \frac{\partial T}{\partial x}(0, t) = 0, \quad \lambda_k\frac{\partial T}{\partial x}(a, t) + h[T(a, t) - T_0] = 0$$

$$E(x, 0) = 0, \quad \begin{cases} E(x_p, t) = 0, & x_p > 0 \\ \frac{\partial E}{\partial x}(0, t) = 0, & x_p = 0 \end{cases}, \quad \frac{\partial E}{\partial x}(a, t) = \frac{\mu_0}{4b}\frac{dI}{dt} \tag{6.40}$$

Here, x_p is the movable coordinate of the current penetration front defined by the relation

$$4b\int_{x_p}^{a} J(x, t)\,dx = \frac{dI}{dt}t$$

The physical meaning of the parameters used was discussed above. Note that the problem (6.39)-(6.40) differs from the previously considered problem (6.1)-(6.6) only in the form of the superconducting composite under study.

In addition to the model (6.39)-(6.40), other approximations to analyze the macroscopic states of the superconducting composite will be used.

Along with the nonstationary system of equations (6.39)-(6.40), the corresponding stationary equations will be considered to describe the thermo-electrodynamic states of a superconducting tape in the approximation of the infinitely low current charging ($dI/dt \to 0$). In other words, the fully penetrated modes will be analyzed in the static one-dimensional approximation. (Physical features of these states are discussed below in detail). In this case, the distribution of the temperature and electric field in the cross-section of the superconducting tape can be determined from the solution of the system of stationary equations of the form

$$\frac{d}{dx}\left(\lambda_k(T)\frac{dT}{dx}\right) + EJ = 0, \quad \frac{d^2E}{dx^2} = 0, \quad 0 < x < a \tag{6.41}$$

$$J = \eta J_s + (1-\eta)J_m, \quad E = E_c[J_s/J_c(T, B)]^n = J_m\rho_m(T, B),$$

$$J_c(T, B) = J_{c0}(B)[T_{cB}(B) - T]/[T_{cB}(B) - T_0]$$

under the boundary conditions

$$\frac{dT}{dx}(0) = 0, \quad \lambda_k\frac{dT}{dx}(a) + h[T(a) - T_0] = 0, \quad \frac{dE}{dx}(0, t) = 0, \quad \frac{dE}{dx}(a, t) = 0 \,.$$

The partial current penetration mode may also be described using the scaling approximation formulated in Chapter 2. Then, the distribution of the electric field and the time variation of the current penetration front can be calculated at $n \ge 10$ as follows

$$E(x,t) = \begin{cases} 0, & 0 \le x \le x_p \\ \dfrac{\mu_0}{4b}\dfrac{dI}{dt}(x - x_p), & x_p \le x \le a \end{cases}$$

(6.42)

$$x_p(t) = a - \left(\frac{n+1}{n}t\right)^{\frac{n}{n+1}}\left(\frac{\mu_0}{4b}\frac{dI}{dt}\right)^{\frac{n-1}{n+1}}\left[\frac{E_c}{\mu_0^n \eta^n J_{c0}^n}\right]^{\frac{1}{n+1}}$$

To analyze the full penetration mode, let us use also the static zero-dimensional model as above. If the matrix resistivity is practically constant, according to section 6.2, the thermo-electrodynamic states of the composite are described by the relations

$$J = \left(\frac{\eta J_{c0}}{E_c^{1/n}}E^{1/n} + \frac{1-\eta}{\rho_m}E\right)\bigg/\left[1 + \frac{J_{c0}}{E_c^{1/n}}\frac{\eta a}{h(T_{cB}-T_0)}E^{(n+1)/n}\right],$$

$$T = T_0 + \frac{Ea}{h}\left(\frac{\eta J_{c0}}{E_c^{1/n}}E^{1/n} + \frac{1-\eta}{\rho_m}E\right)\bigg/\left[1 + \frac{J_{c0}}{E_c^{1/n}}\frac{\eta a}{h(T_{cB}-T_0)}E^{(n+1)/n}\right]$$

(6.43)

taking into consideration that for tape $S/p = a$. The relations (6.43) allow one to find the boundary of the stable current states during full penetration of the transport current into the tape solving the transcendental equation (6.25).

Below, the investigation of the thermo-electrodynamic phenomena occurring in the Ag/Bi2212 composite placed in a magnetic field $B_a = 10$ T is made. The calculations were carried out at $a = 0.19 \times 10^{-3}$ m, $b = 2.45 \times 10^{-3}$ m, $T_0 = 4.2$ K, $h = 10$ W/(m^2 × K), $\rho_m(273$ K)/$\rho_m(4.2$ K$) = 10$, $\rho_m(273$ K$) = 1.48 \times 10^{-8}$ Ω × m, $dI/dt = 10$ A/s, $n = 10$, $\eta = 0.2$, $T_{cB} = 26.1$ K, $J_{c0} = 1.5 \times 10^8$ A/m^2, $E_c = 10^{-4}$ V/m. The thermal conductivity coefficient of the tape in its cross-section is calculated taking into account the Wiedemann-Franz law $\lambda_k(T) = 2.45 \times 10^{-8}T(1-\eta)/(1+\eta)/\rho_m(T, B)$. The heat capacity of the tape and the resistivity of the matrix were calculated in the same way as noted above.

Using the proposed models, let us discuss the main physical regularities governing the formation of the stable and unstable states of superconducting tapes during current charging.

The dependencies of the electric field and temperature as a function of the charged current are shown in Figure 6.33. Here, Curves 1' and 1" describe the corresponding dependencies on the surface and in the center, respectively. Curves 2 are the voltage-current and temperature-current characteristics calculated according to (6.43), that is assuming the uniform electric field and temperature distributions. Accordingly, the electric field E_q, current I_q and temperature T_q defining the stability limit of the spatially homogeneous stationary states of the composite, following from relations (6.25), are shown in Figure 6.33. Curves 3 correspond to the static $E(I)$ and $T(I)$ characteristics of the tape, which were determined on the basis of the spatially nonhomogeneous stationary approximation as mentioned above. In this approximation, the quenching parameters are as follows: the electric field $E_{q,1}$, the current $I_{q,1}$ and the temperature $T_{q,1}$. Curve 4 describes the scaling states. Their calculation was performed on the basis of the approximation (6.42).

It is seen that the partial penetration stage is accompanied not only by a practically uniform temperature distribution but also characterized by a weak deviation from the coolant temperature (Figure 6.33b) despite the nonuniform distribution of the electric field over the cross-section (Figure 6.33a). This is because the main part of the charged current flows through the superconducting core of the tape and therefore, the heat release is insignificant. According to (6.42), one can find the time during which the current will fill the entire cross-section of the tape and also the corresponding electric field on its surface. Their values are equal to

$$t_f = \frac{n}{n+1}\frac{\mu_0 \eta J_{c0}}{E_c^{1/n}}a^{\frac{n+1}{n}}\left(\frac{\mu_0}{4b}\frac{dI}{dt}\right)^{\frac{1-n}{n}}, \quad E_f = \frac{\mu_0}{4b}\frac{dI}{dt}a$$

(6.44)

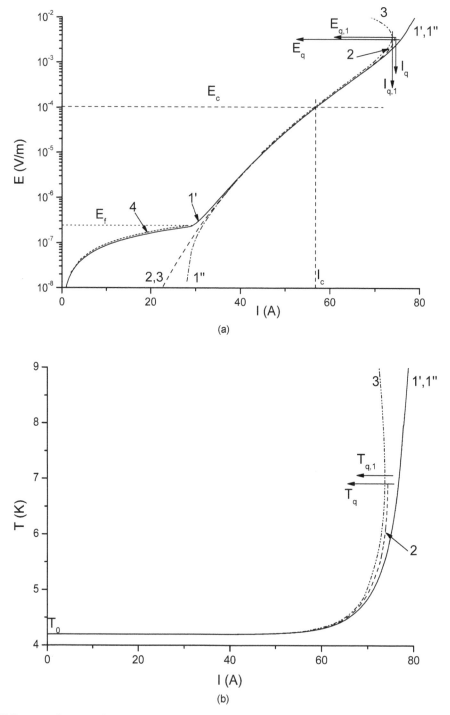

FIGURE 6.33 Static and dynamic change in the electric field (a) and the temperature (b) of the superconducting tape, calculated on the basis of various models: $1'$, $1''$ – the electric field and the temperature on the surface ($1'$) and in the center ($1''$), respectively, numerically determined according to the 1D model (6.39)-(6.40), 2 – zero-dimensional model (6.43) at $\rho_m = \rho_m(T_0)$, 3 – 1D static fully penetrated model (6.41) ($dI/dt \rightarrow 0$), 4 – scaling approximation (6.42).

These formulae show how the rate of the current charging, the geometric parameters of the tape affect the formation of its electrodynamic states in the partial penetration mode, which must be considered when measuring the critical current of a superconductor. In particular, the value of E_f will increase with an increase in the thickness of the tape. As a result, a nonuniform distribution of the electric field over the cross-section of the tape will be observed not only at the initial stage of the full penetration mode but may also occur at the voltage range used to measure the critical current. The higher the current charging rate, the clearer this trend becomes. The scaling approximation makes it possible, first, to estimate the electric voltage limit below which the measurement of the critical current of a superconductor is incorrect. It equals

$$V_0 = 2\int_0^a E dx = \frac{\mu_0 a^2}{2b}\frac{dI}{dt}$$

Second, Figure 6.33 shows that the initial stage of the full penetration mode is nonetheless also characterized by the nonuniform distributed electric field after exceeding V_0. As follows from Figure 6.33a, the distribution of the electric field, which is close to uniform, will take place when the simplest condition is observed $E_f \ll E_c$, that is, for example at $dI/dt \ll 4bE_c/(\mu_0 a)$. The distribution of the electric field over the cross-section of the tape will be uniform only in the limiting case $dI/dt \rightarrow 0$. That is why zero-dimensional approximations will lead to the correct values of the quenching parameters.

In the supercritical electric fields ($E > E_c$), the nonstationary $E(I)$ and $T(I)$ characteristics of the tape (Curves 1′ and 1″) are quantitatively and qualitatively different from dependencies calculated in the stationary approximations (Curves 2 and 3). This feature is associated with its finite stable overheating and the effect of the heat capacity, as discussed above, on the nonstationary modes occurring in the superconducting composites. According to the results discussed above, the rise in dependencies $E(I)$ and $T(I)$ (slope of Curves 1′ and 1″) is always positive and will decrease with increasing of the current charging rate due to a corresponding increase of the heat capacity with temperature.

It should be emphasized that the condition

$$\partial E/\partial J \rightarrow \infty \qquad (6.45)$$

allowing one to find the limiting quenching values E_q, I_q and T_q are not respected in the approximation of a stationary nonuniform distribution of the electric field and temperature, as it follows from Figure 6.33a. In the general case, the current instability border follows from the condition

$$\partial E/\partial I \rightarrow \infty \qquad (6.46)$$

It takes into account the heterogeneity of the macroscopic states of composite tape in the stationary approximation. The difference between Conditions (6.45) and (6.46) is a consequence of the fact that the thermo-electrodynamic states of the superconducting composites depend on the current and temperature distributions in their cross-sections. It follows from the above that the calculated range of the stable currents determined by taking into account the inhomogeneous distribution of the temperature over the cross-section of the composite will be smaller compared to similar calculated values obtained in the framework of the zero-dimensional model. The corresponding quenching parameters calculated in the framework of zero-dimensional and one-dimensional models are shown in Figure 6.33 as noted earlier.

The stable and unstable branches of the static $E(I)$ and $T(I)$ dependencies are presented in Figure 6.34. They are calculated for the surface of the tape under consideration in the framework of the 1D stationary approximation (6.41) by varying the heat transfer coefficient and the critical current density of the superconductor. The results show that the stationary $E(I)$ and $T(I)$ characteristics of the composite superconductors can increase without a catastrophic development of the instability (Curve 1′) under certain conditions. Accordingly, they have a positive slope in the entire range of the variation in the transport current even when the superconductor is heated up to the critical temperature. These conclusions demonstrate how important it is for the analysis of the current states of the composite superconductors, primarily made of HTS, to know their temperature which can stably rise to the critical temperature.

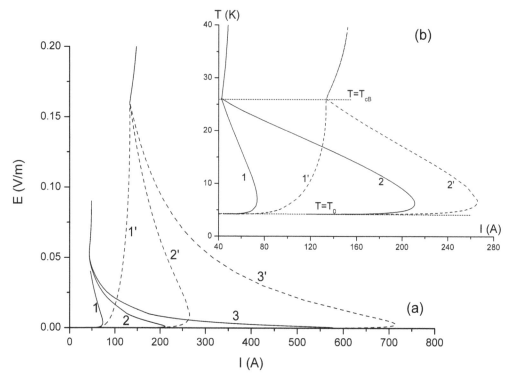

FIGURE 6.34 Influence of the heat transfer coefficient and critical current density on the static voltage-current (a) and temperature-current (b) characteristics: —— - h = 10 W/(m² × K), - - - - - h = 100 W/(m² × K); 1, 1' - J_{c0} = 1.5 × 10⁸ A/m², 2, 2' - J_{c0} = 5 × 10⁸ A/m², 3, 3' - J_{c0} = 15 × 10⁸ A/m².

In general, Figures 6.33 and 6.34 confirm the existence of the stable supercritical modes discussed earlier when the current instability occurs at the electric field and currents exceeding the conventionally set values E_c and J_c. Accordingly, there is a noticeable stable overheating of the superconductor. It leads to the thermal degradation of the current-carrying capacity of the composite superconductors. Indeed, as follows from Figure 6.34, the increase in the critical current density by 3.33 times (transition from State 1 to State 2) at h = 10 W/(m² × K) leads to the increase in the quenching current by 2.85 times. The increase in the critical current density by 3 times (transition from State 2 to State 3) leads to an increase in the quenching current by 2.73 times. Finally, the increase in the critical current density by 10 times (transition from State 1 to State 3) leads to an increase in the quenching current only by 7.81 times. In addition, the degradation also intensifies at the improvement of the heat exchange: the transition from State 2' to 3' (h = 100 W/(m² × K)), when the critical current density increases by 3 times, is accompanied by the increase in the quenching current by 2.69 times, that is less than during corresponding transition from the State 2 to the State 3. Thus, the increase in the quenching current is always less than the corresponding increase in the critical current of the superconductor. The difference is more noticeable, the higher the critical current density of the superconductor or the heat transfer coefficient.

In order to give a general physical formulation to the discussed formation features of the current-voltage and temperature-current characteristics of composite superconductors that will be observed in the full penetration mode, let us perform their analysis in dimensionless approximation. In this case, the stationary distribution of the electric field over the cross-section of the tape becomes uniform as follows from the model (6.41). Introduce the dimensionless variables, namely dimensionless temperature $\theta = (T_{cB} - T)/(T_{cB} - T_0)$, dimensionless spatial coordinate $X = x/a$ and dimensionless heat resistance $H = ha/\lambda_k$, assuming a constant coefficient of thermal conductivity of the tape. Accordingly, let us find a solution to the following boundary problem

$$\frac{d^2\theta}{dX^2} - \theta\frac{E_c}{E_\lambda}\left(\frac{E}{E_c}\right)^{(n+1)/n} - \frac{E^2}{E_\lambda E_\rho} = 0,$$

$$\frac{d\theta}{dX}(0) = 0, \quad \frac{d\theta}{dX}(1) + H\theta(1) = H$$

(6.47)

Here,

$$E_\lambda = \frac{\lambda(T_{cB} - T_0)}{\eta J_{c0}a^2}, \quad E_\rho = \frac{\eta J_{c0}\rho_m}{1-\eta}, \quad E_h = \frac{h(T_{cB} - T_0)}{\eta J_{c0}a}$$

(6.48)

and there are equalities $H = E_h/E_\lambda$ and the Stekly parameter $\alpha = E_\rho/E_h$, as introduced in Section 6.2.

According to (6.47) and (6.48), the electrodynamic states of the composite superconductors are characterized by three characteristic values of the electric field E_λ, E_ρ and E_h, affecting the intensity of temperature rise and therefore, on the steepness of the V-I characteristic rise (characteristic values E_ρ and E_h existing in zero-dimensional approximation have already been partially discussed before). Their values depend on the critical parameters of the superconductor, thermal and electrical properties of the matrix and conditions of the heat transfer to the coolant. Accordingly, there will be various mechanisms for the formation of thermo-electrodynamic states of composite superconductors. Let us discuss them at a qualitative level of rigor.

When $E_\lambda \to \infty$, the temperature distribution over the cross-section of the composite will be uniform. Therefore, E_λ is the characteristic electric field that affects the spatial character of the temperature distribution over cross-section, depending on the amount of the heat removed by the thermal conductivity of the composite. In this case, the parameter determining the variation of E_λ, is the thermal conductivity coefficient.

When $E_\rho \to \infty$, the mechanism of the current sharing between the superconducting core and the matrix will not affect $E(I)$ and $T(I)$ characteristics. So, E_ρ is the characteristic value of the electric field, which determines the boundary of the electrical voltages, after exceeding of which the development of thermal and electrodynamic states will depend on the properties of the stabilizing matrix. The determining parameter influencing the values of E_ρ is the matrix resistivity.

When $E_h \to \infty$, the V-I characteristic of the composite is isothermal ($T = T_0$). Consequently, E_h is the characteristic electric field, which makes it possible to evaluate the influence of the cooling intensity on the character of the V-I characteristic increase and take into account the deviation of the electrodynamic states from states that would exist under ideal cooling conditions.

The variation of the remaining parameters (the critical current density of the superconductor and its critical temperature, filling factor, coolant temperature and composite thickness) will be accompanied by an interconnected variation in the values of E_λ, E_ρ and E_h. Accordingly, there will be an increase or decrease in the role of one or another mechanism for the formation of the V-I characteristic. For example, as the critical current density of superconductor increases, E_ρ increases but E_λ and E_h decrease. As a result, the thermo-electrodynamic states of the composite superconductors will depend less on the resistivity of the matrix. But they will be characterized by a more inhomogeneous temperature distribution, which will have a corresponding effect on the electrodynamic states of composite superconductors. An increase in the filling factor will also lead to an increase in the influence of the temperature and its distribution on the V-I characteristic.

Note the contribution of the thermal mechanisms (conductive and convective) to the rise of the V-I characteristic. Their role will depend on the difference between the values of E_λ and E_h. Indeed, if $E_h \ll E_\lambda$, then a spatially uniform increase in the V-I characteristic in the range of the electric fields $E \ll E_h$ will depend on the conditions of its cooling. Accordingly, the change of the thermo-electrodynamic states will be nonisothermal at $E_h \ll E \ll E_\lambda$. But the nonuniform temperature distribution over the composite cross-section will influence the stationary V-I characteristic at $E \gg E_\lambda$. At the same time, the electrodynamic states of the composite will also depend on the spatial inhomogeneity of the temperature field at $E_\lambda \ll E_h$. In particular, when $E_\lambda \ll E_h \ll E$, the analysis of the electrodynamic states of the superconducting composite must be performed taking into account the inhomogeneous temperature distribution over its cross-section.

Dependencies E_λ, E_ρ and E_h on the temperature for the considered tape are shown in Figure 6.35. It is seen that the following condition takes place $E_q < E_h < E_\rho \ll E_\lambda$ in the case under consideration. It explains why the temperature distribution over its cross-section is almost uniform, the current sharing does not have a noticeable effect on the stable formation of the thermo-electrodynamic states of the tape and the instability occurs at the supercritical parameters.

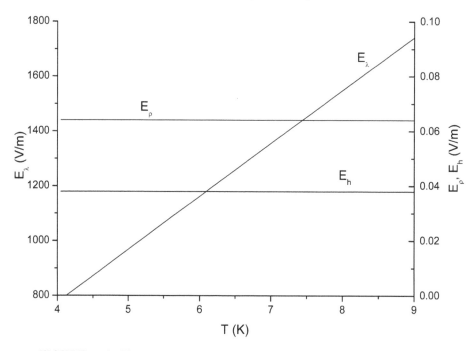

FIGURE 6.35   Temperature effect on the characteristic values of the electric field.

The solution of the boundary problem (6.47) is the function

$$\theta(X) = \begin{cases} \theta_0 ch\sqrt{\gamma}X - q/\gamma, \quad \theta_0 = \dfrac{H(1+q/\gamma)}{\sqrt{\gamma}sh\sqrt{\gamma} + Hch\sqrt{\gamma}}, \quad \theta(0) > 0 \\[4mm] \dfrac{q}{2}(x^2 - 1) + \dfrac{H-q}{H}, \quad \theta(0), \theta(1) \le 0 \end{cases} \tag{6.49}$$

In zero-dimensional approximation, according to (6.43), the uniform temperature distribution over the cross-section of the tape with respect to the used dimensionless variables is described by the following expressions

$$\theta = \begin{cases} (H-q)/(H+\gamma), \quad \theta > 0 \\ (H-q)/H, \quad \theta \le 0 \end{cases} \tag{6.50}$$

Here,

$$\gamma = \frac{E_c}{E_\lambda} \left(\frac{E}{E_c} \right)^{(n+1)/n} \quad \text{and} \quad q = \frac{E^2}{E_\rho E_\lambda}.$$

Correspondingly, there is a connection $q = \varepsilon \gamma^{2n/(n+1)}$, where $\varepsilon = E_c/E_\rho (E_\lambda/E_c)^{(n-1)/(n+1)}$. The parameters γ and q describe the influence of the heat conductive mechanism and the current sharing on the operating states of the composite. In particular, according to (6.49), it is easy to find that at $\gamma \ll 1$, that is at an efficient heat transfer due to the thermal conductivity of the tape ($E \ll E_\lambda$), there will be states with uniform temperature distribution. Accordingly, Expressions (6.49) transit to (6.50).

At
$$\frac{q}{\gamma} = \frac{E_c}{E_\eta} \left(\frac{E}{E_c} \right)^{(n-1)/n} \ll 1,$$

that is at $E \ll E_\rho$, the main part of the current will flow in the superconducting core, and the influence of the matrix will be negligible.

Note that the Formula (6.49) describes only two thermal regimes. First of all, it is the regime when superconductivity remains in the entire volume of the composite at $\theta(0) > 0$. If $\theta(0) < 0$, then the central part of the composite begins to change to the normal state and the superconducting state of the composite is destroyed at $\theta(1) < 0$. The mathematical description of the modes when the superconducting and normal states can stably exist in the composite is very cumbersome and is not given here. At the same time, calculations show that for practically important regimes, namely at $H \ll 1$, the difference in conditions that take into account the existence of both $\theta(0)$ and $\theta(1)$ is insignificant. Therefore, the Formula (6.49) describes the states when it is assumed that when $\theta(0) < 0$, the composite completely changes to the normal state.

According to (6.49) and (6.50), it is easy to find relations when the condition $\theta = 0$ is fulfilled, which describes the boundary between the superconducting and nonsuperconducting states. In the framework of one-dimensional approximation, it is performed at

$$H = \varepsilon \gamma^{\frac{3n-1}{2(n+1)}} sh\sqrt{\gamma}/[1 + \varepsilon \gamma^{\frac{n-1}{n+1}}(1 - ch\sqrt{\gamma})] \tag{6.51}$$

and in the framework of zero-dimensional approximation at

$$H = \varepsilon \gamma^{\frac{2n}{n+1}} \tag{6.52}$$

Expressions (6.51) and (6.52) allow one to estimate the electric field and current after exceeding which the superconducting properties of the composite superconductors will be destroyed without their jump-transition to the normal state. The corresponding values of γ as a function of the thermal resistance are shown in Figure 6.36. The calculations were performed at $E_\lambda = 10^3$ V/m varying E_ρ. It can be seen that for $H \ll 0.1$, the values of γ are not only much smaller than unity but can also be described with satisfactory accuracy by zero-dimensional model. On the other hand, from Figure 6.36 also follows that it is necessary to take into account the nonuniformity of the temperature distribution over the cross-section at $H \gg 0.1$. Accordingly, the effect of nonuniform temperature distribution will have a significant impact on the formation of the thermo-electrodynamic states of the intensely cooled massive current-carrying elements.

In the dimensional form, the stationary nonuniform temperature distribution over the cross-section of the composite during full current penetration is described by the expression

$$T(x) = T_{cB} - (T_{cB} - T_0) \begin{cases} \theta_0 ch\sqrt{\gamma} X - q/\gamma, & T(0) < T_{cB} \\ \dfrac{q}{2}(X^2 - 1) + \dfrac{H - q}{H}, & T(0) \geq T_{cB} \end{cases} \tag{6.53}$$

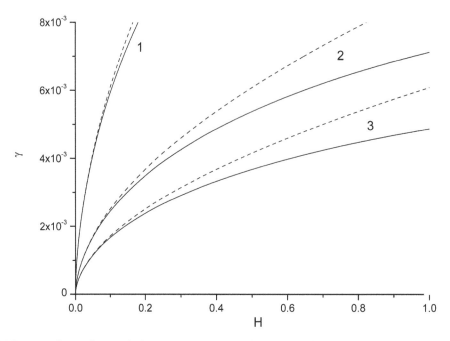

FIGURE 6.36 Dependence of the maximum permissible values of the parameter γ on the thermal resistance of the composite, after exceeding which all current will flow stably only in the matrix: 1 - E_ρ = 0.05 V/m, 2 - E_ρ = 0.01 V/m, 3 - E_ρ = 0.005 V/m —— – one-dimensional approximation, - - - - zero-dimensional approximation.

In this case, the connection between the charged current per unit of the tape width and the electric field, determined with regard to the symmetry of the problem under consideration as

$$I^*(E) = 2\int_0^a J(E)\,dx,$$

is written as follows

$$I^* = \begin{cases} 2\eta J_{c0}a\left(\dfrac{E}{E_c}\right)^{1/n} \dfrac{1+q/\gamma}{\sqrt{\gamma}/th\sqrt{\gamma}+\gamma/H}, & T(0) < T_{cB} \\[4mm] 2\eta J_{c0}a\dfrac{E}{E_\rho}, & T(0) \geq T_{cB} \end{cases} \tag{6.54}$$

The corresponding expressions that follow from zero-dimensional approximation are

$$T = \begin{cases} T_{cB} - (T_{cB}-T_0)\dfrac{H-q}{H+\gamma}, & T < T_{cB} \\[4mm] T_{cB} - (T_{cB}-T_0)\left(1-\dfrac{q}{H}\right), & T \geq T_{cB} \end{cases} \tag{6.55}$$

$$I^* = \begin{cases} 2\eta J_{c0}a\left(\dfrac{E_\lambda}{E_c}\gamma\right)^{1/(n+1)} \dfrac{1+q/\gamma}{1+\gamma/H}, & T < T_{cB} \\[4mm] 2\eta J_{c0}a\dfrac{E}{E_\rho}, & T \geq T_{cB} \end{cases} \tag{6.56}$$

For states that satisfy the condition $\gamma \ll 1$, the Expressions (6.55) and (6.56) allow the writing of more rigorous estimates of the electric field range when the current charging is isothermal or the role of the matrix is negligible. Omitting the obvious transformations, let us formulate them.

The V-I characteristic of a composite superconductor will be close to isothermal if the condition is fulfilled

$$\frac{T_{cB} - 2T_0}{T_0} \frac{E}{E_h} \left(\frac{E}{E_c}\right)^{1/n} + \frac{T_{cB} - T_0}{T_0} \frac{E^2}{E_h E_\rho} \ll 1$$

When $T_{cB} > 2T_0$, it can be written in a more simplified form, namely

$$\frac{E^2}{E_h E_\rho} \ll 1$$

The matrix will not affect the slope of the V-I characteristic when

$$\frac{E}{E_\rho}\left(\frac{E_c}{E}\right)^{1/n} \ll 1$$

(This estimate has already been obtained above). In this case, the temperature of the superconductor will differ slightly from the temperature of the coolant when the additional condition is met

$$\frac{T_{cB} - 2T_0}{T_0} \frac{E}{E_h} \left(\frac{E}{E_c}\right)^{1/n} \ll 1$$

When $T_{cB} > 3T_0$, it is converted to

$$\frac{E}{E_h}\left(\frac{E}{E_c}\right)^{1/n} \ll 1 .$$

The written estimates lead to a practically important conclusion: the smaller the difference between the critical temperature of the superconductor and the temperature of the coolant, the greater the influence of the matrix and the higher the stable heating of the superconductor. It is of particular importance for HTS since they are characterized by high values of the temperature margin.

The dependencies γ and $\Theta = 1 - \theta$ on the dimensionless parameter $\iota = I^*/(2a\eta J_{c0})$ are shown in Figures 6.37 and 6.38, which are the generalized static voltage-current and temperature-current characteristics calculated assuming nonuniform and uniform temperature distribution over the cross-section of the composite. The calculations were based on the use of the Formulae (6.51)-(6.56). In this case, it was assumed that $E_\lambda = 10^3$ V/m based on the results shown in Figure 6.35, and the values E_h and E_ρ are varied. Namely, curves presented in Figure 6.37 were obtained at E_ρ = 0.05 V/m and variation of E_h, and curves depicted in Figure 6.38 were obtained at $E_h = 0.01$ V/m and variation of E_ρ.

The results presented prove that zero-dimensional model allows one to adequately describe the thermal and electrodynamic states observed in experiments if $\gamma \ll 1$ and $H \ll 1$. Besides, the current modes can be stable in the supercritical region ($\iota > 1$). This pattern is more noticeable when there is higher E_h or lower E_ρ. Composite superconductors may have stable stationary voltage-current and temperature-current characteristics with a positive slope in the entire range of their temperature variation up to the critical temperature as already noted. In other words, the transition from the superconducting state to the normal state will occur without an avalanche-like occurrence of the current instability only after reaching the critical temperature of the superconductor.

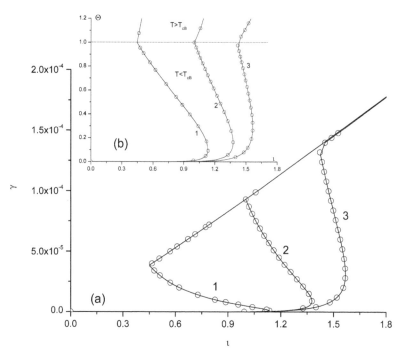

FIGURE 6.37 Stable and unstable branches of the generalized voltage-current (a) and temperature-current (b) characteristics at varying cooling conditions: 1 - E_b = 0.01 V/m, (α = 5); 2 - E_b = 0.05 V/m, (α = 1); 3 - E_b = 0.1 V/m, (α = 0.5); ——— - one-dimensional approximation, o - zero-dimensional approximation.

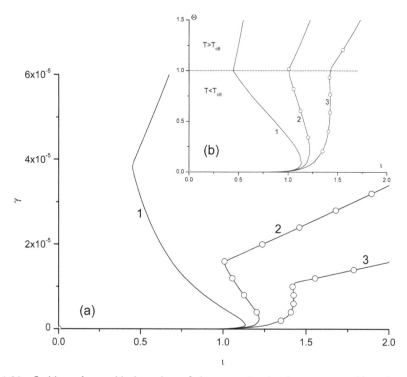

FIGURE 6.38 Stable and unstable branches of the generalized voltage-current (a) and temperature-current (b) characteristics of the superconducting composite at varying E_ρ: 1 - E_ρ = 0.05 V/m, α = 5; 2 - E_ρ = 0.01 V/m, α = 1; 3 - E_ρ = 0.005 V/m, α = 0.5; ——— - one-dimensional approximation, o - zero-dimensional approximation.

According to Condition (6.46), it is easy to find the quenching parameters: $\gamma_{q,1}$, that is the electric field before the current instability and $\iota_{q,1}$ which is the instability (quenching) current. In terms of the dimensionless variables used, the value $\gamma_{q,1}$ in the framework of the one-dimensional model is determined by solving equation

$$H = \dfrac{\dfrac{2n}{n+1}\sqrt{\gamma_{q,1}}\,th\sqrt{\gamma_{q,1}}\left(1+\dfrac{\varepsilon}{n}\gamma_{q,1}^{(n-1)/(n+1)}\right)}{\dfrac{2\sqrt{\gamma_{q,1}}}{sh2\sqrt{\gamma_{q,1}}}-\dfrac{n-1}{n+1}+\varepsilon\gamma_{q,1}^{(n-1)/(n+1)}\left(\dfrac{2\sqrt{\gamma_{q,1}}}{sh2\sqrt{\gamma_{q,1}}}+\dfrac{n-1}{n+1}\right)} \tag{6.57}$$

In zero-dimensional approximation, the value γ_q satisfies the equality

$$H = \gamma_q\dfrac{n+q/\gamma_q}{1+nq/\gamma_q} \tag{6.58}$$

Accordingly, the quenching currents, determined according to one-dimensional and zero-dimensional approximations, are equal to

$$\iota_{q,1}=\left(\dfrac{E_\lambda}{E_c}\gamma_{q,1}\right)^{1/(n+1)}\dfrac{1+q/\gamma_{q,1}}{\sqrt{\gamma_{q,1}}/th\sqrt{\gamma_{q,1}}+\gamma_{q,1}/H},$$

$$\iota_q=\left(\dfrac{E_\lambda}{E_c}\gamma_q\right)^{1/(n+1)}\dfrac{1+q/\gamma_q}{1+\gamma_q/H} \tag{6.59}$$

considering (6.57) and (6.58), respectively.

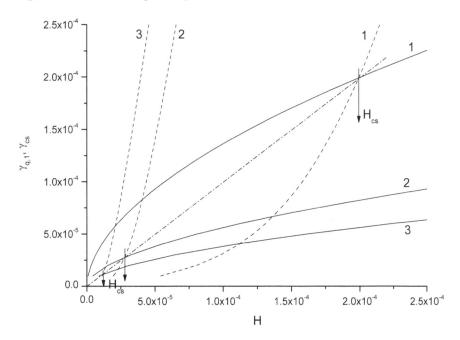

FIGURE 6.39 Influence of dimensionless thermal resistance on the values γ_q (- -) and γ_{cs} (——): 1 - E_p = 0.05 V/m, (α = 5); 2 - E_p = 0.01 V/m, (α = 1); 3 - E_p = 0.005 V/m, (α = 0.5).

Dependence $\gamma_{q,1}$ on the thermal resistance H for different values of E_p is shown in Figure 6.39. The calculations were carried out according to the Formula (6.57) at $E_\lambda = 10^3$ V/m and

$E_h = 0.05$ V/m. The curves $\gamma_{cs}(H)$, calculated according to (6.52) and thus, corresponding to the thermal states of the composite when its temperature becomes equal to the critical one are shown also. The intersection of the curves $\gamma_{q,1}(H)$ and $\gamma_{cs}(H)$ allows one to find the values H and γ, after exceeding of which the composite superconductors would be in a normal condition without the current instability. According to zero-dimensional approximation, the following states $\gamma_{q,1}(H_{cs}) = \gamma_{cs}(H_{cs})$ lead to the condition

$$\gamma = q = H_{cs} \qquad (6.60)$$

and it is shown by the dash-dotted line in Figure 6.39. Regarding the dimensional variables, it is written as follows

$$E_c^2 = E_\rho^{n+1}/E_h^{n-1} .$$

Therefore, the charged currents will stably flow only in the matrix at

$$E_c^2 > E_\rho^{n+1}/E_h^{n-1}$$

As discussed above, such modes are known as fully stable. Let us rewrite this condition in the form

$$\alpha < \left(\frac{E_c^2}{E_\rho E_h} \right)^{1/n} = \left[\frac{a(1-\eta)E_c^2}{\rho_m h(T_{cB} - T_0)} \right]^{1/n} ,$$

Here, α is the Stekly parameter introduced in Section 6.2. This condition shows that the full stability condition will be observed at such values of a, which should be knowingly less than 1 if $E_c < (E_\rho E_h)^{1/2}$. This feature is right for the modes discussed above. Indeed, the full stability is not observed even at $\alpha = 0.5$ as follows from Figures 6.37 and 6.38. At the same time, the full stability will take place at the value a that may exceed 1 at $E_c > (E_\rho E_h)^{1/2}$.

According to (6.58) and (6.59), the boundary value of the current, after which the avalanche-like transition to the normal state does not occur, is equal to $\iota_{q,0} = (E_h/E_c)^{1/(n+1)}$. It is obvious that the stable supercritical modes will take place for all $E_h > E_c$, since in this case $\iota_{q,0} > 1$.

In general, the written criteria show that the probability of the existence of the stable current charging in the supercritical region will be higher, the smaller E_ρ or the greater E_h. It should be emphasized that it is not possible to describe such modes within the framework of the thermal stabilization theory based on the jump-like transition from the superconducting state to the normal one ($n \to \infty$). It may be done only in the framework of the smoothed voltage-current characteristic of the superconductor.

The above-discussed features of the stable state's formation and the conditions of their destruction are based on the size effects affecting the stable performance of the current-carrying elements of superconducting magnetic systems. They lead to the following conclusion (Romanovskii and Watanabe 2006b). If the current-carrying elements have the same cross-sectional area but different transverse dimensions, they will not have identical conditions for the formation and destruction of their superconducting state. In Figure 6.40, the results of the numerical simulation of nonstationary (Curves 1′, 1″, 2′, 2″, 3′ and 3″) and stationary (Curves 1, 2 and 3) operating modes of the above superconducting composite are shown, the transverse dimensions of which were varied, provided that its cross-sectional area and the filling factor remain constant. In the first case, the calculation was carried out on the basis of the problem (6.39)-(6.40) and zero-dimensional model (6.43) was used in the second case. Dependencies described by Curves 1, 1′ and 1″ were obtained at $a = 9.5 \times 10^{-5}$ m, $b = 4.9 \times 10^{-3}$ m, Curves 2, 2′, 2″ at $a = 1.9 \times 10^{-4}$ m, $b = 2.45 \times 10^{-3}$ m and Curves 3, 3′, 3″ at $a = 3.8 \times 10^{-3}$ m, $b = 1.225 \times 10^{-3}$ m. Straight horizontal and vertical lines correspond to the quenching values of the electric field and current determined in the stationary approximation for each considered case.

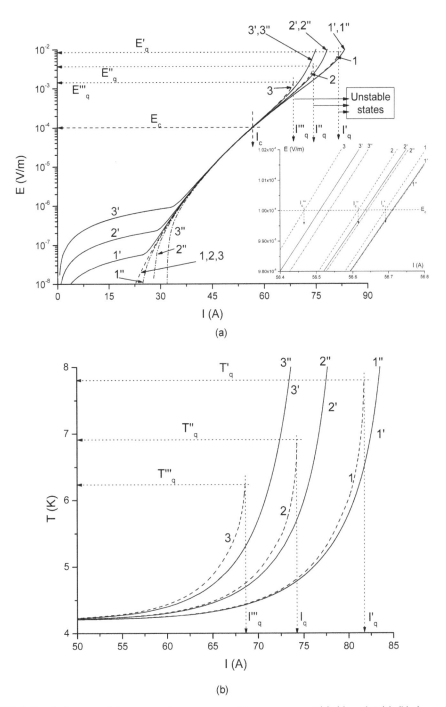

FIGURE 6.40 Influence of the composite transverse dimensions on $E(I)$ (a) and $T(I)$ (b) dependencies that originate on the surface (Curves $1'$, $2'$, $3'$) and in the center ($1''$, $2''$, $3''$) of tape when its cross-sectional area remains constant. Here, Curves 1, 2, 3 are the linear zero-dimensional approximation (6.43); Curves $1'$, $1''$, $2'$, $2''$, $3'$, $3''$ are the nonlinear 1D approximation (6.39)-(6.40).

Although S = const, the following regularities will be observed due to the size effect.

First, the initial formation stages of the full penetration modes may differ noticeably. As a result, the greater the thickness of the tape (and therefore its width is less), the more noticeable is not only the nonuniform distribution of the electric field over the cross-section of the tape but also the deviation of the critical current determined at $E = E_c$ from currents, which will be measured in experiments (see the inset in Figure 6.40a). This significantly increases the value of V_0, which is equal to

$$V_0 = \frac{2\mu_0 a^3}{S} \frac{dI}{dt},$$

and determines the minimum voltage level suitable for measuring the critical current when $S =$ const.

Second, a variation in the temperature of a composite in the stable supercritical region is very sensitive to the variation of the transverse tape dimensions. As a result, there will be a noticeable difference in the steepness of the rise in temperature when a current is charged. That means that the conditions under which the current instability occurs may be significantly different. Indeed, while maintaining the cross-sectional area of the composite, the quenching current calculated for the states described by Curves 1, 1' and 1'' is 16% larger than the instability current of the states described by Curves 3, 3' and 3''. In general, the following tendency is observed: the larger the thickness of the composite, the less its permissible overheating and the instability current at $S =$ const.

Since the appearance of instability is preceded by a stable increase in the temperature of the composite, it leads to a corresponding deviation of the nonstationary dependencies $E(I)$ and $T(I)$ from the stationary ones, to the greatest extent observed after complete current penetration. However, despite the well-conducting silver matrix in which the considered superconductor is located, the nonuniform temperature distribution over the cross-section affects the quenching. To confirm this conclusion, the results of determining the boundary of stable states by the method of finite perturbation of the initial state are presented in Figure 6.41. The numerical experiment was carried out in the framework of the nonstationary zero-dimensional (6.38) and one-dimensional (6.39)-(6.40) approximations for the composite discussed in section 6.6.

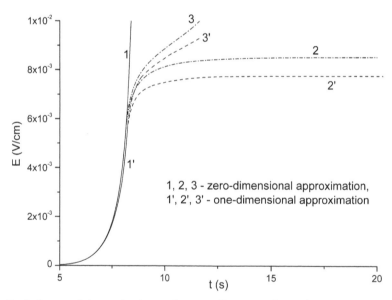

FIGURE 6.41 Influence of the conductive mechanism of heat transfer on the occurrence of the instability current at $dI/dt = 10$ A/s, $a = 9.5 \times 10^{-5}$ m, $b = 0.49 \times 10^{-2}$ m: 1, 1' - continuous current charging, 2 - $I_0 = 81.7$ A, 3 - $I_0 = 81.8$ A, 2' - $I_0 = 82.3$ A, 3' - $I_0 = 82.4$ A.

Thus, the results presented show that the occurrence of the current instability in composite superconductors can occur both at the partial and full penetration of the current into the composite. The highest current-carrying capacity is achieved at a low current charging rate. With its increase, it decreases (Figure 6.5). Under certain conditions, instability starts to occur when the composite cross-section is partially filled with the current. In other words, the current-carrying capacity of superconductors will not be fully realized under these conditions. Moreover, the dependence of $I_m(dI/dt)$ is characterized by a sharp decrease in the instability currents as the current charging rate increases. From the solution of equations (6.25) and (6.44) at $E_q = E_f$, it is easy to find the boundary separating the stability regimes, the violation of which occurs with the full and partial penetration of the current into the composite. The corresponding equation can be written as follows

$$\frac{\eta J_{c0}}{n E_c^{1/n}}\left(\frac{\mu_0 a}{4b}\frac{dI}{dt}\right)^{\frac{1-n}{n}} - \left(\frac{\eta J_{c0}}{E_c^{1/n}}\right)^2 \frac{a}{h(T_{cB}-T_0)}\left(\frac{\mu_0 a}{4b}\frac{dI}{dt}\right)^{\frac{2}{n}} = \frac{1-\eta}{\rho_m}\left[\frac{\eta}{n}\frac{J_{c0}}{E_c^{1/n}}\frac{a}{h(T_{cB}-T_0)}\left(\frac{\mu_0 a}{4b}\frac{dI}{dt}\right)^{\frac{n+1}{n}} - 1\right]$$

In particular, this equation allows one to estimate the value of the current charging rate after exceeding of which the realization of the current-carrying capacity of the superconductor will depend essentially on dI/dt, and instabilities will occur with the current partially penetrating the composite.

The composite superconductors have characteristic values of the electric field, which affect the steepness of the increase in their V-I characteristic. They depend on the current charging rate, the geometric dimensions of the composite, the critical properties of the superconductor, the conditions of heat transfer and the characteristics of current sharing between the superconductor and the stabilizing matrix. They underlie the formation mechanisms of the V-I characteristic of composite superconductors. The written criteria allow one to find the conditions for the existence of the stationary V-I characteristics with a positive slope that stably increase as the superconductor heats up to the critical temperature. With these modes, the spontaneous development of the current instability does not occur since the transport current stably flows from the superconductor to the stabilizing matrix as it increases. In this case, the stable supercritical modes may be observed when the charged current exceeds the conditionally set critical current of a superconductor. The formulated conditions for the occurrence of the current instabilities show that the nonuniform temperature distribution over the cross-section of a composite superconductor decreases the quenching currents. This regularity leads to an increase in the effect of the thermal degradation of the current-carrying capacity.

In general, the nonuniform temperature distribution over the cross-section of a composite superconductor depends not only on its thermal resistance but also on its critical properties whose influence will increase with the improvement of the heat transfer conditions. The introduced dimensionless parameter γ makes it possible to estimate the influence of the critical properties of a superconductor on the formation of its thermal and electrodynamic states. As a result, at $\gamma \ll 1$, the analysis of the thermo-electrodynamic states can be carried out on the basis of zero-dimensional approximation with a good degree of accuracy.

The discussed features of the thermal-electrodynamic state's formation in composite superconductors must be taken into account when measuring their V-I characteristics as well as when determining the conditions for their stable performance. The obtained results are especially important when analyzing the stability conditions of the superconducting state of the intensely cooled current-carrying elements manufactured on the basis of superconductors with high values of the critical current density.

6.8 FORMATION FEATURES OF THE THERMO-ELECTRODYNAMIC STATES DURING AN ALTERNATING CURRENT CHARGING INTO SUPERCONDUCTING COMPOSITES BASED ON HTS

As shown above, the superconducting composites can retain their operability under supercritical regimes under direct current load. Many experiments show that the current modes of HTS composites can be stable when an alternating current is charged into them, the peak values of which can be significantly higher than the conditionally specified critical current of the superconductor. However, the permissible stable modes of HTS composites carrying alternating current, as a rule, in practice are limited to the values of the critical currents determined on the basis of experiments performed for direct current using a priori defined critical value of the electric field of a superconductor. In other words, the stability mechanisms leading to the existence of the stable AC regimes are not taken into account. Let us discuss them when an alternating current charging into an HTS composite.

Let us investigate the formation mechanisms of the current modes in a superconducting composite of rectangular shape ($-a < x < a$, $-b < y < b$, $-\infty < z < \infty$, $b \gg a$) when a sinusoidal current is charged into it. Assume that

- superconductor $Bi_2 Sr_2 CaCu_2 O_8$ uniformly distributed with a filling factor η over the cross-section of the composite with a silver matrix;
- the transport current $I(t) = I_m \sin(2\pi f t)$ of a peak value I_m and frequency f is charged in the longitudinal direction of the axis z;
- the composite is in a constant external magnetic field parallel to its surface and the field has completely penetrated the superconductor;
- the own field of the current is much less than the external magnetic field;
- the critical current density J_c is described by a linear dependence of the form $J_c(T, B) = J_{c0}(T_{cB} - T)/(T_{cB} - T_0)$;
- the distribution of the temperature and electric field in the cross-section is uniform.

According to the model of an anisotropic medium, such operating mode may be described by the following nonstationary zero-dimensional model

$$C_k(T)\,dT/dt = -h(T - T_0)/a + E(t)I(t)/S, \quad T(0) = T_0$$
$$E(t) = E_c[J_s(t)/J_c(T, B)]^n = J_n(t)\rho_n(T, B), \quad E(0) = 0 \tag{6.61}$$
$$I(t) = \eta J_s(t)S + (1 - \eta)J_n(t)S$$

in which the thermal and physical parameters of the composite were defined above, and the physical meaning of the written equations has already been discussed.

Numerical simulations will be carried out at $a = 1.9 \times 10^{-4}$ m, $b = 0.245 \times 10^{-2}$ m, $\eta = 0.2$, $n = 10$, $E_c = 10^{-4}$ V/m, $J_{c0} = 1.52 \times 10^9$ A/m², $T_{cB} = 26.12$ K, $f = 10$ Hz. There were made under nonintensive heat transfer conditions ($h = 10$ W/(m² × K)) by taking the coolant temperature to be $T_0 = 4.2$ K and assuming that the background magnetic field is equal to 10 T.

The time dependencies of the temperature, and electric field for two characteristic peak values of the charged current are shown in Figure 6.42. Their determination was made according to the method of the finite perturbation of initial state. The results presented show that the formation of the stable and unstable states of the composite during an alternating current charging occurs similarly in the formation of the stable and unstable states observed for the operating modes when the transport currents close to the quenching current I_q are charged with a constant rate, when the metastable superconducting state either persists ($I < I_q$) or is destroyed ($I > I_q$). Accordingly, if charging an alternating current, then there is a limiting peak value I_m below which the superconductor retains its superconducting state, and above which it is destroyed. According to this concept, it is possible to determine the stability limit of the alternating current regimes in experiments using two characteristic peak values of the charged current. In this case, the minimum of them will correspond to the upper permissible value of the peak current, which is stably charged into the superconductor despite its high stable overheating and consequently high energy losses. The maximum value will determine the peak value of the current when the modes of the alternating current are unstable.

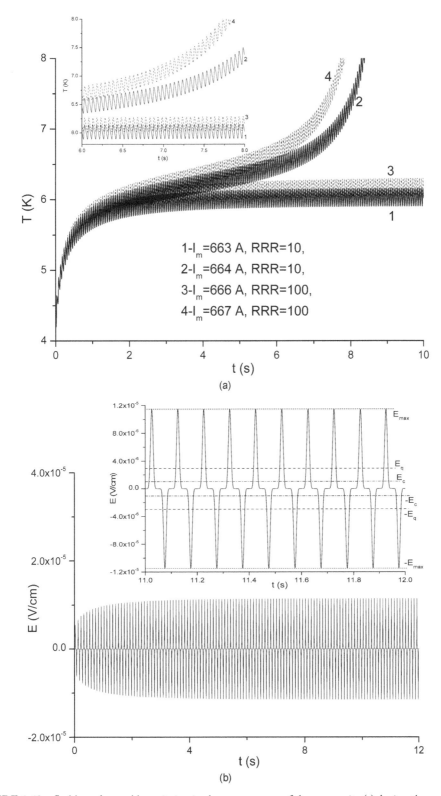

FIGURE 6.42 Stable and unstable variation in the temperature of the composite (a) during the continuous charging of the alternating currents close to the instability currents (in the inset the corresponding curves are shown in more detail); (b) - stable change in time of the electric field at RRR = 10, I_m = 663 A.

The role of the matrix resistance also demonstrates Figure 6.42. Namely, the lower its resistance, the higher the operating current. This regularity is due to the effect of the current sharing between the superconductor and matrix. Therefore, the stable peak temperature at RRR = 100 exceed the corresponding values at RRR = 10.

Let us discuss the mechanisms of sustainable formation of the alternating current modes. The curves describing the stable dynamics of the current, electric field and temperature are shown in Figure 6.43, which is observed near the instability boundary for the states corresponding to Curve 1 in Figure 6.42. In this case, the critical current of the composite is 566 A and the calculations of the limiting quenching parameters lead to the following values $E_q = 3 \times 10^{-4}$ V/m, $I_q = 573$ A, $T_q = 5.9$ K.

First, the existing stable modes of the charged alternating current can be defined as overloaded as follows from the presented results. In this case, the peak values I_m and E_{max} significantly exceeds not only the critical quantities, but also the corresponding values I_q and E_q. In particular, the peak value of the induced electric field is almost 10 times greater than a priori defined critical electric field.

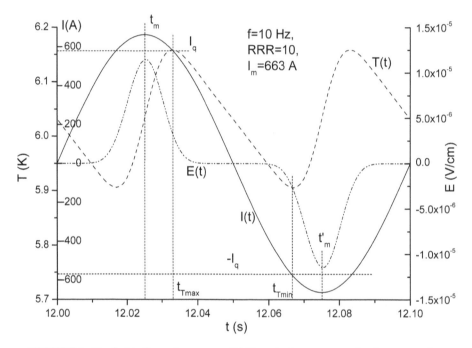

FIGURE 6.43   Stable formation stages of AC modes in a superconducting composite.

Second, there are characteristic stages of the stable development of the AC states in an HTS composite. Let us consider them when changing the current from I_m to $-I_m$, assuming for definiteness that the cycle begins after the peak value I_m which is reached at $t = t_m$ (Figure 6.43).

For the operating mode under consideration, the first stage exists in the interval $t_m < t < t_{Tmax}$, where t_{Tmax} is the time when the composite temperature reaches its maximum value. In this case, the current and electric field decrease almost simultaneously. It is important to emphasize that the charged current is unstable at this stage from the point of view of the stability conditions of the current that constantly flows through the superconductor since it exceeds the quenching current ($I(t) > I_q$). Calculations show that if a current is fixed at this stage, a spontaneous transition of the superconducting composite to a normal state will occur. However, the instability of the charged alternating current at this stage does not occur for the following reasons. First of all, it happens because the current is continuously decreasing. Besides, at $E > E_c$ the development of the electrodynamic states depends on the temperature dependence of the composite heat capacity as it was shown above for DC regimes. It plays a stabilizing role in the permissible increase in the electric field and temperature since the value $C_k(T)$ at this stage increases with increasing temperature.

In the second stage, which starts at $t > t_{T\max}$, the temperature begins to decrease. It is characterized by a decreasing current that is less than the quenching current ($I(t) < I_q$). The end of this stage is $t_{T\min}$. In addition, $I(t) = -I_q$.

At the third stage ($t > t_{T\min}$) the mutual change of the charged current and the induced electric field leads to an increase in the temperature of the composite, as $|I(t)| > I_q$. At this stage, the current is not stable in terms of the instability of direct current. The end time of this stage is $t = t_m'$, when the current and electric field takes their minimal values.

The following stages of the current variation at $t > t_m'$ are repeated as described above but with the opposite sign, and the stable formation of the overloaded cycles continues.

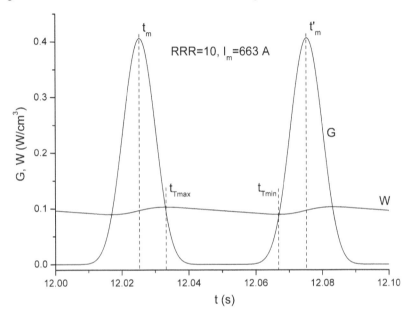

FIGURE 6.44 Stable variation stages of the Joule heat release G and heat flux into the coolant W.

Such formation stages of the stable overloaded alternating current operating modes lead to the corresponding features of the cyclical change of the released and removed heat fluxes (Figure 6.44). In fact, let us rewrite the first equation of the system (6.61) as follows $dT/dt = (G - W)/C_k(T)$. Since $dT/dt > 0$ in the first stage ($t_m < t < t_{T\max}$), then $G > W$. Accordingly, in the second stage $dT/dt < 0$, therefore, $G < W$, and in the third stage $G > W$, because $dT/dt > 0$. These features are necessary to understand the relationship between the values dI/dt and dE/dt, as the formula shows

$$\frac{dE}{dt} = \frac{\dfrac{1}{S}\dfrac{dI}{dt} + \dfrac{G(T) - W(T)}{C(T)}\left[\eta\left|\dfrac{dJ_c}{dT}\right|\left(\dfrac{E}{E_c}\right)^{1/n} + \dfrac{1-\eta}{\rho_n^2}\dfrac{d\rho_n}{dT}E\right]}{\dfrac{1-\eta}{\rho_n} + \dfrac{\eta J_c(T)}{nE}\left(\dfrac{E}{E_c}\right)^{1/n}}$$

which follows from the system (6.61). It demonstrates that the difference between the times of reaching minimax values of dI/dt and dE/dt, that is, the duration of the stages, depends on the frequency of the charged current, heat release and heat exchange, the filling factor, the temperature dependencies of the critical current density $J_c(T)$, matrix resistivity $\rho_n(T)$ and heat capacity of the composite $C_k(T)$. At a certain value of the frequency of charged current, the value of dI/dt becomes to play a prevailing role. In particular, the higher the frequency of the charged current, the shorter the duration of the existence of unstable states during which $|I(t)| > I_q$. In other words, operating AC modes will become more stable with increasing frequency.

Let us formulate stability conditions for overloaded AC modes by analyzing the features of changes in time of the corresponding averaged values of temperature, heat release and cooling power, defining them as

$$T_{av} = \frac{1}{t}\int_0^t T dt, \quad G_{av} = \frac{1}{t}\int_0^t G dt, \quad W_{av} = \frac{1}{t}\int_0^t W dt.$$

They allow one to go from the corresponding instantaneous values to more visual average values.

Figure 6.45 shows the results of the numerical simulations of the stable (Curves 1 and 3) and unstable (Curves 2 and 4) increase in the average temperature of the composite during continuous charged of the alternating currents close to the instability states. Their corresponding instantaneous values are shown in Figure 6.42. It is seen that, there is a maximum allowable increase in the average temperature of the composite $T_{q,av}$. After exceeding this temperature the instability will occur. Thus, the change in time of the average temperature allows one to determine the stability boundary of the overloaded AC modes in accordance with the method of the finite perturbations of initial state. It is obvious that, in the same way, one can find the corresponding average value of the electric field, which is the boundary between stable and unstable states. Let us emphasize that the value $T_{q,av}$ appears as a result of the dynamic conditions of the heat balance between the released and removed heat fluxes. As an illustration of the existence of the stable heat release in a composite carrying alternating current, the corresponding values of G_{av} are presented in Figure 6.46.

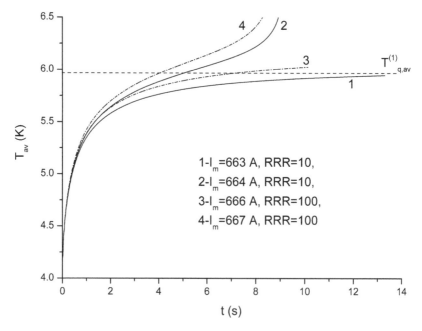

FIGURE 6.45 Change in the averaged values of the temperature of composite with different resistance of the matrix near the instability boundary. Here, $T_{q,av}^{(1)}$ is the temperature boundary between stable and unstable states for regime 1.

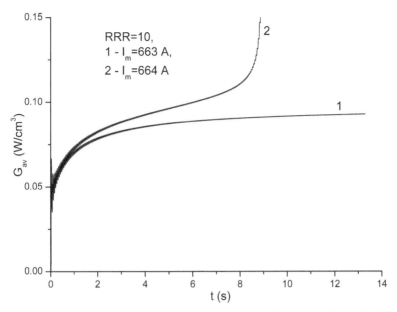

FIGURE 6.46 Dynamics of the averaged heat release in the stable (Curve 1) and unstable (Curve 2) states.

As a result, the determination of the stable state's boundary and the corresponding value $T_{q,av}$ can be performed by analyzing the dependencies $G_{av}(T_{av})$ and $W_{av}(T_{av})$. The results of the corresponding calculations are presented in Figure 6.47. It also shows that the limiting stable regimes of the alternating current arc uniquely determined by the conditions of the dynamic equilibrium between the averaged values of the Joule heat release and heat flux into the coolant. According to these results, the stability condition of the alternating current regimes in zero-dimensional approximation is written as follows

$$G_{av}(T_{q,av}) = W_{av}(T_{q,av}), \; \partial G_{av}(T_{q,av})/\partial T = \partial W_{av}(T_{q,av})/\partial T$$

at $t \to \infty$.

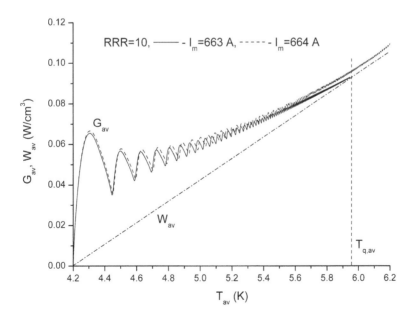

FIGURE 6.47 Averaged Joule heat release G_{av} and heat flux to the coolant W_{av} as a function of the average temperature of the composite T_{av}: —— - steady states, - - - - - unstable states.

The general mechanisms for forming current states and methods for determining the stability of alternating current discussed above allow for a detailed analysis of the stability conditions of superconducting composites, taking into account different cooling conditions, external magnetic field, design parameters of composites and other features. In particular, the results of determining the amplitude of the limiting currents depending on the frequency of the charging current are shown in Figure 6.48 for the considered composite. It is seen that under the conditions of nonintensive cooling, the limiting AC modes will weakly depend on the frequency of the charged current in a wide range of variations in the operating parameters of the composite, which is a consequence of the stabilizing effect of the matrix that effectively dampens the occurrence of the unstable states.

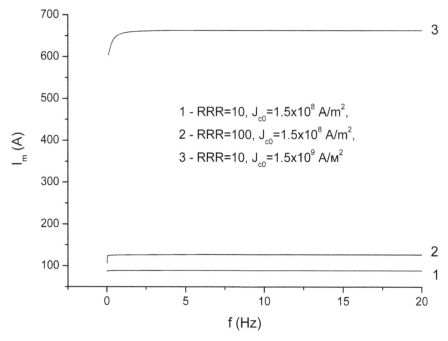

FIGURE 6.48   Influence of the frequency for the current charged into the superconducting composite on its peak limit values.

6.9   COMPARISON OF THEORETICAL RESULTS WITH EXPERIMENTS

To verify the above-formulated regularities of nonisothermal formation of the electrodynamic states of superconducting composites, the theoretical results of analyzing the conditions for the occurrence of current instabilities in the composite based on $Bi_2Sr_2CaCu_2O_8$ in a silver matrix cooled by a cryocooler to the temperature of liquid helium were compared with the experimental results (Romanovskii and Watanabe 2009b, Watanabe et al. 2004). The experiments were performed at High Field Laboratory for Superconducting Materials, Institute for Materials Research, Tohoku University, Japan. The initial parameters are presented in Table 5.

The superconducting composite was produced according to the powder-in-tube method. The sample was made in the form of a single-layer coil consisting of nontouching turns. The coil turns were coated with a thin layer of vacuum grease to mechanically fix the sample. The sample was supplied with potential contacts and resistance thermometers to monitor the sample temperature and measure the developed voltage. The induction of the external magnetic field was parallel to the wide side of the tape. The magnetic field was controlled by a Hall sensor. The critical currents were determined by the level of the electric field equal to 1 μV/cm using the V-I characteristic.

TABLE 5 Ag/Bi2212 composite parameters

Width (mm)	4.9
Thickness (mm)	0.38
Number of wires	54
RRR	110
Ag/Bi2212	4/1
T_c (K)	87
T_0 (K)	4.2
n	11

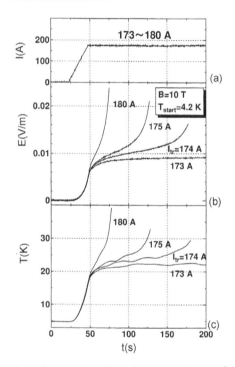

FIGURE 6.49 Experimental dependencies describing the time variation of the charged current (a) electric field (b) and temperature (c) of the composite.

For the experimental determination of the boundary of stable current states, the method of finite perturbation of the initial state was applied (Figure 6.49). In the calculations, the model described by Equations (6.39)-(6.41) was used, taking into account the nonlinear dependence of the critical current density of the superconductor on temperature. According to experimental measurements, the latter was determined by the Formula (1.8) with parameters (6.32). The volumetric heat capacity of a superconductor was described by the Expression (4.34), the matrix heat capacity was calculated according to (Dresner 1993) and the matrix thermal conductivity was determined in accordance with Wiedemann-Franz law by taking into account, as described above, the dependence of the electrical resistivity of silver on the temperature and external magnetic field.

The experimental temperature curves were used as calibrations in terms of setting the effective value of the heat transfer coefficient. In this case, the desired value was equal to $h = 2 \times 10^{-7}$ W/(m^2 × K). Accordingly, it was used in further calculations. As a result, the currents of instability occurrence and the corresponding temperatures which precede them, as functions of the induction of an external magnetic field, were determined. They are compared with the experiments in Figure 6.50.

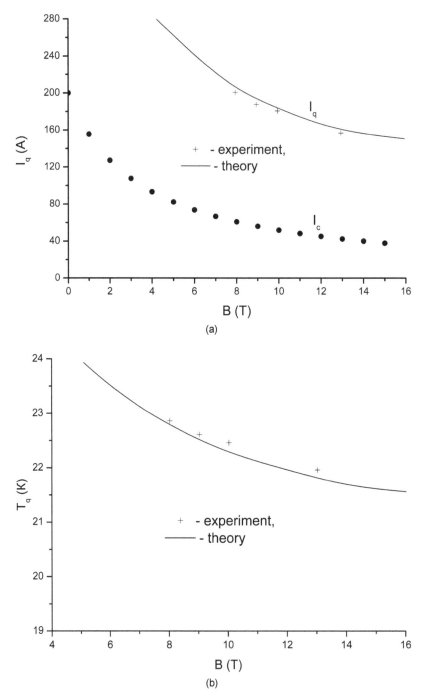

FIGURE 6.50 Dependence on the magnetic field of the instability currents I_q and critical current I_c of the composite (a) and the temperature of the composite (b) before their occurrence.

The presented numerical results satisfactorily describe the experimentally observed dependencies. They confirm that the currents of instability may be supercritical and are accompanied by significant stable overheating of the composite under the nonintensive cooling conditions. Namely, the considered composite retains the current-carrying capacity when overheated by more than 15 K. Note that the experimental confirmation of the dependence of permissible stable overheating on the properties of HTS composites and the conditions for their cooling follows from many experiments. In general, the experimental and theoretical results that take into account the stable increase in the temperature of the composite, which can be observed under stable conditions, significantly expand the range of the stable performance of superconducting composites.

6.10 CONCLUSIONS

The nonisothermal phenomena of the stable current mode formation in superconducting composites and the conditions for their violation are characterized by the existence of the following regularities:

1. The equivalency of the stability conditions for currents charged into the composite superconductors, whose V-I characteristics are described by power or exponential equations, depend on the rise parameters of V-I characteristics. The difference between them increases with the decreasing of parameters of the V-I characteristic rise. The most noticeable difference will be observed in superconductors with a strong creep ($n < 10$, $J_\delta/J_{c0} > 0.1$).
2. The current instabilities can occur both at the partial and full penetration of the charged current. Current instability conditions arising at the full penetration mode do not depend on the current charging rate but depend on the superconductor temperature margin and the features of the current sharing between the superconducting core and matrix. The latter can lead to a significant stable increase in the composite overheating (more than 10 K) before the occurrence of instability.
3. There are characteristic values of the electric field, which determine the influence of the thermal mechanisms of the stable state's formation on the development of thermo-electrodynamic phenomena in superconducting composites.
4. The boundary values of the permissible values of the electric field and current may be both subcritical and supercritical. Their existence depends on the properties of the superconductor and matrix, the conditions of the heat exchange with the coolant and the induction of an external magnetic field. In this case, the supercritical stability modes are characterized by the high permissible overheating of the composite.
5. Due to the inevitable overheating of composites before the instability, there is the effect of the thermal degradation of their current-carrying capacity. As a result, the currents of instability do not increase in proportion to the increase of their critical current. It significantly affects the stability conditions in the subcritical modes.
6. In the supercritical stability region, multistable modes may arise. Their existence is associated with the mechanism of the stable current sharing between the superconducting core and matrix. They are the result of an interconnected change in temperature and external magnetic field of the values $J_c(T)$, $\rho_m(T)$, $|\partial J_c/\partial T|$ and $\partial \rho_m/\partial T$. As a result, at the static V-I characteristic of composites, the stable voltage surges may be observed that do not convert the superconductor to the normal state despite a significant increase in its temperature.
7. A consequence of the existence of multistable states is the occurrence of modes that are stable over the entire range of the temperature variations in the composite from the coolant temperature to the critical temperature of the superconductor.
8. The mechanism of the current sharing between the superconducting core and matrix may have a significant effect on the stability conditions of the charged current when cooling HTS composites by liquid coolant with a high-temperature range of existence of the nucleate boiling regime. In addition, with the use of liquid refrigerants, the occurrence of unstable

states can occur before the boiling crisis occurs due to the violation of the steady increase in the stationary V-I characteristic of the composite.

9. The formation of the supercritical V-I characteristic branch of a superconducting composite (monostable or multistable) substantially depends on the variation of its heat capacity with temperature. As a result, the V-I characteristic of the composite not only has a positive slope at continuous current charging but it decreases with an increase in the rate of current charging due to a corresponding increase in the temperature of the composite in both stable and unstable states. Moreover, the heat capacity of the composite, which increases with temperature, significantly modifies the character of the V-I characteristic increase in which multistable regions exist.

10. When an alternating current is charged into a superconducting composite, there are stable overloaded regimes when the currents of occurrence of unstable states not only significantly exceed the conditionally defined critical parameters but also the quenching currents defined for DC modes.

7

Thermal Phenomena in Composite Superconductors with an Ideal Voltage-Current Characteristic and their Thermal Stabilization Conditions

To provide the stable performance of superconducting magnet systems, it is important to study the conditions of the onset and propagation of the local areas with normal resistivity (the so-called normal zones). This is because superconductors in a normal state have very high electrical resistivity. When a normal zone appears in the SMS, the energy stored by the magnet is released in it heating the current-carrying element. As a result, more and more part of SMS begins to move to the normal state leading to a further increase in its temperature. This phenomenon, known as quench, may lead to the destruction of SMS. In this case, the arising problems are directly related to the analysis of the thermal state of the current-carrying elements under the action of numerous perturbations which are unavoidable under the change of the magnetic field, current charging, etc. Their solution underlies the thermal stabilization theory of the composite superconductor. Let us discuss the main features of this theory.

7.1 ONSET CAUSES OF NORMAL ZONES. TYPES OF THERMAL PERTURBATIONS. DESCRIPTION FEATURES OF THE THERMAL PHENOMENA IN COMPOSITE SUPERCONDUCTORS WITH AN IDEAL VOLTAGE-CURRENT CHARACTERISTIC

A local destruction of the superconducting properties may occur for many reasons. First of all, this happens due to the existence of the critical values of temperature, magnetic field induction and current of superconductors. If one of them is exceeded, a normal zone appears in the SMS. In particular, 2G HTS tapes have a local variation of the critical current density over their thickness and length. As a result, a normal zone may appear at a 'weak spot' that may lead to a subsequent transition of tape to the normal state.

To avoid the exceeding of the critical parameters, one should select the permissible operating characteristics of magnets in the following way. The maximum permissible induction of the operating magnetic field is determined by the intersection of the load line and the curve defining the dependence of the critical current density on magnetic induction at the operating temperature T_0.

As an illustration, Figure 7.1a shows a scheme for determining the permissible operating parameters of a superconducting solenoid. Here, Line 1 is the load line defined by magnetic induction at the center of the solenoid, and the load Line 2 is defined by the maximum magnetic induction at the center of the solenoid at its internal radius. Then, Point 3 is the point of the intersection of the Load Line 2 and the dependence $J_c(T, B)$ at the operating temperature T_0. It corresponds to the maximum permissible magnetic induction B_{max} below which the superconductivity of the solenoid is not destroyed. However, in practice, the operating value B_{max} is often chosen with a margin, shifting down the load line, moving from the mode in the Point 3; for example, to the mode in the Point 4 to avoid the consequences of the possible 'weak point' where the critical current density may be lower than the operating critical density (Figure 7.1b).

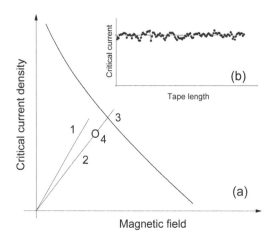

FIGURE 7.1   Determination of the maximum magnetic field in the superconducting solenoid (a) and the critical current distribution along the HTS tape length (b).

Besides, the sources of the current-carrying elements heating may be magnetic flux jumps. According to the results discussed in Chapters 4 and 5, they should be avoided by reducing the transverse size of the superconducting filaments and composites. Another source is losses caused by a variable current or magnetic field. These losses are diminished by reducing the size of superconducting filaments to submicron values. Losses may also occur when the SMS is exposed to high-energy particles.

In general, the noted causes of the normal zone occurrence are predictable, their occurrence conditions may be predicted and thus, one may avoid their appearance. However, other reasons for their premature transition to the normal state may exist. Numerous experiments have led to the conclusion that they are based on the heat release that may occur unpredictably in the winding during current charging or when the magnetic field changes. They may lead to the local heating of the superconductor above its critical temperature. Indeed, the current-carrying elements of SMS carrying the high-density currents are subjected to considerable ponderomotive forces at high magnetic fields. As a result, they experience high mechanical loads. Besides, the difference in the thermal expansion coefficients of the winding components also leads to significant mechanical stresses. As a result, the movement of turns, their friction against each other, plastic matrix deformation and jump-like deformation of superconductors at low operating temperatures may occur in SMS. Such phenomena are accompanied by the release of a certain amount of the heat leading to a local temperature increase of the current-carrying element and the possible appearance of the normal zone. Consequently, numerous perturbations of the mechanical nature acting in SMS are responsible for the degradation of its current-carrying capacity and the training of SMS.

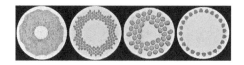

FIGURE 7.2   Composite structure of some technical superconductors.

The undesirable appearance of the normal zones in the SMS led to the development of the thermal stabilization method providing the preservation of their operability. It consists of shunting a superconductor by a nonsuperconducting metal of high conductive properties (Figure 7.2). The composite structure of the superconducting current-carrying element leads to the sharing of the transport current between its superconducting core and nonsuperconducting matrix after a normal zone appears in a superconducting composite for any reason (Figure 7.3a). The heat released in the matrix is discharged to the coolant. Under certain conditions, which are discussed below, the mechanism of the current sharing and the heat sink of the heat released allow the normal zone to disappear. As a result, the superconducting properties of the current-carrying element are restored, and the current returns to the superconducting cores. When the conditions of the thermal stability are violated, the quench (irreversible transition to the normal state) occurs (Figure 7.3b) and the SMS irreversibly passes to the normal state.

current sharing   normal zone

(a)   (b)

FIGURE 7.3 Current sharing in the superconducting composite: (a) – the transport current both in the superconducting core and the matrix; (b) – the transport current flows only in the matrix when a superconductor is in the normal state.

In general, the calculation of the operating modes of SMS made of the superconducting composite wires, which are isolated from each other by nonconducting isolation with the coefficient of the thermal diffusivity three or four orders lower than in the current-carrying element, requires the solution of the multi-dimensional nonstationary equations of the thermal diffusion with the piecewise-continuous coefficients. The models used in this case should take into consideration the design features of the SMS, physical heterogeneity of its internal structure and magnetic field variation in it and so on. However, this is associated with a significant amount of the calculations based on a system of the form (1.12)-(1.13). It significantly limits the formulation of general physical regularities that allow one to understand the features of the phenomena occurring in the current-carrying elements when a normal zone appears in SMS. In this connection, the analysis of the thermal stability conditions and the peculiarities of the irreversible propagation of the thermal instabilities are usually carried out for a single superconducting composite. In this case, the operating modes of the composite are described only by the thermal conductivity equation of the form

$$C_k(T)\frac{\partial T}{\partial t} = \text{div}[\lambda_k(T)\text{grad }T] + G + P \tag{7.1}$$

Here, as above-mentioned, C_k is the volumetric heat capacity of the composite, λ_k is its coefficient of the thermal conductivity, $G = EJ$ is the Joule heat release caused by an instantaneous electric field change by temperature and P is the volumetric power of external heat sources.

The value G may be calculated within the framework of various approximations. In general, the equation of the form (1.12) must be solved. However, as noted in Chapter 5, the development of the thermo-electrodynamic states in composite superconductor depends on the ratio of the characteristic times of the electromagnetic field diffusion to the time of the heat flux diffusion. Accordingly, it leads to the dimensionless parameter $\Lambda_k = \lambda_k\mu_0/C_k\rho_k$, which satisfies the condition $\Lambda_k \gg 1$ for the superconducting composites (Gurevich, Mints, Rachmanov, 1997c). Then the value G can be determined by taking into account the real voltage-current characteristic of the composite in which the electric field and current density are determined by the instantaneous temperature variations of the composite according to the condition $\Lambda_k \gg 1$. Both of these approximations will be used below. Finally, the calculation G can be performed within the framework of the simplest model based on the ideal V-I characteristic of a superconductor when a jump-like S-N transition from the superconducting to the normal state occurs at the critical current. In this case, the differential resistivity of superconductor abruptly varies from zero to an infinitely large value. Let us use this approximation and find the corresponding value G.

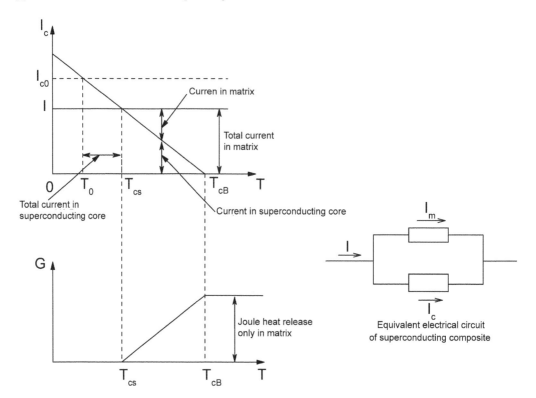

FIGURE 7.4 Equivalent electrical circuit of the superconducting composite, the current sharing mechanism in it and the temperature dependence of the Joule heat release at $\rho_m = $ const and a linear temperature decrease of the critical current.

If a transport current I, which is less than the critical current, flows in the composite, then it flows only in the superconducting core of the composite without energy losses. If, for any thermal reason, the critical current becomes less than the transport current, then a part of the transport current will flow into the matrix (Figure 7.3a). As a result of the current sharing, the current I_m will flow through the matrix and the current which is equal to the corresponding value of the critical current $I_c(T)$ will flow in its superconducting core. The equivalent electrical circuit of the superconducting composite as two parallel-connected conductors is presented in Figure 7.4. According to the Kirchhoff equations, for this circuit the following relations

$$I = I_c + I_m, \quad E = \frac{I_m \rho_m}{(1-\eta)S}$$

take place. Here, ρ_m is the electrical resistivity of the matrix depending in general on temperature and magnetic field and η is the filling factor of the composite with a superconductor. Accordingly, the Joule heat release is equal to

$$G = EJ = EI/S = \frac{I^2 \rho_m}{(1-\eta)S^2}\left(1 - \frac{I_c}{I}\right)$$

At a linear temperature dependence of the critical current, that is as above at $I_c(T) = I_{c0}(T_{cB} - T)/(T_{cB} - T_0)$, it is easy to find

$$\frac{I_c}{I} = \frac{I_{c0}}{I}\frac{T_{cB} - T}{T_{cB} - T_0} = \frac{T_{cB} - T}{T_{cB} - T_{cs}},$$

where T_{cs} is the so-called current sharing temperature after exceeding which a part of the current $I - I_c(T)$ overflows into the matrix because $I > I_c(T)$ and the composite goes to the resistive state. Its value is determined from the condition $I = I_c(T_{cs})$, and it is equal to

$$T_{cs} = T_{cB} - (T_{cB} - T_0)I/I_{c0} \qquad (7.2)$$

Accordingly, all current flows only in the superconducting core without the Joule losses at $T < T_{cs}$ (Figure 7.4). When $T > T_{cB}$, the superconductor goes into a normal state and the transport current flows only in the matrix. As a result, the temperature dependence of the Joule heat release in a superconducting composite is described by the following expression

$$G(T) = \frac{I^2 \rho_m}{(1-\eta)S^2} \begin{cases} 1, & T > T_{cB} \\ \dfrac{T - T_{cs}}{T_{cB} - T_{cs}}, & T_{cs} \leq T \leq T_{cB} \\ 0, & T < T_{cs} \end{cases} \qquad (7.3)$$

within the framework of the jump S-N transition model.

Thus, in the simplest approximation, the power of the heat released in a superconducting composite is described by a piecewise linear temperature dependence (7.3). It changes from zero value at $T_0 < T < T_{cs}$ to the maximum value at $T > T_{cB}$ (Figure 7.4). It should be emphasized that the dependence $G(T)$ is shown in Figure 7.4 that corresponds to the state when ρ_m = const. For well-conducting metals, this condition is satisfied with acceptable accuracy for $T < 20$ K. At the same time, the corresponding temperature dependencies $\rho_m(T)$ must be taken into account at $T > 20$ K.

Many experiments and the reasons discussed above for the occurrence of a normal zone in SMS show that the perturbations that actually may exist in the superconducting winding are characterized by their length (from local to extended) and duration (from short-term perturbations to stationary ones). They may be described using the following thermal models.

If in some volume of winding V_1 a certain amount of the heat release is emitted almost instantly and uniformly, then the temperature of this area rises adiabatically from the coolant temperature T_0 to some temperature T_1. These states may be described using the initial condition of the form

$$T\big|_{t=0} = \begin{cases} T_1, & V \leq V_1 \\ T_0, & V > V_1 \end{cases} \qquad (7.4)$$

If $P = 0$, then the value of the energy that is released during this perturbation is equal to

$$E_1 = \int_{V_1} dV \int_{T_0}^{T_1} C_k(T)dT \qquad (7.5)$$

Let us call such perturbations as *the temperature perturbations*.

If the length of the region where a certain amount of heat is released may be neglected, then the corresponding change in the thermal state of the composite may be described by the boundary condition

$$\lambda_k \frac{\partial T}{\partial \bar{n}}\bigg|_{\bar{r}=0} = \begin{cases} f_1, & t \leq t_1 \\ 0, & t > t_1 \end{cases} \qquad (7.6)$$

Here, f_1 is the heat release per unit of the cross-sectional area released during time $0 < t < t_1$. If $P = 0$ and the initial condition is $T\big|_{t=0} = T_0$, then the total energy released as a result of the action of this type of perturbation, called as *a point impulse perturbation*, is equal to

$$E_1 = 2\int_S dS \int_0^{t_1} f_1 dt \tag{7.7}$$

taking into account the symmetry temperature distribution in the composite.

The operating modes of the superconducting composites under the action of the thermal perturbation of finite duration and length may be described by a model where the spatiotemporal distribution of the volumetric heat release source is determined by the quantity P. Then, provided that the temperature of the composite was equal to the temperature of the coolant at the initial time, the total amount of the energy released in the volume V_1 during t_1 is determined as follows

$$E_1 = \int_{V_1} dV \int_0^{t_1} P dt \tag{7.8}$$

Such perturbations may be called as *volumetric*. They may be both local and extended as well as their duration may be short or long.

These models allow one to describe the thermal phenomena in the superconducting composites under the action of the thermal perturbations of various types making it possible to understand the basic mechanisms responsible for the thermal destruction of the superconducting properties of SMS.

7.2   THERMAL PHENOMENA IN THERMALLY THIN COMPOSITE SUPERCONDUCTORS: QUASI-LINEAR APPROXIMATION

Let us discuss the characteristic features of the thermal mode development in the superconducting composites investigating the change in the thermal state of a cooled composite superconductor under the action of the thermal perturbations. Let us assume that the composite, the boundaries of which $z = \pm l_k$ are maintained at the coolant temperature T_0, is in a constant external magnetic field; perfect thermal and electrical contacts are provided between superconducting filaments and matrix; the condition $hS/(p\lambda_\perp) \ll 1$ takes places ($\lambda_\perp$ is the thermal conductivity coefficient of the composite in the transverse direction) and the nonuniform temperature distribution in the cross-section of the composite may be neglected (the so-called thermally thin composite); the volume fraction of superconducting in the composite (the filling coefficient) is η. Assume that the sources of the thermal perturbation of the first or third types are arranged symmetrically to the boundaries of the composite; the heat transfer coefficient is h and the thermophysical properties of the composite (C_k, λ_k, ρ_m) do not depend on temperature. Averaging Equation (7.1) in the plane of the cross-section, the one-dimensional heat flux diffusion in the longitudinal direction of the composite may be described by the heat conduction equation of the form

$$C_k \frac{\partial T}{\partial t} = \lambda_k \frac{\partial^2 T}{\partial z^2} - \frac{hp}{S}(T - T_0) + \frac{I^2}{S^2}\rho_k(T) + P(z, t) \tag{7.9}$$

considering the following initial and boundary conditions

$$T\big|_{t=0} = \begin{cases} T_1, & 0 \le |z| \le l_1 \\ T_0, & l_1 < |z| \le l_k \end{cases}, \quad T(\pm l_k, t) = T_0 \tag{7.10}$$

Here, l_1 is the half-length of the initial temperature perturbation, on which the composite instantly warms up to a temperature T_1, p is the cooled perimeter, S is the cross-sectional area, I is the transport current flowing along the axis Z and $\rho_k(T)$ is the effective resistivity of the composite described according to (7.3) by the formula

$$\rho_k(T) = \rho_0 \begin{cases} 1, & T > T_{cB} \\ \dfrac{T - T_{cs}}{T_{cB} - T_{cs}}, & T_{cs} \leq T \leq T_{cB} \\ 0, & T < T_{cs} = T_{cB} - (T_{cB} - T_0)\dfrac{I}{I_{c0}} \end{cases} , \quad \rho_0 = \dfrac{\rho_m}{(1-\eta)} = \text{const} \tag{7.11}$$

To carry out the analysis at a qualitative level of rigor, let us introduce the next dimensionless variables $i = I/I_{c0}$, $\theta = (T - T_0)/(T_{cB} - T_0)$, $Z = z/L_z$, $\tau = t/t_x$ using the following values of the characteristic length $L_z = \sqrt{\lambda_k(T_{cB} - T_0)S^2/(I_{c0}^2\rho_0)}$ and characteristic time $t_x = C_k L_z^2/\lambda_k$. Let us estimate these values for a composite superconductor based on Nb-Ti in a copper matrix setting averaged parameters $S \sim 3 \times 10^{-6}$ m^2, $\lambda_k \sim 200$ W/(m $\times$ K), $C_k \sim 10^3$ J/(m$^3 \times$ K), $\rho_0 \sim 10^{-9}$ $\Omega \times$ m, $I_{c0} \sim 10^3$ A, $T_{cB} - T_0 \sim 5$ K. Then, at adiabatic states, the operating modes existing in superconducting composites at the helium temperature level have the following characteristic space-time scales $L_z \sim 3 \times 10^{-3}$ m and $t_x \sim 4.5 \times 10^{-5}$ s.

Regarding the introduced dimensionless variables, the problem (7.9)-(7.11) is rewritten as follows

$$\frac{\partial \theta}{\partial \tau} = \frac{\partial^2 \theta}{\partial Z^2} - W(\theta) + G(\theta) + P_0(Z, \tau) \tag{7.12}$$

$$\theta\big|_{\tau=0} = \begin{cases} \theta_1, & |Z| \leq \xi_1 \\ 0, & |Z| > \xi_1 \end{cases} , \quad \theta(\pm L, \tau) = 0, \quad \xi_1 = l_1/L_z, \quad L = l_k/L_z \tag{7.13}$$

Here,

$$W(\theta) = \theta/\alpha, \ G(\theta) = i^2 r(\theta), \ P_0 = \frac{PL_x^2}{\lambda_k S(T_{cB} - T_0)}, \ r(\theta) = \begin{cases} 1, & \theta > 1 \\ \dfrac{\theta - 1 + i}{i}, & 1 - i \leq \theta \leq 1 \\ 0, & \theta < 1 - i \end{cases} \tag{7.14}$$

$$\alpha = \frac{I_{c0}^2 \rho_0}{hpS(T_{cB} - T_0)} = \frac{\eta^2 J_{c0}^2 \rho_m S}{hp(1-\eta)(T_{cB} - T_0)} \tag{7.15}$$

Equation (7.12) allows to make a quasi-linear analysis of the thermal states of a composite superconductor under the action of the thermal perturbations of various types, varying the dimensionless current ($0 < i < 1$) and the dimensionless parameter $\alpha > 0$, known as the thermal stabilization parameter or the Stekly parameter as mentioned before.

In the framework of the introducing dimensionless variables, the heat release and heat transfer are described by the expressions $G = i^2 r(\theta)$ and $W = \theta/\alpha$; accordingly the dimensionless current sharing temperature is equal to the $1 - i$, the dimensionless critical temperature of the superconductor is the $\theta_c = 1$ and the temperature of the coolant is zero. Therefore, the current begins to share at $1 - i < \theta < 1$ in the area with moving boundaries $\pm\xi_n(\tau)$ and $\pm\xi_c(\tau)$, which are determined from the solution of equations $\theta(\pm\xi_n(\tau)) = 1 - i$ and $\theta(\pm\xi_c(\tau)) = 1$, if they exist.

In the approximation under consideration, the parameter α, which formally appeared as a result of the reduction to dimensionless form of the heat equation (7.9), is equal to the ratio of the characteristic values of the heat release in the matrix where the critical current I_{c0} flows to the heat flux into the coolant when the temperature of the composite is equal to the critical temperature T_{cB}. This parameter is also determined as the ratio between the characteristic values of the electric field affecting the current sharing and the deviation of the composite temperature from the coolant temperature, as it was shown in Section 6.2. The value of α depends on the critical parameters of the superconductor (and hence, the magnetic field induction), the transverse size of the composite, the resistivity of its matrix, filling factor and heat exchange conditions with the coolant. Let us estimate the value α for the composite superconductor with the following parameters $\rho_m \sim 10^{-10}\ \Omega \times$ m, $J_{c0} \sim 10^9$ A/m^2, $T_{cB} - T_0 \sim 5$ K, $\eta \sim 0.5$, $S/p \sim 10^{-3}$m. Then, $\alpha \sim 100$ under the nonintensive conditions of the heat transfer ($h \sim 100$ W/(m$^2 \times$ K)) and $\alpha \sim 10$ under the intensive conditions of the heat transfer ($h \sim 10^3$ W/(m$^2 \times$ K)).

Using the parameter α, Stekly has formulated the cryogenic stability condition for the first time (Kantrowitz, Stekly, 1965). It is written as follows $\alpha \leq 1$ for the constant values of the coefficients contained in the Stekly parameter. If the operating current density has to be high, the coil may be tightly packed, for example, as in accelerator magnets, and the cryostability condition is not applicable since $\alpha > 1$ for these types of SMS. The first cryostable magnet was the Big European Bubble Chamber at CERN. At nominal operating conditions, its current-carrying element was operating at the Stekly parameter of about 0.5. Stekly condition of the cryogenic stability is still used to design the large-scale superconducting magnets to ensure their stability conditions with respect to the large spectrum of the external heat perturbations. In general, the quasi-linear model described by equations (7.12)-(7.15) is widely used to make the qualitative analysis of the main stability features of the resistive states of superconducting composites that carry a constant transport current using the jump-like V-I equation under the change of the thermal stabilization parameter α, the dimensionless current i, and the parameters of the heat disturbancies.

It should be emphasized that the dimensionless variables and parameters introduced above are not interdependent, therefore they do not violate of the invariance of the problem (7.12)-(7.15). In other words, the dimensional quantities contained in the dimensionless parameter α restrict the choice of the characteristic length of the dimensionless transformation (Romanovskii 1987c). Indeed, if one chooses it, for example as $L_h = \sqrt{\lambda_k S/(hp)}$, then the heat conduction equation obtained as a result of this dimensionless transformation does not allow, firstly, to analyze the thermal modes occurring in the thermally insulated superconducting composites since the dimensionless heat conduction equation loses physical sense at $h \to 0$, because of $L_h \to \infty$. Secondly, a change in the stability parameter α may be associated with a variation of one of its dimensional quantities I_{c0}, ρ_m, η, h, p, S, T_{cB} or T_0. However, some of them will also be used when the characteristic length of the form $L_h = \sqrt{\lambda_k S/(hp)}$ is used to provide a transition to a dimensionless heat conduction equation. Therefore, if the parameter α will be varied during the thermal analysis of the operating modes of superconducting composites, then other dimensionless variables must inevitably change due to a corresponding change in any of the values of I_{c0}, ρ_m, η, h, p, S, T_{cB} or T_0 included in the dimensionless parameter α. Thus, the invariance will be lost and the variation of the parameter α with constant values of other dimensionless parameters, for example ξ_1 or P_0 will lead to the physically incomparable results. (This problem is discussed below in more detail). At the same time, an independent variable dimensional parameter is the heat transfer coefficient h if the characteristic length of the form $L_z = \sqrt{\lambda_k (T_{cB} - T_0) S^2 / (I_{c0}^2 \rho_0)}$ is used. In this case, an arbitrary change in the parameter α is possible with constant values of the other dimensionless parameters.

Let us use the model (7.12)-(7.15) and discuss the formation features of the thermal modes of the composite superconductors. The characteristic dependencies of the heat flux to the coolant and the Joule heat release on temperature are presented in Figure 7.5. They were obtained at various approximations. First, it was assumed that ρ_m and h do not change with temperature (Figure 7.5a) and secondly when ρ_m and h depend on temperature (Figure 7.5b). In the latter case, curve $W(T)$ corresponds to the typical steady-state cooling mode at which a boiling crisis occurs. Namely, the initial ascending branch of curve $W(T)$ describes the temperature dependence of the heat flux removed into the coolant during nucleate boiling of the coolant leading to high values of the heat transfer coefficient. When the heat flux exceeds the characteristic value, an abrupt transition to the film boiling regime occurs. In this case, the cooled surface is isolated by a vapor film, and the value of the heat transfer coefficient decreases sharply. Conversely, as the surface temperature decreases, there is an abrupt transition from the film boiling regime to the nucleate boiling regime.

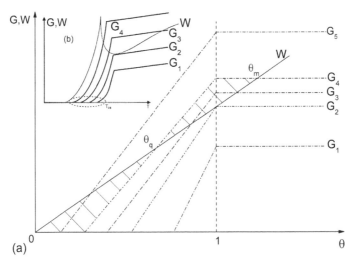

FIGURE 7.5 Temperature dependencies of the Joule heat release G in the composite superconductor and heat transfer W in the quasi-linear (a: $W = \theta/\alpha$, $G = i^2\rho(\theta)$, G_1 - $i < i_r$, G_2 - $i = i_r$, G_3 - $i_r < i < i_s$, G_4 - $i = i_s$, G_5 - $i > i_s$) and nonlinear (b: G_1 - $i < i_r$, G_2 - $i = i_r$, G_3 - $i = i_s$, G_4 - $i > i_s$) approximations of the heat exchange conditions.

In Figures 7.6-7.9, the results of the thermal simulations of superconducting composite states at $\alpha = 4$ and different values of the dimensionless current are given in the case of acting the temperature perturbations. Accordingly, the problem (7.12)-(7.15) was solved numerically at $P_0 = 0$ varying values θ_1 and ξ_1. To avoid the boundary effects, the dimensionless length of the composite was set $L = 1,000 \gg \xi_1$. The results presented show the formation features of the thermal modes in the superconducting composites depending on the transport current, conditions of the thermal stabilization and the temperature perturbation.

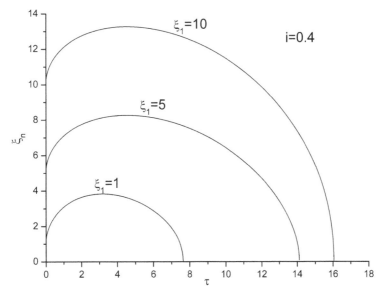

FIGURE 7.6 Fast recovery of superconductivity at $i < i_r$ under the action of the powerful temperature perturbations ($\theta_1 = 10$) of an arbitrary initial length.

At low currents, for example, as at the state G_1 in Figure 7.5 or at the effective thermal stabilization, in particular at high heat transfer coefficients, the specific power of the heat flux to the coolant exceeds the specific power of the heat release in the entire range of the temperature variation of the composite. Accordingly, under the action of the thermal perturbation with arbitrary energy (local or extended spatially, short or long in time), the resistive zones with the current sharing boundary ξ_n will disappear after the termination of the thermal perturbation source. As a result, the temperature of the composite will be equal to the temperature of the coolant, its superconducting properties will be restored and the transport current will flow through the superconducting core again. In this case, the existence of the thermal perturbations is of short duration. They are shown in Figure 7.6.

Such regimes exist up to the state which is determined from the condition of the thermal equilibrium of the Joule heat release and heat flux into the coolant at the critical temperature of the superconductor. In this case, the G and W curves have a common point at $\theta = 1$. In Figure 7.5, they correspond to the state G_2. In the quasi-linear approximation, this state is described by the relation

$$\alpha i^2 = 1 \tag{7.16}$$

It implies the existence of the current i_r that is called the full recovery current and is equal to $i_r = 1/\sqrt{\alpha}$. Then, as already noted in the current range $0 < i < i_r$, the superconducting properties of the composite are restored in a short time after the source of the perturbation ceases to act.

When $i > i_r$ (modes $G_3 - G_5$ in Figure 7.5), an additional heat generating region appears in the temperature range close to the critical temperature of the composite. As a result, new thermal states appear. First of all, let us discuss current modes when the total power of the heat release does not exceed the total power of the heat removal as, for example, during the G_3 mode depicted in Figure 7.5a.

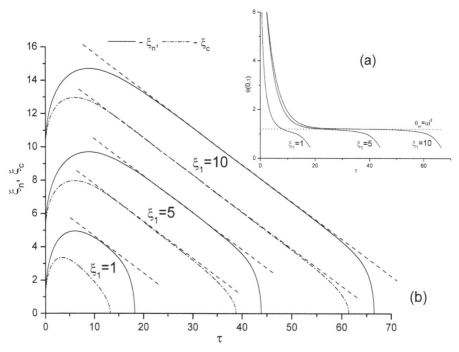

FIGURE 7.7 Stable modes of a superconducting composite under the action of a powerful perturbation ($\theta_1 = 10$) with its arbitrary initial lengths when the reduction of the resistive boundaries may occur at a constant velocity: (a) - change in the maximum temperature of the composite at $Z = 0$, (b) - propagation and disappearance of the fronts ξ_c and ξ_n.

The results of the numerical experiments performed at $i_r < i = 0.55$ and different values of the initial perturbation length are presented in Figure 7.7. It is seen that the temperature of the composite in its most heated part (in the center of the composite) decreases necessarily to its initial unperturbed value after the action of the temperature perturbations of arbitrary energy due to the effective thermal stabilization condition. As a result, the normal zones disappear after some time. However, unlike the modes, which are shown in Figure 7.6, the characteristic feature of their kinetics is the occurrence of the modes when the reduction of ξ_n and ξ_c occurs at almost constant velocity after the action of the extended powerful perturbations. This confirms the same slope of the tangents drawn to them. This conclusion may be interpreted as the occurrence of a thermal autowave moving in an infinitely long superconducting composite from its less heated outer boundary toward the more heated one that is in the direction opposite to the z-axis. Such states are discussed in more detail below.

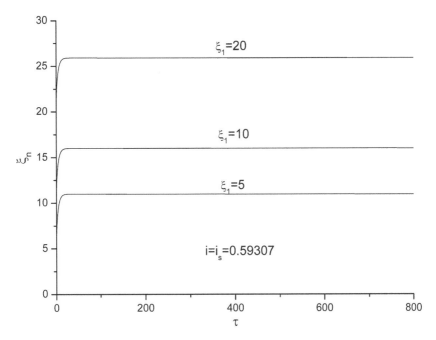

FIGURE 7.8   Equilibrium states of the resistive zone at a minimum propagating current.

It is easy to understand that as the current increases, the velocity of the reduction of the normal and resistive zones will decrease in absolute value, approaching zero due to the corresponding increase in the Joule heat release. In the limiting case, namely at some current i_s, a new mode will occur due to the satisfaction of the condition $G(\theta) = W(\theta)$ in the temperature range $0 < \theta < \theta_m$. It corresponds to the mode G_4 in Figure 7.5a. The areas of shaded regions are equal to each other. Thus, the velocity of any normal zone that has arisen will be zero after some time at current i_s, which is called the minimum propagating current of the normal zone, and the normal zone will be in equilibrium (Figure 7.8). Consequently, a stationary heat generating region will appear in the superconducting winding at this current.

Note that there are several characteristic temperatures of the composite at $i > i_r$ (Figure 7.5a). First, it is the temperature of the ends of composite maintained at the coolant temperature T_0 (in dimensionless form $\theta = 0$). At this temperature, the superconducting composite is in the initial equilibrium state and it will be in that state after the appearance and disappearance of the normal zone in it as, for example, it is shown in Figure 7.6 and 7.7. Secondly, there are temperatures θ_q and θ_m as a consequence of the intersection of the curves G and W. In the quasi-linear approximation, θ_m is the second equilibrium temperature that corresponds to the maximum stable temperature of the composite in the supercritical temperature range. It is not difficult to find that in the quasi-linear approximation $\theta_m = \alpha i^2$. The whole composite will be at this temperature after the action of the thermal perturbation in the case when it propagates over its entire length. (The exception is the outer boundaries if the composite length is finite). When the initial temperature of the composite satisfies the condition $\theta_1 > \theta_q$, the transition of the composite to the normal state will occur under the action of infinitely extended temperature perturbation ($\xi_1 \to \infty$). At the same time, it will return to its initial equilibrium state ($\theta = 0$) at $\theta_1 < \theta_q$ and $\xi_1 \to \infty$. These features are explained by the fact that the heat flux to the coolant exceeds the heat release at $0 < \theta < \theta_q$, and the heat release exceeds the heat flux to the coolant at $\theta_q < \theta < \theta_m$. On the other hand, in the temperature range, which places to the right of θ_m, the heat release is lower than the heat removal. Therefore, the temperature of the composite will tend to θ_m in this region of the temperature perturbations when $\theta_1 > \theta_m$.

Let us find the general condition that allows one to determine the minimum propagating current i_s. To do this, let us use one-dimensional stationary heat conduction equation of the form

$$\frac{d}{dz}\left[\lambda_k(T)\frac{dT}{dz}\right] - W(T) + G(T) = 0 \tag{7.17}$$

Here, as above, G and W are the heat release and heat flux to coolant, respectively.

For an infinitely extended composite, physically obvious conditions will be fulfilled at its boundaries

$$T\big|_{z\to\pm\infty} \to T_0, \quad \frac{\partial T}{\partial z}\bigg|_{z\to\pm\infty} \to 0 \quad \left(\theta\big|_{Z\to\pm\infty} \to 0, \quad \frac{\partial\theta}{\partial Z}\bigg|_{Z\to\pm\infty} \to 0\right). \tag{7.18}$$

Introduce the new variable $\Theta = \lambda_k dT/dz$. Then, taking into account that

$$\frac{d}{dz}\left[\lambda_k(T)\frac{dT}{dz}\right] = \frac{d\Theta}{dz} = \frac{d\Theta}{dT}\frac{dT}{dz}$$

Equation (7.17) is rewritten as

$$\frac{\Theta}{\lambda_k(T)}\frac{d\Theta}{dT} = W(T) - G(T)$$

Integrating it by temperature and taking into account the existence of the two equilibrium temperatures discussed above (in the dimension form T_0 and T_m) between which the temperature of the superconducting composite will change in the stationary mode, one can find

$$\frac{\Theta^2}{2}\bigg|_{\Theta_1}^{\Theta_2} = \int_{T_0}^{T_m} [W(T) - G(T)]\lambda_k(T)\,dT$$

Taking into account the boundary conditions (7.18), the condition of the existence of a stationary state will be described by the equality

$$\int_{T_0}^{T_m} [W(T) - G(T)]\lambda_k(T)\,dT = 0 \tag{7.19}$$

in an infinitely extended composite. It was first formulated in (Maddock et al. 1969) and is called the 'cold-end' equilibrium condition of the normal zone (the Maddock condition). This condition is satisfied by the mode G_3 in Figure 7.5b. In the quasi-linear approximation ($\lambda_k = $ const), the condition (7.19) takes the form

$$\int_{T_0}^{T_m} [W(T) - G(T)]\,dT = 0 \tag{7.20}$$

known as the theorem of the 'equal areas' due to the obvious interpretation of the integral (7.20) as the area between curves G and W. It is observed for the G_4 mode in Figure 7.5a and leads to the equality of the areas of the shaded areas in the temperature range from 0 to θ_m.

The relations (7.19) and (7.20) make it easy to find the minimum propagating currents. (It is also called the 'cold-end' recovery current). In particular, in the quasi-linear approximation, they satisfy the condition

$$\alpha i^2 + i - 2 = 0 \tag{7.21}$$

which leads to the next value of the minimum propagating current

$$i_s = \frac{\sqrt{8\alpha + 1} - 1}{2\alpha} \tag{7.22}$$

When the transport current flows in the composite above the value i_s, the thermal equilibrium conditions (7.19) or (7.20) will be violated as the total amount of the heat released in the temperature range from 0 to θ_m will exceed the total heat flux to the coolant. Modes G_5 in Figure 7.5a and G_4 in Figure 7.5b correspond to these states. In this case, the superconducting state becomes metastable since an unbalanced amount of the heat released may lead to the irreversible propagation of the normal zone throughout the composite.

Stable and unstable modes are depicted in Figure 7.9, which shows the time variation of the normal zone length (Figure 7.9a), the temperature of the most heated part of the composite (Figure 7.9b) and the velocity of the normal zone fronts (Figure 7.9c). It may be seen that depending on the character of the thermal perturbation, for example, its length or intensity, the composite may either restore its superconducting properties or lose them. In the latter case, as discussed above, the temperature at the hottest point of the composite (at $Z = 0$) tends to the equilibrium value $\theta_m = \alpha i^2$ after transient time. It is important to emphasize that a characteristic feature of the irreversible transition of the composite to a normal state is the occurrence of the states when the normal zone begins to move almost at a constant velocity.

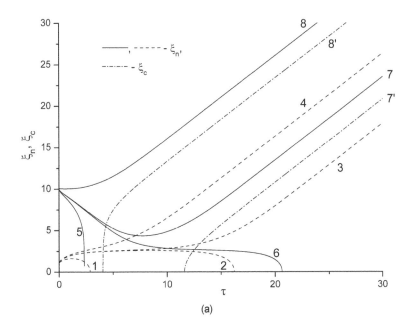

(a)

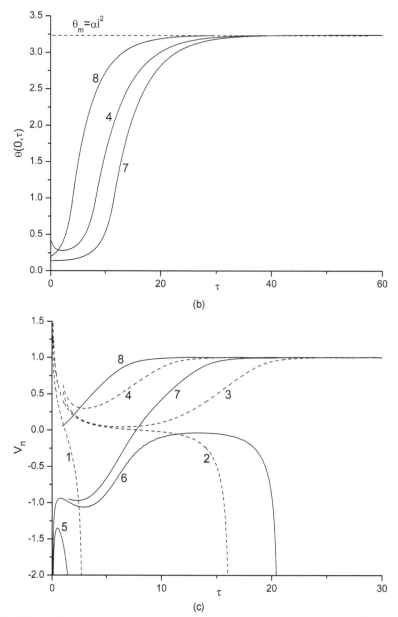

FIGURE 7.9 Metastable states of a superconducting composite at $i = 0.9$ ($i_s < i < 1$) under the action of the local and extended perturbations: 1 - $\xi_1 = 1$, $\theta_1 = 0.3$; 2 - $\xi_1 = 1$, $\theta_1 = 0.37$; 3 - $\xi_1 = 1$, $\theta_1 = 0.371$; 4 - $\xi_1 = 1$, $\theta_1 = 0.4$; 5 - $\xi_1 = 10$, $\theta_1 = 0.13$; 6 - $\xi_1 = 10$, $\theta_1 = 0.139$; 7, 7' - $\xi_1 = 10$, $\theta_1 = 0.14$; 8, 8' - $\xi_1 = 10$, $\theta_1 = 0.2$. Here, (a) – normal zones recovery and propagation, (b) – temperatures rise in the center of the composite, (c) – kinetics of the normal zones.

The presented results lead to important conclusions. Firstly, when $i_s < i < 1$, there is a boundary value of the thermal perturbation energy. The emerging normal zones do not destroy superconductivity below that value. When that value is exceeded, the normal zone spontaneously propagates throughout the composite. In other words, there is *critical energy* of the thermal perturbation. It is also called as the minimum quench energy (E_{MQE}). As follows from Figure 7.9, the method of the finite perturbation of initial state allows one to find the critical energy of any perturbations. Secondly, the irreversible transition of the composite from the superconducting state to the normal one is a self-sustaining autowave mode. These states are characterized by the occurrence of the thermal wave propagating with a constant velocity ($\partial \xi_n / \partial \tau = \partial \xi_c / \partial \tau = \text{const}$) in an extended superconducting composite which, as follows from Figure 7.9, does not depend on the nature of the initial thermal perturbation. Let us discuss these characteristic aspects of the thermal stabilization problem of superconducting composites in more detail.

7.3 THERMAL STABILITY CONDITIONS OF SUPERCONDUCTING COMPOSITES

As follows from the previous discussion, the composite superconductors may lose their superconducting properties due to the onset of the thermal instability. Let us perform a quasi-linear analysis of the thermal stability conditions of the composite superconductors, based on the model (7.12)-(7.15), determining the boundary of the stable states for the temperature perturbations. In this case, it should be taken $P_0 = 0$ in Equation (7.12). Then, taking into account the symmetry of the states considered, the general critical energy of the temperature perturbation in the terms of the utilized dimensionless variables is defined as follows $\varepsilon_q = E_q/E_x = 2\xi_1\theta_q$. Here, according to (7.5),

$$E_q = 2l_1 S \int_{T_0}^{T_q} C_k(T)\,dT = 2l_1 S C_k (T_q - T_0),$$

where T_q is the maximum permissible perturbation temperature T_1, θ_q is its dimensionless value, l_1 (or ξ_1) is its half-length of perturbation, $E_x = L_z S C_k(T_{cB} - T_0)$ is the characteristic value of the perturbation energy. For a composite superconductor with averaged parameters $S \sim 3 \times 10^{-6}\,\text{m}^2$, $\lambda_k \sim 200\,\text{W/(m} \times \text{K)}$, $C_k \sim 10^3\,\text{J/(m}^3 \times \text{K)}$, $\rho_0 \sim 10^{-9}\,\Omega \times \text{m}$, $I_{c0} \sim 1000\,\text{A}$, $T_{cB} - T_0 \sim 5\,\text{K}$, one can find $E_x \sim 4.5 \times 10^{-5}\,\text{J}$.

The dependencies of the critical energy on the transport current under the action of a local temperature perturbation in the case of a change in the thermal stabilization parameter in a wide range are shown in Figure 7.10. (The physical sense of the local nature of the perturbation is discussed below). The dashed vertical straight lines correspond to the currents that follow from (7.22), that are the currents at which the thermal equilibrium of the normal zone occurs according to the Maddock condition. The presented calculation results make it possible to formulate general regularities of the change in the thermal stability conditions of superconducting composites.

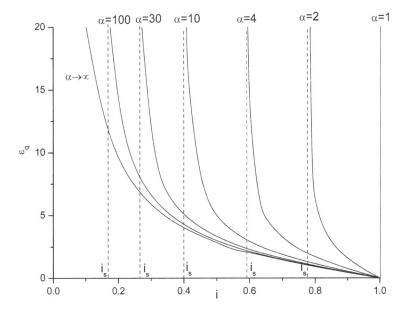

FIGURE 7.10 Critical energies of the local temperature perturbation ($\xi_1 = 1$) as a function of the current for various parameters of the thermal stabilization.

The critical energies decrease monotonically with transport current. They take an infinitely large value at the 'cold-end' recovery current i_s (let us call it as the full stabilization current) and is zero at the critical current ($i = 1$). The explanation of the infinitely large value of the critical energy at i_s follows from Figure 7.8. During this mode, under the action of any thermal perturbation in the composite, the emerged normal zone stabilizes after the termination of the perturbations action. This peculiarity means that the composite does not completely go into the normal state. On the other hand, at the critical current, any infinitesimal thermal perturbation will lead the composite superconductor to a normal state. Therefore, $\varepsilon_q = 0$. Thus, the current i_s divides the entire range of transport currents into two areas. In the first area, that is at $0 < i < i_s$, the superconducting state of the composite is stable for arbitrary thermal perturbations, and the energy of the external thermal perturbations is not limited, but the critical energies have the finite value at $i_s < i < 1$. As a consequence, the thermal stabilization conditions are called as the full thermal stabilization conditions at the currents $0 < i < i_s$, and it is the partial thermal stabilization at $i_s < i < 1$.

The critical energies decrease with increasing of the thermal stabilization parameter α. This is due to a decrease in the heat transfer coefficient. In the limiting case $\alpha \to \infty$ ($h = 0$), the critical energies limit the stability of the superconducting state of composite in the entire range of the variation of transport current ($0 < i < 1$) since $i_s = 0$. In other words, heat-insulated superconducting composites have only partial thermal stabilization conditions. As follows from Figure 7.10, the thermal stability of the composite in a wide range of the transport current variation approaches the adiabatic stability conditions at $\alpha > 30$. A more general statement is valid. The permissible perturbations depend very little on the stabilization parameter, that is on the cooling conditions at $\alpha t^2 \gg 1$. Thus, the condition $\alpha t^2 \gg 1$ corresponds to the practically adiabatic thermal stability condition of the superconducting composites.

In another limiting case $\alpha = 1$, the superconducting properties of composites will always remain up to the critical current under the action of external thermal perturbations with arbitrary energy. In this case, the superconducting composites are in the full thermal stabilization mode. For the first time, this stability condition was formulated by Stekly using the following idea. Let us assume that the cooled superconducting composite carrying the critical current is in a normal state at the critical temperature of the superconductor. If one needs to provide stability of this mode, the heat released in the matrix must be completely removed to the coolant. Under this assumption, Stekly formulated the full stability condition as follows $\alpha \leq 1$. From a formal point of view, the condition of the full thermal stabilization of a superconducting composite with a transport current that is equal to the critical one ($i = 1$) follows from the expressions (7.16) and (7.21). This means that the transport current flowing in the composite is equal to both full recovery current and minimum propagating current at $\alpha = 1$.

The effect of the initial length of the temperature perturbation on the thermal stabilization conditions is shown in Figure 7.11. In Figure 7.11a, the dashed and dash-dotted curves describe the change in the maximum permissible value of the initial perturbation temperature (in the general case, the energy density of the short-time perturbation) for two values of the thermal stabilization parameter corresponding to 'poor' cooling ($\alpha = 100$) and 'good' cooling ($\alpha = 2$). Here, the vertical dashed straight lines show the values i_s, where $\theta_q \to \infty$. Besides, in Figure 7.11a, the solid curves show the limiting values θ_q for an infinitely extended temperature perturbation. In the framework of this approximation, to determine them, one should neglect the conductive heat flow in the longitudinal direction in the appropriate heat conduction equation, which leads to zero-dimensional model. Then, the corresponding values θ_q can be found as the intersection point of the temperature dependencies G and W. As an illustration, this value is given for G_4 mode in Figure 7.5a. It is easy to find

$$\lim_{\xi_1 \to \infty} \theta_q = \frac{\alpha i (1 - i)}{\alpha i - 1} \tag{7.23}$$

The corresponding values $\theta_q(\alpha, i)$ are shown in Figure 7.11b. Here, the maximum value θ_q is limited by the critical temperature of composite ($\theta_q \leq 1$). According to (7.23), it is easy to find that $\theta_q = 1$ at $\alpha i^2 = 1$. Therefore, in Figure 7.11, the ranges of the full thermal stabilization that follow from one-dimensional and zero-dimensional approximations do not coincide. However, as follows from Figure 7.11a, zero-dimensional model satisfactorily describes the stability boundary under the action of extended temperature perturbations in the region of the transport currents that are not too adjacent to the full stabilization current i_s.

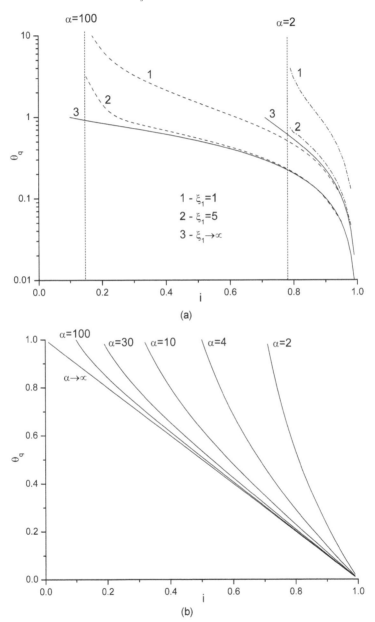

FIGURE 7.11 Maximum permissible temperature perturbation as a function of its length: (a) - from local to infinitely extended perturbations, (b) - for infinitely extended perturbation.

The dependence of the thermal stabilization conditions on the length of the temperature perturbation is shown in Figure 7.12. As it follows from Figure 7.12a, in most cases, the critical energies are minimal under the action of the point perturbations ($\xi_1 \ll 1$). Moreover, in this case, the stability conditions weakly depend on their length. The exceptions are operating modes with effective heat removal of the Joule losses. Namely, at $\alpha = 2$ and $i = 0.8$, that is near the full stabilization current ($i_s = 0.78$) and the values $\varepsilon_q(\xi_1)$ have a minimum under the action of relatively extended perturbations. In general, the critical energies asymptotically approach values (Figure 7.12b) that correspond to instantaneous heating by an infinitely long temperature perturbation in the area of extended perturbations ($\xi_1 > 5$), which are described by the Formula (7.23). It is easy to understand that this pattern is associated with a decrease in the efficiency of the heat conduction mechanism.

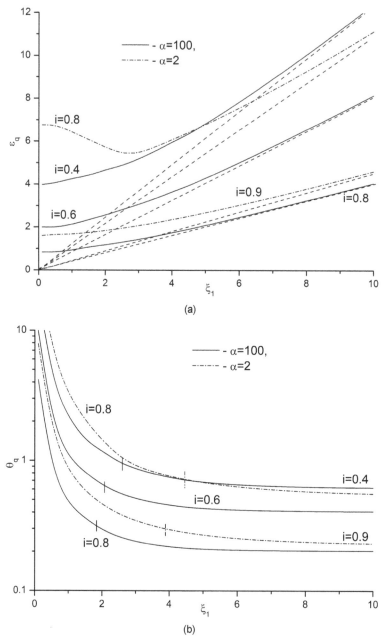

FIGURE 7.12 Critical energies (a) and maximum permissible perturbation temperatures (b) as a function of its length at various values of the current and thermal stabilization parameter.

Let us discuss the features of the thermal stabilization conditions in superconducting composites under the action of the volumetric thermal perturbation of finite length of various durations in the framework of the formulated above problem (7.12)-(7.15). Assume that the temperature of a composite at the initial time is equal to the temperature of the coolant, and the external thermal impulse in time has a rectangular shape. Accordingly, the problem (7.12)-(7.15) will be solved at $\theta_1 = 0$ and

$$P_0 = \begin{cases} P_1 = \text{const}, 0 < \tau \leq \tau_1, |Z| \leq \xi_1 \\ 0, \quad \tau > \tau_1, -L \leq Z \leq -\xi_1, \xi_1 < Z \leq L \end{cases}$$

varying the values of τ_1. Then, the total energy released by the action of this perturbation type in a dimensionless form is equal to $\varepsilon_1 = 2P_1\xi_1\tau_1$ under (7.8) taking into account the symmetric temperature distribution along the length of the composite. To avoid the boundary effects, the dimensionless length of the composite was assigned as before $L = 1,000 \gg \xi_1$. To determine the critical energies, the method of the finite perturbation of initial state was used.

The dependencies of the critical energy on the duration of the local perturbation for various values of the stabilization parameter and current are shown in Figure 7.13. The results depict that the critical energies are minimal under the action of the short-time heat pulses as a rule ($\tau_1 \ll 1$). As in the case of temperature perturbations, the exception is the regimes with an effective heat sink of the Joule losses. So, the minimum at the dependencies $\varepsilon_q(\tau_1)$ shifts toward longer pulses at $\alpha = 2$ and $i = 0.8$.

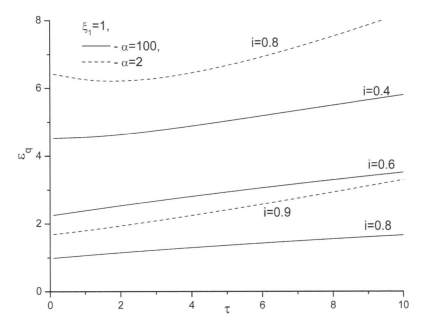

FIGURE 7.13 Dependence of the critical energies of a local impulse perturbation on its duration.

It should be noted that the specified rectangular pulse shape does not reduce the generality of the results obtained at $\tau_1 \ll 1$. Physically, this is because soon after the cessation of the short-time perturbations, the temperature distribution arising along the composite is the same for perturbations of various shapes but of the same energy. This effect is due to the influence of the longitudinal thermal conductivity of the composite on the redistribution of the heat flux.

In the extreme case, when the perturbation is distributed throughout the entire length of the composite, it is not difficult to find the condition within the framework of zero-dimensional model that describes the boundary of the stable superconducting states under the act of the infinitely extended thermal pulse of arbitrary shape. Taking into account that the permissible increase in the temperature of the composite must not exceed the value defined by the Formula (7.23) in this case, the corresponding expression for determining the stability limit under the act of arbitrary thermal pulse may be written in the form

$$\int_{\tau_i}^{\tau} P_0(t) \exp \frac{(\alpha i - 1)(\tau_i - t)}{\alpha} dt = \frac{1 - i}{\alpha i - 1}, \quad \alpha i^2 > 1 \qquad (7.24)$$

Here, τ_i is the dimensionless time of the heating the composite to the temperature $(1 - i)$ at which current begins to share. It follows from the solution of the integral equality

$$\int_{0}^{\tau_i} P_0(t) \exp\left(\frac{t - \tau_1}{\alpha}\right) dt = 1 - i$$

In particular, the boundary of the stable states for a rectangular pulse is described by the following criterion

$$\frac{\alpha P_0}{1 - i}\left[1 - \exp\left(-\frac{\alpha i - 1}{\alpha i^2}\tau\right)\right] = 1, \quad \alpha i^2 > 1$$

These formulae allow one to determine the maximum allowable duration of the extended heat pulse for a given value P_0 or find the limit value of the perturbation power density for a given time of its action.

The results show that local short-time thermal impulses are usually the most dangerous. Let us define the local nature of perturbations.

Let us estimate the characteristic length of the normal zone at which the heat release in the area with the normal conductivity is equal to the conductive heat flux in the adiabatic conditions. As above-mentioned, let us consider the symmetric redistributing of the heat flux in a thin superconducting composite. Let the transport current flow through it with a density ηJ_{c0}. Place the origin of the coordinates in the center of the composite assuming that its central part of the length $2l_n$ has heated to the critical temperature T_{cB} and the remainder of the composite is at a temperature T_0. Then, the amount of the heat released in the area, which is in the positive direction relative to the coordinate system, where the composite in the normal state is equal to $Sl_n\rho_0\eta^2 J_{c0}^2$. The temperature gradient in each area of composite $0 < |z| < l_n$, which is in the normal state, is determined by the temperature decrease on the length l_n. For the equilibrium of the normal zone, the total allocated amount of the Joule heat release $2Sl_n\rho_0 J_{c0}^2$ must be equal to the total amount of the heat transferred by the heat conduction in both positive and negative directions, that is $2S\lambda_k(T_{cB} - T_0)/l_n$. As follows from Figure 7.9a, this is a state of unstable equilibrium toward which the perturbed thermal states tend. Then, from the equality $2Sl_n\rho_0\eta^2 J_{c0}^2 = 2\lambda_k S(T_{cB} - T_0)/l_n$, one may find the characteristic value of the half-length of the normal zone. It equals

$$l_n = \sqrt{\lambda_k(T_{cB} - T_0)/(\rho_0\eta^2 J_{c0}^2)}$$

This value coincides with the characteristic length L_z, introduced above to get the heat conduction Equation (7.9) in the dimensionless form. Thus, this estimate determines the physical meaning L_z: taking into account the symmetry condition, the value $2\sqrt{\lambda_k(T_{cB} - T_0)/(\rho_0\eta^2 J_{c0}^2)}$ corresponds to the total length of the normal zone in the case of its unstable equilibrium in the adiabatic mode. If it is exceeded, the thermal balance conditions between the Joule heat release and the heat flux transferred by the thermal conductivity of the composite are violated. It is called the Minimum Propagating Zone (MPZ). Let us discuss its influence on the formation of stable and unstable states in more detail.

As follows from Figure 7.9a, at temperature perturbations with the energy close to the critical, the development of the thermal states of the composites, first, has a bifurcation character; second, they tend to the unstable equilibrium state with some characteristic length ξ_m. It should be emphasized that these regularities are observed during the action of the perturbations of an arbitrary length. Indeed, the thermal mode under the temperature perturbation with the parameters $\xi_1 = 1$ and $\theta_1 = 0.371$ (Curve 3), leading to an irreversible transition of the composite to the normal state, are preceded by the stable modes when, first, there is a growth of the initial perturbation length and then, second, its subsequent reduction occurs (Curves 1 and 2). The stable and unstable perturbations described by Curves 2 and 3 exist due to the bifurcation of these modes that are close to the unstable equilibrium state. Therefore, the condition $\xi_n < \xi_m$ exists during stable states. The perturbations may be defined as local if $\xi_1 < \xi_m$. At a temperature perturbation with the length $\xi_1 = 10$ ($\xi_1 > \xi_m$), the restoration of superconducting properties by the composite (Curves 5 and 6) is provided by the mandatory decrease in the length of the heat-releasing area below the same characteristic value ξ_m. If the normal zone exceeds this value after a finite period of action time of a perturbation, then such nonlocal perturbation ($\xi_1 > \xi_m$) leads to an unstable state. During stable states, the normal zone disappears even at $\xi_1 > \xi_m$.

Thus, there always exists the upper limit of the permissible energies for the thermal perturbations of an arbitrary length. As it follows from the Formulae (7.23) and (7.24), they are finite, even under the action of infinitely extended perturbations. Physically, this is primarily due to the heat removal of the Joule heat release to the coolant. The thermal stability of superconducting composites is also ensured by the thermal conductivity of the matrix, the contribution of which depends on the length of the initial perturbation (Figure 7.11a). As a result, regardless of the initial perturbation length in the entire range of the stabilization parameter, the allowable overheating of the composite is always nonzero at $i_s < i < 1$.

In the framework of the problem (7.12)-(7.15), the boundary of the resistive area ξ_m limited by the isotherm $1 - i$ will be determined from the solution of the corresponding stationary equation (Romanovskii 1985a). Solving it for an infinitely long composite, one may find

$$\xi_m = \begin{cases} \xi_0 + \sqrt{\dfrac{\alpha}{\alpha i - 1}} \arccos \dfrac{(\alpha i - 1)\sqrt{1 - \alpha i^3} - \alpha i^2 + 1}{\alpha i(1 - i)}, & \alpha i^3 < 1 \\[3mm] \sqrt{\dfrac{\alpha}{\alpha i - 1}}\left(\pi - \arccos \dfrac{1}{\sqrt{\alpha i}}\right), & \alpha i^3 \geq 1 \end{cases} \tag{7.25}$$

where

$$\xi_0 = \frac{\sqrt{\alpha}}{2} \ln \frac{\alpha i^2 - 1 + \sqrt{1 - \alpha i^3}}{\alpha i^2 - 1 - \sqrt{1 - \alpha i^3}}.$$

The written expressions have no physical meaning at $i = 1$ since this state is not stable to the thermal perturbations. When $\alpha \to \infty$, the Formula (7.25) is reduced to the form

$$\lim_{\alpha \to \infty} \xi_m = \frac{\pi}{2\sqrt{i}} \tag{7.26}$$

which is written in a dimensional form as follows

$$\lim_{h \to 0} l_m = \frac{\pi}{2} S \sqrt{\frac{\lambda_k (T_{cB} - T_0)}{\rho_0 I I_{c0}}} = \frac{\pi}{2} \sqrt{\frac{\lambda_k (T_{cB} - T_0)}{\rho_0 \eta J_{c0} J}}$$

It differs from the above-written value l_n by a numerical factor. It demonstrates the dependence of the characteristic length of the quasi-equilibrium state on the transport current. However, this value of l_m slightly differs from the value l_n obtained on the basis of the simplest estimation at $I = I_{c0}$. Let us use this formula to evaluate l_m. For example, for uncooled Nb-Ti in a copper matrix

1.5 mm in diameter, with a filling factor of 50% in a magnetic field of 5 T, the total length of the quasi-equilibrium state area is approximately 17 mm, 8 mm and 6 mm for currents of 100 A, 500 A and 1,000 A. In this case, the following averaged values are used: $\lambda_k \sim 200$ W/(m × K), $\rho_0 \sim 10^{-9}\,\Omega \times$ m, $I_{c0} \sim 1200$ A, $T_{cB} - T_0 \sim 5$ K. When the composite is cooled, these values increase.

In Figure 7.12b, the marker "|" denotes the corresponding values ξ_m for each of the considered modes. It is seen that the value of the permissible overheating depends on the initial length of the perturbation when $\xi_1 \ll \xi_m$. However, the critical energy has an almost constant value at $\xi_1 \ll \xi_m$ (Figure 7.12a) and vice versa, the permissible perturbation temperatures (in general, it is the volume perturbation density) is practically independent of ξ_1 at $\xi_1 \gg \xi_m$. As mentioned above, they are close to their asymptotic values, which follow from the Formula (7.23), which was obtained without taking into account the mechanism of the thermal conductivity. Therefore, the critical energies increase almost linearly with the length of the perturbation when $\xi_1 \gg \xi_m$.

In the dimensional form, the characteristic half-length of the normal zone may be rewritten as follows $l_m = \xi_m(\alpha, i)\sqrt{\lambda_k(T_{cB} - T_0)/(\rho_0\eta^2 J_{c0}^2)}$. Here, $\xi_m(\alpha, i) > 1$, as follows from (7.25), (7.26) and Figure 7.12b, including adiabatic conditions. Therefore, the condition $l_m > l_n$ takes place, that is the actual length of the quasi-equilibrium state is greater than l_n and increases with decreasing of α and i.

For the first time, the concept of the characteristic length of the normal zone as the maximum allowable length of the normal zone was formulated in (Martinelli and Wipf 1972). It was postulated that when it is exceeded, the superconductivity is destroyed. The main conclusions of the MPZ theory are formulated using the solution of the stationary equations. Meanwhile, the above-mentioned thermal stability conditions appeared as a consequence of the solution of the nonstationary heat conduction equation using the method of the finite perturbation of the initial state with the obligatory determination of the final thermal state of the composite after thermal impulse. In the stationary MPZ approximation, the stability of superconducting states is limited to the critical energy, which is equal to

$$E_{\mathrm{MPZ}} = \int\limits_{T_0}^{T_{\mathrm{MPZ}}} \int\limits_{V_{\mathrm{MPZ}}} C_k(T)\,dT\,dV \approx V_{\mathrm{MPZ}} \int\limits_{T_0}^{T_{\mathrm{MPZ}}} C_k(T)\,dT,$$

where V_{MPZ} is the volume occupied by the stationary MPZ temperature distribution T_{MPZ} changing, in general, in the space. It is claimed, formally, that any states are stable if the energy (enthalpy) of the temperature profile created at that is lower than E_{MPZ}. Note that the choice of this value E_{MPZ} as the critical value of the external perturbation energy is not strictly justified. Moreover, the MPZ theory does not allow to determine the dependence of the permissible perturbations on any length at $\xi_1 < \xi_m$, that is at local perturbations. The MPZ theory also cannot answer the influence of the duration of a point impulse perturbation on the thermal stability conditions.

It should be emphasized that in determining ξ_m the mechanism of the current sharing plays an essential role. If it is not taken into account, as it is usually done in the MPZ studies performed, the resulting values ξ_m will not only be greatly underestimated but also the limit transition $\lim\limits_{\alpha\to\infty} \xi_m = 0$ will take place at $\alpha \to \infty$ (Romanovsky 1985a). To avoid this uncertainty and estimate the thermal stability conditions under the adiabatic conditions, the multidimensional models are used in the MPZ-theory. Besides, as calculations show (Romanovsky 1985a), the values ξ_m calculated according to the jump-like model of the transition from the superconducting state to the normal one at currents close to critical ones, are very low and tending to zero at adiabatic conditions. At the same time, if the current sharing is taken into account, the value ξ_m is always finite, including value under the adiabatic conditions, as follows from (7.25) and (7.26). This means that the efficiency of the heat conduction mechanism is significantly underestimated due to the small values of ξ_m within the model of a jump-like transition. Thus, for instantaneous point perturbations, the MPZ theory allows one to obtain a very rough estimation of the character size of the unstable states. Moreover, as it will be shown below, the true values of the critical energies may be either less or more than the enthalpy of the temperature MPZ-profile.

The results presented in Figure 7.13 show that the transition from the region of stable states in which the critical energies practically do not depend on the duration of the thermal pulse τ_1 to the region, where the critical energies approach their asymptotic values, depends on the conditions of the heat exchange and transport current. In general, curves presented in Figure 7.13 allow one to conclude that the permissible energies of the thermal point perturbations are practically independent on the impulse duration τ_1 when the simplest condition is met

$$\tau_1 \ll \tau_m = \frac{\alpha(1-i)}{\alpha i^2 + i - 2} \tag{7.27}$$

If the composite is not cooled, then this condition is written as follows

$$\tau_1 i^2 \ll 1 - i$$

In the dimensional variables, it takes the form

$$t_1 \ll \frac{C_k(T_{cB} - T_0)S^2}{\rho_0 I^2} \tag{7.28}$$

This estimate can also be obtained by comparing the characteristic propagation length $l_i \sim \sqrt{\lambda_k t_1 / C_k}$ of the isotherm T_{cB} during thermal impulse and the characteristic unstable equilibrium length of the normal zone l_n. When a current I flows through the composite, according to the estimate obtained above, it may be defined as $l_n(I) \sim \sqrt{\lambda_k(T_{cB} - T_0)S^2/(\rho_0 I^2)}$. Then, the condition (7.28) follows from the estimation $l_i \ll l_n$. The estimate (7.27) may be generalized using the condition $l_i \ll l_m$, where l_m is the length of the unstable equilibrium of the normal zone of cooled composite described in the dimensionless form by the Formula (7.25).

Thus, to find the minimum value of the critical (minimum quench) energy (MQE) for a superconducting composite under given cooling conditions and carrying a transport current exceeding the full stabilization current i_s (minimum propagating current), it is necessary to consider the thermal stability problem of a superconducting composite under the action of a point ($\xi_1 \ll \xi_m$) short-time ($\tau_1 \ll \tau_m$) perturbation. It is in this case, they are the minimum energies under which the composite will go to a normal state. Otherwise, they are not minimal and increase with increasing length and duration of the perturbation. Their existence allows one to explain such phenomena as the degradation and training of SMS. Indeed, if the unwanted regular heat perturbations take place in the winding, then it is impossible to achieve the critical current of a short sample in it (degradation effect). On the other hand, during the mechanical ordering of wires in a magnetic system, the value of the thermal perturbation energy may decrease in the successive transitions of SMS to a normal state. Therefore, the subsequent currents of the transition of the winding to the normal state may increase (the training effect).

Along with the numerical analysis of the thermal stability conditions of the superconducting state, the analytical formulae have, undoubtedly, practical meaning. These formulae make it possible to easily and quickly estimate the critical energies and, first of all, for instantaneous point thermal perturbations which, as discussed above, are the most unfavorable for stable SMS performance. In the framework of the model of a jump-like transition of a superconducting composite to a normal state (Gurevich et al. 1997c) and assuming that it occurs at the dimensionless temperature of $1 - i$, the following formula was obtained for calculating the critical energy

$$\varepsilon_q = \frac{2.3}{i} \frac{(1-i)^{3/2}}{\sqrt{1 - 2(1-i)/(\alpha i^2)}} \tag{7.29}$$

Its dimensionless value is written in terms of the dimensionless variables introduced above.

According to the results shown in Figure 7.10-7.13, the critical energies taking into account the current sharing can be determined using the approximate formula

$$\varepsilon_q = 2\frac{\alpha i(1-i)}{\alpha i - 1}\left[\xi_m + \frac{(ai-1)(1-i)^2}{\alpha i^2 + i - 2}\right], \quad i_s \le i \le 1 \tag{7.30}$$

Here, the value ξ_m is described by the Formula (7.25).

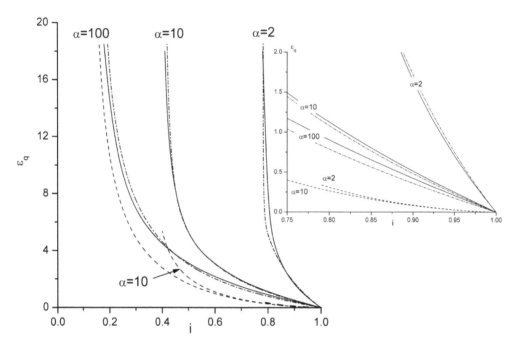

FIGURE 7.14 Dependence of the minimum critical energies on the current at various values of the thermal stabilization parameter: ————- - numerical calculation according to the method of the finite perturbation of the initial state, - - - - - - calculation by the Formula (7.29), - · - · - · - - calculation by the Formula (7.30). In the inset, the critical energies are shown for currents close to the critical current.

In Figure 7.14, the critical energies for the point instantaneous perturbations at different values of the stabilization parameter calculated in the framework of different approximations are compared with each other. It is seen that the jump-transition model underestimates the critical energies. In this case, the deviation is the more noticeable, the smaller the stabilization parameter, that is the error increases with the improvement of the heat transfer conditions. At the same time, the approximate Formula (7.30), which takes into account the current sharing, describes the dependence $\varepsilon_q(\alpha, i)$ with a good degree of accuracy. The error of the jump-transition model is explained by the fact that a jump from the superconducting state to the normal state is supposed at the temperature at which the current begins to share. Therefore, in the temperature range $1 - i < \theta < 1$, this approximation significantly overestimates the power of the Joule losses compared to the heat release that occurs when the current is shared.

Within the framework of the MPZ theory, the following formula was obtained for the calculation of MPZ energies (Dresner 2002b)

$$\varepsilon_d = \frac{E_{MPZ}}{C_k S(T_{cB} - T_0)L_h} =$$

$$= \begin{cases} 2\dfrac{\alpha i(1-i)}{\alpha i-1}\left[1+\dfrac{1}{\sqrt{\alpha i-1}}\left(\arcsin\dfrac{\alpha i^2-1}{(1-i)\sqrt{\alpha i}}+\arcsin\dfrac{1}{\sqrt{\alpha i}}\right)\right]+ \\[2mm] \alpha i^2 \ln\dfrac{\alpha i^2-1+\sqrt{1-\alpha i^3}}{\alpha i^2-1-\sqrt{1-\alpha i^3}}-2\sqrt{1-\alpha i^3}\,\dfrac{\alpha i}{\alpha i-1}, \quad \alpha i^3<1 \\[2mm] 2\dfrac{\alpha i(1-i)}{\alpha i-1}\left[1+\dfrac{1}{\sqrt{\alpha i-1}}\left(\dfrac{\pi}{2}+\arcsin\dfrac{1}{\sqrt{\alpha i}}\right)\right], \quad \alpha i^3\geq1 \end{cases}$$

(7.31)

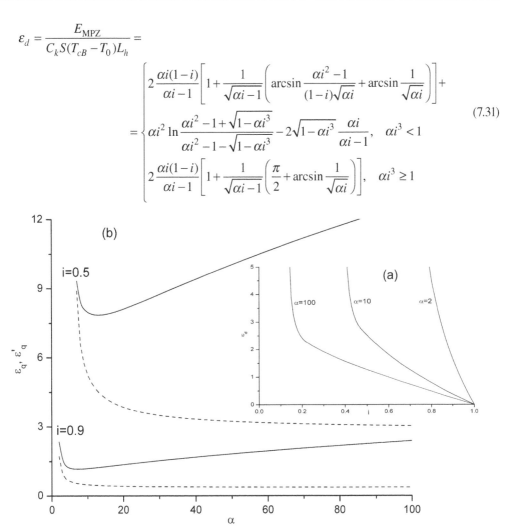

FIGURE 7.15 Critical energies of the point instantaneous perturbations calculated by different models: (a): calculation $\varepsilon_d(\alpha, i)$ according to the Formula (7.31), (b): (- - - -) - calculation by the Formula (7.30) and (——) - calculation $\varepsilon_q' = \varepsilon_d \sqrt{\alpha}$ according to the Formula (7.31).

It is written in the framework of the dimensionless model when the characteristic length $L_h = [\lambda_k S/(hp)]^{1/2}$ was used instead of $L_z = \sqrt{\lambda_k(T_{cB} - T_0)S^2/(I_{c0}^2\rho_0)}$, as it was done in section 7.2. In Figure 7.15a, the dependencies $\varepsilon_d(\alpha, i)$ as a function of the current for some values of the stabilization parameter are shown. The depicted dimensionless values $\varepsilon_d(\alpha, i)$ are finite at $\alpha \to \infty$ in this approximation. However, if one turns to dimensional values, then the MPZ-energy will be unlimited for the uncooled superconducting composite since $\lim_{h\to 0} E_{MPZ} = \varepsilon_d E_x = \varepsilon_d L_h S C_k(T_{cB} - T_0) \to \infty$ due to $\lim_{h\to 0} L_h \to \infty$. As noted above, the use of L_h leads to physically incomparable results in the transition from a dimensionless model to a dimensional one due to the violation of the invariance in the used dimensionless variables. To avoid this, the quantity L_z should be used to get the correct dimensionless model. Considering that $L_h/L_z = \sqrt{\alpha}$, let us compare the values $\varepsilon_d(\alpha, i)$, which follow from the Formula (7.30), and the corresponding values $\varepsilon_q' = \varepsilon_d \sqrt{\alpha}$. These values as functions of the stabilization parameter are shown in Figure 7.15b. It is seen that the Formula (7.31) allows one to correctly describe the conditions of the thermal stabilization only at $\alpha < 5$. At $\alpha > 10$, it leads to a physically incorrect result when the conditions of the heat stabilization are improved with the increase of α, that is at the deterioration of the heat transfer conditions. That is why in the MPZ-model the limiting transition $\lim\limits_{h\to 0} E_{MPZ} \to \infty$ is observed.

Thus, for a correct quantitative analysis of the thermal stability conditions of superconducting composites, one should take into account the current sharing between the superconductor and matrix and also taking into account the kinetics of the normal conductivity zones. The conclusions formulated above are of a general physical nature and may be used to study more complex modes. Using them, the features of the thermal stabilization of the composite superconductors based on LTS and HTS will be discussed below.

7.4   AUTOWAVE PROPAGATION OF THE THERMAL PERTURBATIONS

As it was shown above, the quasi-stationary states may arise in a superconducting composite when the propagation velocities of the normal zones, which are the result of the thermal perturbation, are constant. They may be both negative (Figure 7.7b) and positive (Figure 7.9c) and are called the autowaves. Thermal autowaves transfer the composite from one stable state to another one restoring superconducting properties or destroying them.

The characteristic dependencies of the normal zone velocities on the current for various values of the stabilization parameter are shown in Figure 7.16. The calculation was carried out on the basis of the numerical solution of the problem (7.12)-(7.15) under the assumption of the powerful temperature perturbation action. Here, the corresponding values i_r and i_s are given for each value α.

The presented curves show that the currents range with the negative velocities ($i_r < i < i_s$) is expanding, and range with the positive velocities ($i_s < i < 1$) is decreasing at decreasing α (at an increase of the heat transfer coefficient). In the limiting case $\alpha = 1$, the Curve $V_n(i)$ degenerates into a point because in this case $i_r = i_s = 1$. This means that the thermal autowave does not exist at $\alpha = 1$ and $0 < i < 1$. As a result, the normal zone will necessarily disappear after the termination of any external thermal perturbation. In other words, the superconducting state is completely stable under the action of arbitrary thermal perturbations at $\alpha = 1$. Earlier, Stekly got a similar conclusion but, as already noted, on the basis of other physical assumptions.

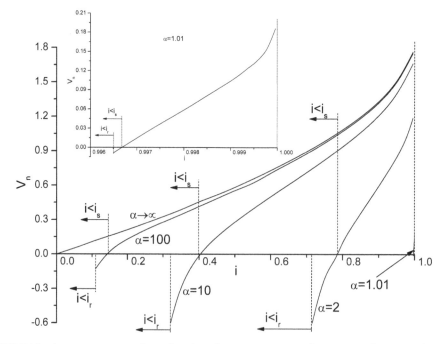

FIGURE 7.16 Autowave propagation velocities of temperature perturbations as a function of current. In the inset, the dependence $V_n(i)$ at $\alpha = 1.01$ is shown in more detail.

The solution of equation similar (7.12) in the form of an autowave can be found in the same way as it was done in (Kolmogorov et al. 1937). Let us find the solution of equation (7.12) at $P_0 = 0$ in the region $-\infty < Z < \infty$. In this case, the first equilibrium temperature occurs at $Z \to \infty$, that is $\theta|_{Z\to\infty} \to 0$ and the second one at $Z \to -\infty$, that is, $\theta|_{Z\to-\infty} \to \theta_m = \alpha i^2$, as discussed above. Introduce the new variable $\chi = Z - V_n\tau$, $V_n = \partial\xi_n/\partial\tau = \text{const}$. Considering that $\partial/\partial Z = d/d\chi$ and $\dfrac{\partial}{\partial\tau} = -V_n\dfrac{d}{d\chi}$, Equation (7.12) is converted into the form

$$\frac{d^2\theta}{d\chi^2} + V_n\frac{d\theta}{d\chi} - \frac{1}{\alpha}\theta + i^2 r(\theta) = 0 \tag{7.32}$$

which must satisfy the conditions

$$\theta_{\chi\to-\infty} = \theta_m, \quad \theta_{\chi\to\infty} \to 0, \quad \frac{\partial\theta}{\partial\chi}\bigg|_{\chi\to\pm\infty} \to 0 \tag{7.33}$$

according to the existing equilibrium states arising from the irreversible propagation of the normal zone.

Multiplying Equation (7.32) by $d\theta/d\chi$ and integrating it over χ in the range from $-\infty$ to ∞, one finds

$$\frac{1}{2}\left(\frac{d\theta}{d\chi}\right)^2\bigg|_{-\infty}^{\infty} + V_n\int_{-\infty}^{\infty}\left(\frac{d\theta}{d\chi}\right)^2 d\chi + \int_{\theta_m}^{0}\left[i^2 r(\theta) - \frac{1}{\alpha}\theta\right]d\theta = 0$$

Taking into account the boundary conditions, it is easy to write

$$V_n = \frac{\displaystyle\int_0^{\theta_m}\left[i^2 r(\theta) - \frac{1}{\alpha}\theta\right]d\theta}{\displaystyle\int_{-\infty}^{\infty}\left(\frac{d\theta}{d\chi}\right)^2 d\chi} \tag{7.34}$$

This formula shows that the normal zone in a quasi-stationary state depends only on α and i. It also leads to the 'equal area' theorem (7.20) at $i = i_s$ when the propagation velocity of the normal zone is zero. Besides, it is not difficult to understand that the velocity of the normal zone will decrease at decreasing current or parameter α. When $i < i_s$, the numerator in (7.34) takes negative values. Therefore, the arisen normal zone will disappear over time.

In general, the integrals in the Formula (7.34) depend on the velocity of the normal zone due to its corresponding connection with the temperature of the composite. Therefore, it implicitly describes the kinetics of the normal zone. To find the value $V_n(\alpha, i)$, let us place the origin of coordinates in the point where the current sharing begins. Then, in the quasi-linear approximation, the problem (7.32)-(7.33) leads to the solution of the following system of the stationary ordinary differential equations

$$\frac{d^2\theta_1}{d\chi^2} + V_n\frac{d\theta_1}{d\chi} - \frac{1}{\alpha}\theta_1 + i^2 = 0$$

$$\frac{d^2\theta_2}{d\chi^2} + V_n\frac{d\theta_2}{d\chi} - \frac{1}{\alpha}\theta_2 + i(\theta_2 + i - 1) = 0 \tag{7.35}$$

$$\frac{d^2\theta_3}{d\chi^2} + V_n\frac{d\theta_3}{d\chi} - \frac{1}{\alpha}\theta_3 = 0$$

under the following conditions

$$\theta_1(-\infty) = \theta_m,$$

$$\theta_1(-\chi_c) = \theta_2(-\chi_c) = 1, \quad \frac{d\theta_1}{d\chi}(-\chi_c) = \frac{d\theta_2}{d\chi}(-\chi_c),$$

$$\theta_2(0) = \theta_3(0) = 1 - i, \quad \frac{d\theta_2}{d\chi}(0) = \frac{d\theta_3}{d\chi}(0), \tag{7.36}$$

$$\theta_1(\infty) = 0$$

This boundary value problem describes the temperature distribution along the composite in three areas: in the area of the normal conductivity ($-\infty < \chi < -\chi_c$) in which the temperature changes from $\theta_m = \alpha i^2$ to 1, in the current sharing area ($-\chi_c < \chi < 0$) where the temperature varies from 1 to $1 - i$ and in the superconducting part of the composite ($0 < \chi < \infty$). Due to the cumbersome form of writing the expressions obtained, its solution, which is reduced to solving a transcendental equation, is not written out. At the same time, from (7.35)-(7.36) there are two limiting values V_n at $i = i_r$ and $i = 1$. Namely, in the framework of the used dimensionless variables, they are equal

$$V_n\big|_{i \to i_r} = -2\sqrt{(\sqrt{\alpha} - 1)/\alpha}, \quad V_n\big|_{i \to 1} = 2\sqrt{(\alpha - 1)/\alpha} \tag{7.37}$$

and $\partial V_n / \partial i \to \infty$ at $i = i_r$ and $i = 1$ (Lvovsky 1984, Gurevich et al. 1997c). Let us discuss the specifics of these values.

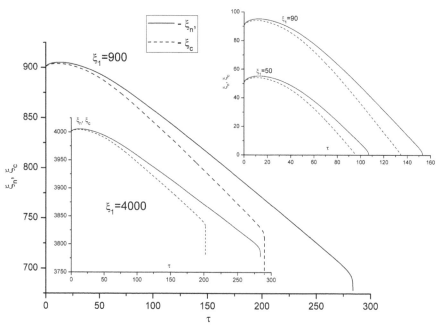

FIGURE 7.17 Propagation and disappearance of the powerful temperature perturbation at the full recovery current ($i = i_r$) during temperature perturbation with various initial length

The results of the numerical solution of the problem (7.12)-(7.15) at $\alpha = 10$, $\theta_1 = 10$ and $i = i_r$ are presented in Figure 7.17. The length of the computational domain was assumed to be $L = 5,000$ and the initial length of the temperature perturbation varied. It is seen that in contrast to the results shown in Figure 7.7, the velocities of the fronts ξ_c ($\theta(\xi_c, \tau) = 1$) and ξ_n ($\theta(\xi_n, \tau) = 1 - i$) at $i = i_r$ are different throughout the lifetime of the thermal perturbation under the action of the perturbations of any length. Moreover, $|\partial \xi_c/\partial \tau| > |\partial \xi_n/\partial \tau|$ in the area of the negative values, that is the front ξ_c 'runs away' from ξ_n. As a consequence, the thermal autowave is not formed at $i = i_r$ and the value $V_n(\alpha, i_r)$ just limits the corresponding values $V_n(\alpha, i)$ for a given value of the thermal stabilization α.

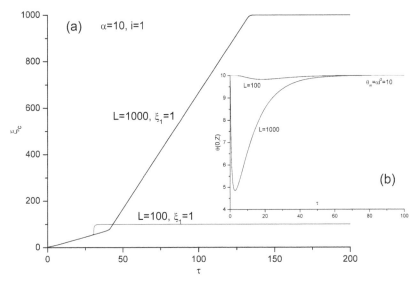

FIGURE 7.18 Propagation of the temperature perturbation at the critical current depending on the length of the composite (a) and the change in temperature at its center (b).

The change over time of the normal zone length in the composite carrying the critical current ($i = 1$) is shown in Figure 7.18. The numerical calculations were carried out for the local temperature perturbation at $\alpha = 10$, $\theta_1 = 10$, and the length of the computational domain varied. The presented curves show the following formation features of the thermal autowaves in the case when a critical current flows in the composite. First, if its length is relatively small, then only one value of the normal zone velocity exists according to the second Formula (7.37). However, the second value of the propagation velocity may exist when the length of the composite increases.

The noted features may be explained by analyzing the corresponding $G(T)$-$W(T)$ diagrams.

The value θ_m (the second value of the stable thermal equilibrium as discussed above) must appear as the intersection point of Curves $G(T)$ and $W(T)$ but it is missing at full recovery current (Figure 7.19). Thus, the uncertainty arises in the Formula (7.34). Therefore, the autowave state is not formed. This conclusion may be interpreted differently. For the existence of the autowave states, it is necessary the existence of the heat release. At the full recovery current, as it follows from Figure 7.19, this condition is not met.

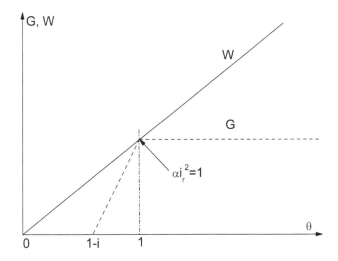

FIGURE 7.19 $G(T)$-$W(T)$ diagram for the full recovery current.

If the critical current flows in the composite, then the transition from the superconducting state to the normal state will occur at any thermal perturbation at $\alpha > 1$. As follows from (7.34), the normal zone velocity is influenced not only by the value of θ_m but also by the temperature gradient change along the entire length of the composite. If the value of θ_m is set almost immediately in the relatively short composite ($L = 100$, Figure 7.18b) then the value of V_n is unique and equal to $V_n = 2\sqrt{(\alpha - 1)/\alpha}$ when the length of the composite is sufficient for its formation. However, when the setting of θ_m is preceded by its significant drop due to the heat conduction mechanism ($L = 1,000$, Figure 7.18b) that leads to the corresponding temperature gradient distribution, then the normal zone may begin to propagate at a different velocity after the redistribution of the temperature gradient when the quantity θ_m begins to increase. From a formal point of view, the second value of V_n is a consequence of the multivalue solution of problem (7.35)-(7.36), which belongs to the class of the Sturm-Liouville problems (Lvovsky 1984).

In addition to the directly solving problem (7.35)-(7.36), several analytical formulae were obtained for calculating the velocity of the normal zone. In deriving these formulae, the assumption of a jump-like transition of the composite from the superconducting state to the normal state was used, that is neglecting the current sharing mechanism. Let us introduce the temperature θ_j, after which a jump in the effective resistivity of the composite occurs from 0 to 1. Then, neglecting the second equation and the corresponding conjugation conditions in the system (7.35), it is easy to find a solution for the transformed system and as a result, the value of the propagating velocity of the normal zone. It is equal to

$$V_n(\alpha, i) = \frac{\alpha i^2 - 2\theta_j}{\sqrt{\alpha \theta_j (\alpha i^2 - \theta_j)}} \tag{7.38}$$

This formula includes all previously obtained expressions for different values θ_j. Let us give them. If one takes $\theta_j = 1$, that is the jump of the effective resistivity of the composite occurs at the critical temperature of the superconductor, then the velocity of the normal zone is described by the formula (Cherry and Gittelman 1960)

$$V_n(\alpha, i) = \frac{\alpha i^2 - 2}{\sqrt{\alpha(\alpha i^2 - 1)}} \tag{7.39}$$

Assuming that the jump occurs at the temperature of the beginning of the current sharing, that is at $\theta_j = 1 - i$, the Formula (7.38) is reduced to (Keilin et al. 1967)

$$V_n(\alpha, i) = \frac{\alpha i^2 + 2i - 2}{\sqrt{\alpha(1 - i)(\alpha i^2 + i - 1)}} \tag{7.40}$$

If to take $\theta_j = 1 - i/2$ as the value of the jump transition temperature of the composite from the superconducting state to the normal state, then one gets (Dresner 1979a)

$$V_n(\alpha, i) = \frac{\alpha i^2 + i - 2}{\sqrt{\alpha(1 - i/2)(\alpha i^2 + i/2 - 1)}} \tag{7.41}$$

TABLE 7.1

Formula	θ_j	i_r	i_s	V_{min}	V_{max}	$V_n\|_{\alpha \to \infty}$
7.39	1	$1/\sqrt{\alpha}$	$\sqrt{2/\alpha}$	$-\infty$	$\dfrac{\alpha-2}{\sqrt{\alpha(\alpha-1)}}$	i
7.40	$1-i$	$\dfrac{\sqrt{1+4\alpha}-1}{2\alpha}$	$\dfrac{\sqrt{1+2\alpha}-1}{\alpha}$	$-\infty$	∞	$\dfrac{i}{\sqrt{1-i}}$
7.41	$1-i/2$	$\dfrac{\sqrt{1+16\alpha}-1}{4\alpha}$	$\dfrac{\sqrt{1+8\alpha}-1}{2\alpha}$	$-\infty$	$\dfrac{2(\alpha-1)}{\sqrt{\alpha(2\alpha-1)}}$	$\dfrac{i}{\sqrt{1-i/2}}$
7.42	$-$	$1/\sqrt{\alpha}$	$\dfrac{\sqrt{1+8\alpha}-1}{2\alpha}$	$-\infty$	$2\sqrt{\dfrac{\alpha-1}{\alpha}}$	$2(1-\sqrt{1-i})$

Analyzing the solutions of the system (7.35)-(7.36), the formula for calculating the velocity of the normal zone propagation was proposed in (Turk 1980) taking into account the current sharing. In dimensionless form, it is written as

$$V_n(\alpha, i) = \frac{\alpha i^2 + i - 2}{\sqrt{\alpha(\alpha i^2 - 1)}} + \sqrt{\frac{\alpha-1}{\alpha}}\left(2 - 2\sqrt{\frac{1-i}{1-i_s}} - \frac{i-i_s}{1-i_s}\right) \qquad (7.42)$$

The results of the numerical calculations of V_n are compared with the results of the calculation of the velocities of the normal zone propagation by the Formulae (7.39)-(7.42) in Figure 7.20.

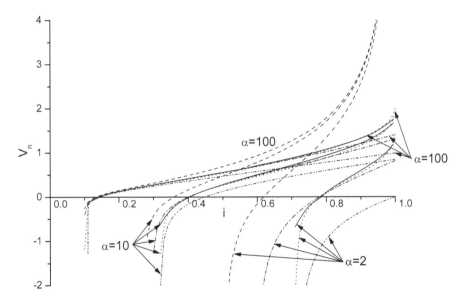

FIGURE 7.20 Dependence of the normal zone velocity on the current for various parameters of the thermal stabilization: ———— - numerical calculation, - ·· - ·· - - calculation by the Formula (7.39), - - - - - calculation by the Formula (7.40), - · - · - · - - calculation by the Formula (7.41), · · · · · · - calculation by the Formula (7.42).

The curves presented show that the most accurate formula describing the dependence $V_n(\alpha, i)$ is the Formula (7.42). The Formula (7.41) loses accuracy under intense cooling conditions.

The Formulae (7.39) and (7.40) allow one to obtain only the lower and upper estimates of the thermal autowave velocity, respectively. These formulae have a significant error at $i > 0.6$, which is most noticeable in the neighborhood of the critical current and under intensive cooling. In general, all formulae distort the functional dependence $V_n(\alpha, i)$ at currents close to i_r. This is confirmed by Table 7.1. Here, the corresponding dependencies $V_n(\alpha, i)$ for the uncooled composite are also presented. They allow easily estimating the propagation velocity of the normal zone in tight windings of SMS, albeit with varying degrees of accuracy but without resorting to the numerical calculations. In particular, the above formulae show that in this case $V_n \sim i$ at low currents.

To go to the dimensional values, the following characteristic velocity should be used

$$V_x = \frac{L_z}{t_x} = \frac{I_{c0}}{SC_k}\sqrt{\frac{\lambda_k \rho_0}{T_{cB} - T_0}} \approx \frac{\eta J_{c0}}{C_k}\sqrt{\frac{\lambda_m \rho_m}{T_{cB} - T_0}}$$

according to the above characteristic values L_z and t_x and taking into account that $\lambda_k \approx (1-\eta)\lambda_m$. According to its physical meaning, V_x determines the characteristic propagation velocity of the normal zone under adiabatic conditions when the critical current flows in the composite. Using the averaged values of the composite parameters given above, it is easy to estimate that $V_x \sim 100$ m/s for LTS. The value V_x is directly proportional to the critical current density. The main influence of the thermophysical properties on the normal zone velocity determines the heat capacity of the composite since the product $\lambda_m \rho_m$ satisfies the Wiedemann-Franz law.

Above, a superconducting composite without taking into account the insulating coating of its surface was considered. The presence of insulation will undoubtedly lead to a change in its thermal state. Let us study the effect of a nonconductive coating on the irreversible propagation velocity of a normal zone along a cooled composite superconductor. For ease of analysis, let us assume that the boundary surfaces of the composite are flat; its thickness is d_k and the thickness of the coating d_i are small; there is no contact thermal resistance between the insulation and the composite; the thermal and electrophysical parameters do not depend on temperature. Then, the change in the temperature of the composition 'superconducting composite + insulation' in the longitudinal direction may be described by one-dimensional unsteady heat conduction equation

$$C_e \frac{\partial T}{\partial t} = \lambda_e \frac{\partial^2 T}{\partial z^2} - \frac{h_e p}{S}(T - T_0) + \frac{I^2}{S^2}\rho_e(T)$$

with averaged coefficients $C_e = C_k d_k/d + C_i d_i/d$, $\lambda_e = \lambda_k d_k/d + \lambda_i d_i/d$, $1/h_e = 1/h + d_i/\lambda_i$, $1/\rho_e = d_k/\rho(T)d + d_i/\rho_i d \approx d_k/\rho_k(T)d$ because $\rho_i \gg \rho_0$ where $\rho_k(T)$ and ρ_0 are described by the Formula (7.11). Here, C_k and C_i are the volumetric heat capacities of the composite and insulation, respectively, λ_k and λ_i are their thermal conductivity coefficient, $d = d_k + d_i$. The physical meaning of the other parameters remained unchanged. Using the above entered dimensionless variables, it is easy to get the following dimensionless equation of the thermal conductivity for a composite with an insulating coating

$$\left(1 + C\frac{\Delta_i}{\Delta_k}\right)\frac{\partial \theta}{\partial \tau} = \left(1 + \Lambda\frac{\Delta_i}{\Delta_k}\right)\frac{\partial^2 \theta}{\partial Z^2} - \frac{\theta}{\alpha_e} + i^2 r(\theta)$$

Here, $C = C_i/C_k$, $\Lambda = \lambda_i/\lambda_k$, $\Delta_{k,i} = d_{k,i}/L_x$, $\alpha_e = \alpha + 1/\omega$ ($\omega = \Lambda/\Delta_k\Delta_i$).

The influence of the insulation on the temperature of the composite is negligible if $C\Delta_i/\Delta_k \ll 1$, $\Lambda\Delta_i/\Delta_k \ll 1$, $\alpha \gg 1/\omega$. It is also easy to understand what role the thermophysical parameters of insulation and its thickness play in the kinetics of the normal zone. It can be shown that in the case of $\Delta_i/\Delta_k \ll 1$, which is satisfied with the parameters of real current-carrying parameters, the specificity of the thermal mode development in them will determine by the value of the dimensionless complex

$\Delta_k \Delta_i / \Lambda$ which is inversely proportional dimensionless thermal insulation resistance ω. In other words, taking the heat capacity of insulation into account in calculating the normal zone velocity will have little effect on the final results for a wide range of practical applications. The validity of this conclusion is strictly proved below.

Let us estimate the possible variation range of the introduced dimensionless parameters. For Nb-Ti superconductor in a copper matrix with an organic coating, the averaged values of the initial coefficients are equal to $C_k \sim 5 \times 10^3$ J/(m³ × K), $\lambda_k \sim 100$ W/(m × K), $d_k \sim 10^{-3}$ m, $C_i \sim 10^4$ J/(m³ × K), $\lambda_i \sim 0.1$ W/(m × K), $d_i \sim 5 \times 10^{-5}$ m, $\rho_0 \sim 5 \times 10^{-10}$ W × m, $T_{cB} - T_0 \sim 5$ K, $I_{c0} \sim 10^3$ A. Then, it leads to $C \sim 2$, $\Lambda \sim 10^{-3}$, $\Delta_k \sim 1$, $\Delta_i \sim 5 \times 10^{-2}$.

According to the Formula (7.42), the dimensionless velocity of the normal zone propagating along the composite with an insulating coating after the formation of the thermal autowave is equal to

$$V_n = \frac{\sqrt{1 + \Lambda \Delta_i / \Delta_k}}{1 + C \Delta_i / \Delta_k} \left[\frac{\alpha_e i^2 + i - 2}{\sqrt{\alpha_e (\alpha_e i^2 - 1)}} + \sqrt{\frac{\alpha_e - 1}{\alpha_e}} \left(2 - 2\sqrt{\frac{1-i}{1-i_s}} - \frac{i - i_s}{1 - i_s} \right) \right]$$

In Figure 7.21a, the dashed lines show the dependence of the normal zone velocity versus current for different insulation thicknesses calculated at $\alpha = 2$, $C = 1$, $\Lambda = 10^{-3}$, $\Delta_k = 1$. Here, for comparison, the values $V_n(\alpha, i)$ (the solid curves) are given, corresponding to the composite superconductor without an insulating coating. These values are calculated for three characteristic values of the stabilization parameter: $\alpha \to \infty$ - thermal insulation mode (Curve 1), $\alpha = 100$ – 'poor' cooling mode (Curve 2), $\alpha = 2$ – 'good' cooling mode (Curve 3).

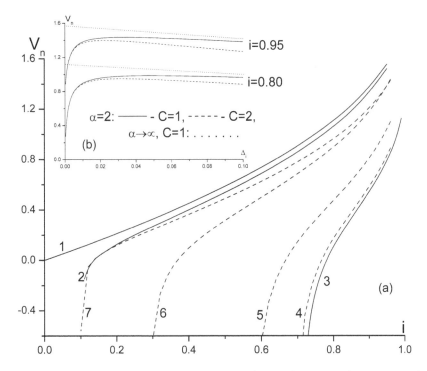

FIGURE 7.21 Dependence of the propagation velocity of normal zone on the current with different insulation thicknesses (a: $\Delta_i = 0$: 1 - $\alpha \to \infty$, 2 - $\alpha = 100$, 3 - $\alpha = 2$; $\alpha = 2$, $\Delta_i \neq 0$: 4 - $\Delta_i = 10^{-4}$, 5 - $\Delta_i = 10^{-3}$, 6 - $\Delta_i = 10^{-2}$, 7 - $\Delta_i = 10^{-1}$) and on the insulation thickness for different values of current (b).

Comparing the results presented, one may notice that even insignificant insulation in its thickness can significantly modify the thermal stabilization conditions of the intensively cooled superconducting composite. First of all, it should be noted that the coating thickness has a significant effect on the range of currents at which the full stability of the superconducting state is preserved. So, the irreversible destruction of superconductivity occurs in the current interval $0.7808 = i_s < i < 1$ at $\Delta_i = 0$ and $\alpha = 2$. At the same time, it almost doubles at $\Delta_i = 10^{-2}$, and the thermal state of the intensively cooled composite is slightly different from the adiabatic conditions at $\Delta_i = 10^{-1}$.

Attention should also be paid to the special nature of change in the velocity of the normal zone in 'well' cooled composite superconductors in the region of currents that are close to the critical under increasing insulation thickness. The corresponding calculation results are plotted in Figure 7.21b. In this case, the dependencies $V_n(\Delta_i)$ have a maximum, which is shifted toward lower values Δ_i if the current increases as follows from Figure 7.21b.

Thus, when using current-carrying elements manufactured on the basis of a composite superconductor with an insulating coating, special attention should be paid to the existence of a thermal barrier, which is formed by isolation. The results discussed above show that the thermal resistance of insulation can almost eliminate the effectiveness of measures providing, as it would seem, intensive cooling of the superconducting winding. This should be taken into account when designing large SMS, for which the stable performance requires intensive cooling of current-carrying elements.

7.5 IRREVERSIBLE PROPAGATION OF THE THERMAL INSTABILITY IN THE CROSS-SECTION OF A MULTILAYER SUPERCONDUCTING COMPOSITION 'COMPOSITE SUPERCONDUCTOR + INSULATION'

As it was shown above, the thermal instability arising in a superconducting composite as a result of an external perturbation with an energy exceeding the critical one (supercritical thermal perturbation) is accompanied by its irreversible propagation along the composite when a thermal autowave moving at a constant velocity is formed in it. Against its background, the destruction of the superconducting properties of SMS occurs. As a rule, the critical energies and velocities of the thermal autowaves are investigated for a single composite superconductor (superconductor + matrix). The proposed models used a priori to describe the modes occurring in multiturn SMS in which the thermal diffusivity of its structural elements differs by three to four orders of magnitude. However, as the analysis carried out in the previous section shows, the presence of the isolation may not only quantitatively but also qualitatively modify the kinetics of the normal zone. This limits the use of the conclusions of the thermal stabilization theory formulated for a single composite.

Let us discuss the features of the irreversible propagation of thermal perturbations in the cross-section of a multilayer region. Assume that it is formed by N of the thermally thin slab-shaped composite superconductors, the width of them is much greater than the thickness, which are separated from each other by the insulation (Figure 7.22). Let us focus on the mechanisms affecting the characteristic features of the destruction of superconducting state in the cross-section of this composition, and determine the velocity of the transverse propagation of the thermal instability depending on the thermal properties and thickness of the insulation. Suppose also that the constant current I flows in all composite superconductors cooled to a temperature T_0, and they are in a constant external magnetic field. Let the thermal instability be initiated at the initial time by an extended supercritical temperature perturbation as a result of which one or several superconducting layers ($n_i = 1, 2, \dots$) are instantly going to the normal state. In this case, the longitudinal thermal conductivity of the matrix may be neglected. To simplify the analysis, let us also assume that the temperature of each composite in its cross-section is constant, and the temperature variation does not affect the thermal and electrophysical properties of all the composition elements.

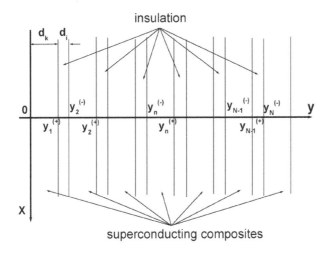

FIGURE 7.22 Scheme of the calculated area.

Under these assumptions, the three-dimensional problem of determining the temperature in a multicomponent composition is reduced to the system of the conjugate one-dimensional heat conduction equations taking into account the conditions of the heat exchange between its elements. Then, the temperature change in the multilayer superconducting composition without taking into account the longitudinal mechanism of the thermal conductivity can be described by the following system of the heat conduction equations (for example: Funaki et al. 1986, Rusinov 1994)

$$C_k \frac{\partial T_1}{\partial t} = \frac{I^2}{S^2} \rho_k(T_1) + \frac{\lambda_i}{d_k} \frac{\partial T_{1,i}}{\partial y}\bigg|_{y_1^{(+)}} - \frac{hp}{S}(T_1 - T_0), \quad n = 1$$

- - - - - - - - - - - - - - - - - - - -

$$C_k \frac{\partial T_n}{\partial t} = \frac{I^2}{S^2} \rho_k(T_n) + \frac{\lambda_i}{d_k}\left(\frac{\partial T_{n,i}}{\partial y}\bigg|_{y_n^{(+)}} - \frac{\partial T_{n-1,i}}{\partial y}\bigg|_{y_n^{(-)}} \right) - \frac{hp}{S}(T_n - T_0), \quad n = 2, ..., N-1$$

- - - - - - - - - - - - - - - - - - - -

$$C_k \frac{\partial T_N}{\partial t} = \frac{I^2}{S^2} \rho_k(T_N) - \frac{\lambda_i}{d_k} \frac{\partial T_{N-1,i}}{\partial y}\bigg|_{y_N^{(-)}} - \frac{hp}{S}(T_N - T_0), \quad n = N$$

$$C_i \frac{\partial T_{n,i}}{\partial t} = \lambda_i \frac{\partial^2 T_{n,i}}{\partial y^2}, \quad n = 1, ..., N$$

$$T_{n,i}\big|_{y_n^{(+)}} = T_n, \quad T_{n,i}\big|_{y_{n+1}^{(-)}} = T_{n+1}$$

with initial conditions

$$T_{n,i}\big|_{t=0} = T_0, \quad T_n\big|_{t=0} = \begin{cases} T_d, & n = n_i, \\ T_0, & n \neq n_i \end{cases}$$

Here, n and n_i are the current numbers of the superconducting composite and the insulation layer in the composition; C_k and $C_i \sim$ const are the heat capacities of the composite and insulation, respectively; λ_k, $\lambda_i \sim$ const are their coefficients of the thermal conductivity, h is the heat transfer coefficient, p is the cooled perimeter of each layer, S is the cross-sectional area of the superconducting composite layer, T_d is the temperature perturbation of the layer/layers and $\rho_k(T)$ is the effective resistivity of the composite varying with temperature according to (7.11).

As above-mentioned, to simplify the analysis, let us introduce the following dimensionless variables

$$Y = y/L_z, \quad \tau = \lambda_k t/(C_k L_z^2), \quad \theta_n = (T_n - T_0)/(T_{cB} - T_0), \quad \Theta_n = (T_{n,i} - T_0)/(T_{cB} - T_0)$$

Then, the change in the dimensionless temperature in the cross-section of this multilayer composition will be determined on the basis of the numerical solution of the problem in the form

$$\frac{d\theta_1}{d\tau} = i^2 r(\theta_1) - \frac{1}{\alpha}\theta_1 + \frac{\Lambda}{\Delta_k}\frac{\partial\Theta_1}{\partial Y}\bigg|_{Y_1^{(+)}}$$

$- -$

$$\frac{d\theta_n}{d\tau} = i^2 r(\theta_n) - \frac{1}{\alpha}\theta_n + \frac{\Lambda}{\Delta_k}\left(\frac{\partial\Theta_n}{\partial Y}\bigg|_{Y_1^{(+)}} - \frac{\partial\Theta_n}{\partial Y}\bigg|_{Y_1^{(-)}}\right) \quad (7.43)$$

$- -$

$$\frac{d\theta_N}{d\tau} = i^2 r(\theta_N) - \frac{1}{\alpha}\theta_N - \frac{\Lambda}{\Delta_k}\frac{\partial\Theta_{N-1}}{\partial Y}\bigg|_{Y_N^{(-)}}$$

$$C\frac{\partial\Theta_n}{\partial\tau} = \Lambda\frac{\partial^2\Theta_n}{\partial Y^2},$$

$$\Theta_n(Y_n^{(+)}, \tau) = \theta_n(\tau), \Theta_n(Y_{n+1}^{(-)}, \tau) = \theta_{n+1}(\tau)$$

with initial conditions

$$\theta_n(Y, 0) = 0, \quad \theta_n(0) = \begin{cases} \theta_d = (T_d - T_0)/(T_{cB} - T_0), n = n_i \\ 0, n \neq n_i \end{cases}$$

Here, the quantities L_z, a, i, $r(\theta)$, C, Λ and Δ_i, Δ_k were introduced earlier for a single composite.

As in the analysis of the effect of insulation on the longitudinal velocity of the instability, the original dimensionless insulation parameters can be combined by the dimensionless parameter $\omega = \Lambda/\Delta_k\Delta_i$ introduced above. According to its physical meaning, it corresponds to the dimensionless thermal resistance of insulation.

In Figures 7.23-7.26, the results of the numerical experiments performed for uncooled ($\alpha \to \infty$) composition with varying insulation parameters at $\Delta_k = 1$ are presented. (Features of the transient modes in the cooled superconducting multicomponent composition are discussed in the next section). To reduce the influence of the boundary effect due to a finite number of elements in the composition, the number of them was set equal to $N = 50$. In most cases, it was assumed that the instability is initiated in the first layer at the initial time, setting its initial temperature equal to $\theta_d = 10$ (Curves 1-3 in Figure 7.23). However, in some cases (and, first of all, at the low currents), it turns out that such heat impulses may not lead to the irreversible destruction of the superconducting properties of the entire composition. For these modes, the total number of the composites was increased in which the sections with normal conductivity appeared at the initial time (Curves 4-6 in Figure 7.23).

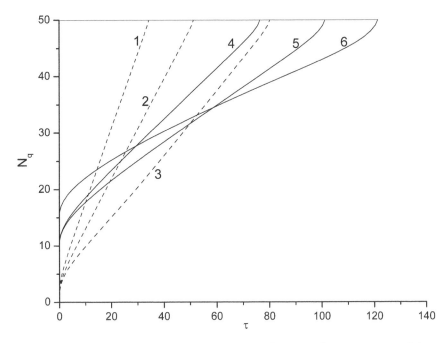

FIGURE 7.23 Transition kinetics of the temperature perturbation in the cross-section of the multilayer superconducting composition at $C = 1$, $\Delta_i = 10^{-3}$, $\Lambda = 10^{-3}$ and various currents: 1 - i = 0.9, 2 - i = 0.7, 3 - i = 0.5, 4 - i = 0.4, 5 - i = 0.3, 6 - i = 0.2.

The results presented in Figure 7.23 allow one to formulate the main regularities that determine the characteristic features of the irreversible propagation of thermal perturbation in the cross-section of a superconducting multilayer composition.

First, during normal zone propagation, there are periods when the growth of the transverse dimensions of the resistive region occurs with unsteady velocity. The latter takes place in the initial and final stages of change in time dependences $N_q(\tau)$. As a rule, they are not long enough. However, the smaller the current, the power and length of the external disturbance or the worse the heat exchange conditions between the elements of the composition, the greater the time of their existence. Second, in the intermediate times, the velocity of the increase in the size of the resistive region takes an almost constant value which does not depend on the character of the initial perturbation. (More exactly, it asymptotically approaches a constant value in the framework of the computational model used). Therefore, thermal autowave modes may appear the existence conditions of which depend only on the properties of the composition elements.

In Figure 7.24, the results of calculating the dimensionless values of the transverse autowave velocity V_y of the thermal instability as a function of current are shown. The simulation was performed for the insulation with the dimensionless thermal conductivity coefficient $\Lambda = 10^{-3}$ at various dimensionless values of its thickness and heat capacity. The values V_y were determined by the differentiation of curves $N_q(\tau)$ in those parts of their growth when the development of the transition mode asymptotically approaches the autowave one ($\partial N_q/\partial \tau \sim$ const). From a formal point of view, these values V_y are determined with a good degree of accuracy by the numerical differentiation of the central parts of curves $N_q(\tau)$. The given dependencies show the very insignificant role of the heat capacity insulation in changing the velocity of the irreversible propagation of perturbations. Therefore, taking into account the heat capacity of the insulation has a small effect on the values of the thermal instability velocity (Figure 7.25).

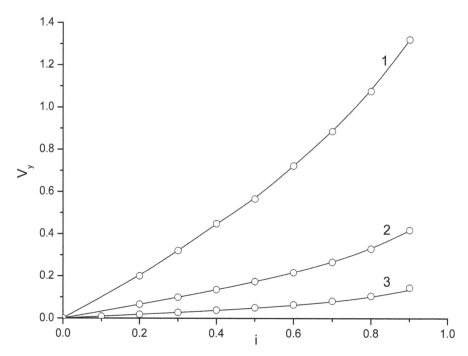

FIGURE 7.24 Dependence on the current of the dimensionless propagation velocity of normal zone in the cross-section of the uncooled multilayer composition: o - $C = 0$, ——— - $C = 1$: 1 - $\Delta_i = 10^{-3}$ ($\omega = 1$), 2 - $\Delta_i = 10^{-2}$ ($\omega = 0.1$), 3 - $\Delta_i = 10^{-1}$ ($\omega = 0.01$).

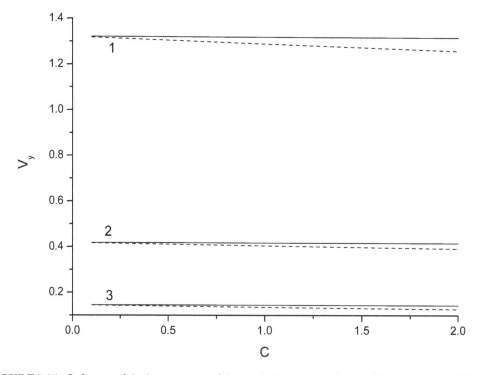

FIGURE 7.25 Influence of the heat capacity of the insulation on the velocity of the thermal instability at $i = 0.9$: ——— - $\Delta_i = 0.01$, – – – – $\Delta_i = 0.1$; 1 - $\Lambda = 0.1$, 2 - $\Lambda = 10^{-2}$, 3 - $\Lambda = 10^{-3}$.

The study of functional dependence of the transverse propagation velocity of the thermal instabilities on current is of undoubted interest from the point of view of assessing the effect of the discrete nature of the superconducting composition on it. From the results shown in Figure 7.24, it follows that with increasing insulation thickness, the change V_y with current takes a more linear character, reducing the range of currents in which the dependence V_y is nonlinear. Therefore, the relationship between the propagation velocities of the normal zone in the longitudinal and transverse directions of a multicomponent superconducting composition will differ from that which can be formulated in the approximation of an anisotropic medium model with averaged parameters. According to the model (7.43), this approximation will take place only at $\omega \to \infty$, for example in the case of negligible insulation thickness. (The interrelation of the thermal instability propagation in a multilayer composition in the longitudinal and transverse directions is discussed in more detail below).

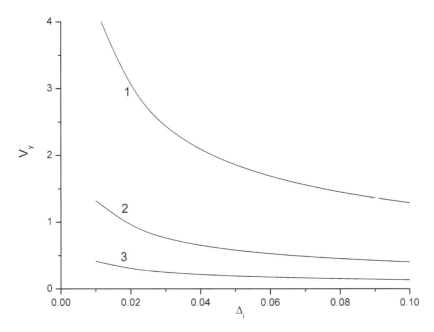

FIGURE 7.26 Influence of the insulation thickness on the transverse autowave velocity of the thermal instability at $i = 0.9$ and $C = 1$: $1 - \Lambda = 10^{-1}$, $2 - \Lambda = 10^{-2}$, $3 - \Lambda = 10^{-3}$.

In Figure 7.26, the results of the calculation of the transverse propagation velocity of thermal instability depending on the insulation thickness for different values of the dimensionless thermal conductivity of the insulation at a current close to the critical are shown. The presented results depict that the insulation thickness strongly affects the velocity of the normal zone in the cross-section compared with the heat capacity of the insulation. The most noticeable effect of the thickness of the insulating layer occurs in the region of small values Δ_i.

The weak dependence of the transverse propagation velocity of thermal instability on the heat capacity of the insulation leads to the conclusion that a more simplified model may be used to calculate it. In this case, the thermal interaction of the superconducting composites with each other is described only taking into account the thermal insulation resistance, which depends on the dimensionless parameter ω. Correspondingly, the operating mode of the multilayer superconducting composition may be described using zero-dimensional approximation that may be written as follows

$$\frac{d\theta_n}{d\tau} = i^2 r(\theta_n) - \frac{1}{\alpha}\theta_n + \begin{cases} \omega(\theta_1 - \theta_2), & n = 1 \\ \omega(2\theta_n - \theta_{n-1} - \theta_{n+1}), & n = 2, \ldots, N-1 \\ \omega(\theta_N - \theta_{N-1}), & n = N \end{cases} \qquad (7.44)$$

$$\theta_n(0) = \begin{cases} \theta_d, n = n_i \\ 0, n \neq n_i \end{cases}$$

(The derivation and formulation of equations of the form (7.44) is discussed in the next section).

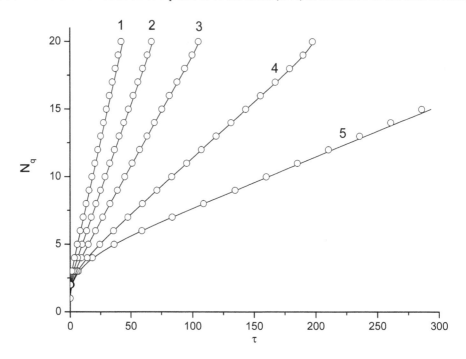

FIGURE 7.27 Development of the transition modes in the cross-section of a multilayer region with 'thin' insulation at $\alpha = 100$: o - calculation by the model (7.43) ——— - calculation by the model (7.44); 1 - $i = 0, 9$, 2 - $i = 0, 7$, 3 - $i = 0, 5$, 4 - $i = 0, 3$, 5 - $i = 0, 2$.

In Figure 7.27, the markers 'o' show the results of the numerical simulation of the irreversible kinetics of thermal instability in the cross-section of the composition with parameters $N = 20$, $C = 10^{-2}$, $\alpha = 100$, $\Delta_k = 1$, $\Delta_i = 10^{-4}$, $\Lambda = 10^{-5}$. They were obtained according to the model (7.43) for different current at $\theta_d = 10$. The solid curves correspond to the calculations by the model (7.44). The calculations were made at $\theta_d = 10$, $n = 1$ and $\omega = 0.1$. The calculation results of the transverse velocity of thermal perturbation as a function of current performed for various conditions of thermal interaction of the composites with each other ($\omega = 0.01$ ($\Lambda = 10^{-6}$) and $\omega = 0.1$ ($\Lambda = 10^{-5}$)) are given for this composition in Figure 7.28.

The results, which are shown in Figure 7.27 and 7.28, demonstrate a very satisfactory agreement between the calculations of the transverse propagation velocity of thermal instability performed within the framework of various thermophysical models of a multilayer superconducting composition. Therefore, to assess the effect of the isolation on the irreversible propagation of instability in SMS, a more simplified approximation may be used in which the thermal connection between the turns is described by the thermal resistance of the insulation layer (in dimensionless form by parameter ω).

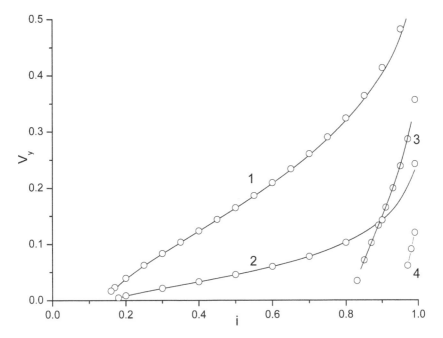

FIGURE 7.28 Dependence on current of the propagation velocity of thermal instability in the cross-section of a multilayer superconducting composition: o - calculation by the model (7.43), ———— - calculation by the model (7.44); 1 - α = 100, ω = 0.1, 2 - α = 100, ω = 0.01, 3 - α = 2, ω = 0.1, 4 - α = 2, ω = 0.01.

Thus, in the whole range of the transport current change, the transverse autowave propagation velocity of the thermal instability primarily depends on the value of the insulation thermal resistance. The influence of its heat capacity on the character of the normal zone propagation in superconducting windings manufactured on the basis of the materials currently used is not significant. Therefore, to determine the propagation velocities of the thermal instabilities in the superconducting multicomponent composition with a good degree of accuracy, one can use a simplified model. Within that model, the thermal interaction between superconducting turns and insulation is described by the model using the concept of thermal resistance.

7.6 QUASI-TWO-DIMENSIONAL KINETICS OF THE IRREVERSIBLE TRANSITION TO THE NORMAL STATE OF A MULTICONDUCTOR SUPERCONDUCTING COMPOSITION

As already noted, many theoretical results of the analysis of transients in SMS, as a rule, usually follow from the anisotropic continuum model. In this case, SMS is considered as a continuum bulk with average characteristics. As a result of this approximation, the spatio-temporal development of the transition modes in the winding is described by an expanding ellipsoid model (Wilson 1983). According to this model, the ellipsoid axes are formed by the velocity vectors of the normal zone penetrating from turn to turn and from layer to layer. In the framework of this approximation, between the velocities in the transverse (v_y) and longitudinal (v_x) directions, with which the ellipsoid increases the existence of the following connection $v_y = v_x \sqrt{\lambda_y / \lambda_x} \, C_w / C_k$ is a priori assumed, where C_w and C_k are the averaged values of the heat capacity of the winding and the composite, respectively, λ_x and λ_y are the thermal conductivity coefficients in the corresponding directions. Moreover, v_x is determined from the solution of the corresponding one-dimensional problem of the transition to the normal state of a single superconducting composite. However, such an approximation does not allow to fully describe the features of the transient modes in SMS taking into account the discrete nature of its internal structure since the characteristic mechanisms responsible for the propagation of a normal zone in multiturn areas, in particular the effect of isolation are omitted from attention.

In this regard, let us discuss the irreversible propagation mechanisms of the normal zone, which is initiated by the supercritical temperature perturbation, in the longitudinal and transverse directions in a cooled superconducting composition with a discrete (multicomponent) structure composed, as discussed above, from N straight-line superconducting composites (Figure 7.22). To simplify the analysis, let us assume that the current I in each element of the composition is constant, and their thermal and electrophysical properties do not depend on the temperature and magnetic field. For the same purpose, let us also neglect the temperature change in the cross-section of each element of the composition. Let us place the origin of coordinates in the center of the initial temperature perturbation and describe the symmetric temperature distribution in the interconnecting composite superconductors separated from each other by final thermal resistance R using the following system of the heat conductivity equations

$$C_k \frac{\partial T_n}{\partial t} = \lambda_k \frac{\partial^2 T_n}{\partial x^2} - \frac{hp}{S}(T_n - T_0) + \frac{I^2}{S^2} \rho_k(T_n) - \begin{cases} \dfrac{P}{SR}(T_1 - T_2), & n = 1 \\[2mm] \dfrac{P}{SR}(2T_n - T_{n-1} - T_{n+1}), & n = 2, \ldots, N-1 \\[2mm] \dfrac{P}{SR}(T_N - T_{N-1}), & n = N \end{cases} \quad (7.45)$$

with initial-boundary conditions

$$T_n(x, 0) = \begin{cases} T_d, & 0 \leq x \leq l_1, & n = n_i, & i = 1, 2, \ldots \\ T_0, & l_1 < x \leq l_k, & n = n_i \\ T_0, & 0 < x \leq l_k, & n \neq n_i \end{cases}$$

$$\frac{\partial T_n}{\partial x}(0, t) = 0, \quad T_n(l_k, t) = T_0 \qquad (7.46)$$

Here, $n = 1, \ldots, N$ is the number of the element in the composition, C_k is the volumetric heat capacity of the composites, λ_k is the thermal conductivity coefficient of them in the longitudinal direction, h is the heat transfer coefficient, p is the cooled perimeter, S is the cross-sectional area, P is the contact perimeter between two neighboring composites, R is the thermal contact resistance between composites, I is the transport current in each composite, T_0 is the coolant temperature, T_d is the initial temperature of the thermal perturbation of a length l_1 and $\rho_k(T_n)$ is the effective resistivity of the superconducting composite taking into account the existence of the current sharing in the n-th composite and is described by the Formula (7.11).

The problem (7.45)-(7.46) summarizes the description of the modes discussed above that occur in the longitudinal direction of a single composite superconductor of small thickness and in the transverse direction of a multilayer superconducting composition. Accordingly, it describes the collective modes of quasi-two-dimensional symmetric temperature redistribution in a multicomponent composition consisting of small cross-section elements after temperature perturbation with the temperature T_d and length l_1. That perturbation may cause the local normal zone in the composites with the numbers $n_i = 1, 2, \ldots$. To determine the temperature distribution in all elements of the composition and the corresponding propagation velocity of the normal zones in all directions, the numerical method was used. In this case, only supercritical temperature perturbations were considered, which necessarily cause an irreversible transition of the superconducting composite to the normal state in the form of a thermal autowave.

Reduce the total number of varied parameters using, as above-mentioned, the following dimensionless variables

$$X = x/L_z, \ \tau = t/t_x, \ i = I/I_{c0}, \ \theta_n = (T_n - T_0)/(T_{cB} - T_0),$$

where

$$L_z = \sqrt{\lambda_k (T_{cB} - T_0)S^2/(I_{c0}^2 \rho_0)} \ \text{and} \ t_x = C_k L_z^2/\lambda_k.$$

In this case, thermal modes in the dimensionless terms may be described as follows

$$\frac{d\theta_n}{d\tau} = \frac{\partial^2 \theta_n}{\partial X^2} - \frac{1}{\alpha}\theta_n + i^2 r(\theta_n) + \begin{cases} \omega(\theta_1 - \theta_2), & n = 1 \\ \omega(2\theta_n - \theta_{n-1} - \theta_{n+1}), & n = 2, ..., N-1 \\ \omega(\theta_N - \theta_{N-1}), & n = N \end{cases} \quad (7.47)$$

$$\theta_n(X, 0) = \begin{cases} \theta_d = (T_d - T_0)/(T_{cB} - T_0), & 0 \le X \le X_1, \quad n = n_i, \quad i = 1, 2, ... \\ 0, & X_1 < X \le L = l_k/L_x, \quad n = n_i \\ 0, & 0 \le X \le L, \quad n \ne n_i \end{cases}$$

$$\frac{\partial \theta_n}{\partial X}(0, \tau) = 0, \quad \theta_n(L, \tau) = 0$$

Here,

$$r(\theta_n) = \begin{cases} 1, & \theta_n > 1 \\ (\theta_n - 1 + i)/i, 1 - i \le \theta_n \le 1, \\ 0, & \theta_n < 1 - i \end{cases} \quad \alpha = \frac{I_{c0}^2 \rho_0}{hpS(T_{cB} - T_0)}, \quad \omega = \frac{PS(T_{cB} - T_0)}{RI_{c0}^2 \rho_0}$$

As noted above, the dimensionless parameters α and ω take into account the heat transfer to the coolant and the thermal connection between the composites, respectively, and $r(\theta_n)$ describes their effective resistivity in a dimensionless form according to approximation (7.14).

This simplified description of the problem of SMS transition to the normal state allows one to carry out a quasi-linear analysis of the basic regularities of thermal modes occurring during the destruction of the superconducting properties of composites contacting each other. This analysis may be done using the generally accepted terms of the thermal stabilization theory. Neglecting the longitudinal thermal conductivity, which is negligible under the action of an extended perturbation, the system (7.47) goes into (7.44). The results of the numerical experiments reflecting the qualitative regularities governing the transition of a superconducting discrete composition to a normal state are discussed below. To reduce the influence of the boundary effect due to the specification of the boundary condition at $X = L$, the length of the computational domain was taken to be $L = 500$ during calculations. Without loss of generality, in most cases, it was assumed that the normal zone occurs in the first wire ($n_i = 1$) at the initial time as a result of the powerful local overheating assuming $\theta_1 = 10$.

First of all, let us discuss the formation features of the normal conductivity core in a superconducting composition of a discrete structure.

In Figure 7.29, Curves 1-6 describe the change in time of the dimensionless velocity $dX_{q,n}/d\tau$ of isotherm $\theta_n(X_{q,n}, \tau) = 1$ with which it moves along the composite with the number $n = 1$, when the normal zone was initiated. The calculation was carried out at $\omega = 0.2$ for two characteristic values of the stability parameter corresponding to 'good' ($\alpha = 2$) and 'poor' ($\alpha = 100$) cooling at different values of the current, total number of composites and length of perturbation. For the case $\alpha = 100$, the values of the propagation velocity of the normal zone along the elements of the composition with $N = 51$ are presented in Figure 7.29 (Curves 7-9). In Figure 7.30, the influence of the number of composite elements on the formation of a core with normal conductivity in a superconducting discrete composition is shown. Initial parameters were set equal to $\alpha = 100, i = 0.5, \omega = 0.1, X_1 = 10$.

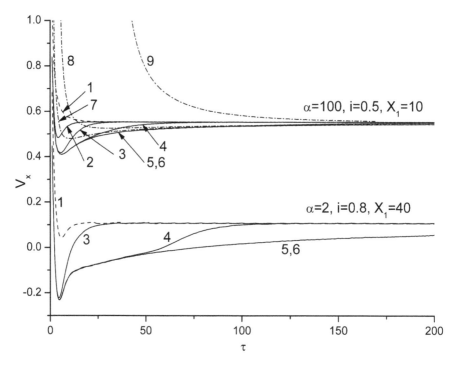

FIGURE 7.29 Kinetics of the normal zone front in the longitudinal direction of a multicomponent composition with different numbers of elements: 1 - $N = 1$, $n = 1$ (single composite); 2 - $N = 2$, $n = 1$; 3 - $N = 3$, $n = 1$; 4 - $N = 5$, $n = 1$; 5 - $N = 10$, $n = 1$; 6 - $N = 51$, $n = 1$; 7 - $N = 51$, $n = 2$; 8 - $N = 51$, $n = 3$; 9 - $N = 51$, $n = 10$.

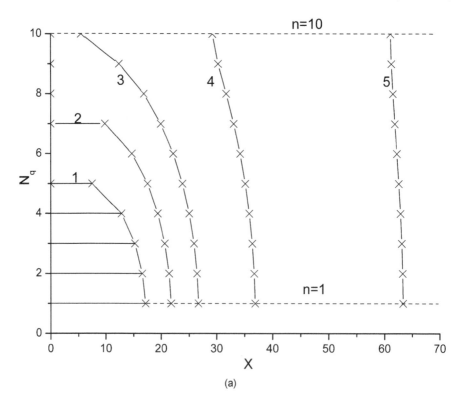

(a)

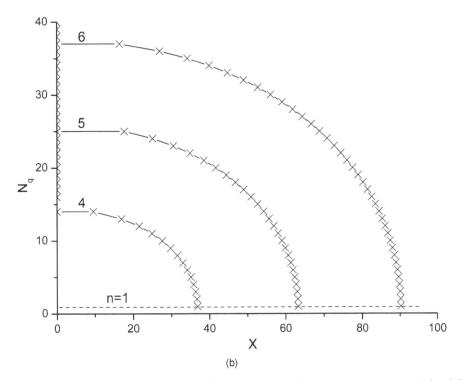

(b)

FIGURE 7.30 Kinetics of the normal conductivity core in a multicomponent composition with a different number of superconducting elements (a - N = 10, b - N = 51) initiated by thermal instability in the element with n_i = 1: 1 - τ = 10, 2 - τ = 20, 3 - τ = 30, 4 - τ = 50, 5 - τ = 100, 6 - τ = 150. Here, the front of the normal zone in each element of the composition is depicted by marker '×'.

The results show the existence of two characteristic modes that occur when the normal zone propagates along a multicomponent composition. As noted above, at the initial stage of the formation of normal conductivity core, the change in the velocity of the normal zone is of a substantially nonstationary character. Its existence time depends on cooling conditions, current load, the thermal connection between composites and number of them. In the second stage, the velocities of the normal zones in each element are alternately stabilized and take constant values equal to the corresponding value of the propagation velocity of the normal zone in a first composite (Figure 7.29). This corresponds to the so-called quasi-stationary states when the thermal instability propagates along the composite superconductor in the form of the thermal autowave. For understandable reasons, in the approximation under consideration, the achievement of the steady-state normal zone velocity in each element of the composition is asymptotic. This mode is primarily achieved in the composite where the normal zone was initiated. The propagation of the temperature perturbation in the remaining elements of the composition has not only the more nonstationary character but may also occur at higher velocities than in a similar single composite. As a result, in the first element of the composition, the velocity of the normal zone approaches its asymptotic limit from below. In other elements, this process has a different type of convergence, namely both from below and from above. Within the model of an expanding ellipsoid, a similar description of the kinetics of the normal zones in the multicomponent composition cannot be fulfilled.

It is important to emphasize that an increase in the size of a normal conductivity core in a superconducting composition with a discrete structure is characterized by the existence of some regularities that are a direct consequence of the jump-transition change in their properties. First, the boundary of a normal core has the shape of a truncated oval with a piecewise continuous outer boundary. The discrete character of the region with normal conductivity is shown in Figure 7.30a for the state of the described Curve 1. The formation of such a normal core is due to the finite transition time of each superconducting element of the composition to the normal state. Obviously, the worse the thermal connection between the elements, the more piecewise-truncated shape will have a normal core. Second, after the transition to the normal state of all composition elements in the cross-section, there is a gradual flattening of the boundary separating the nonsuperconducting region from the superconducting one, which ultimately transforms into a straight front. It is easy to understand that the flattening of the boundary will occur the faster, the better the heat exchange conditions between the elements of the composition, that is the higher the ω. This effect is due to the equalization of the temperature of all composition elements that have turned into the normal state. As a result, the heat flux in the cross-section practically disappears. As a consequence, the last term in the heat conduction equation (7.47) ceases to influence the thermal state of the composition which is, thus, described by an equation close to the heat conduction equation describing the thermal mode of a single composite.

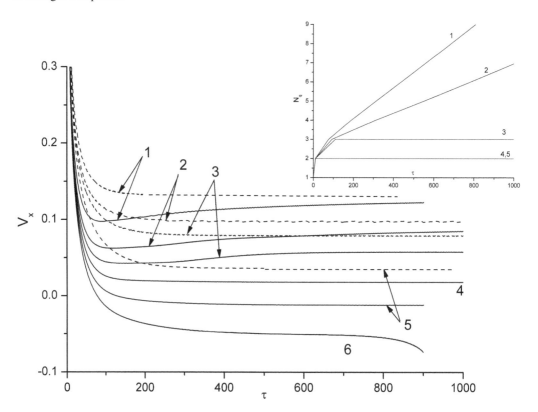

FIGURE 7.31 Time dependence of the longitudinal propagation velocity of the normal zone in the composite with the number $n = 1$ at $N = 51$ near the full stabilization current (i_s): $1 - i = 0.2$; $2 - i = 0.18$; $3 - i = 0.17$; $4 - i = 0.16$; $5 - i = 0.15$; $6 - i = 0.14$.

Finally, let us note another limitation of the applicability of the anisotropic medium model. As follows from the simulation results presented in Figure 7.31 and 7.32, for the currents close to the full stabilization current i_s, which is calculated in the approximation of a single composite, it is possible such modes under which an irreversible transition to the normal state of the entire superconducting composition of a discrete structure does not occur. In Figure 7.31, the propagation velocities of the isotherm $\theta_n(X_{q,n}, \tau) = 1$, which propagates along with the first element of the composition, where the normal zone appears at $\alpha = 100$, $X_1 = 10$ and $\omega = 0.1$ are depicted. The current change range is $i = 0.2 - 0.14$ that slightly exceeding the stationary stabilization current $(i_s = 0.1356)$. Here, the dashed curves show the corresponding velocity of the isotherm $\theta_n(X_{q,n}, \tau) = 1$ in a single composite. The kinetics of the normal zone in the central part of the cross-section of the composition is shown in the inset. It is seen that the above-discussed regularities of the irreversible transition of the entire superconducting composition to the normal state are fully observed at $i \geq 0.18$ (Curves 1 and 2). At the same time, at $i < 0.18$, it is possible the modes when only a limited number of the composition elements go to the normal state. So, areas with the normal conductivity appear only in three elements of the composition at $i = 0.17$, and only two elements go to the normal state at $i = 0.16$ and $i = 0.15$. The quasi-stationary normal states may not occur during a further decrease of current. The existence of such modes leads to the fact that the propagation velocity of the normal zone in the discrete composition may be lower than the corresponding velocity value of the normal zone in a single composite. Moreover, this difference increases at decreasing the current.

It is important to emphasize that the shape of the core with normal conductivity is usually a priori assumed to be ellipsoid regardless of the stage of transition mode. However, an ellipse as a continuous curve has two focus points relative to which is the sum of distances to any point on the ellipse curve is a constant value, which is equal to the length of the central axis. However, this law is not the case for a superconducting composition of a discrete structure. As a result, within the framework of the expanding ellipsoid, it is not possible to correctly describe the features noted above.

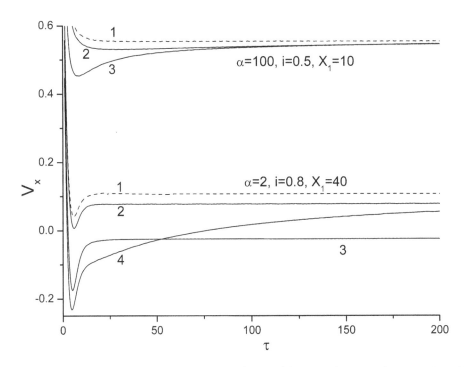

FIGURE 7.32 Change in time of the propagating velocity of the normal zone in the composite with the number $n = 1$, depending on the heat exchange conditions between the elements of the composition ($N = 51$): 1 - single composite; 2 - $\omega = 0.01$; 3 - $\omega = 0.1$; 4 - $\omega = 0.2$.

The existence of these states also depends on the character of the thermal connection between the elements of the composition. Curves describing the kinetics of the normal zone in the composition with different values of ω are plotted in Figure 7.32. It can be seen that when $\alpha = 100$, $i = 0.5$, the value of ω affects the transition time only. However, when $\alpha = 2$, $i = 0.8$ (the full stabilization current for a single composite is equal to $i_s = 0.7808$), the conditions for the occurrence and propagation of a normal zone in a discrete superconducting composition may change. So, only one composite turns into the normal state at $\omega = 0.01$, two of them turn into the normal state at $\omega = 0.1$ and the entire composition turns into the normal state at $\omega = 0.2$. Therefore, only in the latter case, the velocity of the normal zone in the first composite will begin to asymptotically approach the velocity of the normal zone in a single composite.

Let us now discuss the propagation regularities of a thermal perturbation in the cross-section of a superconducting discrete composition.

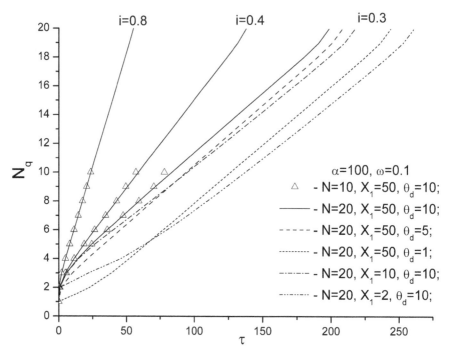

FIGURE 7.33 Influence of the total number of composites in a superconducting discrete composition, the temperature of the initial perturbation and its length on the propagation of the normal zone in its central section.

The results of the numerical simulation of the normal zone propagation in the central section of the composition at $\alpha = 100$ and $\omega = 0.1$ are shown in Figure 7.33. The calculations were performed for various current, temperature and length of the initial perturbation and the number of elements in the composition. The boundary of the normal zone, as above-mentioned, was determined by solving equation $\theta_n(X_{q,n}, \tau) = 1$. The time occurrence of the normal zone in n element of the composition was determined from the solution of equation $\theta_n(0, \tau_{q,n}) = 1$. This allows one to approximate the discrete mode of new area formation with normal conductivity by the corresponding continuous dependencies.

The results, first, confirm the basic regularities determining the characteristic features of the irreversible propagation of thermal perturbation in the cross-section of a discrete structure superconducting composition formulated in section 7.5. Second, they demonstrate the peculiarities of the irreversible destruction of the superconducting properties of the entire composition under the action of the temperature perturbations with a different energy. In particular, it should be noted that the development of a transient mode in the cross-section of the composition may depend on the length of the initial temperature perturbation. To avoid this situation, the necessary studies of the transients occurring at low currents must be carried out with an increased number of the composition elements ($N > 20$). This feature must be considered during relevant experiments. Besides, the model of the extended perturbations, considered in the previous section and described by equations (7.44), allows one to simplify the calculations. At the same time, the influence of the energy of the extended initial perturbation on the kinetics of the transition mode is practically absent if it is supercritical. Taking into account these features, the results of the numerical experiments performed at $N = 20$, $X_1 = 50$, $\theta_1 = 10$ are discussed below.

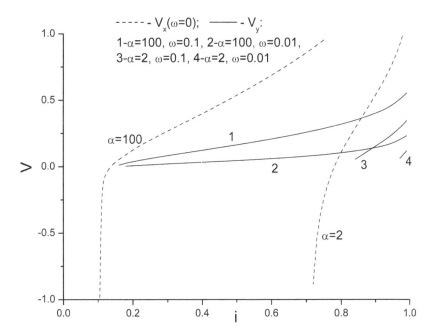

FIGURE 7.34 Dependence of the steady-state values of the propagating velocity of the normal zone on the current: (——) - V_y, (- - - - -) - V_x.

In Figure 7.34, the solid curves describe the transverse propagation velocity of the normal zone as a function of current for various values of the dimensionless thermal resistance between the elements of the composition. The calculations were carried out for two characteristic values of the stability parameter. The depicted curves were calculated under the assumption that the instability at the initial time occurs only in the first composite. Here, for clarity, the dashed curves show the corresponding dependencies of the longitudinal propagating velocity of the normal zone in a single composite when the irreversible destruction of its superconducting properties occurs against the background of a thermal autowave moving at a constant velocity. Formally, this mode corresponds to the case $\omega = 0$.

From Figure 7.34, it follows that the deterioration of the thermal coupling between the elements of the composition not only reduces the transverse propagation velocity of the normal zone but also reduces the range of the currents at which the complete transition of the composition to the normal state will take place in its cross-section. In other words, in SMS with well-isolated turns, its transition will mainly depend on the longitudinal propagation velocity of the normal zone. To the greatest extent, this effect is observed in the SMS when, from the point of view of the thermal stabilization, its elements are 'well' stabilized. Besides, the dependencies $V_y(i)$ can be approximated by a linear dependence in a wide range of the current variation. They acquire a nonlinear character in the region of the full stabilization current and near the critical current as well as with an increase in ω. It follows from the above that for a discrete superconducting composition, the ratio of the steady-state velocity of the normal zone in a cross-section to its velocity value in the longitudinal direction, strictly speaking, is not a constant throughout the entire range of the current variation as it is assumed in the expanding ellipsoid model. Besides, the model of the expanding ellipsoid also leads to the dependence of the transverse propagation velocity of the normal zone on the thermal conductivity of a single composition element in its longitudinal direction. Such a connection is evident for a continuum, the physical properties of which are not discontinuous. At the same time, in the case of a superconducting discrete composition, the propagation of the normal zone in the transverse direction primarily depends on the thermal properties of the insulating layers and the temperature of the composition in its most heated part. Its value is mainly determined by the Joule heat release power and the conditions for its removal to the coolant and to adjacent layers. Therefore, the longitudinal mechanism of the thermal conductivity of each composition element will have a weak effect on the conditions for the occurrence and propagation of the normal zone in the transverse direction.

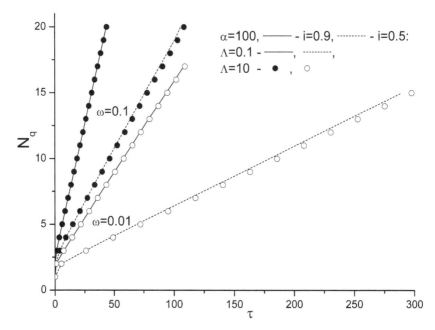

FIGURE 7.35   Effect of the longitudinal thermal conductivity on the kinetics of the transition mode in the cross-section of a superconducting discrete composition.

To illustrate this conclusion, the results of the numerical experiment simulating the thermal instability propagation in compositions with various thermal conductivity coefficients in the longitudinal direction are shown in Figure 7.35. They differ from each other by two orders of magnitude. It was based on the solution of the problem (7.47) in which the $\partial^2 \theta_n / \partial X^2$ term was replaced by the more general expression $\Lambda \partial^2 \theta_n / \partial X^2$ where Λ is the dimensionless value of the thermal conductivity coefficient nth element in the longitudinal direction. When solving a model problem, the stability parameter was set to $\alpha = 100$, the current and the conditions of the thermal coupling between the composites were varied. The calculations demonstrate the very insignificant influence of the mechanism of the longitudinal thermal conductivity on the kinetics of the perturbation penetration in the cross-section of a multicomponent composition.

Thus, the discussed results show that the irreversible propagation of the normal zone in a discrete superconducting composition has several features that are not described by the expanding ellipsoid model. There are the following regularities:

1. The core with normal conductivity which is formed as a result of an irreversible transition to the normal state of superconducting composition elements that was composed of superconducting composites separated by a final thermal resistance is of a truncated oval shape.

2. The propagation velocity of the normal zone in the longitudinal direction of the composition element where the normal zone originated is less than the corresponding value of the propagation velocity of the normal zone in a single composite. The latter is the asymptotic limit for all values of the velocity with which the normal zone will be propagated in other elements of the composition. This limiting case is reached only after a transition of all elements of the composition to the normal state when the moving front separating the superconducting area from the nonsuperconducting one becomes flat.

3. For the current modes close to the currents of the full stabilization of a single composite, there may be states under which not all elements of the composition break down their superconducting properties. As a result, within a superconducting composition with a discrete structure, a limited number of the subdomains of normal conductivity can exist in which the propagation of a normal zone with a constant velocity takes place.

4. Near the full stabilization current, the character of the transient modes largely depends on the heat exchange conditions between the elements of the composition. Therefore, if there are insulating layers with relatively high values of the thermal resistance in a multicomponent composition, the appearance of a normal zone in a single element may not be accompanied by a transition of the entire composition to a normal state. To the greatest extent, this effect is observed in 'well' stabilized wires.

5. The propagation velocity of the normal zone in the cross-section of a multicomponent superconducting composition is almost independent of the longitudinal thermal conductivity of its elements. First of all, it is determined by thermal conductivity and thickness of the insulation.

7.7 THERMAL STABILIZATION OF SUPERCONDUCTING COMPOSITES TAKING INTO ACCOUNT THE SIZE EFFECT

The basic principles of the thermal stabilization theory for single superconducting composites formulated above demonstrate the features of the conditions for their stable operability under the action of external thermal perturbations but when the heat flux propagates only in the longitudinal direction. This approximation not only simplifies the analysis but also provides practically important criteria for the thermal stability of superconducting composites. However, taking into account even the nonisothermal state of the superconducting filament in its cross-section leads to an increase in the power of the heat release, hence to a corresponding change in the conditions of the thermal stabilization (Wilson 1983). Therefore, an appropriate analysis of the thermal stability should be carried out taking into account the multidimensional character of the heat flux propagation when developing large SMS with massive current-carrying elements. In this connection, let us study the influence of size effect on the conditions for the occurrence and propagation of the thermal instabilities in superconducting composites initiated by temperature perturbations.

Let us consider within the framework of the anisotropic continuum and the above-made assumptions, a composite superconductor of finite length, having a cylindrical shape with radius r_0 carrying a transport current I, uniformly distributed over its cross-section and cooled initially to a coolant temperature of T_0. Let the entire volume of the composite be instantly warmed up to the temperature T_1 on the length $-l_1 \leq z \leq l_1$ at the initial time. Describe in a two-dimensional approximation the change in its temperature $T(z, r, t)$ in the longitudinal ($-l_k \leq z \leq l_k$) and transverse ($0 \leq r \leq r_0$) directions by the heat conduction equation of the form

$$C_k \frac{\partial T}{\partial t} = \lambda_z \frac{\partial^2 T}{\partial z^2} + \lambda_r \frac{1}{r} \frac{\partial}{\partial r}\left(r \frac{\partial T}{\partial r} \right) + \frac{I^2}{S^2} \rho_k(T) \qquad (7.41)$$

with initial-boundary conditions

$$T(z, r, 0) = \begin{cases} T_1 = \text{const}, & 0 \le |z| \le l_1, 0 \le r \le r_0 \\ T_0, & l_1 < |z| \le l_k \end{cases} \tag{7.42}$$

$$T\big|_{z=-l_k} = T\big|_{z=l_k} = T_0, \quad \lambda_r \frac{\partial T}{\partial r} + h(T - T_0)\bigg|_{r=r_0} = 0 \tag{7.43}$$

Here, C_k, λ_z and λ_r are the average values of the volumetric heat capacity of the composite and its thermal conductivity coefficients in the corresponding directions. The physical meaning of the remaining parameters was discussed above.

Reduce the number of the variable dimensional variables by going to dimensionless ones

$$Z = z/L_z, R = r/L_r, i = I/I_{c0}, \tau = \lambda_z t/C_k L_z^2, \theta = (T - T_0)/(T_{cB} - T_0) \tag{7.44}$$

assuming that the filling factor and the current density remain constant when the outer radius changes. Then, the problem (7.41)-(7.43) takes the form

$$\frac{\partial \theta}{\partial \tau} = \frac{\partial^2 \theta}{\partial Z^2} + \frac{1}{R} \frac{\partial}{\partial R}\left(R \frac{\partial \theta}{\partial R}\right) + i^2 r(\theta)$$

$$\theta\big|_{\tau=0} = \begin{cases} \theta_1 = \text{const}, & 0 \le |Z| \le \xi_1, 0 \le R \le R_0 \\ 0, & \xi_1 < |Z| \le L \end{cases} \tag{7.45}$$

Here,

$$L_{z,r} = \left[\frac{\lambda_{z,r}(T_{cB} - T_0)}{\eta^2 J_{c0}^2 \rho_0}\right]^{1/2}, \beta = \frac{hL_r}{\lambda_r}, L = \frac{l_k}{L_z}, R_0 = \frac{r_0}{L_r}, \xi_1 = \frac{l_1}{L_z}, \theta_1 = \frac{T_1 - T_0}{T_{cB} - T_0}$$

Let us note the features of the multi-dimensional character of the thermal phenomena in the composite superconductors. In the one-dimensional approximation, the conditions of the thermal stability of the superconducting state primarily depend on the Stekly parameter α described by the Formula (7.15). When analyzing the conditions of the thermal stability with regard to the nonuniform temperature distribution over the cross-section of the composite, this parameter is absent. In this case, the thermal stabilization will depend on the dimensionless parameter

$$\beta = \frac{hL_r}{\lambda_r} = \frac{hS}{I_{c0}}\sqrt{\frac{T_{cB} - T_0}{\lambda_r \rho_0}}$$

which is the ratio of the conductive heat flux in the transverse direction to the heat flux into the coolant. In the heat conduction theory, it is known as the Bio parameter. It is easy to find that there is a relationship,

$$\alpha\beta = \frac{S}{pL_r} \tag{7.46}$$

that is this quantity is the thermal stabilization parameter of the superconducting composites in the multi-dimensional approximation.

Since the parameter β characterizes the relationship between the temperature of the composite and the heat transfer conditions, it is easy to get the simplest condition describing the nonuniform character of the temperature in the cross-section of the composite. It will be observed when its thermal resistance is many times greater than the thermal resistance of the heat transfer. In the framework of the introduced dimensionless variables, the condition for violation of uniform temperature distribution in the cross-section of a cylindrical composite is written as follows $2\beta R_0 \gg 1$.

The transition to dimensionless variables (7.44) allows one not only to simplify the analysis but also to understand the physical features of the multi-dimensional nature of the thermal phenomena in superconducting composites. In particular, the dimensional analysis makes it possible to estimate energies required for the destruction of superconductivity in the simplest form. Indeed, according to (7.5), the characteristic energy of the instantaneous temperature perturbation is equal to

$$E_x' = C_k S L_z (T_{cB} - T_0) = \frac{C_k S}{\eta J_{c0}} \sqrt{\frac{\lambda_z}{\rho_0} (T_{cB} - T_0)^3}$$

in the one-dimensional approximation. In the two-dimensional approximation, relative to the variables adopted above, it is defined as follows

$$E_x'' = \frac{\lambda_r}{\lambda_z} C_k L_z^3 (T_{cB} - T_0) = \frac{C_k \lambda_r}{\rho_0 \eta^3 J_{c0}^3} \sqrt{\frac{\lambda_z}{\rho_0} (T_{cB} - T_0)^5}$$

These values are related to the following relation $E_x'' = E_x' L_r^2 / S$. It shows that the calculated values of the critical values obtained in the framework of the two-dimensional theory will monotonously decrease with respect to the corresponding values calculated in the one-dimensional approximation with the cross-section increasing. A more accurate quantitative response can be obtained after an appropriate analysis of the critical energies.

Within the framework of the two-dimensional approximation, one can determine the full stabilization current i_s which, as discussed above, separate the region of the full stability ($\varepsilon_q \to \infty$) from partial (ε_q is finite) stability conditions. To do this, let us use the method of the finite perturbation of initial state. In this case, numerically solving equation (7.45) for given values β and R_0, it is possible to look for such current, below which the temperature of the composite in its most heated part always becomes over time lower than the current sharing temperature ($\theta(0, 0, \tau)$ < 1 − i) after the action of any powerful extended temperature perturbation. In one-dimensional approximation, the condition of the full stabilization is described by expressions (7.21) or (7.22).

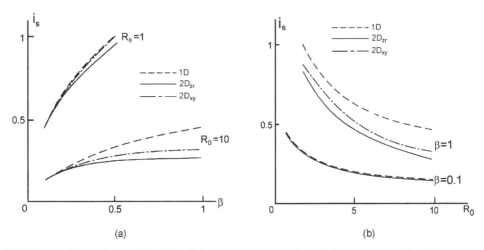

FIGURE 7.36   Dependence of the full stabilization currents on the stability parameter (a) and the transverse size of the composite (b) with various cross-sectional shapes: 1 D is the one-dimensional approximation, $2 D_{zr}$ is the composite having a cylindrical shape and $2 D_{xy}$ is the composite having a square shape.

The simulation results of the full stabilization condition performed in the framework of one-dimensional and two-dimensional approximations are presented in Figure 7.36. In the first case, the full stabilization current was determined according to the condition (7.21) taking into account the relationship (7.46). Figure 7.36a shows the effect of the stability parameter β on i_s for the thin

($R_0 = 1$) and massive ($R_0 = 10$) composites and Figure 7.36b depicts the effect of the transverse size of the composite on the full stabilization currents under various conditions of its cooling: $\beta = 0.1$ is the nonintensive cooling and $\beta = 1$ is the intensive cooling. The dashed curves are described the result of calculations in one-dimensional approximation. The solid curves describe the values i_s for a cylindrical composite, and the dash-dotted curve corresponds to the composite of the square cross-section with a cross-sectional area equal to the corresponding area of the cylindrical composite. In the latter case, in the heat conduction equation of the problem (7.45), when the differentiation was made with respect to the spatial coordinates changed in accordance with the transition from the cylindrical to rectangular coordinates. The results obtained describe the conditions for the full stabilization of the composite, the two parallel surfaces of which are thermally insulated, and cooling is performed through the two other surfaces.

The presented results show that one-dimensional theory gives a noticeably overestimated boundary of their full stabilization conditions for the massive composite superconductors. The full stabilization current determined in the framework of one-dimensional theory is slightly different from the corresponding values calculated in the two-dimensional approximation if the transverse size of the composite is small or it is under nonintensive cooling conditions (more strictly if $2\beta R_0 \ll 1$). The values of i_s are also influenced by the value of the cooled perimeter as follows from (7.46). That is why at all other things being equal, the full stabilization currents of the square-shaped composite superconductors are higher than those of the cylindrical composites.

Let us formulate the condition of the full stabilization in general form taking into account the nonuniform distribution of the temperature, current and electric field in the cross-section of the composite. Let us integrate equation (7.1) over spatial variables. Taking into account the Green formula and the boundary condition on the surface of the composite

$$\lambda_k \frac{\partial T}{\partial n} + h(T)(T - T_0) \bigg|_p = 0$$

describing the convective heat transfer to the coolant along the perimeter p, one may get

$$\iint_S C_k \frac{\partial T}{\partial t} dS = -\int_p h(T - T_0) dp + \iint_S EJ dS$$

Then, under the condition

$$\iint_S EJ dS \le \int_p h(T - T_0) dp \tag{7.47}$$

the left side of the previous equation is not positive. Therefore, any external thermal perturbation will always fade after the cessation of its action. Criterion (7.47) is a general form of the full stabilization condition, that is a generalization of the Stekly condition. It can be written in a more simplified form if $E, J \sim$ const over the cross-section of the composite, and the heat transfer coefficient does not depend on temperature. Using the parameter Stekly α, after simple transformations one may get

$$\alpha i^2 \frac{T_{cB} - T_0}{\dfrac{1}{p} \displaystyle\int_p (T - T_0) dp} \le 1$$

The formulated criterion shows that the condition of the full stabilization depends on the thermal state of the surface of the current-carrying element. Moreover, the true value of the full stabilization current is always lower than the corresponding value calculated under the assumption that the temperature distribution in its cross-section is uniform. The difference between them increases not only with an increase in the transverse dimensions of the current-carrying elements but also with the improvement of the conditions of their heat exchange with the coolant. This condition should be taken into account when developing large magnetic systems using massive current-carrying elements. Namely, if the conditions for the complete maintenance of the superconducting properties by the winding are calculated on the basis of one-dimensional theory, then the premature irreversible transition due to thermal instability development in the magnetic system with the parameters set in this way may lead to the destruction of the SMS in the real operating conditions.

Let us discuss in the framework of the formulated problem (7.45) the influence of the transverse size of a composite on the stability conditions of its superconducting state under the action of a local temperature perturbation ($\xi_1 = 1$) using the finite perturbation method of initial state.

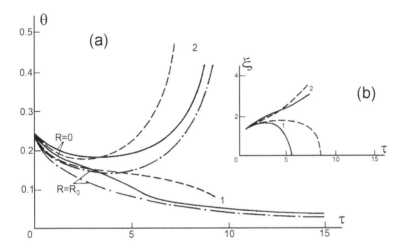

FIGURE 7.37 Temperature change of the composite in the center and on the surface (a) and the length of the normal zone (b) under the action of the perturbations that are close to the critical: 1 - $\theta_1 = 0.24$, 2 - $\theta_1 = 0.25$; ———— - two-dimensional model, – – – – – - one-dimensional model.

In Figure 7.37, the restoration and violation of superconducting properties by the composite under the action of the subcritical and supercritical perturbations are shown in case when $\beta = 0.1$, $R_0 = 3$, $i = 0.9$. Curves describing the kinetics of the normal zone (Figure 7.37b) prove the local character of the initial perturbation. Comparing one-dimensional and two-dimensional calculations of the composite temperature, it is easy to see that the development of the instability is determined by the thermal state of the central part of the composite for the considered perturbation. It is seen that under the action of a supercritical perturbation the temperature increase in the center of the composite occurs more intensively than on its surface. Moreover, as follows from the calculations, the increase β leads to a more noticeable difference between the temperatures of the composite in its central part and on the surface.

The dimensionless critical energies $\varepsilon_q = \varepsilon_s \pi R_0^2$ calculated at $i = 0.9$ are plotted in Figure 7.38 as a function of the radius of the composite. Here, $\varepsilon_s = 2\xi_1\theta_q$ is the dimensionless perturbation energy per unit of the cross-sectional area. The dashed curves show the corresponding dependencies of the maximum permissible value of the initial perturbation temperature. The results presented demonstrate the existence of the minimum in the $\varepsilon_q(R_0)$ dependence (the solid curves). The latter is due to a monotonous decrease in the permissible temperature perturbation (in the general case in the perturbation energy density) at a corresponding increase in the transverse size of the composite. With an increase of β (for example, with improved cooling conditions), the falling branch of curve $\varepsilon_q(R_0)$ shifts to the area of large values R_0.

In general, the calculations show that the stability of the superconducting state of the composites with small transverse size depends on the longitudinal thermal conductivity of the composite. But its effect is small for massive composites. In this case, the role of the transverse thermal conductivity of the composite is more significant. Qualitatively, these conclusions are supported by the characteristic values of E'_x and E''_x introduced above which show that $E'_x \sim \sqrt{\lambda_z}$ and $E''_x \sim \lambda_r\sqrt{\lambda_z}$.

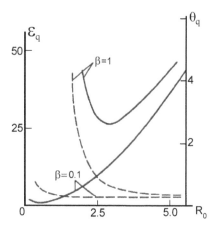

FIGURE 7.38 Dependence of the critical energy (———) and the maximum permissible perturbation temperature (- - - -) on the radius of the composite.

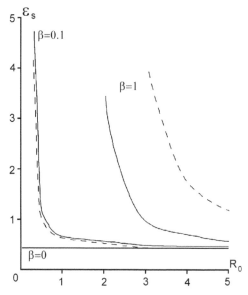

FIGURE 7.39 Dependence of the critical energy density on the radius for different values of the stabilization parameter: ——— two-dimensional model, - - - - one-dimensional model.

In Figure 7.39, one-dimensional and two-dimensional calculations of the dimensionless critical energy density ε_s as a function of the radius of the composite are compared with each other. Calculations were carried out for the local temperature perturbations ($\xi_1 = 1$) at current close to the critical ($i = 0.9$) and various cooling conditions. It is seen that the critical energy of the thermally insulated composite ($\beta = 0$) does not depend on its transverse size, and two-dimensional calculations completely coincide with one-dimensional ones. For cooled composites, the difference in the calculated values of the critical energies obtained from one-dimensional and two-dimensional models depend on β (Figures 7.39 and 7.40). In the area of the small values β (for example, at the nonintensive cooling), the calculated two-dimensional values ε_s slightly exceed the corresponding values calculated in the framework of the one-dimensional model over the entire range of change R_0. At relatively high values of β, there is a significant discrepancy between one-dimensional and two-dimensional calculations. It turns out that there are such values of the parameter β when the improvement of the cooling conditions in the massive composites will very poorly affect the values of the critical energies (Figure 7.40).

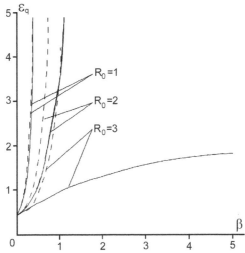

FIGURE 7.40 Dependence of the critical energy on the stabilization parameter at different values of the composite radius: ———— - two-dimensional model, - - - - - one-dimensional model.

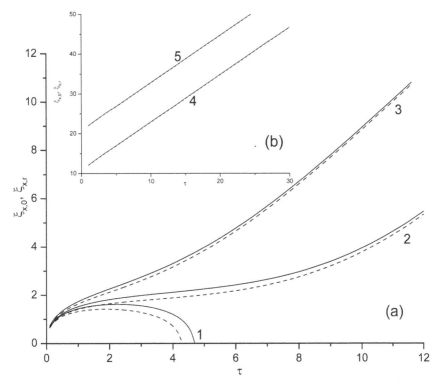

FIGURE 7.41 (a): propagation of the normal zone along the axis of the composite $R = 0$ (————) and on its surface $R = R_0$ (- - - -) under the action of sub- and supercritical point temperature perturbations ($\xi_1 = 1$): 1 - $\theta_1 = 1.6$, 2 - $\theta_1 = 1.7$, 3 - $\theta_1 = 2$; (b): autowave propagation of the normal zone along the axis of the composite $R = 0$ (————) and on its surface $R = R_0$ (- - - -) under the action of the extended supercritical temperature perturbations: 4 - $\xi_1 = 10$, 5 - $\xi_1 = 20$.

Let us investigate the kinetics of the irreversible propagation of the thermal instability in the framework of the multi-dimensional problem (7.45). Correspondingly, the velocity of the thermal autowave will be determined according to the numerical solution, which appears as a result of a supercritical perturbation action. As an illustration, Figure 7.41 shows the time variation of the normal zone length in the central part of the composite ($R = 0$, $\xi_{z,0}$) and on its surface ($R = R_0, \xi_{z,r}$) determined from the solution of equations $\theta(Z, R, \tau)\big|_{R=0, R=R_0} = 1$ at various initial temperature perturbations (subcritical and supercritical). The calculation was carried out at $\beta = 0.1$, $R_0 = 2$, $i = 0.9$. It is seen that, in this case, the thermal instability becomes irreversible at $\theta_1 > 1.7$. As a consequence, the thermal autowave modes are formed.

It is important to emphasize that as during the formation of a thermal autowave in a multicomponent composition, the thermal autowave that appears in a massive superconducting composite is also characterized by the formation of a practically flat front separating the nonsuperconducting part of the composite from the superconducting. These features confirm the results of the calculations presented in Figure 7.41.

The dimensionless autowave velocities of the thermal instability propagation as a function of the transport current are presented in Figure 7.42 for different cooling conditions and two values of the radius of composite: thin ($R_0 = 1$, Figure 7.42a) and massive ($R_0 = 10$, Figure 7.42b). Here, the dashed curves show the results of the corresponding calculations for one-dimensional model.

As expected, the thermal autowave velocity in the thermally insulated composite ($\beta = 0$) does not depend on its transverse size since the thermal state of the composite is uniform over its cross-section. At the same time, the difference in the autowave velocity, calculated from one-dimensional and two-dimensional models for cooled composites, depends on the value β. As one would expect, the dependencies $V_n(i)$ for a thin composite in two-dimensional approximation slightly exceed the corresponding values obtained from one-dimensional model in a wide range of the current variation. However, massive composites lead to a significant discrepancy between one-dimensional and two-dimensional calculations of the thermal autowave propagation. This difference may be not only quantitative but also qualitative due to a significant increase in the calculated velocity of the normal zone in two-dimensional approximation. As a result of the transition from one-dimensional model to two-dimensional model, the currents range with the negative velocity corresponding to the stable superconducting states is significantly reduced. Therefore, if transients in a massive composite were studied on the basis of one-dimensional models, in particular its stable states under arbitrary perturbations are determined (in areas where $V_n < 0$), then it may be in the metastable state ($V_n > 0$) under these parameters in real conditions.

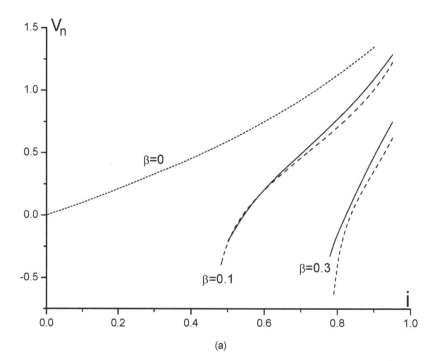

(a)

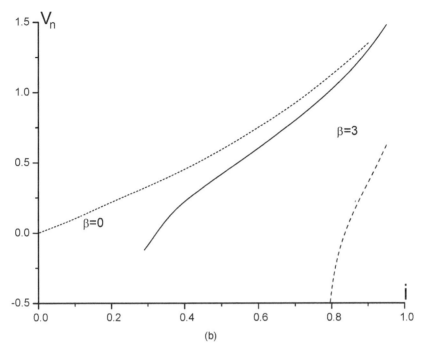

(b)

FIGURE 7.42 Dependence of the normal zone velocity on the current calculated under nonintensive (a) and intensive (b) cooling conditions using one-dimensional (- - - -) and two-dimensional (————) models.

The dependence of the autowave velocity of the normal zone on the radius of the composite for different values of the parameter β and $i = 0.9$ are shown in Figure 7.43. They show that the difference between one-dimensional and two-dimensional calculations increases with improved thermal stabilization conditions. This discrepancy may be qualitative for several modes, velocities differ not only in magnitude but also in the sign (Figure 7.44). In other words, if the thermal instabilities do not pose a danger within the framework of the one-dimensional model since they fade out with time, then they may be unstable in real operating conditions.

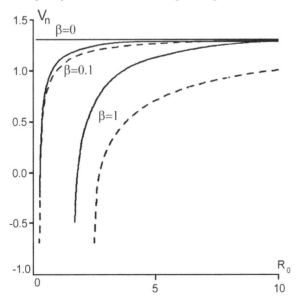

FIGURE 7.43 Dependence of the autowave velocity of the normal zone on the radius under different cooling conditions: – – – – – one-dimensional model, ———— - two-dimensional model.

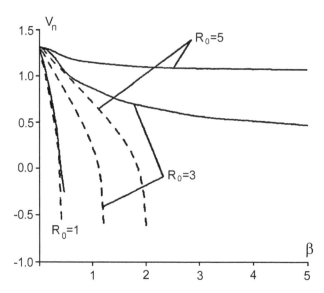

FIGURE 7.44 Dependence of the autowave velocity of the normal zone on the stabilization parameter: - - - - - one-dimensional model, ——— - two-dimensional model.

The influence of the size effect on the thermal stabilization of the superconducting state of the composites is explained as follows. When $\beta = 0$, the temperature distribution over their cross-section is uniform since there is no heat flux to the coolant. Therefore, one-dimensional and two-dimensional calculations are the same. The thermal state of the composite changes when β increase: the temperature of its surface decreases and it rises in the central part (Figure 7.37a). As a result, the average temperature of the composite rises. Let us prove this conclusion more strictly by estimating the temperature distribution in the cross-section of a cylindrical composite when it is in a normal state. In this case, one can restrict to solving the following equation

$$\frac{1}{R}\frac{d}{dR}\left(R\frac{d\theta_{max}}{dR}\right) + i^2 = 0$$

with boundary condition

$$\frac{d\theta_{max}}{dR} + \beta\theta_{max}\bigg|_{R=R_0} = 0$$

Finding the quantity θ_{max}, it is easy to estimate the average temperature of the composite in its most heated part. It is equal to

$$\langle\theta\rangle = \frac{1}{S}\int_S \theta_{max}\,ds = (\alpha + R_0^2/8)i^2$$

This expression shows that compared with the corresponding value of αi^2 which, according to one-dimensional approximation, corresponds to the maximum stable heating of the composite in the supercritical temperature range, the size effect leads to its monotonic increase with the increasing transverse size of the composite. As a result, the corresponding change in the thermal stabilization conditions takes place.

7.8 THERMAL STABILIZATION OF THE TECHNICAL SUPERCONDUCTORS UNDER ADIABATIC CONDITIONS

As follows from the above-discussed results, the thermal perturbations (especially point impulses) can cause a premature SMS transition to a normal state. These perturbations occur due to high mechanical loads on the structural elements of the magnetic system. As already noted, they may lead to the so-called training and degradation phenomena when the currents charged into the winding are below the critical currents of short samples. Therefore, the analysis of the stability of the current-carrying elements of the magnetic system, the increase in the threshold of their stable states is a necessary step in the development of SMS. In this regard, let us discuss the features of ensuring the conditions of the thermal stability of the technical superconductors (superconducting composites based on the LTS and HTS) under the most unfavorable adiabatic conditions.

As in the quasi-linear analysis made above, let us assume that a local part of the superconducting composite becomes instantly in the normal state under the action of the temperature perturbation. The boundaries of the composite $z = \pm l_k$ are maintained at the temperature of the coolant T_0. Let us describe the change in its temperature in the longitudinal direction by the nonlinear heat conduction equation of the form

$$C_k(T)\frac{\partial T}{\partial t} = \frac{\partial}{\partial z}\left[\lambda_k(T)\frac{\partial T}{\partial z}\right] + \frac{I^2}{S^2}\rho_k(T) \qquad (7.48)$$

with initial-boundary conditions

$$T\big|_{t=0} = \begin{cases} T_1 = \text{const}, & 0 \le |z| \le l_1 \\ T_0, & l_1 < |z| \le l_k \end{cases}, \quad T(\pm l_k, t) = T_0 \qquad (7.49)$$

Problem (7.48)-(7.49) is similar to the previously set problem (7.9)-(7.11). The difference between them is that in Equation (7.48) the temperature dependence of the thermo-electrical parameters of the composite is taken into account. Besides, when calculating the critical temperature of a superconductor, the formula $T_{cB} = T_{c0}\sqrt{1 - B/B_{c0}}$ will be used. The critical current of the superconductor is described by the relation $I_c(T, B) = A_0(1 - T/T_{cB})/B$. Here, T_{c0}, B_{c0} and A_0 are the given parameters depending on the type of superconductor.

The maximum allowable temperature perturbation T_q will be determined according to the finite perturbation method of initial state. Then, as above, the critical energy of the perturbation, taking into account the symmetry of the temperature perturbation, is

$$E_q = 2l_1 S \int\limits_{T_0}^{T_q} C_k(T)\,dT.$$

Within the framework of this problem, let us study the thermal stability conditions of the superconducting state of the composites based on Nb-Ti and Nb$_3$Sn. The length of the composites will be set to $l_k = 10$ cm, and the initial local perturbation length is $l_1 = 0.05$ cm.

The heat capacity of the niobium-titanium composite is determined according to the rule of the mixture, that is as follows $C_k = \eta C_1 + (1 - \eta)C_2$, where η is the filling factor of the composite by superconductor. The corresponding values equal

$$C_1[\text{J/(cm}^3 \times \text{K)}] = \begin{cases} 0.812\times10^{-3}\text{T} + 1.29\times10^{-5}\text{T}^3, & T > T_{cB} \\ 0.812\times10^{-3}\text{T}\dfrac{B}{B_{c0}} + 4.273\times10^{-5}\text{T}^5, & T \le T_{cB} \end{cases} \quad - \text{ for Nb-Ti,}$$

$$C_2[\text{J/(cm}^3 \times \text{K)}] = 8\times10^{-6}\text{T}^3 - \text{ for cooper} \qquad (7.50)$$

The thermal conductivity of the composite will be determined under the Wiedemann-Franz law

$$\lambda_k \rho_0 = 2.45 \times 10^{-8} T, \; W \times \Omega/K \tag{7.51}$$

accepting for copper $\rho_m [\Omega \times cm] = (2.13 + 0.605 B) \times 10^{-8}$, where as before, $\rho_0 = \rho_m/(1-\eta)$.

The stability analysis of the superconducting state of the niobium-tin composite is performed for different matrix. First of all, consider the composite with the bronze matrix. In this case, its thermal conductivity is defined by the Formula (7.51) and $\rho_0 = \rho_s/\eta_1$ setting $\rho_s = 7.8 \times 10^{-6}\,\Omega \times cm$. Here, η_1 is the volume fraction of bronze in the composite. Besides, it is consider the niobium-tin composite with cooper and bronze matrixes. In this case: $C_k = C_1\eta + C_2\eta_2 + C_3\eta_1$, where η_2 is the filling factor of cooper in niobium-tin composite. Correspondingly, the heat capacity of Nb_3Sn is calculated as follows $C_1 [J/cm^3/K] = 2.27 \cdot 10^{-5} T^3$ if $T < T_{cB}$ and $C_1 [J/cm^3/K] = 0.988 \cdot 10^{-3} T + 1.388 \cdot 10^{-5} T^3$ if $T > T_{cB}$. The heat capacity of bronze is calculated as follows $C_3 [J/cm^3/K] = 10.324 \cdot 10^{-5} T + 1.024 \cdot 10^{-5} T^3$; C_2 is the heat capacity of cooper.

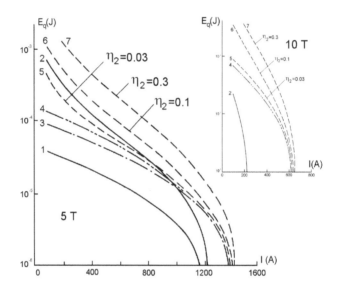

FIGURE 7.45 Dependencies of the critical energies on current for LTS at various values of the external magnetic field: 1, 3 - heating to T_{cB}, 2 and 4-7 - heating to T_q. Here, 1, 2 - Nb-Ti in the copper matrix, 3, 4 - Nb_3Sn in a bronze matrix, 5-7 - Nb_3Sn in a matrix 'bronze + copper'.

In Figure 7.45, the dependencies of the critical energies on the current calculated for composites of 0.15 cm in diameter based on Nb-Ti in the copper matrix (Curves 1, 2) and Nb_3Sn in a bronze matrix (Curves 3, 4) are compared. The Curves 1 and 3 correspond to the energies

$$E_c = 2l_1 S \int_{T_0}^{T_{cB}} C_k(T)\,dT$$

needed to heat the uncooled composites to the critical temperature of the superconductors T_{cB}. It defines the stability conditions of the superconducting state to the infinitely extended perturbation $(l_1 \to \infty)$ determined without taking into account the effect of the thermal conductivity of the matrix. Calculations show that under adiabatic conditions the temperature perturbations of more than 1 cm in length are almost infinitely extended. In this case, the permissible increase in temperature of the composite slightly differs from T_{cB} (formally $T_q \to T_{cB}$ at $l_1 \to \infty$). It is easy to find the change in the temperature of the composite with time at $T_1 > T_{cB}$. According to equation (7.48), it is described by the expression

$$\int_{T_1}^{T} \frac{C_k(T)S^2}{I^2 \rho_k(T)} dT = t$$

Summarizing it for the cooled composite, one can get

$$\int_{T_1}^{T} \frac{C_k(T)}{\Psi(T)} dT = t, \quad \Psi(T) = \frac{I^2}{S^2} \rho_k(T) - \frac{h(T)p}{S}(T - T_0)$$

where the parameters used were previously introduced.

Curves 2 and 4-7 in Figure 7.45 describe the energy boundary of stable states to the local temperature perturbation of the composite. The critical energies for Nb_3Sn composite, the matrix of which consisted not only of bronze but also of copper, are also presented in Figure 7.45 (the dashed Curves 5-7). In this case, taking into account the corresponding change in all filling factors, the resistivity of the composite was calculated as follows $1/\rho_0 = \eta_2/\rho_m + \eta_1/\rho_s$, neglecting as above, the contribution of Nb_3Sn to the resistivity of the superconducting composite. The parameters of superconductors used in the calculations are presented in Table 7.2.

TABLE 7.2

Superconductor	Nb-Ti	Nb$_3$Sn
Filling factor, η	0.5	0.15
Critical temperature, T_{c0}	9.5 K	15.4 K
Critical magnetic field, B_{c0}	14 T	25 T
Parameter A_0	1.38×10^4 A$\times$T	1.04×10^4 A$\times$T

Let us compare the results of these calculations with the results that follow from the quasi-linear analysis. First of all, it may be emphasized that the simplest estimation of the characteristic value of the perturbation energy given in Section 7.3, which is

$$E_x = \frac{C_k S}{\eta J_{c0}} \sqrt{\frac{\lambda_k}{\rho_0}(T_{cB} - T_0)^3},$$

satisfactorily describes the average value of the critical energies under adiabatic conditions. This estimation can be improved when the Formula (7.30) will be used. According to quasi-linear analysis, the dependencies $E_q(I)$ under adiabatic conditions exist in the entire range of the current variation, monotonously decreasing to zero at the critical current. At the same time, the critical energies turn out to be insignificant even at very low currents as follows from Figure 7.45. For clarity, let us note that the energy of the order of 10^{-3} J will be released when a body weighing 10^{-4} kg is dropped from a height of 1 m and weight of 10^{-6} kg is enough for energy release of the order of 10^{-5} J.

The results of the numerical experiments presented in Figure 7.45 also show that the values E_q for the Nb-Ti composite significantly exceed the corresponding values E_c. At the same time, for the Nb_3Sn composite in a bronze matrix, this difference is insignificant. This conclusion follows from the comparison of Curves 1 and 2 for the Nb-Ti composite and Curves 3 and 4 for the Nb_3Sn composite. It is explained by the influence of the matrix thermal conductivity on the stability conditions of the superconducting state under the action of the local perturbations during adiabatic conditions. In the first case, there is an intensive redistribution of the heat flux along the composite while the contribution of the thermal conductivity of the bronze matrix to the thermal stability of the composite is small. However, for the Nb_3Sn composite in a bronze matrix, the values E_c are significantly higher than those of the Nb-Ti composite for all operating currents, and the corresponding values E_q for local perturbation may be higher than the ones of the Nb-Ti composite at high operating currents. This feature explains an unexpected at first glance, the result when in laboratory SMS based on multifilament niobium-tin composites in a bronze matrix, currents close to those of a short sample are achieved. In this case, due to the 'poor' electrophysical properties of bronze, it is quite expected that the current-carrying capacity will degrade compared to composites with a copper matrix. However, this effect is not observed.

It should be underlined that there is a range of the transport currents where the stability of the superconducting state of the Nb-Ti composite is higher despite its lower critical properties than that of the Nb_3Sn composite under the action of the local perturbation. This effect is due to the stabilizing role of the thermal conductivity mechanism of the copper matrix. However, with an increase in the induction of an external magnetic field, the range of increased thermal stability of the Nb_3Sn composite increases.

To estimate the stabilizing effect of copper on the thermal stability of the Nb_3Sn composite in the bronze matrix, the critical energies were calculated under the assumption that there are 3, 10 and 30% copper in the total cross-section of the matrix. The corresponding results are presented in Figure 7.45 by the dashed curves. It can be seen that already with the addition of 10% copper, the thermal stability of the Nb_3Sn composite over the whole range of the transport currents and magnetic field variation is higher than that of the Nb-Ti composite with 50% copper. This result indicates the possibility of a significant increase in the thermal stability of the SMS, almost without reducing the operating current density.

This comparison gives grounds for choosing the magnetic field induction in combined SMS based on Nb-Ti and Nb_3Sn at which the transition from Nb-Ti composites to Nb_3Sn is expedient. The analysis made also makes it possible to substantiate the choice of the type of superconductor for manufacturing magnetic systems with a high level of mechanical stresses, for example, dipoles. Figure 7.46 shows the calculation result of the magnetic field induction where the transition from the Nb-Ti composite to the Nb_3Sn with equal critical energies under the action of a local perturbation is appropriate. Here, the region above the dependence $B(I)$ corresponds to greater stability of the Nb_3Sn without copper, and the Nb-Ti composite with 50% copper has better thermal stability conditions under this curve.

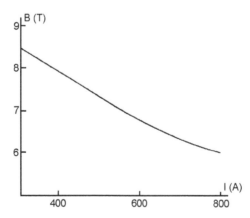

FIGURE 7.46 Dependence of the induction of external magnetic field on the current at which the critical energies of composites based on Nb-Ti in a copper matrix and Nb_3Sn in a bronze matrix are equal to each other under the action of a local perturbation.

Let us discuss possible ways to increase the thermal stability of the composite superconductors, namely, lowering the temperature of the coolant and using superconductors with an increased current-carrying capacity. For this study, in addition to the composites investigated above, which will be conventionally designated as NT (niobium-titanium composite in a copper matrix) and NS1 (niobium-tin composite in a copper-bronze matrix with a copper filling factor $\eta_2 = 0.17$), the latter niobium-tin composite with a higher current-carrying capacity, designated as NS2, will be considered setting $A_0 = 1.5 \times 10^4$ A × T for it. To estimate the effectiveness of lowering the coolant temperature from 4.2 K to 1.8 K as well as the effectiveness of increasing the critical current, let us introduce the coefficient of the relative increase in the critical energy as follows $\delta_E = (E_{q,2} - E_{q,1})/E_{q,1}$.

The relative increase of the critical energies under the action of a local temperature perturbation is shown in Figure 7.47, which will take place when the coolant temperature decreases (the quantities $E_{q,1}$ are calculated at $T_0 = 4.2$ K, and $E_{q,2}$ are defined at $T_0 = 1.8$ K). It is seen that lowering the temperature of the coolant is most advisable for Nb-Ti composites in the considered range of changes in the induction of an external magnetic field. In this case, the rise in permissible perturbation energies increases with increasing of the magnetic field and operating current density. So, for the niobium-tin composite, the increase in the critical energies does not exceed 50% while for the niobium-titanium composite the change in the critical energies can be 100%.

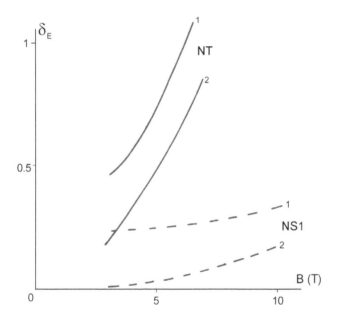

FIGURE 7.47 Relative increase in the critical energy when the coolant temperature decreases from 4.2 K to 1.8 K: 1 - I = 1,100 A, 2 - I = 1,000 A.

In Figure 7.48, the calculating results of the relative increase in the critical energy with an increase of the critical current density of the superconductor in niobium-tin composites are presented. Here, the quantity $E_{q,1}$ corresponds to the composite NS1, and the quantity $E_{q,2}$ are defined for the composite NS2. The solid and dashed curves show δ_E under the local perturbation of composites cooled to 4.2 K and 1.8 K, respectively. The dash-dotted curves show the relative increase of the critical energies for an infinitely extended temperature perturbation at $T_0 = 4.2$ K. Correspondingly, for the perturbations of arbitrary but finite length, the required values δ_E will lie in the region bounded by the solid and dash-dotted curves.

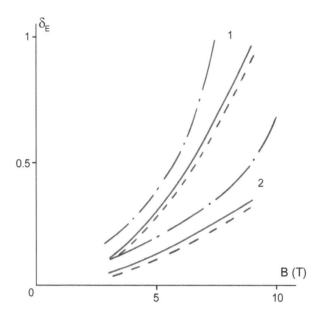

FIGURE 7.48 Relative increase of the critical energy in the transition to superconductors of increased current-carrying capacity: 1 - I = 1,100 A, 2 - I = 1,000 A. - - - - - T_0 = 1.8 K, ——, —. —. — - T_0 = 4.2 K.

It is easy to see that an increase of the current-carrying capacity of Nb_3Sn leads to a slight increase in the critical energies in cases when the composite is in the area of 'weak' magnetic fields (B < 5 T). However, in the area of 'strong' fields, this effect is very significant. Therefore, for improving the thermal stability of SMS, superconductors with an increased current-carrying capacity must be used in high-field high-current magnets.

The ways considered above for increasing the stability of the superconducting state of LTS to the thermal perturbations have shown the real possibilities of high-current superconducting composites at the helium temperature level. The HTS can solve this problem more efficiently because the use of liquid nitrogen is more economical. In this regard, let us estimate the change in the conditions of the thermal stabilization of superconducting current-carrying elements during variation of the so-called temperature margin, which is equal to $\Delta T = T_{cB} - T_0$.

TABLE 7.3

T_0, K	4.2	78
C_κ, J/(cm³ × K)	10^{-3}	1
λ_k, W/(cm × K)	1	2
ρ_0, Ω × cm	6×10^{-8}	4×10^{-7}

Let us use the above-formulated problem (7.48)-(7.49) in which the resistivity of the composite is defined as follows

$$\rho_k(T) = \rho_0 \begin{cases} 1, & T > T_{cB} \\ 0, & T < T_{cB} \end{cases} \tag{7.52}$$

for ease of analysis, and the parameters of the composite are approximated by the temperature-independent estimated values given in table 7.3. This simplest model allows one to qualitatively understand the influence of ΔT on the thermal stability of the superconducting state of both LTS and HTS.

The temperature margin dependencies of the critical energies per unit of the cross-sectional area of the composite, which are equal to $E_{q,s} = 2l_1 C_k (T_q - T_0)$, are shown in Figure 7.49. They are obtained on the basis of the numerical solution of the problem (7.48)-(7.49), (7.52) for the local perturbation ($l_1 = 0.05$ cm) according to the method of the finite perturbation of initial state. The results of the numerical experiment show that when passing from liquid helium coolant to nitrogen coolant, the permissible energies increase of almost three orders of the magnitude and higher. For a physical interpretation of this conclusion, let us use the results of a quasi-linear analysis, according to which the characteristic value of the perturbation energy density is

$$E_{x,s} = \frac{C_k}{\eta J_{c0}} \sqrt{\frac{\lambda_k}{\rho_0}} \Delta T^3 .$$

Then, the possible change in the permissible energies can be estimated as follows

$$\frac{E_{x,s}^{(78)}}{E_{x,s}^{(4.2)}} \sim \frac{J_{c0}^{(4.2)}}{J_{c0}^{(78)}} \frac{\left(C_k \sqrt{\lambda_k/\rho_0}\right)\big|_{78\,\mathrm{K}}}{\left(C_k \sqrt{\lambda_k/\rho_0}\right)\big|_{4.2\,\mathrm{K}}}$$

at other conditions being equal. Here,

$$J_{c0}^{(4.2)}/J_{c0}^{(78)} > 1 \text{ and } C_k\sqrt{\lambda_k/\rho_0}\big|_{78\,\mathrm{K}} \Big/ C_k\sqrt{\lambda_k/\rho_0}\big|_{4.2\,\mathrm{K}} \sim 0.55 \times 10^3 .$$

Therefore, this estimate explains such a change in the critical energies and shows the parameters of the composite that primarily affect the thermal stability conditions of the superconducting state.

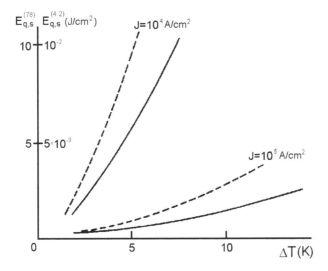

FIGURE 7.49 Dependence of the critical energy density on the temperature reserve for two values of current density: $- - - - - T_0 = 4.2$ K, $———- T_0 = 78$ K.

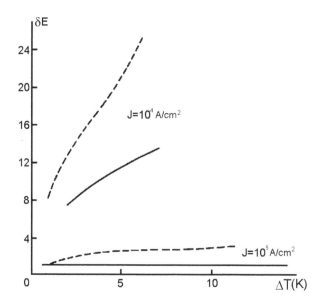

FIGURE 7.50 Effect of the temperature margin on the difference between the critical energies of local and extended perturbations: $- - - - -$ $T_0 = 4.2$ K, $——-$ $T_0 = 78$ K.

From Figure 7.49 it also follows that the benefit from an increase in the temperature margin to increase the critical energies is more noticeable at liquid helium temperatures than at liquid nitrogen temperatures. For the thermal stabilization of superconducting current-carrying elements, this conclusion confirms the advisability of using HTS at temperatures below liquid nitrogen temperatures.

As it was repeatedly discussed above, the thermal stability conditions of a superconducting state under the action of the local perturbations essentially depend on the thermal conductivity of the matrix. Let us consider this problem from the point of view of the temperature margin influence on the thermal stability condition by comparing the critical energies per unit of the cross-sectional area of the composite $E_{q,s}$ of the local temperature perturbation with the energy density $E_{q,c} = 2l_1 C_k(T_{cB} - T_0)$ required to heat the composite at a given perturbation length to a critical temperature T_{cB}. The results of the calculations of $\delta E = E_{q,s}/E_{q,c}$ as a function of the temperature margin are presented in Figure 7.50. They show that to ensure the thermal stability conditions, the role of the thermal conductivity of the matrix is more significant at the liquid helium temperatures than at the liquid nitrogen temperatures. Increasing the transport current will lead to a reduction in the difference between $E_{q,s}$ and $E_{q,c}$. Moreover, this feature is more noticeable in HTS. In general, at high operating current densities, an increase in the temperature margin has a weak effect on improving the thermal stability conditions of superconducting composites to local perturbations. In the HTS, this effect is practically absent. Based on the physical meaning of the parameter δE, the following general conclusion is valid: at the liquid nitrogen level of temperatures, the mechanism of the heat conduction at high currents is not significant ($E_{q,s} \sim E_{q,c}$), and the thermal stability, first of all, depends on the heat capacity of the composite.

The variation of the temperature margin will also affect the characteristic development times of the thermal modes in composites. Indeed, according to the quasi-linear analysis, the characteristic thermal time under adiabatic conditions is equal to $t_x = \dfrac{C_k \Delta T}{\rho_0 \eta^2 J_{c0}^2}$. Then, its possible change can be estimated as

$$\frac{t_x^{(78)}}{t_x^{(4.2)}} \sim \frac{J_{c0}^{(4.2)}}{J_{c0}^{(78)}} \frac{(C_k/\rho_0)|_{78K}}{(C_k/\rho_0)|_{4.2K}}$$

Here, $J_{c0}^{(4.2)}/J_{c0}^{(78)} > 1$ and $C_k/\rho_0|_{78K}/C_k/\rho_0|_{4.2K} \sim 10^2$, that is the diffusion phenomena in the superconducting composites at liquid nitrogen temperatures will occur much slower than it may be at liquid helium temperatures. This feature directly affects the velocities of the normal zone, in particular its irreversible propagation along the composite.

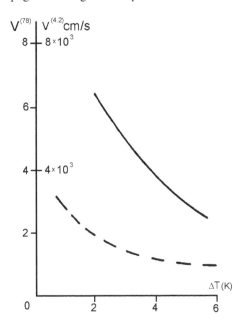

FIGURE 7.51 Dependence of the normal zone velocity on the temperature margin at $J = 10^5$ A/cm²: – – – – – $T_0 = 4.2$ K, ——— - $T_0 = 78$ K.

The corresponding calculations of the thermal autowave velocity as a function of the temperature margin are presented in Figure 7.51. They confirm the conclusion that the velocity of the normal zone is substantially reduced in the transition from liquid helium to liquid nitrogen temperatures. It also follows from them that at the liquid nitrogen level of temperatures the propagation of the thermal autowave also inevitably slows down when using in SMS of the superconductors with high values of the temperature margin.

Thus, the discussed features of the adiabatic thermal stability conditions of the technical superconductors show that:

- The conditions of the thermal stabilization of low- and high-temperature superconductors are significantly different from each other. Firstly, due to the significantly large difference in the values of critical energies and secondly, due to a noticeable decrease in the propagation velocities of the thermal instabilities.
- The role of the thermal conductivity of the matrix is more significant at the liquid helium temperatures than at the liquid nitrogen temperatures.
- There are areas of the magnetic fields where increase in the critical current density of the superconductor or decrease in the temperature of the coolant has little effect on the thermal stability of the superconducting states of the composites.
- It is possible to choose the induction of the magnetic field in which the transition from Nb_3Sn to Nb-Ti composites is expedient in the combined SMS.

7.9 CHARACTERISTIC REGULARITIES OF A RESISTIVE REGION FORMATION IN SPATIALLY LIMITED MULTICONDUCTOR SUPERCONDUCTING COMPOSITIONS

As noted above, the theoretical analysis of the transient modes in SMS widely uses the expanding ellipsoid model. This model with no change is extended to the case when the increase in the size of the resistive region is limited in any spatial dimension due to the finite size of the real winding (Wilson 1983). Let us discuss the formation mechanisms of a resistive core in a superconducting multiconductor composition based on Nb-Ti in a copper matrix in more detail while limiting its spatial dimensions in one direction.

As in the previous investigation based on the problem (7.45)-(7.46), let us describe the variation of the temperature T_n in a rectilinear uncooled composition of N thin superconducting composites that are in contact with each other, solving the following system of the quasi-conjugate equations of the form

$$C_k(T)\frac{\partial T_n}{\partial t} = \frac{\partial}{\partial x}\left(\lambda_k(T)\frac{\partial T_n}{\partial x}\right) + \frac{I^2}{S^2}\rho_k(T_n) - \begin{cases} \dfrac{P}{SR}(T_1 - T_2), & n = 1 \\[2mm] \dfrac{P}{SR}(2T_n - T_{n-1} - T_{n+1}), & n = 2, ..., N-1 \\[2mm] \dfrac{P}{SR}(T_N - T_{N-1}), & n = N \end{cases}$$

in which unlike (7.45), the dependence of the thermo-electrophysical properties of each element on temperature is taken into account. (The physical meaning of the problem and the initial parameters are similar to the previously discussed). Let us use the following temperature dependencies:

$$C_k\left[\frac{J}{cm^3 \times K}\right] = 10^{-6}\begin{cases} 104\,T^{2.5} - 197\,T, & 4.2 \le T < 10\ \text{K} \\ 366\,T^2 + 180\,T, & 10 \le T < 40\ \text{K} \\ 25.5\,T^{4,02 - 0,35\ln T}, & 40 \le T < 300\ \text{K} \end{cases},$$

$$\rho_0[\Omega \times cm] = 2.93 \times 10^{-8}\begin{cases} 0.968 + 0.0076\,T, & 4.2 \le T < 20\ \text{K} \\ 1.71 \times 10^9\,T^{2.35\ln T - 14.1}, & 20 \le T < 35\ \text{K} \\ 8.8 \times 10^{-4}\,T^{2.2} + 0.12, & 35 \le T < 100\ \text{K} \\ 0.37\,T - 14.8, & 100 \le T < 300\ \text{K} \end{cases}$$

$$\lambda_k(T) = 2.45 \times 10^{-8}\,T/\rho_0, \ \text{W/(cm} \times \text{K)}$$

setting the following initial parameters

$$I = 500\ \text{A},\ T_d = 10\ \text{K},\ l_1 = 1\ \text{cm},\ P = 0.01\ \text{cm},\ S = 0.012\ \text{cm}^2,\ T_0 = 4.2\ \text{K},\ T_{cB} = 8.5\ \text{K}.$$

In Figure 7.52 and 7.53, the results of the numerical experiments simulating the propagation of a normal zone in a multiwire superconducting composition are presented. It is supposed that the operating region is bounded either in the transverse (Figure 7.52) or in the longitudinal (Figure 7.53) directions, and the normal zone is initiated in the first element of the composition ($n_i = 1$). In the first case, a 'long' superconducting composition with a relatively small number of the superconducting composites in the transverse direction was considered ($l_k = 500$ cm, N = 11) and in the second case, it was considered a 'short' composition in the longitudinal direction with a relatively large number of elements ($l_k = 50$ cm, N = 51). The value of the thermal resistance between the elements was varied.

The results presented demonstrate the formation features of the normal conductivity core in spatially limited superconducting composition (in general, in SMS). Their distinctive feature is the mandatory flattening of the resistive core boundary after it reaches one of the outer limits of the composition, which after this begins to propagate in a direction that is free from restrictions. In this case, the flattening of the boundary occurs the faster, the better the heat transfer conditions between the elements of the composition. Obviously, with the limitation of the increase in the size of the resistive core in all directions, the increase in the size of the region of normal conductivity will be completed after reaching all boundaries and the entire composition goes into a normal state.

These features should be considered when calculating the total resistance of the SMS during transition modes. This value determines the correct determination of the current decay character in a circuit composed of active-inductive sections. Moreover, after the transition of a composition into a normal state in any direction, an increase in its dimensions in another direction with a good degree of accuracy can be approximated by a simple model with an ever-increasing resistive region with a moving flat front. It should be considered that the propagating velocity of the latter, as it follows from Figure 7.52, increases at the thermal resistance decreasing. These effects are not taken into account in the propagating ellipsoid model since it is a priori based on the velocity of the irreversible propagation of the normal zone calculated for a single wire.

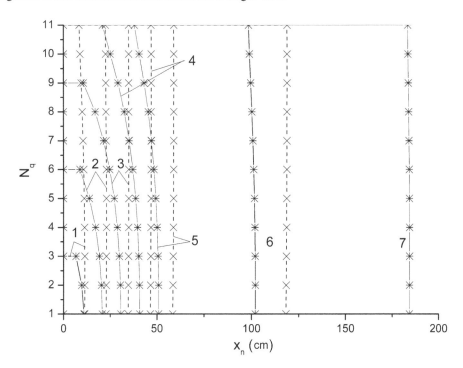

FIGURE 7.52 Formation of the boundary of normal conductivity core $(T_n(x_n, t) = T_{cB})$ if the propagation area is limited in the transverse direction: —— - $R = 1$ cm$^2 \times$ K/W, - - - - - $R = 0.1$ cm$^2 \times$ K/W, 1 - $t = 0.01$ c, 2 - $t = 0.02$ c, 3 - $t = 0.03$ c, 4 - $t = 0.04$ c, 5 - $t = 0.05$ c, 6 - $t = 0.1$ c, 7 - $t = 0.18$ c.

The increase in the resistive core depends on two characteristic times: the propagation time of the normal zone in the longitudinal direction t_x and the heating time of the composite up to a temperature above the critical temperature t_k. If $t_x \sim l_k/V_x$ and $t_x \sim C_k SR/P$ are taken as the simplest estimates $(V_x = I_{c0}\sqrt{\lambda_k \rho_0/(T_{cB} - T_0)}/SC_k$ is the characteristic propagating velocity of a normal zone along a single wire), then it becomes obvious that the resistive core in a longitudinally bounded superconducting composition will have a flat front extending in the transverse direction at $t_x \ll t_k$ (Figure 7.53). Conversely, at $t_x \gg t_k$ in a superconducting multiwire composition bounded in the transverse direction, a resistive core will be formed penetrating with a straight front in the longitudinal direction after a certain time (Figure 7.52).

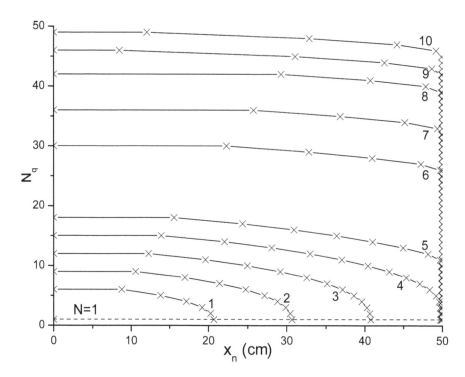

FIGURE 7.53 Formation of the boundary of normal conductivity core $(T_n(x_n, t) = T_{cB})$ at limitation in the longitudinal direction at $R = 1$ cm$^2 \times$ K/W: $t = 0.01$ c, 2 - $t = 0.02$ c, 3 - $t = 0.03$ c, 4 - $t = 0.04$ c, 5 - $t = 0.05$ c, 6 - $t = 0.06$ c, 7 - $t = 0.1$ c, 8 - $t = 0.12$ s, 9 - $t = 0.14$ c, 10 - $t = 0.15$ c, 11 - $t = 0.16$ c.

Thus, when making experiments to determine the propagation velocities of a normal zone in a multiconductor superconducting current-carrying element in both longitudinal and transverse directions, it is necessary to take into account the formation peculiarities of its resistive core by taking into account the spatial boundedness of the superconducting region, first of all, in the transverse direction. This conclusion should also be taken into account when describing transients in superconducting magnets, since in general the expanding ellipsoid model is not correct for superconducting composition with a discrete structure.

7.10 CONCLUSIONS

Thermal processes occurring in superconducting composites, as at the magnetic and current perturbations, are characterized by bifurcation of stable and unstable states near the instability boundary. The analysis of the thermal phenomena regularities in composite superconductors with ideal V-I characteristic and in multicomponent structures based on them, carrying a constant transport current allows to understand the main mechanisms underlying the preservation and destruction of superconducting properties of current-carrying elements of SMS under the action of the thermal perturbations differing in length and duration. Within the framework of the anisotropic continuum model, the criteria of the full thermal stability of composite superconductors are written in one-dimensional and two-dimensional approximations when superconducting properties of composites are preserved under the action of the thermal perturbations with arbitrary energy. For conditions of the partial thermal stabilization, which are characterized by the existence of the critical energies of external thermal disturbances that determine the upper limit of the allowable impulse heat dissipation, the dependence of the critical energies on the length and duration of the thermal disturbance is shown. To determine the critical energies, the method of the finite perturbation of initial state allows one to correctly find the thermal stability conditions regardless of the nature of thermal perturbation.

The thermal modes in the composite superconductors are characterized by the existence of a quasi-equilibrium state to which the temperature field of the composite tends after the action of any type perturbations with energy close to critical. It has a characteristic linear size ξ_m, which is determined by the length of the resistive area of the composite in a quasi-equilibrium state. Correspondingly, the perturbation is almost point-like if its initial length is much less than ξ_m. In this case, the contribution of the heat conductivity mechanism of the matrix to the thermal stability of the superconducting composite is significant. When the initial length of the temperature perturbation is much greater than ξ_m, the efficiency of the thermal conductivity mechanism of the matrix is small. For these modes, the thermal stability conditions of a superconducting composites approach to the corresponding conditions, which exist at the action of infinitely extended perturbation.

The most dangerous perturbations are short-time point-like heat impulses. Under there action, the thermal stabilization is characterized by the value of total perturbation energy. In this case, the critical energy is practically independent of the perturbation parameters. Under the action of the extended or long-time thermal perturbations, the value determining the thermal stability of composite superconductors is the density of the perturbation energy. In these cases, the critical energies increase almost linearly with the increase in the length or duration of the disturbance approaching their asymptotic values. These conclusions show that the assumption of the MPZ-theory about the existence of the maximum allowable length of the initial disturbance is erroneous. The possibility of the restoration of superconducting properties by the composite regardless of the length of initial perturbation is primarily due to the efficiency of the Joule losses diversion into the coolant arising from the appearance of a normal zone.

At a performance of the condition $\alpha i^2 \gg 1$, the critical energy of the thermal perturbations rather weakly depends on the cooling conditions. Thus, this condition is essentially a condition of the adiabaticity of both perturbations and the thermal state of the composite superconductors. The thermal resistance of the insulation can significantly reduce the efficiency of the intensive cooling conditions of the winding. Taking into account the size effect shows that the inhomogeneous temperature distribution in the cross-section of a superconducting composite leads to an increase in its average temperature. Therefore, the massive superconducting current-carrying elements have a considerable divergence between the results of the analysis of thermal modes obtained in the framework of one-dimensional and two-dimensional approximations. In particular, two-dimensional calculation of the conditions of full thermal stabilization shows that one-dimensional theory leads to the overestimated estimates of the full stabilization current and significantly overestimated critical energies for massive intensively cooled current-carrying elements. At the same time, one-dimensional calculation of the thermal instability kinetics along the cooled composite underestimates the velocity of the irreversible propagation of normal zone.

The existence of the critical energies of thermal perturbations may provide a basis for the choice of magnetic field induction at which the transition from one superconductor to another is expedient in combined magnets.

The dependence of the stability conditions of composite superconductors on the energy of external thermal disturbances explains the existence of the effects of degradation and training. Correspondingly, if in a magnetic system there is constant heat dissipation, for example the heat released during movement of the turns their friction against each other, cracking of the compound, etc., then it is impossible to achieve the critical current of a short sample (degradation effect) if the energy of the active disturbances is higher than the critical one. On the other hand, the energy of thermal perturbations can be reduced during the mechanical ordering of the turns in the magnetic system. Therefore, in the course of successive transitions of SMS to a normal state, there is a possibility of increasing the current transition of the magnetic system to a normal state (the training effect).

When the critical energy is exceeded, superconducting composites go irreversibly to the normal state. These modes are characterized by the formation of a thermal autowave propagating spontaneously at a constant velocity. The propagation of the thermal perturbations in the superconducting multiconducting composition is based on the nontrivial thermal mechanisms of spatial formation of the core with normal conductivity in longitudinal and transverse directions, which are not described by the model of expanding ellipsoid for a superconducting composition with a discrete structure. At the irreversible propagation of the thermal instability in a multiconductor composition, the normal conductivity core has the form of a truncated oval with a piecemeal-continuous boundary. The velocity of the thermal instability propagation in the direction of its main axis is less than the corresponding value of the normal zone propagation velocity in a single superconducting composite. The latter is the asymptotic limit for all velocities with which the normal zone will propagate in each element of the composition.

In the case of the irreversible propagation of thermal perturbation in the spatially bounded superconducting region, the front of the normal zone flattened after it reached one of the external boundaries. After this, the normal core propagates with a flat front in a direction free of restrictions. The given destruction mechanism of superconductivity is observed also during the multidimensional formation of the area with normal conductivity in a massive superconducting current-carrying element.

The development of transients in superconducting multiconducting compositions depends to a large extent on the thermal resistance between their elements. As a consequence, within a superconducting composition with a discrete structure, there may be a limited number of subdomains with normal conductivity in which the normal zone propagates at a constant velocity.

The influence of insulation heat capacity on the velocity of the normal zone propagation in the transverse direction is not significant. Its value also practically does not depend on the coefficients of longitudinal heat conductivity of the multiconductor composition elements.

In general, the determination of the thermal stability conditions of composite superconductors and multiconductor compositions should be carried out on the basis of the analysis of emerging area dynamics with normal conductivity. It is of a general nature and can be used in more complex case studies.

8

Electrophysical Features of Thermal Instabilities Occurrence in Technical Superconductors with Various Nonlinearity Types of V-I Characteristics

In the previous chapter, the main provisions of the thermal stabilization theory were formulated within the framework of a model suggesting a jump-like *S-N* transition from the superconducting state to the normal, that is in the case of an ideal V-I characteristic. As a result, the operating current range of technical superconductors is limited by the critical current of the superconducting composite. Correspondingly, as already discussed in Chapter 7, the Joule heat release in a composite superconductor begins only when its temperature exceeds the temperature of the resistive transition T_{cs} (in other words, the current sharing temperature) at which the transport current is equal to the critical current of the superconducting composite I_{c0} at cooling temperature T_0. When it is exceeded, the transport current begins to share between the superconducting core and the matrix, and it flows only through the matrix after reaching the critical temperature of the superconductor T_{cB}. However, both LTS and HTS composites have continuously increasing V-I characteristics for many reasons, as discussed in Chapter 1. Their account allows one to describe the thermal phenomena occurring in superconducting composites with real V-I characteristics (see, for example, Paasi et al. 2000, Majoros et al. 2004, Ishiyama et al. 2005). In this connection, let us perform a comparative analysis of the thermal stabilization conditions for superconducting composites having the ideal and the power V-I characteristics and formulate the stability criteria of a superconducting state with respect to the thermal perturbations for composite superconductors with continuously increasing V-I characteristic.

8.1 RESISTIVE STATES OF COMPOSITE SUPERCONDUCTORS: QUASI-LINEAR APPROXIMATION

Let us consider, as above, the thermally thin superconducting composite of finite length ($z = \pm l_k$) with the cross-sectional area S carrying constant transport current of density J and placed in a static external magnetic field B_a. Let the local part $l_1 < z < l_1$ ($l_k \gg l_1$) of the composite cooled to a coolant temperature T_0 is instantly heated to the temperature of T_1 at the initial time. Assume that the distributions of the temperature T and the electric field E over the cross-section of the composite are uniforms; there are an ideal thermal and electrical contacts between the superconducting filaments and the matrix; the volume fraction of the superconductor in the composite is η; superconductor V-I characteristic is described by the power equation $E(J) = E_c[J/J_c(T, B)]^n$ where $J_c(T, B)$ is the critical current density of the superconducting filaments determined at a priori-defined value of the electric field E_c and n is the increasing parameter of the V-I characteristic (n-value). In this section, the variation of the $J_c(T, B)$ with temperature and magnetic induction is described, as above-mentioned, by the linear dependence of the form $J_c(T, B) = J_{c0}(B)(T_{cB}(B) - T)/(T_{cB}(B) - T_0)$.

Let us place the origin of the coordinates in the center of the initial temperature perturbation and write the one-dimensional heat conduction equation in the form

$$C_k \frac{\partial T}{\partial t} = \lambda_k \frac{\partial^2 T}{\partial z^2} - \frac{hp}{S}(T - T_0) + \begin{cases} G_c(T) \\ G(T) \end{cases} \tag{8.1}$$

with the initial-boundary conditions

$$T\big|_{t=0} = \begin{cases} T_1, & 0 \le |z| \le l_1 \\ T_0, & l_1 < |z| \le l_k \end{cases}, \quad T(\pm l_k, t) = T_0 \tag{8.2}$$

Here, the physical meaning of the initial parameters is the same as in Equation (7.9). The values G_c and G are the Joule heat release in a composite calculated either in the framework of the ideal V-I characteristic or taking into account its continuous rise, respectively. In the first case, the value $G_c = EJ$ follows from (7.11) and is defined as follows

$$G_c(T) = \frac{J^2 \rho_m(T, B)}{1 - \eta} \begin{cases} 1, & T > T_{cB} \\ (T - T_{cs})/(T_{cB} - T_{cs}), & T_{cs} \le T \le T_{cB} \\ 0, & T < T_{cs} = T_{cB} - (T_{cB} - T_0)J/(\eta J_{c0}) \end{cases} \tag{8.3}$$

In the case of a continuously increasing V-I characteristic, to calculate $G = EJ$ one may use the model of two parallel-connected conductors. Then, the instantaneous values of the electric field and current densities in the superconducting core J_{sc} and matrix J_m satisfy the system of Equations (6.12) and (6.13). Taking into account the linear temperature dependence $J_c(T, B)$, it is easy to get the following formula

$$G(T) = \begin{cases} \eta J_{c0} E \left(1 - \dfrac{T - T_0}{T_{cB} - T_0}\right) \left(\dfrac{E}{E_c}\right)^{1/n} + (1 - \eta)\dfrac{E^2}{\rho_m}, & T < T_{cB} \\ (1 - \eta)\dfrac{E^2}{\rho_m}, & T \ge T_{cB} \end{cases} \tag{8.4}$$

In this case, the instantaneous value of the electric field E for the given values of the current density J and temperature T is the solution of the transcendental equation

$$\eta J_{c0} \left(1 - \frac{T - T_0}{T_{cB} - T_0}\right) \left(\frac{E}{E_c}\right)^{1/n} + (1 - \eta)\frac{E}{\rho_m} - J = 0 \tag{8.5}$$

at $T < T_{cB}$, and it is described by the equality

$$E = J\rho_m/(1 - \eta) \tag{8.6}$$

at $T \ge T_{cB}$.

Let us perform, first of all, a quasi-linear analysis of the resistive state formation of a superconducting composite within the framework of the formulated models to understand the qualitative features of the development of electric modes in superconducting composites with real V-I characteristics. To do this, assume that the value of the matrix resistivity is constant. Introduce the following dimensionless variables: $i = J/(\eta J_{c0})$ – dimensionless current, $e = E/E_c$ – dimensionless electric field, $\theta = (T - T_0)/(T_{cB} - T_0)$ – dimensionless temperature, $g_c = G_c/(\eta J_{c0} E_c)$ – dimensionless heat release calculated within the framework of the model of jump-like S-N transition,

$g = G/(\eta J_{c0}E_c)$ – dimensionless heat release in the composite with continuously increasing V-I characteristic. Then, the Expressions (8.3) and (8.6) are converted into the form

$$g_c(\theta) = i^2\varepsilon_1 \begin{cases} 1, & \theta > 1 \\ (\theta - 1 + i)/i, & 1 - i \leq \theta \leq 1, \\ 0, & \theta < 1 - i \end{cases} \tag{8.7}$$

for the ideal V-I characteristic and

$$g(\theta) = \begin{cases} (1-\theta)e^{1+1/n} + e^2/\varepsilon_1, & \theta < 1 \\ e^2/\varepsilon_1, & \theta \geq 1 \end{cases} \tag{8.8}$$

for the power V-I characteristic. According to (8.7) and (8.8), the corresponding dimensionless V-I characteristics of the composite in the nonisothermal approximation are described by the relations

$$e_c = i\varepsilon_1 \begin{cases} 1, & \theta > 1 \\ (\theta - 1 + i)/i, & 1 - i \leq \theta \leq 1 \\ 0, & \theta < 1 - i \end{cases} \tag{8.9}$$

and

$$i = \begin{cases} (1-\theta)e^{1/n} + e/\varepsilon_1, & \theta < 1 \\ e/\varepsilon_1, & \theta \geq 1 \end{cases} \tag{8.10}$$

for the composite with ideal or power V-I characteristics, respectively.

The voltage-current characteristics (8.9) and (8.10) allow one to find the dimensionless effective resistivity of a composite as a function of temperature if the equalities $e_c = i\varepsilon_1 r_c$ and $e = i\varepsilon_1 r_k$ will be used for each of the models under consideration. Here, $\varepsilon_1 = E_\rho/E_c$ and $E_\rho = \eta J_{c0}\rho_m/(1-\eta)$, which were introduced in section 6.2 when analyzing the stability of the current states of a superconducting composite in zero-dimensional approximation. As already noted, according to their physical meaning, the values E_ρ and ε_1 describe the influence of the matrix resistance on the current sharing within an approximation that takes into account the real V-I characteristic of a superconductor. According to the estimates given, the following characteristic values $E_\rho \sim 1$ V/m and $\varepsilon_1 \sim 10^4$ may be obtained. At the same time, it follows from (8.9) that the current sharing, hence the quantity r_c do not depend on ε_1 at a jump-like S-N transition, that is on the matrix resistance since, as noted above, the beginning and the development of the current sharing between the superconducting core and the matrix depends only on the decrease of the critical current of the superconductor with temperature. According to (8.10), with a continuous increase in the V-I characteristic, the higher ε_1 (that is, the higher the matrix resistance), the greater the current in the superconducting core, and most of the current will flow through the superconducting core with a slight overheating of the composite at $1 \ll \varepsilon_1$. In other words, the influence of the matrix resistance on the current sharing in composite superconductors can be correctly described only within the framework of a model that takes into account the continuous increase of the V-I characteristic.

To illustrate the role of ε_1 on the current state formation of the composite superconductors, the results of the current sharing simulation between the superconducting core and the matrix are compared with each other in Figure 8.1. They are obtained in the framework of both models at $i = 0.8$. The calculation of the current modes of the composite with the power V-I characteristic was performed according to (8.10) at $n = 25$ and various values of ε_1. Within the model of the jump-like S-N transition (8.9), the dimensionless currents in superconductor i_{sc} and matrix i_m were calculated as $i_{sc} = 1 - \theta$, $i_m = i - 1 + \theta$ in the temperature range $1 - i < \theta < 1$. Figure 8.1 demonstrates the difference in the formation of the current modes described by various equations of V-I characteristics and their corresponding dependence on ε_1 that occurs only for the composite with the continuous increase of the V-I characteristic.

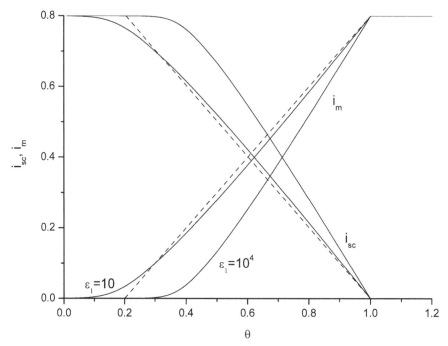

FIGURE 8.1 Effect of the parameter ε_1 on the current sharing between the superconducting core and the matrix: —— - model with the power V-I characteristic, - - - - model with the ideal V-I characteristic.

Let us rewrite the problem described by Equations (8.1)-(8.2) in a dimensionless form by introducing the dimensionless length $X = x/L_z$ and the dimensionless time $\tau = t/t_x$ using the characteristic quantities $L_z = [\lambda_k(T_{cB} - T_0) S/(E_c I_{c0})]^{1/2}$, $t_x = C_k L_z^2/\lambda_k$, $I_{c0} = \eta J_{c0} S$. Then, one gets the following initial-boundary problem

$$\frac{\partial\theta}{\partial\tau} = \frac{\partial^2\theta}{\partial Z^2} - \frac{1}{\alpha_E}\theta + \begin{cases} g_c(\theta) - \text{for an ideal V-I characteristic} \\ g(\theta) - \text{for a power V-I characteristic} \end{cases} \qquad (8.11)$$

$$\theta(Z, 0) = \begin{cases} \theta_1, & |Z| \le \xi_1 \\ 0, & \xi_1 < |Z| < L \end{cases}, \quad \theta(\pm L, \tau) = 0, \quad L = l_k/L_z, \quad \xi_1 = l_1/L_z$$

Here, $\alpha_E = E_c I_{c0}/[hp(T_{cB} - T_0)]$ is the parameter of the thermal stabilization of the composite superconductor with the continuously increasing V-I characteristic when a priori-defined value E_c is used. To avoid the influence of the boundary condition at $Z = L$ on the character of the spatial temperature distribution in the composite, it is assumed that $\xi_1 \ll L = 1,000$ in the numerical simulations.

The problem described by Equations (8.7), (8.8) and (8.11) allows one to perform a comparative analysis of the main characteristics of the dissipative states of composite superconductors with a constant transport current, varying the thermal stability parameter α_E, dimensionless current i and temperature perturbation parameters. It should be emphasized that the introduced dimensionless variables and parameters are not interdependent. They were introduced in the same way as in Section 7.2, to the retention of the invariance of the problem described by Equations (8.7), (8.8) and (8.11) when the initial parameters are changed. As a consequence, the results obtained at $n \to \infty$ follow the limiting transition from a model with a power V-I characteristic to a model with an ideal V-I characteristic.

For the convenience of the comparison of simulation results obtained below, the Stekly parameter is used instead of the thermal stabilization parameter α_E taking into account the connection $\alpha = \varepsilon_1 \alpha_E$. The latter relation indicates that the Stekly parameter depends on the dimensionless parameter ε_1 while the thermal stabilization parameter α_E does not depend on it.

The dependencies of the effective resistivity of superconducting composite on the dimensionless temperature calculated according to (8.9) and (8.10) are presented in Figures 8.2 and 8.3. In the framework of the approximation (8.10), the calculations were performed at $n = 25$ and different values of the parameter ε_1 and the transport current i. Figure 8.2 demonstrates the influence of ε_1 on the values $r_k(\theta)$ at a current lower than the critical value. The inset to Figure 8.2 shows the variation in the effective resistivity of composite in the small overheating area. It is seen that there is a common point in this region where the curves $r_c(\theta)$ and $r_k(\theta)$ intersect. In Figure 8.3, the temperature dependencies of the composite effective resistivity at currents exceeding the critical current of the composite are shown. In the framework of the jump-like S-N transition, all current will flow only through the matrix at $i \geq 1$ because the superconducting state is destroyed. The differencies between values $r_c(\theta)$ and $r_k(\theta)$ due to the influence of the n-value are shown in Figure 8.4. On the inset, these values are shown in more detail in the overheating area close to the temperature of the resistive transition. The presented results allow one to formulate the following conclusions.

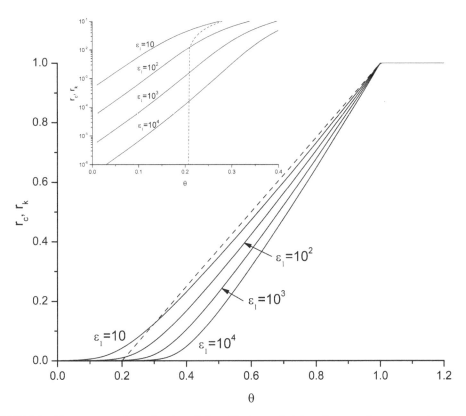

FIGURE 8.2 Temperature dependencies of the effective resistivity of composite superconductor at $i = 0.8$ and different values ε_1: —— model with the power V-I characteristic, - - - - model with the ideal V-I characteristic.

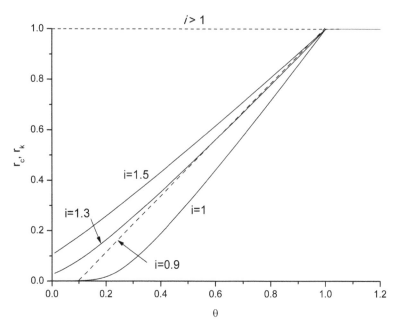

FIGURE 8.3 Temperature dependencies of the effective resistivity of composite superconductor at $\varepsilon_1 = 10^4$ and supercritical currents: ——— - model with the power V-I characteristic, - - - - model with the ideal V-I characteristic.

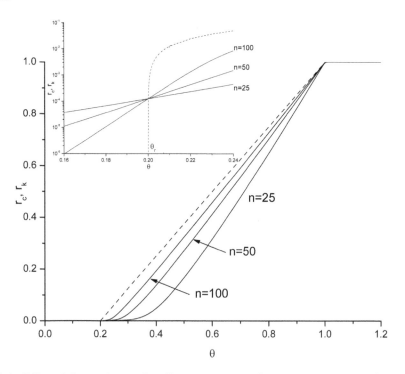

FIGURE 8.4 Effect of the n-value on the effective resistivity of composite at $i = 0.8$ and $\varepsilon_1 = 10^4$: ——— - model with the power V-I characteristic, - - - - model with the ideal V-I characteristic.

In the temperature range from 0 to $\theta = 1 - i$, the values $r_k(\theta)$ differ from zero while the values $r_c(\theta)$ are zero. The difference between them increases if the transport current increases or if n-value and ε_1 decrease. This difference is due to the finite value of the electric field induced in a composite superconductor with a real V-I characteristic.

In the temperature dependencies $r_k(\theta)$, the three characteristic regions may be distinguished. First, there is a temperature θ_r. The resistivities of the composite calculated in the framework of the two formulated approximations are equal to each other at this temperature. This quantity is shown on the inset to Figure 8.4. It is not difficult to find that $\theta_r = 1 - i + 1/\varepsilon_1$ (in dimensional form, $T_r = T_{cs} + (T_{cB} - T_0)/\varepsilon_1$). In the temperature range from 0 to θ_r, the values $r_k(\theta)$ described in the framework of the model with a power V-I characteristic are higher than the corresponding values $r_c(\theta)$. Therefore, the values $G(T)$ will also be higher than $G_c(T)$ in this temperature range. The difference between them increases with increasing the current and decreasing n-value and ε_1. This difference is due to the electric field induced in the superconductor with a real V-I characteristic when any current flows through it. A distinctive feature of the value θ_r is its independence from the n-value as shows the inset to Figure 8.4. Because $e = e_c = 1$ at $\theta = \theta_r$, that is $E = E_c$ at $T = T_r$, as follows from the Formulae (8.9) and (8.10), then the value θ_r separates the subcritical region of the electric fields from the supercritical one even at the subcritical currents. Second, after exceeding the temperature θ_r, the opposite tendency takes place. Namely, in the supercritical region of the electric fields, the effective resistivity of the composite calculated according to the model with an ideal V-I characteristic exceeds the corresponding values of the effective resistivity of composite with a continuous V-I characteristic. It should be emphasized that this feature can be retained for currents exceeding the critical current of the composite. It is not obvious at first glance. As a result, the supercritical modes $(i > 1)$ may be possible. Moreover, it will be shown below that they may be characterized not only by the stable flow of the currents through the composite, as was shown in Chapter 6, but also by states that are stable to external thermal perturbations. These features are a direct consequence of the current sharing mechanism that depends on the character of the increase of the V-I characteristic and the matrix resistivity in composite superconductors. Thirdly, in the area of the supercritical overheating $(\theta > 1$, that is at $T > T_{cB})$, the effective resistivity calculated according to both (8.9) and (8.10) do not differ from each other in the framework of the used quasi-linear approximation. In this case, the Joule losses will be determined by the losses only in a matrix for which the value ρ_m is assumed to be constant.

Let us estimate the temperature growth of the effective resistivity of composite with a power V-I characteristic in the temperature range $0 < \theta < 1$. To begin, let us consider the initial area of the temperatures $(\theta \ll 1)$. In this case, all current in the composite virtually flows through the superconducting core and therefore, one may find $i\varepsilon_1 r_k(\theta) \cong [i/(1-\theta)]^n$ according to (8.10). Logarithmic this equation, one gets the following relation $\ln i\varepsilon_1 r_k(\theta) = n[\ln i - \ln(1-\theta)]$. Expanding the term $\ln(1-\theta)$ into the power series, the latter equality may be rewritten in the form $\ln i\varepsilon_1 r_k(\theta) = n[\ln i + \theta + \theta^2/2 + ...]$. Then

$$r_k(\theta) = \frac{i^{n-1}}{\varepsilon_1} \exp[n(\theta + \theta^2/2 + ...)] .$$

Therefore, the increase in the dependence of $r_k(\theta)$ with the temperature is exponential in the initial temperature area. Furthermore, the higher the n-value, the greater the rate of increase $r_k(\theta)$. These conclusions confirm the insets to Figure 8.2 and 8.4.

To estimate $r_k(\theta)$ near the critical temperature, let us introduce a new function $u = 1 - r_k$, which according to (8.10) will lead to equation

$$1 = \frac{1-\theta}{i}[(1-u)i\varepsilon_1]^{1/n} + 1 - u$$

Expanding the factor $(1 - u)^{1/n}$ in the power series, one may get

$$1 = \frac{1-\theta}{i}(i\varepsilon_1)^{1/n} \left(1 - \frac{u}{n} - \frac{n-1}{2n^2}u^2 + ...\right) + 1 - u$$

Then, considering only the linear approximation in temperature, one may find

$$r_k(\theta) = 1 - \frac{1}{i}(i\varepsilon_1)^{1/n} + \frac{\theta}{i}(i\varepsilon_1)^{1/n}$$

that is the increase in the effective resistivity of a composite with temperature is almost linear near the critical temperature of a superconductor with a power V-I characteristic and the linear temperature dependence of J_c. As it follows from Figure 8.4, the higher the n-value, the better this regularity is observed. The transition from the exponential growth of the effective resistivity of composite with the temperature increase to a linear one demonstrates the stabilizing role of the matrix in the thermal stabilization of composite superconductors with a real V-I characteristic. These results cannot be obtained within the model with an ideal V-I characteristic.

The regularities discussed above are based on the features of the variation in the differential resistivity of a superconductor depending on the character of its V-I characteristic increase both at subcritical ($e < 1$) and supercritical ($e > 1$) electric fields. It is easy to find that the values $r_k(\theta)$ calculated in the framework of the model with a power V-I characteristic satisfies the limiting transition to an ideal V-I characteristic at $n \to \infty$. However, according to Figure 8.4, even at high but finite n-value, for example at $n = 100$, in the temperature dependencies of the effective resistivity of a composite with a power V-I characteristic also take place the above-noted features of their increase. The difference between the calculated values for both models increases with decreasing n-value. As a result, a model with a power V-I characteristic leads to more lowered values of the heat release in the temperature range $\theta_r < \theta < 1$. It is important to emphasize that these regularities will also occur in current-carrying elements based on LTS for which $n > 50$. Therefore, a theoretical analysis of the thermal stability conditions of superconducting composites performed within the model of a jump-like S-N transition will inevitably lead to the underestimated critical energies of external thermal perturbations and to overestimated values of the irreversible propagation velocity of thermal instabilities. These features are discussed in more detail below.

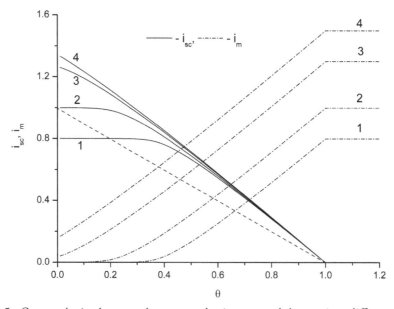

FIGURE 8.5 Current sharing between the superconducting core and the matrix at different values of the transport current: 1 - $i = 0.8$, 2 - $i = 1$, 3 - $i = 1.3$, 4 - $i = 1.5$. Here, curves (——) and (–·–·–) depict the currents in composite with a power V-I characteristic and line (– – –) shows the temperature dependence of the critical current of a superconductor.

Figure 8.5 demonstrates the features of the current sharing between the superconductor and the matrix at subcritical and supercritical currents at $n = 25$ and $\varepsilon_1 = 10^4$. It is seen that even at a subcritical current (Curve 1), the intense current sharing in the composite with a real V-I characteristic is not observed when its temperature exceeds the temperature of the resistive transition ($\theta = 0.2$). Moreover, the current in the superconducting core of the composite remains above the critical current of the composite superconductor even after a significant excess of the temperature of the resistive transition. More than that, under the supercritical current modes, the current in the superconducting core is always higher than the critical current of the composite up to its temperature increase to a value equal to the critical temperature of a superconductor. Consequently, the temperature of the resistive transition T_{cs}, after which the current sharing begins according to the model of the jump-like S-N transition, does not have a physical meaning for composite superconductors with a real V-I characteristic. As can be expected, as the temperature rises for a given current, the continuous current sharing in composite superconductors depends on the rise rate with the temperature of the V-I characteristic and matrix resistivity. This feature is discussed more detail in Sections 8.3 and 8.4.

8.2   STABILITY CONDITIONS AND FEATURES OF RESISTIVE STATES DESTRUCTION OF COMPOSITE SUPERCONDUCTORS

The results presented in Figure 8.1-8.5 lead to the conclusion that the thermal and electrical modes of the composite superconductors calculated in terms of ideal and real V-I characteristics may significantly differ from each other. The characteristic temperature dependencies of the Joule heat release in the composite and the heat flux into the coolant are compared in Figure 8.6. The inset shows some of them in more detail. It demonstrates how the area of the exponential growth of the $g(\theta)$ depends on the current. The calculation was carried out at $n = 25$, $\alpha = 1$, $\varepsilon_1 = 10^4$ and various currents (both subcritical and supercritical). The solid curves correspond to the values $g(\theta)$ calculated according to (8.8). The dashed curves show the heat release $g_c(\theta)$, which are described by the Expression (8.7) and were calculated at $i = 0.8$ and $i = 1.095$. The dash-dotted line indicates the heat flux to the coolant ($w(\theta) = \varepsilon_1\theta/\alpha$). The simulation results allow one to understand the characteristic qualitative features of stable and unstable states formation of superconducting composites with a real V-I characteristic.

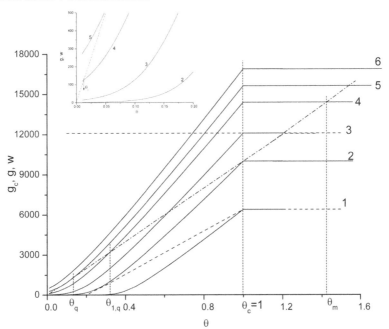

FIGURE 8.6   Possible thermal operating modes of a composite superconductor at various currents: 1 - i = 0.8, 2 - i = 1, 3 - i = 1.095, 4 - i = 1.2, 5 - i = 1.255, 6 - i = 1.3; ――――― - model with the power V-I characteristic, - - - - - - model with the ideal V-I characteristic.

First, note that modes 1 and 2 have only one equilibrium temperature θ_i. This temperature is shown on the inset to Figure 8.6 for Mode 4. The equation for its definition is written below. This state does not occur when the ideal V-I characteristic is used to simulate the operating modes. Mode 2 corresponds to the limiting value of the current i_r after exceeding which additional thermal equilibrium points will appear. The value i_r follows from the obvious condition $\alpha i_r^2 = 1$, which describes the full recovery current, as discussed in Chapter 7. In the framework of the approximations under consideration, its value does not depend on the used V-I characteristics. Accordingly, the temperature of the composite with the real V-I characteristic after the action of any thermal perturbation will be equal to θ_i for all $0 < i < i_r$ after a specific time. As a result, the superconducting composite does not irreversibly go to the normal state at $i < i_r$. This conclusion is also valid for supercritical current modes. In this case, for any current $i > 1$, there always exists a boundary value of the thermal stabilization parameter, which is equal to $\alpha_r = 1/i^2$, when the superconducting properties of the composite carrying the supercritical current are preserved for all $\alpha < \alpha_r$. For these thermal stabilization parameters, the final temperature of the composite after the action of arbitrary thermal perturbation will not exceed the critical temperature of the superconductor since $\theta_i < 1$. As an illustration of the existence of these stable supercritical modes, curves $g(\theta)$ and $w(\theta)$ calculated at $i = 1.5$ and $\alpha = 0.444$ are shown in Figure 8.7. On the inset to the figure, the dependencies $g(\theta)$ and $w(\theta)$ are shown in the temperature range close to θ_i. Let us emphasize that it is not possible to describe such modes within the framework of a model based on a jump-like S-N transition (Fig. 7.5a).

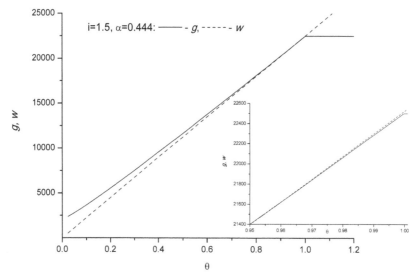

FIGURE 8.7 Stable dissipative regime of the superconducting composite during supercritical current mode when its equilibrium temperature θ_i is close to the critical temperature at $\alpha i^2 < 1$ and $i > 1$ (the destruction of superconducting state does not occur since $\theta_i < 1$, as the inset shows.

At currents greater than i_r, for example, for the current Modes 3 and 4 in Figure 8.6, there may be three equilibrium temperatures. They are depicted for Mode 4 and describe the following states. First, in the region of small overheating ($\theta_i < 1$), the temperature of the stable equilibrium θ_i also takes place. Second, there is a maximum temperature of the stable equilibrium θ_m ($\theta_m > 1$). (In dimensional form, T_m, $T_m > T_{cB}$). In Chapter 7, it was discussed that the temperature of the composite will tend to this temperature, in particular, as a result of an irreversible transition to the normal state. Since the Joule losses are determined by the heat release only in the matrix at $\theta > 1$, the value θ_m does not depend on the V-I characteristics used. Therefore, in the framework of the considered quasi-linear approximation, it is equal to $\theta_m = \alpha i^2$. Third, the temperature of the unstable equilibrium $\theta_{1,q}$ exists. As discussed in Chapter 7, it is corresponds to the critical value of the initial temperature θ_1 of infinitely extended perturbation ($Z_1 \to \infty$). As follows from the calculations, the value $\theta_{1,q}$ will depend on the type of nonlinearity of the V-I characteristic at intensive cooling (in the general case, at relatively small values of α). Note that the condition $\theta_m = \theta_{1,q} = 1$ may be satisfied in the limiting case. A similar state is depicted in Figure 8.8.

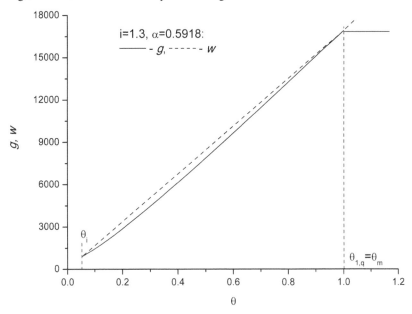

FIGURE 8.8 Dependencies of g and w on the dimensionless temperature at $\theta_m = \theta_{1,q} = 1$ at $\alpha i^2 < 1$ and $i > 1$.

The values θ_i and $\theta_{1,q}$ follow from the obvious equality $w = g$. It leads to the transcendental equation

$$\alpha i^2 = \alpha i (\varepsilon_1 / \alpha i)^{1/n} (1 - \theta) \theta^{1/n} + \theta \tag{8.12}$$

which satisfies both these values. For $\theta \ll 1$, the estimate $\theta_i = \alpha i^{n+1} / \varepsilon_1$ follows from it.

Supercritical Mode 3 shown by the solid curve in Figure 8.6 was calculated for a superconductor with a power V-I characteristic based on the theorem of the equal areas, which allows one to find the full stabilization current (minimum current of the normal zone propagation, according to Chapter 7) i_s. It is easy to understand that its value depends on the used V-I characteristics due to the corresponding difference in the values $g_c(\theta)$ and $g(\theta)$. Moreover, a noticeable difference will be achieved at intensive cooling. As discussed in Chapter 7, in the current range $i_r < i < i_s$, the superconducting state of a composite is stable to the thermal perturbations with arbitrary initial energy. However, the thermal stability of a superconducting state depends on the thermal perturbation energy at higher currents ($i_s < i$).

Modes 5 and 6 only have one equilibrium temperature θ_m. It is not shown in Figure 8.6. The Mode 5 corresponds to the instability current (quenching current) i_q, which, according to Chapter 6, determines the maximum allowable value of the transport current that may stably flow through the composite in the absence of external perturbations. When it is exceeded (Mode 6), the superconducting state of the composite is unstable. The current i_q is characterized by the stable finite values of the electric field e_q and overheating of composite θ_q. According to (Polak, Hlasnik and Krempasky, 1973), and as follows from Figure 8.6, the quenching parameters θ_q, i_q and e_q in the absence of the external thermal perturbations can be determined from the condition $w = g$ and $dw/d\theta = dg/d\theta$ at $\theta = \theta_q$ in the absence of external thermal perturbations. Accordingly, these conditions lead to the relations

$$(\varepsilon_1\theta_q)^2[\alpha\theta_q(\theta_q+n-1)]^{n-1}[(n+1)\theta_q-1]^{n+1}=(n\theta_q)^{2n},$$

$$i_q=\sqrt{\frac{n\theta_q(1-\theta_q)}{\alpha[(n+1)\theta_q-1]}+\frac{\theta_q}{\alpha}},\quad e_q=\frac{\varepsilon_1\theta_q}{\alpha i_q} \tag{8.13}$$

As discussed in Chapter 6, the existence of θ_q, i_q and e_q shows that no instability may occur at a priori-defined critical current in superconducting composites with a real V-I characteristic. For the quasi-linear approximation under consideration, the quenching current is equal to $i_q=1.255$, which is 25% higher than a priori defined critical current despite the finite value θ_q before instability. Moreover, the model of the jump-like S-N transition does not allow correctly formulating the limiting condition of the full thermal stabilization of a composite superconductor when an arbitrary initial perturbation does not move composite into a normal state in the entire range of the current variation. Let us write this condition using (8.13) taking into account that the temperature of the composite before the instability is equal to the critical temperature of the superconductor ($\theta_q=1$), and the transport current will stably flow only in the matrix at the thermal stabilization parameter α_f and current i_f, under which the condition of the full thermal stabilization will be observed. Correspondingly, the conditions $i_f=i_q$ and $\alpha_f i_f^2=1$ must be executed, as discussed above. Then, the value of this thermal stabilization parameter follows from the equality

$$\alpha_f^{n-1}=1/\varepsilon_1^2 \tag{8.14}$$

In this case, the limiting value that is equal to

$$i_f=\varepsilon_1^{1/(n-1)} \tag{8.15}$$

describes the current, after which the composite will be in a stable nonsuperconducting state at $\alpha<\alpha_f$. In the dimensional variables, the condition of the limiting full thermal stabilization (analog of the Stekly cryostability condition at any finite n-value) will be provided if the operating parameters satisfy the conditions

$$\left(\frac{\eta J_{c0}}{E_c^{1/n}}\right)^2\left[\frac{S}{hp(T_{cB}-T_0)}\right]^{1-1/n}\left(\frac{\rho_m}{1-\eta}\right)^{1+1/n}\le 1,\quad J\le\eta J_{c0}\left[\frac{\eta J_{c0}\rho_n}{(1-\eta)E_c}\right]^{1/(n-1)} \tag{8.16}$$

Previously, these criteria were formulated in section 6.2 when analyzing the conditions for the occurrence of the current instability. Note that despite the difference of Equations (8.13) from Equations (6.27) and (6.28) obtained earlier, they lead to the identical results. Besides, relations and criteria (8.12)-(8.16) comply with the limiting transition to the model of a jump-like S-N transition. Actually, it is not difficult to find

$$\theta_{1,q}=\frac{\alpha i(1-i)}{\alpha i-1},\quad \theta_i=0,\quad \alpha_f=1,\quad i_f=1$$

at $n\to\infty$. They are the parameters of the full thermal stabilization of the composite with an ideal V-I characteristic discussed in Chapter 7, that is, these parameters lead to the Stekly cryostability condition of the composite superconductor for the currents that may stably flow only in the range $0<i=I/I_{c0}<1$.

Thus, the criteria (8.14)-(8.16) allow one to formulate correctly the effect of the stabilizing role of the matrix of a composite superconductor with a real V-I characteristic on the thermal stability conditions taking into consideration the supercritical current modes when they can be in the stable modes at currents exceeding a priori defined critical current of the composite ($i>1$).

Let us do the quantitative analysis of the stability of the thermal conditions of the composite superconductor numerically solving problem (8.11) at $n=25$ and $\varepsilon_1=10^4$.

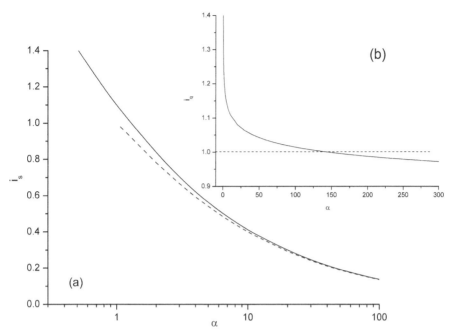

FIGURE 8.9 Dependence of the full stabilization current (a) and quenching current (b) on the thermal stabilization parameter: (————) - the power V-I characteristic, - - - - - the ideal V-I characteristic. The corresponding limiting values are equal to $\alpha_f = 0.4642$ and $i_f = 1.4678$.

The simulation results of the full stabilization i_s and quenching i_q currents, as a function of the stability parameter, are shown in Figure 8.9. The solid curve in Figure 8.9a describes the values i_s calculated within the framework of the model with the power V-I characteristic, and the dashed curve was obtained according to the model with an ideal V-I characteristic. In the first case, the quantity i_s was determined numerically according to the equality

$$\int_0^{\alpha i_s^2} \left[g(\theta) - \frac{\varepsilon_1}{\alpha}\theta \right] d\theta = 0$$

which follows from the theorem of equal areas. In the case of jump-like S-N transition, the full stabilization current is described by the relation (7.22).

The curves presented show that a noticeable difference between the values i_s calculated for both models of the V-I characteristics will be observed for 'well' stabilized composites ($\alpha < 10$). It increases as the quantity α approaches the condition of the full thermal stabilization. As it is proved above, the parameter α_f can be less than 1. For the given initial parameters, the value of the full thermal stabilization parameter is equal to $\alpha_f = 0.4642$. The quenching current that is calculated at $\alpha = 1$ clear shows the existence of the stable supercritical currents. In other words, the thermal instabilities in the form of the autowave can occur when $\alpha < 1$ and $i > 1$ for composite superconductors with real V-I characteristic breaking the Stekly condition. It should be emphasized that the values i_s, i_q and α_f depend on ε_1 in the general case, that is the resistivity of the matrix. These features cannot be formulated in the framework of the jump S-N transition model.

The dashed horizontal line in Figure 8.9b corresponds to the dimensionless value of the critical current. It may be seen that the quenching currents may be supercritical or subcritical, as discussed above. The latter will be observed in the 'poorly' stabilized composites, in particular at their nonintensive cooling ($\alpha \gg 1$). Note that there is tend $i_q \to 0$ in the limiting case $\alpha \to \infty$ (thermally insulated composite) according to (8.13). This feature is explained by the fact that the maximum permissible current is a consequence of the thermal equilibrium of the Joule heat release with the heat flux into the coolant for composites with a real V-I characteristic. Therefore, the self-heating of the thermally insulated composite will lead, strictly speaking, to an unstable state because of the finite electric field exists in composite superconductor due to the real V-I characteristic. Thus, the results presented in Figure 8.9b, prove once again that the concept of the critical current has no physical meaning even for composites based on hard superconductors with strong pinning.

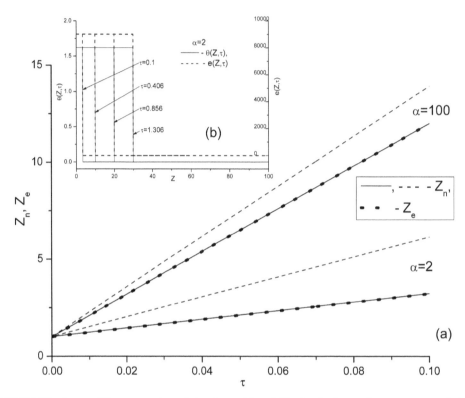

FIGURE 8.10 Irreversible propagation of the thermal instability in the form of temperature and electric waves at different values of the thermal stabilization parameter (a) and change over time the spatial distribution of the temperature and electric field in superconducting composite (b).

   In Figure 8.10, the kinetics of the 'normal zone' (Figure 8.10a), and the spatial-temporal variation of the temperature and electric field in the composite (Figure 8.10b) are presented. The subcritical current $i = 0.9$ for two characteristic values of the thermal stabilization parameter is considered: $\alpha = 2$ – for 'well' stabilized composite and $\alpha = 100$ – for 'poorly' stabilized composite. The parameters of the initial temperature perturbation were set as $Z_1 = 1$ and $\theta_1 = 1$. These parameters correspond to a supercritical temperature perturbation, that is a perturbation that leads to the irreversible destruction of superconductivity. The dashed curves in Figure 8.10a correspond to the calculations made on the basis of the model with the ideal V-I characteristic, the solid curves and curves marked by a short dashed marker '·' were obtained using the model with the power V-I characteristic. When modeling the thermal states of superconducting composites with a continuous V-I characteristic, the concept of a normal zone is absent due to the absence of the boundary of jump-like S-N transition. Therefore, the variation in its 'length' over time was determined by analyzing the kinetics of the point with temperature $\theta = 1 - i$ (in the dimensional form, it is the temperature of the resistive transition $T = T_{cs}$), as well as the point with the small electric field $e = 10$. In the first case, the calculation results of the propagation of given isotherm is described by the solid curves, in the second case, the propagation of the given equipotential is presented by the marker '-'. It can be seen that both approximations lead to the equivalent results at determining the velocity of the irreversible propagation of thermal instability. Therefore, the velocity of the 'normal zone' was determined on the kinetics of the given value of electric field during supercritical current modes ($i > 1$).

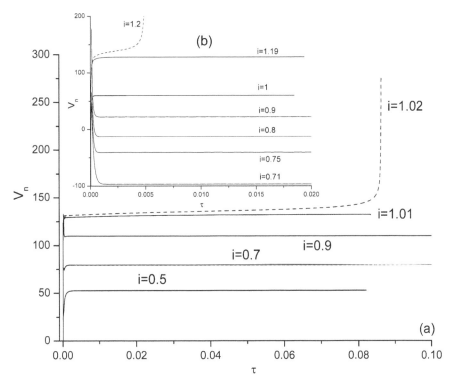

FIGURE 8.11 Formation of the autowave (——) and quenching (- - -) modes at $\alpha = 100$ (a) and $\alpha = 2$ (b) and different currents.

It follows from Figure 8.10 that after the action of the thermal perturbation with energy exceeding the critical value, the thermal and electric autowaves are formed in the composite in the range of currents $i_r < i < i_q$ propagating along the composite at the same constant velocity, as discussed in Chapter 7. Correspondingly, the autowave velocities are negative at $i_r < i < i_s$ and positive at $i_s < i < i_q$ (Figures 8.11 and 8.12). These conclusions are also valid for the currents exceeding a priori defined critical current (Figure 8.12). According to Figure 8.6, the formation of a thermo-electric autowaves will not occur at $i < i_r$ and $i > i_f$. In these operating modes, the disappearance or the irreversible propagation of the initial temperature perturbation is of a substantially nonstationary character due to a noticeable difference between the values $w(\theta)$ and $g(\theta)$ in the temperature range $0 < \theta < \theta_m$ as noted in Chapter 7.

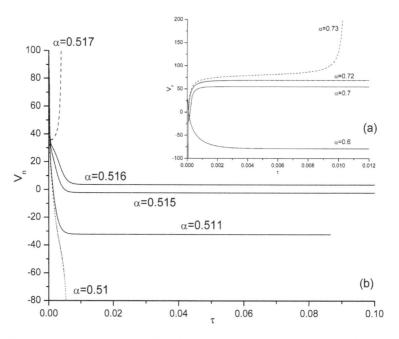

FIGURE 8.12 Autowave (——) and quenching (- - -) modes at supercritical currents (a - i = 1.3, b - i = 1.4) and different values of the thermal stabilization parameter (α_f = 0.4642 and i_f = 1.4678).

As an illustration, the thermal state formation of the 'well' stabilized composite (a = 0.45) under the temperature perturbation with the parameters Z_1 = 10 and θ_1 = 8 carrying the supercritical current i = 1.5 exceeding the full stabilization current, which is equal to i_f = 1.4678, is shown in Figure 8.13. It is seen that the composite loses its superconducting properties rather quickly reaching the equilibrium supercritical temperature θ_m = 1.0125, as discussed above.

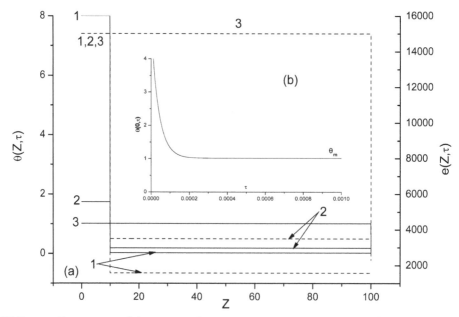

FIGURE 8.13 Destruction of the superconducting properties of the composite (a) and the temperature variation in the center of the composite (b) at $i > i_f$. Here, curves (——) shows the temperature and curves (- - -) shows the electric field distributions for different values of the dimensionless time: 1 - τ = 10^{-5}, 2 - τ = 10^{-4}, 3 - τ = 3.29 × 10^{-3}).

The simulation results of the critical energies for the most dangerous local instantaneous perturbations and autowave velocities as a function of the current obtained in the framework of both models of the V-I characteristics for different values of the thermal stabilization parameter are compared in Figure 8.14 and 8.15. The dashed curves in Figure 8.10a corresponds to the calculations made on the framework of the ideal V-I characteristic, the solid curves and the marker '-' are depicted the calculations based on the power V-I characteristic. The limiting condition $\varepsilon_q \to \infty$ takes place (Figure 8.14), and the autowave velocities of the thermal instability are zero (Figure 8.15) at the full stabilization currents $i_s(\alpha)$. The critical energies are zero at the quenching currents and the velocities of the autowave are maximum. The vertical straight line at $i = 1$ in Figure 8.15 shows the thermal stabilization boundary existing according to the Stekly condition. As noted above, the limiting full stabilization parameters in the framework of the considered quasi-linear approximation for the composite with the power V-I characteristic are equal to $\alpha_f = 0.4642$ and $i_f = 1.4678$ according to the Formulae (8.14) and (8.15), that is, the range of allowable currents is almost 50% larger than it follows from the model with the ideal V-I characteristic. In general, the presented curves show that the difference between the used V-I characteristics the higher, the lower the thermal stabilization parameter or the closer the transport current to a priori defined critical value. The most noticeable difference occurs when calculating the propagation velocities of the thermal instabilities, even though the full recovery currents do not differ from each other as noted above. As a result, the velocities of the thermal instabilities take maximum values regardless of the thermal stabilization parameter at $i = 1$ in the model with an ideal V-I characteristic. At the same time, the currents at which the velocities of the thermal instabilities may be maximum for composite superconductors with the real V-I characteristics are supercritical. Thereby, Figures 8.14 and 8.15 strongly prove that the supercritical current modes significantly expand the range of the thermal stability conditions of composite superconductors with real V-I characteristic.

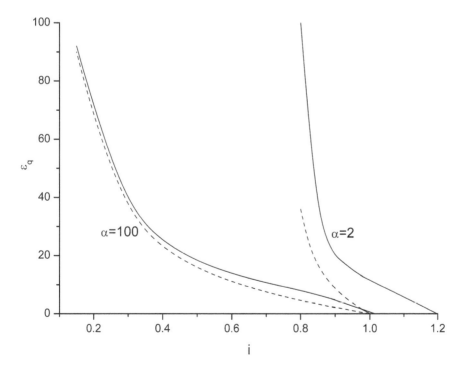

FIGURE 8.14 Dependence of the critical energies on the current at local temperature perturbation ($Z_1 = 1$) and different values of the thermal stabilization parameter: $\varepsilon_q \to \infty$ at $i \to i_s(\alpha)$ and $\varepsilon_q = 0$ at $i = i_q(\alpha)$. Here, (——) - the power V-I characteristic, (- - - -) - the ideal V-I characteristic.

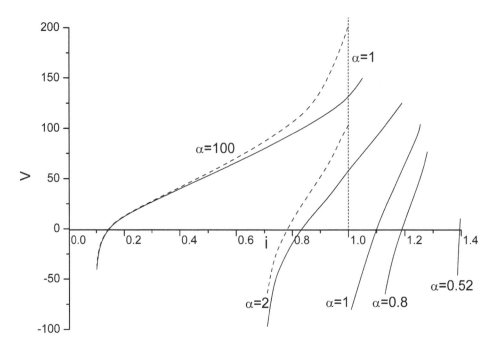

FIGURE 8.15 Thermal autowave propagation velocities as a function of the current for different values of the thermal stabilization parameter ($\alpha_f = 0.4642$ and $i_f = 1.4678$): (———) - the power V-I characteristic, (- - - -) - the ideal V-I characteristic. Here, $V = 0$ at $i = i_s(\alpha)$. The values of $V(\alpha, i)$ are minimum at $i = i_r(\alpha)$ and maximum at $i = i_q(\alpha)$ for each value α.

It should be noted that the power V-I equation that is typical for the superconductors with the multiple crystal lattice defects was used to describe the V-I characteristic of a superconducting composite in this Chapter. Along with the power equation, real V-I characteristics can be described by exponential equations if superconductors have point defects, as noted in Chapter 1. However, a noticeable difference between the thermo-electrodynamic states of superconductors with the power and exponential equations of the V-I characteristics will occur in cases when $n < 10$, as it was shown in Chapter 2. Modern technologies for the production of composites on the basis of HTS make it possible to obtain conductors whose n-value of the V-I characteristic have higher values. Therefore, the conditions discussed above for the thermal stabilization of superconductors with a power V-I characteristic are general, which determine the features of the thermal stabilization of composite superconductors with a continuously increasing V-I characteristic including LTS.

Thus, the qualitative quasi-linear analysis of the thermal stabilization conditions of the composite superconductors with a real V-I characteristic, as a first step of the thermal stability investigation of composite superconductors with real V-I characteristic, showed that they differ not only quantitatively from the corresponding conditions obtained in the Stekly-like approximation with an ideal V-I characteristic but also are characterized by qualitative differences. First of all, the critical current of the composite defined on the basis of a jump-like approximation of its transition from the superconducting state to the normal one, does not correspond to the maximum permissible current that can stably flow in the composite superconductor with a real V-I characteristic. Its real value may be either higher or less than this false value. (Earlier, these important conclusions were formulated in Chapter 6 when analyzing the conditions for the occurrence of the current instabilities). Along with this, the temperature of the resistive transition after exceeding which the transport current begins to be shared between the superconducting core and the matrix in the framework of the model with an ideal V-I characteristic does not have a physical meaning. In the composite superconductors with the continuously increasing V-I characteristics, the current sharing becomes noticeable only at temperatures significantly higher than the temperature of the resistive transition. However, even in this case, the current in the superconducting core of the composite is not equal to the critical current in the whole range of the temperature variation as it is assumed in the theory based on the jump-like S-N transition. Moreover, it exceeds the critical current. As a result, within the framework of the model with an ideal V-I characteristic, the mechanism of the current sharing between the superconducting core and the matrix depends only on the drop of the critical current of the superconductor with the temperature and magnetic field. For superconductors with a continuously increasing V-I characteristic, this mechanism depends on the character of the increase in the V-I characteristic and the matrix resistance with temperature.

In a wide temperature range, the Joule losses in composites with a real V-I characteristic are smaller than those calculated within the model with an ideal V-I characteristic. As a result, the stable supercritical current states are possible in composite superconductors with a real V-I characteristic. Therefore, there are better conditions for the stability of the superconducting state with respect to external thermal perturbations. However, the calculated velocities of the irreversible propagation of the thermal instability along the composite is determined by taking into account the continuous increase of the V-I characteristic are less than the corresponding values calculated within the model of the jump-like S-N transition. In general, the thermal instability propagates along the composite in the form of a thermo-electric autowave both in the subcritical and supercritical current modes.

The above-formulated condition for the full thermal stabilization of composite superconductors takes into account the nonlinear continuous character of the V-I characteristic growth. It significantly expands the range of the currents that are stable to external thermal perturbations during intensive cooling. As a result, the Stekly cryostable condition $\alpha = 1$, according to which the superconducting composite is completely stable to the thermal perturbations of arbitrary energy, does not allow to correctly determine the parameters of the composite with a real V-I characteristic, which ensure its full thermal stabilization. There are possible stable supercritical current modes.

8.3 JOULE DISSIPATION IN COMPOSITE SUPERCONDUCTORS: NONLINEAR APPROXIMATION

The models formulated above allow one to investigate dissipative phenomena in more general cases, for example, in the composite superconductors with significantly nonlinear thermo-physical properties and nonlinear dependence of the critical current density cooled by liquid coolant. Let us consider, as above, a superconducting composite based on $Bi_2Sr_2CaCu_2O_8$ in the silver matrix (Ag/Bi2212) cooled by liquid helium or hydrogen and located in a magnetic field with induction B_a = 15 T and the following parameters: $n = 10$, $S = 1.23 \times 10^{-6}$ m^2, $p = 0.47 \times 10^{-2}$ m^2, $\eta = 0.263$, $E_c = 10^{-4}$ V/m. In this case, one may use the data given before. Accordingly, the critical current density $Bi_2Sr_2CaCu_2O_8$ will be calculated using the Formulae (1.8) and (1.9). They were also used to estimate the corresponding values J_{c0} and T_{cB} needed to describe $J_c(T, B)$ in the linear approximation. As a result of this approximation, one may find $J_{c0} = 1.736 \times 10^9$ A/m^2 and $T_{cB} = 26$ K at $B_a = 15$ T. In this case, the critical current $I_{c0} = \eta J_{c0} S$ is equal to 561 A at $T_0 = 4.2$ and $I_{c0} = 154$ A at $T_0 = 20$ K.

The heat flux $q(T)$ from the surface of the composite to the coolant is described by a jump-like transition from nucleate to film boiling regime after its overheating by ΔT_{cr} using the Formulae (6.37) and (6.33).

To calculate the electrical resistivity of silver, let us use, as above, the data given in (Seeber 1998) accepting $\rho_m(273 \text{ K}) = 1.48 \times 10^{-6} \Omega \times$ m. The results discussed below are obtained at RRR $= \rho_m(273 \text{ K})/\rho_m(4.2 \text{ K}) = 10$.

According to (8.3), the electric field induced in a superconducting composite with the ideal V-I characteristic is described by the formula

$$E(T) = \frac{J\rho_m(T,B)}{1-\eta} \begin{cases} 1, & T > T_{cB} \\ (T-T_{cs})/(T_{cB}-T_{cs}), & T_{cs} \leq T \leq T_{cB} \\ 0, & T < T_{cs} = T_{cB} - (T_{cB}-T_0)J/(\eta J_{c0}) \end{cases} \tag{8.17}$$

The corresponding value in the composite superconductor with the power V-I characteristic is described by relations (8.5) and (8.6).

In Figures 8.16 and 8.17, the solid and dashed curves show the temperature dependencies of the Joule heat release $G = EJ$ at different values of the transport current $I = JS$. The dependencies resulting from the model with the power V-I characteristic are depicted by the solid curves. The dashed curves correspond to the heat release calculated within the model with the ideal V-I characteristic. The dash-dotted lines describe the temperature dependence of the heat flux density $W(T) = q(T)p/S$ removed to liquid helium. Figure 8.16 presents the dependencies $G(T)$ and $W(T)$ for currents that lead to the Joule losses comparable to the heat flux to the coolant. The heat release at the currents leading to a significant excess of $G(T)$ over the values of $W(T)$ in a wide range of the temperature variations of Ag/Bi2212, including the currents above the critical current is shown on the Figure 8.17. The presented results demonstrate the influence of nonlinear properties of Ag/Bi2212 composite on the energy dissipation in it. These results can modify some generally accepted principles of the quasi-linear theory of thermal stabilization with a jump-like S-N transition not only quantitatively but also qualitatively. Let us discuss them.

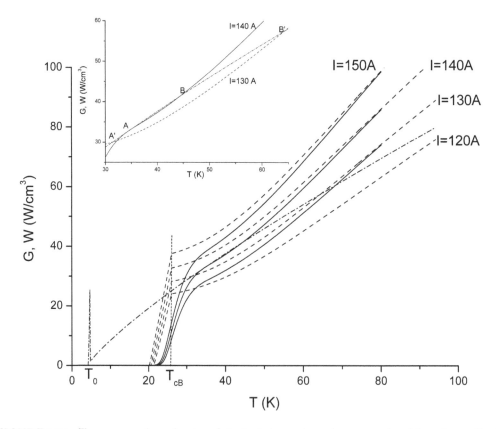

FIGURE 8.16 Temperature dependencies of the Joule heat release (——, - - - -) and heat flux to liquid helium (– · – · –) at currents leading to the nonintensive heat releases: —— - the power V-I characteristic, - - - - the ideal V-I characteristic.

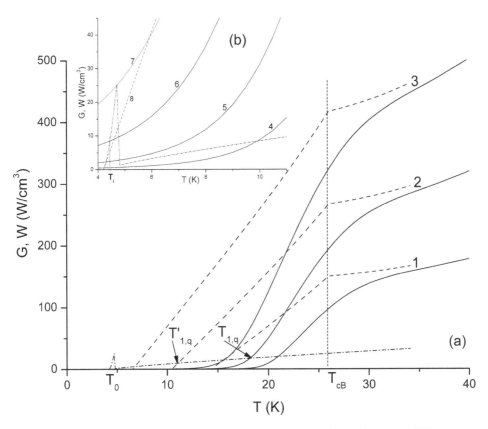

FIGURE 8.17 Temperature dependencies of the Joule heat release (—— - the power V-I characteristic, - - - - - the ideal V-I characteristic) and heat flux into liquid helium (– · – · –) at subcritical (a) and supercritical (b) currents. The power V-I characteristic: 1 - I = 300 A, 2 - I = 400 A, 3 - I = 500 A, 4 - I = 700 A, 5 - I = 800 A, 6 - I = 900 A, 7 - I = 1,009 A; the ideal V-I characteristic: 8 - I = 0.99I_{c0}. Here, $T'_{1,q}$ and $T_{1,q}$ are the temperatures of unstable equilibrium under the infinitely long temperature perturbation.

As noted above, it is necessary, first of all, to take into account the existence of the currents that determine the thermal instabilities occurrence and their development when analyzing conditions of the thermal stabilization of current-carrying elements of SMS. Indeed, the range of operating currents in the composite superconductors is divided into three characteristic areas (see, for example, Figures 7.16, 7.19 and 8.15). They are limited by the full recovery current I_r, the full stabilization current I_s and the maximum permissible current (the critical current I_{c0} for composite with an ideal V-I characteristic or the quenching current I_q for composite with a continuously increasing V-I characteristic). In each of these areas, the development of the thermal perturbation occurs according to a specific scenario. The values of these currents depend on the interrelated variation of the Joule heat release and heat flux into the coolant under the temperature variation.

As shown in Chapter 7, in the currents range $0 < I < I_r$, the superconducting properties of a composite superconductor quickly recover under the action of an arbitrary external thermal perturbation and the composite returns to a stable equilibrium state after the termination of the thermal perturbation. As already discussed, the equilibrium temperature of a composite superconductor is equal to the temperature of the coolant T_0 in the framework of the model with an ideal V-I characteristic because of $G_c(T) = 0$ at $T < T_{cs}$. For a composite superconductor with a continuous V-I characteristic, the corresponding value of the equilibrium temperature for a given transport current is the temperature T_i, at which the equality $W(T_i) = G(T_i)$ in the temperature range close to T_0, is performed since $G(T_i)$

> 0 at $T < T_{cs}$. In the case under consideration, a stable state with a temperature of T_i exists during nucleate boiling regime as shown in Figure 8.17b. Here, the temperature of the thermal equilibrium T_i is depicted for supercritical current $I = 900$ A (Curve 6). In general, the value T_i and therefore, the value $G(T_i)$, depending on the transport current (Figure 8.17b).

Current I_r is determined from the condition $W(T) \geq G(T)$ for all temperatures exceeding the corresponding value of the thermal equilibrium temperature T_0 (for an ideal V-I characteristic) or T_i (for a continuous V-I characteristic). As a result, a new point of the stable equilibrium should appear at current I_r. For example, in the framework of a model with an ideal V-I characteristic, it must exist at $T = T_{cB}$ as shown in Chapter 7. However, according to Figures 8.16 and 8.17, for the composite Ag/Bi2212 with both ideal and power V-I characteristics the condition $W(T) \geq G(T)$ is not performed due to the intensive increase in the resistivity of the stabilizing matrix with temperature. As a result, the self-heating of Ag/Bi2212 will lead to an excess of the Joule heat generation over the heat flux into liquid helium even at low currents after exceeding a certain temperature. (In Figure 8.16, the condition $W(T) \geq G(T)$ is not fulfilled at $T > 100$ K for the current $I = 120$ A). It is clear that the higher the transport current, the more noticeable is the destabilizing role of temperature dependence ρ_m.

Such change with temperature of the dependencies $W(T)$ and $G(T)$ lead to the conclusion that the stable equilibrium temperature stabilizing the unlimited growth of the temperature of composite in the quasi-linear approximation may not exist in composite with the nonlinear properties at $T > T_{cB}$. Therefore, the burnout of the composite even cooled by liquid helium is inevitable after some time due to its self-heating under the supercritical thermal perturbations. It depends on the magnitude of the transport current and the temperature dependence of the matrix resistivity. (Parameters of the cooled superconducting composite, which affect its burnout, are discussed in more detail below). This conclusion is also valid for LTS composites.

In general, to determine the critical energy of the temperature perturbation, a numerical solution of the corresponding heat conduction equation is necessary. However, to find the boundary of stable states during the action of extended perturbations ($l_1 \to \infty$), one can use the diagrams $G(T) - W(T)$, since the mechanism of longitudinal thermal conductivity of a composite superconductor has no stabilizing effect on the thermal stabilization conditions. They allow one to find the temperature of unstable equilibrium $T_{1,q}$ (for composite with a real V-I characteristic) or $T'_{1,q}$ (for composite with an ideal V-I characteristic). As discussed above, they are the maximum allowable temperatures for an infinitely extended temperature perturbation after exceeding which the superconducting state is destroyed. As an illustration, the corresponding values $T_{1,q}$ and $T'_{1,q}$ defined in the framework of both models of V-I characteristics are shown in Figure 8.17a for the current Mode 2.

It should also be noted that consideration of the nonlinear rise of matrix resistivity with temperature can lead to the multistable thermal states. These modes are shown on the inset to Figure 8.16. In this case, they occur at a current of 130 A (in the temperature range between points A' and B'), if the ideal V-I characteristic is used and at a current of 140 A (in the temperature range between points A and B) for composite with the power V-I characteristic. Here, Points A and A' correspond to stable states, and points B and B' are unstable ones. Similar states also occur when liquid hydrogen is used as a coolant.

The full stabilization current I_s, as discussed in Chapter 7 in the framework of the quasi-linear approximation, determines the current at which the velocity of the thermal instability after a certain time becomes zero. Accordingly, the autowave velocity of the thermal perturbation propagation along the composite is negative at $I_r < I < I_s$ (Figure 8.15). Therefore, the superconducting state must be stable to arbitrary thermal perturbations at $I < I_s$. The value I_s follows from the theorem of equal areas and is determined from the integral equality

$$\Sigma(T_s, I_s) = \int_{T_0}^{T_s} [G(T, I_s) - W(T)]\lambda_k(T)dT = 0 \qquad (8.18)$$

where T_s is the maximum equilibrium temperature of the superconducting composite, which satisfies the conditions $d\Sigma(T_s, I_s)/dT = 0$ (Figure 8.18).

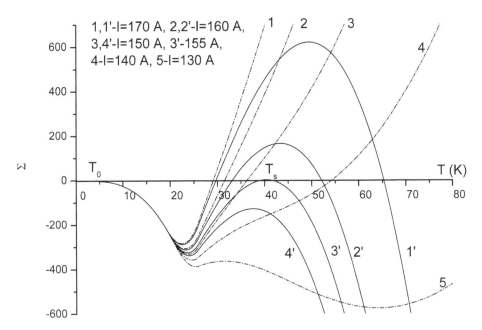

FIGURE 8.18 Determination of the full stabilization current of Ag/Bi2212 with the ideal V-I characteristic, which is cooled by liquid helium: —— - $\rho_m = \rho_m(4.2\ K)$, – · – · – - ρ_m depends on temperature.

The values Σ calculated at different currents as a function of the temperature are shown in Figure 8.18 for liquid helium cooling mode. The simulations are performed in the framework of the ideal V-I characteristic. The dash-dotted curves are obtained under the assumption that the matrix resistivity depends on the temperature, and the solid curves describe the values of $\Sigma(T, I)$ when value ρ_m is constant and equal to $\rho_m(4.2\ K)$. It is seen that for Ag/Bi2212 cooled by liquid helium, the condition (8.18) is not satisfied if ρ_m continuously increases with temperature. As a result, the value of T_s is missing. At the same time, the condition (8.18) is satisfied at $I \approx 155$ A if $\rho_m = $ const. When $\rho_m = $ const, there also exists the full recovery current I_r. The full stabilization current is also absent for Ag/Bi2212 with the power V-I characteristic, as it follows from Figures 8.16 and 8.17.

Thus, the consideration of the nonlinear dependencies $\rho_m(T)$ and $q(T)$ can lead to such modes when the autowave velocity of thermal instabilities is only positive and monotonically increase with the current. Underline that according to the results shown in Figure 8.17, the following inequality $G(T) \gg W(T)$ takes place for currents with high density in the temperature range $T > T_{cB}$ due to the transition from the nucleate boiling regime to film boiling one. Therefore, the temperature of Ag/Bi2212 cooled by liquid helium will increase under practically adiabatic conditions in its hottest area after the occurrence of the thermal instability. This effect will be more noticeable, the higher the current which, however, is limited by the quenching current.

Figure 8.17b shows that when using liquid helium as a coolant, the thermal instability of Ag/Bi2212 occurs after heating to a temperature at which a transition from nucleate boiling regime of liquid helium to film boiling one occurs at $G(T) > W(T)$ for all $T > T_0$. Therefore, the temperature boundary of stable modes is described by the equality $T_q = T_0 + \Delta T_{cr}$. As a result, there are stable supercritical current states despite the finite overheating of the superconductor, which is equal to ΔT_{cr}. In particular, the destruction of a stable resistive state of the Ag/Bi2212 under consideration occurs at a current $I_q = 1010$ A, which is 1.8 times higher than the critical current. In this case, it turns out that the allowable increase in the electric field is substantially higher than a priori set value E_c. These states arise due to the intensive cooling of Ag/Bi2212 by liquid helium.

The full recovery and stabilization currents, which are absent when Ag/Bi2212 is cooled by liquid helium, arise when it is cooled by liquid hydrogen (Figures 8.19 and 8.20), which provides a more effective heat removal. Indeed, as follows from Figure 8.19a, in the framework of models with a power V-I characteristic (the solid curves) and an ideal V-I characteristic (the dashed curves), there are states for which the condition $W \geq G$ will be fulfilled at $T > T_i$ or $T > T_0$, respectively. Therefore, the full recovery current I_r is nonzero in this case and Ag/Bi2212 will return to the equilibrium state that exists at T_i or T_0 in the mode of nucleate hydrogen boiling for all $0 < I < I_r$ after the termination of any thermal perturbation. Besides, the calculation of the values $\Sigma(T, I)$ obtained according to (8.18) and presented in Figure 8.20 show the existence of the full stabilization current I_s in the hydrogen film boiling regime ($I_s = 126.2$ A for the power V-I characteristic and $I_s = 117.6$ A for the ideal V-I characteristic). Accordingly, the autowave values of the thermal disturbance velocities will be negative at $I_r < I < I_s$ ($I_r = 94$ A for the power V-I characteristic and $I_r = 61$ A for the ideal V-I characteristic) after the action of any thermal disturbancies since the total heat release in Ag/$Bi_2Sr_2CaCu_2O_8$ will not exceed the total heat flux to the coolant. In other words, Ag/$Bi_2Sr_2CaCu_2O_8$ will be stable at $0 < I < I_s$ after the termination of any thermal disturbance. The absolute value of this velocity maximum at I_r and it becomes zero at I_s. It is easy to understand that the differencies in the values I_r and I_s, calculated on the basis of both models of V-I characteristic depend on the change character with the temperature of the critical current density.

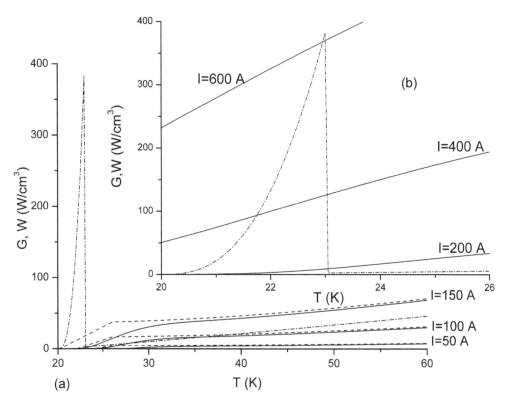

FIGURE 8.19 Temperature dependencies of the Joule heat release (——, - - - -) and heat flux to liquid hydrogen (– · – · –) at subcritical (a) and supercritical (b) currents: —— - the power V-I characteristic; – – – – the ideal V-I characteristic.

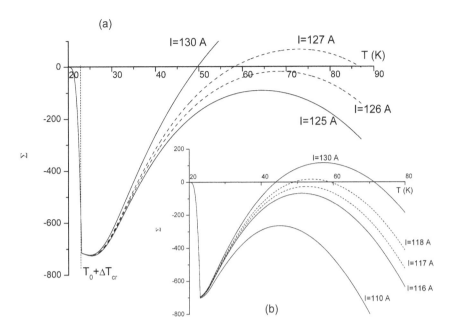

FIGURE 8.20 Calculation of the full stabilization current when Ag/Bi2212 is cooled by liquid hydrogen: (a) - the power V-I characteristic, (b) - the ideal V-I characteristic.

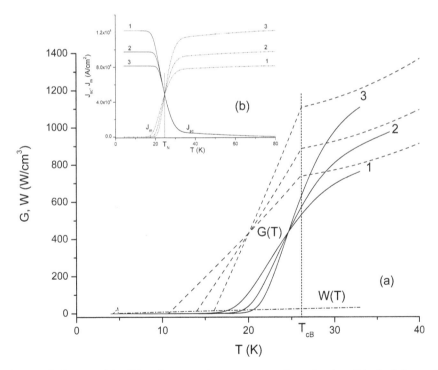

FIGURE 8.21 Influence of the filling factor on the temperature dependencies of the Joule heat release at I = 600 A (a) and variation with temperature of the current density in the superconductor J_{sc} and the matrix J_m (b): 1 - η = 0.4, I_{c0} = 854 A, 2 - η = 0.5, I_{c0} = 1067 A, 3 - η = 0.6, I_{c0} = 1280 A; ——, – ·· – ·· – the power V-I characteristic, - - - - the ideal V-I characteristic.

At $I > I_s$ and under the supercritical perturbations, the thermal instabilities will irreversibly propagate along the composite superconductor with the positive autowave velocity, the value of which will increase with increasing current up to quenching current. This current for Ag/Bi2212 cooled by liquid hydrogen is equal to $I_q = 607$ A. For liquid hydrogen, it is also determined by the transition from nucleate boiling regime to film boiling one (Figure 8.19b) as during cooling of Ag/Bi2212 by liquid helium. It is 3.9 times higher than the corresponding value of the critical current for Ag/Bi2212 since the nucleate boiling regime of liquid hydrogen provides an efficient heat removal of the Joule energy losses.

Thus, a priori set values E_c and I_c have no physical meaning as the critical parameters for the composite superconductors cooled by liquid helium and especially liquid hydrogen. Previously, this conclusion was formulated for cryocooling conditions in Chaptyer 6.

The simulation results presented in Figures 8.17 and 8.19 lead to the conclusion that burnout of technical superconductors becomes more possible at increasing their critical current. This conclusion is vividly confirmed by the dependencies $G(T)$ and $W(T)$ shown in Figure 8.21a for Ag/Bi2212 cooled by liquid helium. They show the variation of the Joule energy losses with temperature when the filling factor varies resulting in a corresponding variation of the critical current. The calculations were performed within the framework of both models of V-I characteristic: ideal (the dashed curves) and power (the solid curves). It is seen that the heat transfer to liquid helium depicted by the dash-dotted curve has almost no effect on its thermal state of the composite in the most heated parts in the temperature range $T > T_{cB}$ (in film boiling mode) since the values of $G(T)$ are significantly higher than $W(T)$ at high critical currents. In this case, the higher the critical current, the higher the difference. Accordingly, the probability of rapid burnout of composite superconductors increases.

Let us note the existence of the characteristic temperature of T_N. The heat release in a composite superconductor does not depend on the filling factor at that temperature. It follows from the relationship $J = \eta J_{sc} + (1 - \eta)J_m$, according to which the condition $J_{sc} = J_m = J$ is observed both for superconductors with ideal and power V-I characteristics. This mode is shown in Figure 21b for superconducting composite with the power V-I characteristic. To determine the value T_N, one may use equation $E_c[J/J_c(T_N, B)]^n = J\rho_m(T_N, B)$. The temperature T_N appears due to different influences of the filling factor on the current density in the superconducting core and matrix (Figure 21b), hence on the total heat release in a composite superconductor. Correspondingly, an increase in the filling factor is accompanied by a decrease in the Joule losses in the temperature range $T < T_N$ due to the corresponding decrease in J_m. Conversely, the heat release increases with an increase in the filling factor at $T > T_N$ since the current in the stabilizing matrix increases.

The results presented in Figures 8.16-8.21, demonstrate the inevitable differencies in the calculations between the models based on ideal and real V-I characteristics. As noted above, they are primarily based on the specifics of the differential resistivity variation of a composite superconductor. Indeed, in the framework of the model with an ideal V-I characteristic, it jumps from zero to an infinitely large value. In superconducting composites with a real V-I characteristic, their differential resistivity varies continuously depending on the n-value and the temperature dependence J_c. Besides, the current sharing depends on the resistivity of the stabilizing matrix. These regularities lead to the characteristic thermal effects affecting the formation of real dependencies $E(T)$, therefore on the current sharing. Figures 8.22 and 8.23 demonstrate these features.

Figure 8.22 shows the results of the electric field calculation as a function of the temperature obtained for Ag/Bi2212 cooled by liquid helium at subcritical current $I = 500$ A and various n-values. On the inset to Figure 8.22, the temperature dependencies $E(T)$ are presented in more detail in the low overheating area. In Figure 8.23, the corresponding curves showing the features of the current sharing between the superconducting core and the matrix are depicted. One can see that the dependence $E(T)$ calculated within the model with the ideal V-I characteristic almost linearly increases with temperature at $T_0 < T < T_{cB}$ because in this case $\rho_m \sim$ const at $T < 20$ K. At $T = T_{cB}$, the dependence $E(T)$ has a kink since the critical current is zero in this case and the whole transport current flows in the matrix. Therefore, the electric field varies in accordance with the temperature dependence of its matrix resistivity at $T > T_{cB}$. The electric field calculated in the framework of the model with the power V-I characteristic gradually increases with temperature in accordance with the continuous change in the differential resistivity of the superconductor and matrix. However, its value turns out to be less than the electric field determined in the framework of the model with an ideal V-I characteristic. Accordingly, the current in the superconducting core I_{sc} calculated in the framework of the model with the power V-I characteristic at temperatures above the resistive transition temperature T_{cs} is always higher than the critical current of Ag/Bi2212 (Figure 8.23). Therefore, the currents in the stabilizing matrix I_m calculated within the framework of both models differ accordingly. These features are observed even at very high n-values, for example at $n = 100$ that corresponds to the practically jump-like transition of the superconductor to the normal state. In general, the difference between the models increases markedly in the temperature range $T_{cs} < T < T_{cB}$ with decreasing n-value. Consequently, a model with the power V-I characteristic will always lead to underestimated values of the electric field, hence the lower values of the heat release in a wide range of the temperature variations ($T > T_{cs}$) of composite superconductors. As a result, a theoretical analysis of the thermal stability conditions performed in the framework of the model with an ideal V-I characteristic will inevitably lead to overestimated values of the autowave velocity of thermal instability. At the same time, since the difference between the computational models decreases in the temperature region $T > T_{cB}$, they will lead to close estimates of the burnout conditions of superconducting composites.

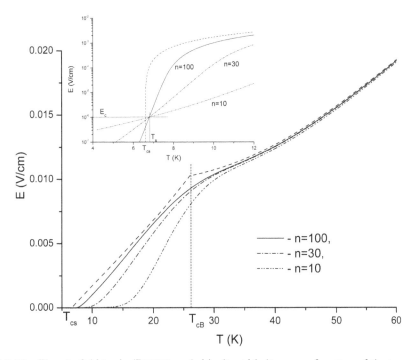

FIGURE 8.22 Electric field in Ag/Bi2212 cooled by liquid helium as a function of the temperature for different n-values: —, – · – · –, – ·· – ·· – the power-law V-I characteristic, – – – - the ideal V-I characteristic.

It should be noted that the composite superconductors with the power V-I characteristic have the characteristic temperature T_s. The curves $E(T)$ intersect independently of the n-value at this temperature, as shown on the inset to Figure 8.22. It is easy to find that $E = E_c$ in this case. Then, the temperature at which this condition is fulfilled follows from the solution of the transcendental equation $J = \eta J_c(T_s) + (1 - \eta)E_c/\rho_m(T_s, B)$. Temperature T_s', at which the condition $E = E_c$ fulfillment for a superconductor with an ideal V-I characteristic, is determined by solving equation,

$$E_c = \frac{J\rho_m(T_s', B)}{1-\eta} \frac{T_s' - T_{cs}}{T_{cB} - T_{cs}}.$$

Therefore, at temperatures $T > T_s$ (for superconductor with ideal V-I characteristic at $T > T_s'$), the current sharing between the superconducting core and the stabilizing matrix occurs at the supercritical values of the electric field even at low overheating relative to the temperature of the resistive transition. This conclusion proves the existence of stable supercritical states in technical superconductors.

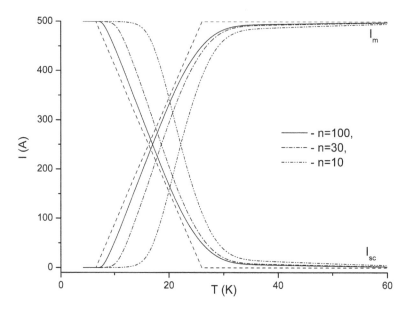

FIGURE 8.23 Effect of n-value on the current sharing between the superconducting core and the matrix: ——, – · – · –, – ·· – ·· – the power V-I characteristic, - - - - the ideal V-I characteristic.

To understand the character of the variation in dependencies $E(T)$, let us evaluate them. At low overheating, the current in the stabilizing matrix is practically absent (Figure 8.21b), that is $J \sim \eta J_{sc}$. Therefore, there is the following relation

$$\frac{E}{E_c} \cong \left(\frac{J}{\eta J_0 (1 - T/T_c)^\gamma \beta_1} \right)^n$$

according to (6.12) and (1.8). Finding the logarithm of this equality, one may obtain

$$\ln\frac{E}{E_c} = n\left[\ln\frac{J}{\eta J_0 \beta_1} - \gamma \ln\left(1 - \frac{T}{T_c}\right) \right]$$

Since $T/T_c \ll 1$, one can expand $\ln(1 - T/T_c)$ into the power series and get

$$\ln\frac{E}{E_c} = n\left[\ln\frac{J}{\eta J_0 \beta_1} + \gamma\left(\frac{T}{T_c} + \frac{T^2}{2T_c^2} + \dots \right) \right]$$

Then,

$$E(T) = E_c \left(\frac{J}{\eta J_0 \beta_1} \right)^n \exp\left[n\gamma\left(\frac{T}{T_c} + \frac{T^2}{2T_c^2} + \dots \right) \right]$$

Since β_1 slightly depends on the temperature at $T/T_c \ll 1$ (this follows from (1.8)), the initial section of the dependence $E(T)$ increases exponentially with temperature. These conclusions are confirmed by the insert to Figure 8.22.

To evaluate $E(T)$ at $T \sim T_{cB}$, let us take into account that in this case $\rho_m \sim$ const. Therefore, the following equality $i = j_c(\theta)e^{1/n} + e/\varepsilon$ may be written in the dimensionless form according to the (6.12). Here, $e = E/E_c$, $i = J/(\eta J_0)$, $\theta = T/T_c$, $\varepsilon = \eta J_0 \rho_m / ((1-\eta)E_c)$, $j_c(\theta) = (1-\theta)^\gamma \beta_1$. The substitution $u = 1 - e/(i\varepsilon)$ leads to equation

$$j_c(\theta) = \frac{i}{(i\varepsilon)^{1/n}} \frac{u}{(1-u)^{1/n}}$$

Since $u \ll 1$, then one can expand the factor $(1 - u)^{-1/n}$ into the power series and obtain

$$j_c(\theta) = \frac{i}{(i\varepsilon_1)^{1/n}} \left(u + \frac{u^2}{n} + \frac{n+1}{2n^2} u^3 + ... \right)$$

Then, in the linear approximation with respect to u, one may find

$$\frac{E}{E_c} = i\varepsilon \left[1 - \frac{(i\varepsilon)^{1/n}}{i} j_c(\theta) \right] = i\varepsilon \left[1 - \frac{(i\varepsilon)^{1/n}}{i} (1-\theta)^\gamma \beta_1 \right] \tag{8.19}$$

Using the expansion in the power serious of θ, it may be written as follows

$$\frac{E}{E_c} = i\varepsilon \left[1 - \frac{(i\varepsilon)^{1/n}}{i} \beta_1 (1 - \gamma\theta + ...) \right]$$

Considering the nonlinear expansion terms in u allows one to estimate the change in the dependence $E(T)$ more accurately. In particular, the quadratic approximation can be written as follows

$$\frac{E}{E_c} = i\varepsilon \left[1 - \frac{(i\varepsilon)^{1/n}}{i} j_c + \frac{(i\varepsilon)^{2/n}}{ni^2} j_c^2 \right] \tag{8.20}$$

Therefore, the values of $E(T)$ increase almost in proportion to the temperature decrease of the critical current density at $T_0 \ll T \leq T_{cB}$, namely,

$$E(T) \approx \left[J - \left(\frac{J\rho_m}{(1-\eta)E_c} \right)^{1/n} \eta J_c(T) \right] \frac{\rho_m}{(1-\eta)} \tag{8.21}$$

as the temperature of the composite superconductor increases. In the limiting case $n \to \infty$, the Formula (8.21) describes the values $E(T)$, which coincide with the values that follow from the model with an ideal V-I characteristic. If J_c decreases linearly with temperature, then the electric field will increase with the temperature almost linearly, as discussed above.

Taking into account (8.21), it is easy to find the variation of the current density in the superconducting core with the temperature. The corresponding formula may be written as follows

$$J_{sc}(T) = \left(\frac{J}{(1-\eta)J_m} \frac{E}{E_c} \right)^{1/n} \eta J_c(T)$$

It justifies the results shown in Figure 8.23. Namely, the current in a superconductor with a real V-I characteristic is always greater than its critical current except for the limiting case $n \to \infty$, when $J_{sc} \to \eta J_c$ due to the observance of the limiting transition to the model with an ideal V-I characteristic.

In the framework of the model with an ideal V-I characteristic, the current flows only in the stabilizing matrix at $T > T_{cB}$. Therefore, the dependence $E(T)$ will change in accordance with the temperature dependence $\rho_m(T)$. A similar conclusion, namely $E \sim J\rho_m/(1 - \eta)$ follows from (8.21) if the condition is met

$$J \gg \left(\frac{J\rho_m}{(1-\eta)E_c} \right)^{1/n} \eta J_c(T)$$

This condition makes it possible to estimate the temperature of a composite superconductor with a power V-I characteristic when the main part of the current will flow in the matrix.

Thus, the Joule heat release in composite superconductors with a real V-I characteristic will change as follows at increasing its temperature. At small overheating ($T \ll T_{cB}$), the dissipated energy exponentially increases with temperature and the main part of the current flows through the superconductor. As the temperature increases, the transport current gradually goes to a nonsuperconducting matrix (Figure 8.23), and the $G(T)$ dependence begins to increase with temperature according to a law which practically complies with a proportional decrease in the critical current density with temperature. This regularity is due to the stabilizing effect of a nonsuperconducting matrix avoiding burnout of composite superconductors in the temperature range up to T_{cB}. At further increase in temperature ($T > T_{cB}$), the character of the variation $G(T)$ mainly depends on the temperature relationship of the resistivity of stabilizing matrix. At high overheating, the dependence $\rho_m(T)$ for silver-coated superconductors can be described by a linear dependence with good accuracy. Therefore, the dependence $G(T)$ increases with the temperature almost linearly.

The analysis performed shows that the full recovery and full stabilization currents determining the range of currents in which the superconducting state is stable to arbitrary thermal perturbations may be absent in the intensely cooled current-carrying elements by liquid coolant. This feature of the thermal stabilization conditions will be observed due to the intensive increase of the stabilizing matrix resistivity with temperature when the increase in the Joule heat release exceeds the temperature increase of the heat flux to the coolant at temperatures above the critical temperature of the superconductor.

The maximum permissible currents of superconductors with a real V-I characteristic cooled by liquid helium or hydrogen follow from the violation condition of the nucleate boiling regime. As a result, when the composite superconductors are cooled with liquid helium or hydrogen, the stable values of the currents are several times higher than a priori set value of the critical current.

An analysis of the interconnected increase in the Joule heat release and heat flux into the coolant shows that a change in the thermal state of technical superconductors cooled by liquid coolant may occur under conditions close to the adiabatic after instability has occurred. This feature leads to the burnout possibility of technical superconductors. In this case, the probability of its burnout increases with an increase in the critical current density of a superconductor.

The formation of the thermal modes of technical superconductors with real V-I characteristic is a direct consequence of the differential resistivity change of superconductors that leads to the corresponding development regularities of dissipative phenomena in them. Therefore, the current in the superconducting core with a real V-I characteristic is always greater than its a priori set critical current when the current shares between the superconducting core and the stabilizing matrix. At the same time, the model of a superconductor with an ideal V-I characteristic leads to the overestimated values of the currents flowing in the stabilizing matrix at temperatures higher than the temperature of the resistive transition. Therefore, the Joule losses calculated in the framework of a model with a continuously rising V-I characteristic are always less than the corresponding values determined according to the model with an ideal V-I characteristic in a wide range of temperature variations of technical superconductors.

8.4 NONLINEAR EFFECTS OF THE THERMAL MECHANISMS OF SUPERCONDUCTIVITY DESTRUCTION

Let us discuss the characteristic features of the irreversible propagation of thermal perturbations along the superconducting composite based on $Bi_2Sr_2CaCu_2O_8$ in a silver matrix and in a magnetic field $B_a = 15$ T within the framework of the linear and nonlinear approximations formulated in the preceding section. Let us set the total length of the composite equal to $2l_k = 1$ m and determine the temperature change over time and along length of the composite ($t > 0$, $-l_k < z < l_k$), from whose surface the heat flow $q(T)$ is removed to the coolant with a temperature T_0, by solving the nonlinear one-dimensional heat conduction equation of the form

$$C_k(T)\frac{\partial T}{\partial t} = \frac{\partial}{\partial z}\left[\lambda_k(T)\frac{\partial T}{\partial z}\right] - q(T)\frac{p}{S} + EJ \tag{8.22}$$

with the initial-boundary conditions

$$T(x,0) = \begin{cases} T_1 = \text{const}, & |z| \leq l_1 \\ T_0 = \text{const}, & l_1 < |z| < l_k \end{cases}, \quad \frac{\partial T}{\partial z}(0,t) = 0, \quad T(l_k,t) = T_0 \tag{8.23}$$

considering the symmetric temperature distribution in the composite.

Here, the electric field in a superconducting composite with an ideal V-I characteristic is described by the Formula (8.17) at the linear temperature dependence of the critical current density used above. The electric field distribution for a superconductor with a power V-I characteristic follows from (8.5) and (8.6).

In Equation (8.22), the heat capacity of the composite will be defined as follows $C_k(T) = \eta C_s(T) + (1 - \eta) C_m(T)$ taking into account the heat capacity of the superconductor C_s and matrix C_m. Here, the temperature dependence of the heat capacity $Bi_2Sr_2CaCu_2O_8$ was calculated using the first Formula (4.34), and the heat capacity of silver C_m was calculated in the same way as it was done in (Dresner 1993b). The coefficient of the thermal conductivity in the longitudinal direction of the composite will be described by the formula $\lambda_k(T) = \eta\lambda_s(T) + (1 - \eta)2.45 \times 10^{-8}T/\rho_m(T, B)$ taking into account the Wiedemann-Franz law. The thermal conductivity coefficient of $Bi_2Sr_2CaCu_2O_8$ will be approximated by the second formula in the expression (4.34).

To describe the heat flux to the coolant, different modes of cooling consider. First, assume that liquid helium or hydrogen at atmospheric pressure are used as a coolant. In these cases, the temperature dependencies $q(T)$ are described by the Formulae (6.37) and (6.33), respectively. Second, the thermal states of Ag/Bi2212 are also investigated for nonintensive cooling conditions that arise when cryocoolers or gaseous coolant are used, and also under indirectly cooling current-carrying elements due to the heat flux redistribution between them. In this case, the heat flux from the surface of the composite into the coolant is described by the formula

$$q(T) = h(T - T_0) \tag{8.24}$$

It was used to analyze the thermal states of the composite under consideration at $h = 10$ W/(m² × K) when the coolant temperature is set equal to $T_0 = 4.2$ K or $T_0 = 20$ K.

According to the results discussed above, the irreversible propagation of thermal instability in the composite occurs in cases when the energy of the initial perturbation exceeds the critical value. Therefore, the initial perturbation parameters l_1 and T_1 were set so as to ensure the formation of unstable states during calculations.

The calculation results of the instability velocity $v_n(t)$ during its irreversible propagation along the Ag/Bi2212 composite cooled by liquid helium are presented in Figure 8.24. Calculations of these dependencies were carried out on the basis of the kinetic analysis of temperatures T_{cs} and $T_0 + 1$ as well as an equipotential E_c. In Figure 8.24a, the change in time of the propagation velocities of isotherms T_{cs} (———) and $T_0 + 1$ (. . . .) for both subcritical and supercritical currents are shown. For the current $I = 500$ A, calculations were performed for perturbations with different initial temperatures. It follows from Figure 8.24 that the thermal instability after the action of the supercritical temperature perturbation propagates along the composite with the constant velocity at the expiration of the transition period as in the quasi-linear approximation, that is the autowave mode is formed in this case. Curves describing time variation of the propagation velocity of the isotherm T_{cs} (———) and equipotential E_c (. . . .) are presented in Figure 8.24b. The results of these calculations show that the used approximations are equivalent to determine the velocity of the autowave.

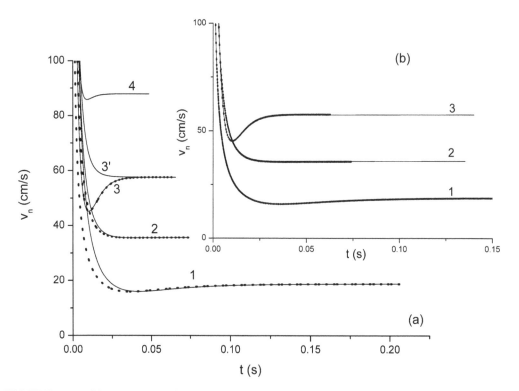

FIGURE 8.24   Change in time of thermal instability velocity in the composite cooled by liquid helium at the local temperature perturbations ($l_1 = 5 \times 10^{-3}$ m) with various initial perturbation temperature T_1: 1 - $I = 300$ A, $T_1 = 30$ K; 2 - $I = 400$ A, $T_1 = 30$ K; 3 - $I = 500$ A, $T_1 = 20$ K; 3′ - $I = 500$ A, $T_1 = 30$ K; 4 - $I = 600$ A, $T_1 = 20$ K. Here, (a) - the calculation of the propagation velocity of isotherms T_{cs} (———) and $T_0 + 1$ ($\cdots$); (b) - calculation of the isotherm propagation velocity T_{cs} (———) and equipotential E_c ($\cdots$).

Figure 8.25 demonstrates the kinetics of the isotherm T_{cs} in the case of the intensive and nonintensive cooling of the composite at $T_0 = 4.2$ K and subcritical currents ($I_{c0} = 561$ A). Comparing the results presented in Figure 8.25, it is easy to see that the autowave velocities of thermal instabilities do not differ significantly from each other even if there is a significant difference in the intensity of the cooling modes. This leads to an important conclusion that liquid helium has no effective stabilizing effect on the irreversible propagation of thermal instability along the composite at a high density of the transport current. This feature of the thermal regime development was already mentioned in the previous section.

Figure 8.26a depicts the features of the thermo-electric autowaves formation in a composite with a nonlinear dependence of the resistivity on temperature when it is cooled by liquid helium. It is seen that the temperature and the electric field are constantly increasing in the center of the perturbation. This means that the thermal equilibrium temperature of the composite is absent in the most heated part during its transition to the normal state. As follows from Figure 8.17, this is explained by the fact that the heat release significantly exceeds the heat transfer to the coolant. Therefore, the thermal state of the composite in the most heated part approaches the adiabatic one. This regularity explains the burnout possibility of the intensively cooled composite if one does not begin to remove timely the current from it. Similar modes arise when the composite is cooled by liquid hydrogen (Figure 8.26b). However, in this case, the increase in the temperature of the composite is not so intense for understandable reasons although it is constantly increasing.

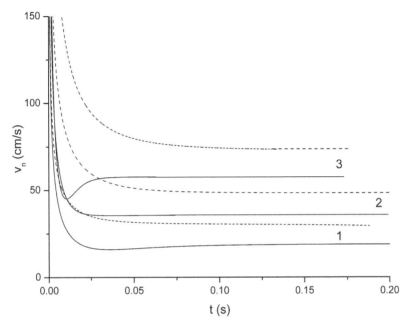

FIGURE 8.25 Change in time of the isotherm T_{cs} velocity at the local temperature perturbation with parameters $l_1 = 5 \times 10^{-3}$ m and $T_1 = 30$ K (1 - $I = 300$, 2 - $I = 400$ A, 3 - $I = 500$ A) under various cooling conditions: —— - liquid helium, - - - - - nonintensive cooling.

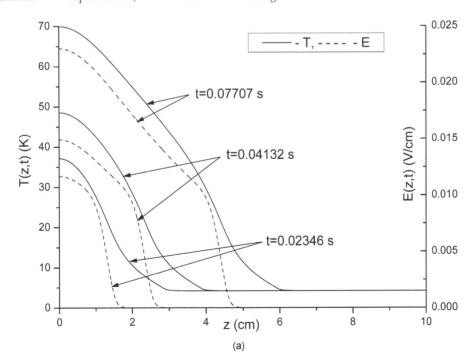

(a)

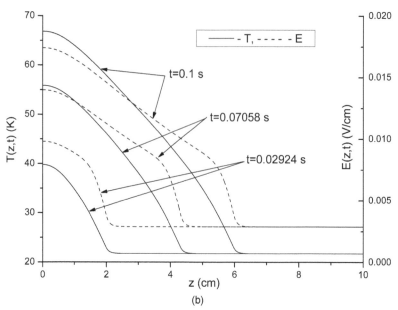

(b)

FIGURE 8.26 Formation of a thermo-electric autowave as a result of the local temperature perturbation of the composite cooled by liquid helium (a - I = 500 A, l_1 = 5 × 10⁻³ m, T_1 = 20 K) and liquid hydrogen (b - I = 400 A, l_1 = 5 × 10⁻³ m, T_1 = 25 K).

Thus, according to the nonlinear temperature approximation, the occurrence and development of the thermal instabilities in technical superconductors have some specific features that must be taken into account, both in the theoretical and experimental analysis of the thermal stabilization conditions. This results in the following laws, which are characteristic for the propagation of thermal instabilities.

In Figure 8.27, the dependencies of the autowave propagation velocity of the thermal instabilities along the composite cooled by liquid helium on the current are presented. The same dependencies are shown in Figure 8.28 for the composite cooled by liquid hydrogen. In both cases, the simulations were performed both for the ideal V-I characteristic at $0 < I < I_c$ and for the power V-I characteristic at subcritical ($0 < I < I_c$) and supercritical currents ($I_c < I < I_q$). For the composite under consideration, which is cooled by liquid helium or liquid hydrogen, the quenching currents are I_q = 1,010 A and I_q = 608 A, respectively, as noted above. The insets to each figure show the corresponding dependencies in detail for a region of small currents. It should be noted that the determination of the instability propagation velocities in a composite cooled by liquid helium at currents less than 100 A leads to a significant increase in the calculation time since the autowave formation time increases significantly at low currents. Nevertheless, let us point out that Curves 1 and 2 in Figure 8.27 have a common beginning at $I = 0$ since the calculations performed according to the equal area theorem show that the full stabilization current I_s and full recovery current I_r is zero despite the use of liquid helium as a coolant. In other words, there is no full stabilization region in this cooling mode.

To demonstrate the influence of the nonlinear character of the matrix resistivity increase and its thermal conductivity with temperature on the velocity of the instability propagation, in Figure 8.27, the calculation results of $V(I)$ performed for different temperature dependencies ρ_m and λ_k are presented. Curves 1, 2 and 5 correspond to the calculation under the models in which the temperature dependencies ρ_m and λ_k were taken into account. Curve 3 was obtained at ρ_m(4.2 K) and Curve 4 was calculated under the condition that the values ρ_m and λ_k do not depend on temperature and are assumed to be equal to ρ_m(4.2 K) and λ_k(4.2 K). In these cases, numerical experiments were performed within the framework of an approximation with the ideal V-I characteristic.

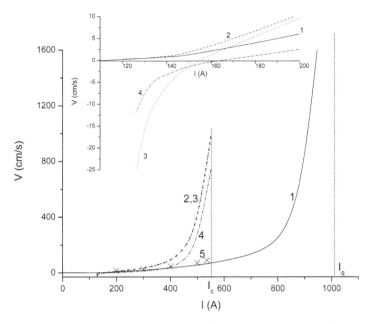

FIGURE 8.27 Propagation autowave velocities of the thermal perturbation as functions of the current in a composite cooled by liquid helium (Curve 1 – V-I characteristic with the power equation; Curves 2, 3, 4 – the ideal V-I characteristic) and under the nonintensive cooling (Curve 5 – the power V-I characteristic).

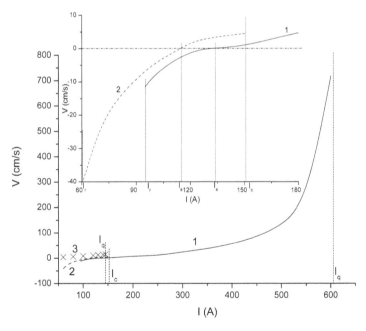

FIGURE 8.28 Autowave velocity as a function of current when the composite is cooled by liquid hydrogen (Curve 1 – V-I characteristic with the power equation; Curve 2 – the ideal V-I characteristic); and at nonintensive cooling when $T_0 = 20$ K (Curve 3 – the power V-I characteristic).

The presented results show that the autowave propagation velocity of the thermal instabilities in a superconducting composite with a silver matrix cooled by liquid helium is not negative over the whole range of the transport current variation (curves 1, 2 in Figure 8.27) if the nonlinear rise of the electrical resistivity of a matrix with temperature is taken into consideration. As noted above, the basis of this regularity is an intense increase in the Joule heat release with temperature compared to an increase in the heat flux with temperature removed into liquid helium at the film boiling regime ($T > 4.8$ K), which exists in the hottest part of the composite. As a result, the heating of the composite is close to an adiabatic mode in this area during instability development. Moreover, as noted above, the value of I_s in this case is zero, which is also typical for uncooled superconducting composites. Consequently, the conclusion about the adiabatic character of the modes in the range of the maximum heating, which are observed in the composite cooled by liquid helium, is valid up to a priori set value of its critical current. In other words, in a wide range of current variations, liquid helium does not provide effective thermal stabilization of composites carrying transport current with high density. In this case, burnout of them is highly likely that is well-known from the application practice.

It should also be noted that the calculation difference between the models with an ideal and a power V-I characteristics increases with increasing current and becomes quite noticeable (Figure 8.27) at currents close to the critical current of the composite since the current instabilities may not occur at significantly greater currents than a priori set value of I_c when a real V-I characteristic is taken into account.

When using liquid hydrogen as a coolant, there is a noticeable difference in the values of the full recovery currents I_r (see inset to Figure 8.28) determined within both V-I characteristic models due to the corresponding difference in the calculated dependencies of the critical current density on temperature. Besides, the range of the currents stably flowing in the composite differs significantly. As a result, there is a significant range of the supercritical currents where the transition to the normal state is not sharp but takes the form of the autowave, which propagates at high velocity due to the occurrence of states close to adiabatic. At the same time, the calculated values of the autowave velocity determined within the framework of both models of the V-I characteristic in the current range $I_r < I < I_c$ are slightly different from each other.

In Figure 8.29a, the depicted curves describe the time change of the composite temperature cooled by liquid helium in its most heated part under the action of a local temperature perturbation ($l_1 = 10^{-2}$ m) with different temperatures of the initial heating of the composite. The simulation was carried out for both subcritical and supercritical currents. In the first case, it was performed for the ideal V-I characteristic and in the second case, simulations were made for the power V-I characteristic. The results show that the transition of the composite to the normal state at the subcritical currents (Curves 3, 3′, 4, 4′) may be characterized by an intense increase in its temperature. Therefore, there is a time to avoid burnout of the composite removing current from it. The calculations lead to this conclusion for both models of the V-I characteristics. However, the probability of a rapid burnout of composite increases when the transport currents with high-density flow in it, especially those that close to the quenching current (Curve 5). At the same time, when liquid hydrogen is used as a coolant, the heating of the composite based on $Bi_2Sr_2CaCu_2O_8$ is not as intense even at supercritical currents. Therefore, there is a reasonable time for removing the current to avoid burnout. This follows from Figure 8.29b, where curves describing the temperature variation over time in the most heated point under the action of a local temperature perturbation ($l_1 = 10^{-2}$ m, $T_1 = 30$ K) and supercritical currents ($I_{c0} = 154$ A) are shown.

The results presented in Figure 8.29a also lead to the conclusion that the calculation of the critical energies at intensive cooling of superconducting composite will depend on the choise of the computational model of V-I characteristic. Indeed, according to the model with the ideal V-I characteristic, a local temperature perturbation transforms the composite into a normal state (Curve 1′) at initial heating temperature $T_1 = 20$ K and current $I = 300$ A. At the same time, the calculation of the thermal mode of composite made in the framework of the model with the power V-I characteristic shows that it retains superconducting properties even when it is locally heated up to a temperature of $T_1 = 25$ K (Curve 2), that is at substantially more overheating. It should be noted that this distinction will be also observed under the perturbations of arbitrary length.

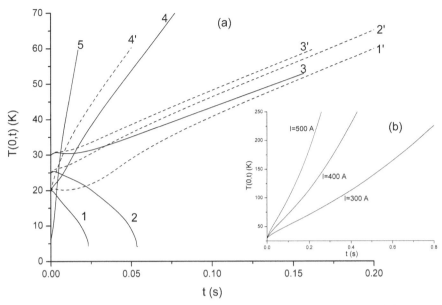

FIGURE 8.29 Temperature change in the center of a superconducting composite cooled by liquid helium (a) and liquid hydrogen (b). Here, (——) - the power V-I characteristic, (– – –) - the ideal V-I characteristic; 1, 1' - $I = 300$ A, $T_1 = 20$ K; 2, 2' - $I = 300$ A, $T_1 = 25$ K; 3, 3' - $I = 300$ A, $T_1 = 30$ K; 4, 4' - $I = 500$ A, $T_1 = 20$ K; 5 - $I = 900$ A, $T_1 = 6$ K.

To estimate the maximum heating temperature of composite superconductors during thermal instability development, one may use a more simplified zero-dimensional model neglecting the heat conduction mechanism. In Figure 8.30a, the simulation results of the time variation in the temperature of a superconducting composite with subcritical current $I = 500$ A cooled by liquid helium under the temperature perturbation $T_1 = 20$ K of different length are compared. Curve 1 describes the heating of the composite at a local perturbation, and Curve 2 corresponds to at an extended perturbation, respectively. They were obtained in the framework of the one-dimensional model. Calculation of Curve 3 was based on the numerical solution of the heat balance equation of the form

$$C_k(T)\frac{dT}{dt} = EJ - \frac{q(T)p}{S}, \quad T(0) = T_1 \tag{8.25}$$

where the electric field was determined according to (8.5) and (8.6), and the critical current density was described by the Formulae (1.8) and (1.9). The solution of this equation can be written as follows

$$\int_{T_1}^{T} \frac{C_k(T)dT}{EJ - q(T)p/S} = t \tag{8.26}$$

Equation (8.25) describes the homogeneous thermal states of the composite, which arise with infinitely extended initial temperature perturbation ($l_1 \to \infty$). Therefore, the results of the heating temperature calculation of a superconducting composite based on the solution of the initial-boundary value problem (8.22)-(8.23) will always lead to more realistic estimates of its burnout conditions and the critical energies of the initial perturbation compared to the estimates following from (8.25). At the same time, its usage to calculate the temperature of the composite superconductor leads to the conclusion that the burnout (in general, the temperature at the most heated point) does not practically depend on the propagation velocity of the thermal instability along the superconductor. The latter affects the size of the burnout area (Figure 8.30b) only. For the upper estimate of the temperature and time of the burnout, one may use the solution (8.26) as follows from Figure 8.30a.

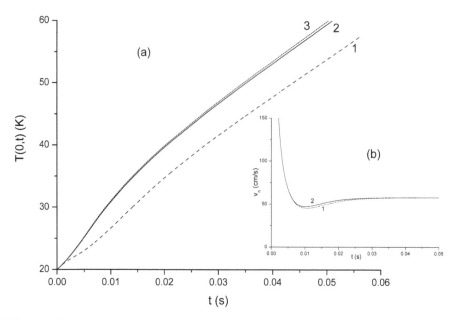

FIGURE 8.30 Temperature increase in the center of the superconducting composite cooled by liquid helium (a) and the time variation of the thermal instability velocity (b) at perturbation with different initial lengths: 1 - l_1 = 5 × 10⁻³ m; 2 - l_1 = 8 × 10⁻² m; 3 - approximation in the framework of zero-dimensional model (8.25).

Note that the well-known 'hot spot' model follows from (8.25) and (8.26) at $EJ \gg q(T)p/S$. It makes to perform a burnout analysis of the superconducting composite under the action of an infinitely long perturbation for the most unfavorable conditions for removing of the Joule heat released in composite when both its heat conductivity and heat transfer to the coolant are not taken into account in the whole volume of composite. At the same time, this conclusion proves the fallacy of the widely held view that the 'hot spot' model describes the heating of a composite superconductor when all stored energy is released at the point of instability. In fact, the 'hot spot' model describes the thermal states of composite superconductors when the stored energy is released in its entire volume.

Thus, one can use the solution (8.26) for an upper estimate of the burnout temperature and time, as follows from Fig. 8.29a. In this relationship, the integral in its left part is a function only of the temperature for the given properties of the composite and the refrigerant. Therefore, the solution (8.26) leads to the simple method allowing to find the burnout time by calculating left part of the dependence (8.26) in the temperature range from T_1 to the melting temperature of the matrix taking into account that all current will flow only through the matrix at $T \gg T_{cB}$. For more simplified estimate, room temperature can be used as the upper limit of integration. It is easy to understand that the estimate $EJ \gg q(T)p/S$ allows one to determine the temperature, after exceeding which the operating conditions close to adiabatic occur. Note that this simplified approach can be used to estimate burnout temperature and time of composites based on both LTS and HTS.

Equation (8.25) also makes it possible to understand the causes of the occurrence for a possible catastrophic temperature increase in intensely cooled composite superconductors at irreversible propagation of the thermal instability. To do this, let use the model with the ideal V-I characteristic and rewrite the right-hand side of Equation (8.25) taking into account the variation peculiarities with temperature of the Joule heat release and the heat flux removed to the coolant. First, note that the resistivity of silver when it is heated up to a temperature about T_r = 20 K slightly depends on temperature but it increases almost linearly at $T_1 > T_r$. Therefore, its value can be written in a simplified form as follows $\rho_m(T) = \rho_r + (T - T_r)d\rho_m/dT$ at $T_1 > T_r$. Here, ρ_r is the residual resistivity, and

$d\rho_m/dT$ = const is the rise rate of resistivity with temperature and T_r is the temperature after which the resistivity of the matrix begins to increase. Second, the heat flux to liquid coolant at high overheating, that is at the film boiling regime, with a satisfactory degree of accuracy can be approximated by the formula $q(T) = h(T - T_0)$, where h = const is the effective value of the heat transfer coefficient during film boiling regime. As a result, the variation in the thermal state of a composite superconductor at $T > T_{cB}$ can be described by the equation of the form

$$C_k(T)\frac{dT}{dt} = \begin{cases} \dfrac{J^2\rho_r}{1-\eta} - \dfrac{hp}{S}(T-T_0), & T_{cB} \leq T \leq T_r \\[3mm] \dfrac{J^2}{1-\eta}\left[\rho_r + \dfrac{d\rho_m}{dT}(T-T_r)\right] - \dfrac{hp}{S}(T-T_0), & T > T_r \end{cases}$$

It may be rewritten as follows

$$C_k(T)\frac{dT}{dt} = \begin{cases} -\dfrac{hp}{S}T + q_1, & q_1 = \dfrac{J^2\rho_r}{1-\eta} + \dfrac{hp}{S}T_0, & T_{cB} \leq T \leq T_r \\[3mm] \left(\dfrac{J^2}{1-\eta}\dfrac{d\rho_m}{dT} - \dfrac{hp}{S}\right)T + q_2, & q_2 = \dfrac{J^2}{1-\eta}\left[\rho_r - \dfrac{d\rho_m}{dT}T_r\right] + \dfrac{hp}{S}T_0, & T > T_r \end{cases} \tag{8.27}$$

Even in this simplest description of dependencies $\rho_m(T)$ and $h(T)$, Equation (8.27) belongs to a class of equations describing diffusion phenomena in an active medium with reproduction, which under certain conditions are of a chain character of explosive type at $T > T_r$. For example, the value T describes the concentration of neutrons by analogy with the phenomenon of the heavy nuclei fission. Then,

$$\gamma_1 = \frac{J^2}{(1-\eta)}\frac{d\rho_m}{dT} = \text{const}$$

is the coefficient of their birth, $\gamma_2 = hp/S$ = const is the absorption coefficient, q_2 is the density of a volume source of neutrons. Accordingly, if

$$\gamma = \gamma_1 - \gamma_2 = \frac{J^2}{(1-\eta)}\frac{d\rho_m}{dT} - \frac{hp}{S} > 0$$

then the generating neutrons mode will prevail over their absorption. Accordingly, the increase in temperature of composite will have an avalanche-like character under this condition. In particular, when C_k = const, the solution of Equation (8.27) in this simplest case is written as follows

$$T = \left(T_r + \frac{q_2}{\gamma}\right)\exp\left(\frac{\gamma}{C_k}t\right) - \frac{q_2}{\gamma}$$

at $T > T_r$, that is an exponential increase of its temperature may occur at the self-heating of a composite superconductor. In general, such intensively rising modes depend on the values J, $d\rho_m/dT$, ρ_r, C, h, η and p/S. In particular, it is the more likely, the higher the value of $\dfrac{J^2}{(1-\eta)}\dfrac{d\rho_m}{dT}$. However, the heat transfer coefficient and heat capacity will have a stabilizing effect on the uncontrolled temperature increase. As a consequence, an avalanche-like increase in the temperature of the composite is inevitable under the condition $\gamma_1 \gg \gamma_2$. It is under this condition that a composite superconductor cooled by liquid helium will exist when its most heated part is in the film boiling regime as a result of self-heating.

It should be emphasized that the discussed features of a burnout occurrence describe both V-I characteristics used. They demonstrate that the burnout effect is based on a trivial increase of the resistance of stabilizing matrix with temperature, which leads to a violation of the thermal equilibrium conditions of a composite superconductor and first all, carrying a high-density current.

The characteristic regularities of the unstable thermal states formation of Ag/Bi2212 discussed above are also observed when analyzing the conditions of the thermal stabilization of superconducting tapes based on YBa$_2$Cu$_3$O$_7$ (Y123). In particular, in Figures 8.31 and 8.32, the results of the numerical simulation of the thermal instability propagation in the nonintensively cooled Y123 tape with two stabilizing coatings of silver and copper are presented. Its geometric parameters (width b, the thickness of superconductor a_s, the thickness of silver a_{ag} and thickness of copper a_{cu}) were assigned as follows $b = 2 \times 10^{-3}$ m, $a_s = 10^{-6}$ m, $a_{ag} = 17 \times 10^{-6}$ m, $a_{cu} = 45 \times 10^{-6}$ m, $l_k = 0.5$ m. The density of the superconductor critical current was described by a linear dependence $J_c = J_{c0}(T_{cB} - T)/(T_{cB} - T_0)$ with the critical parameters $J_{c0} = 6 \times 10^{13}$ A/m^2 and $T_{cB} = 55$ K in the external magnetic field of $B_a = 10$ T and $T_0 = 15$ K. The parameters of the V-I characteristic were assigned as $E_c = 10^{-4}$ V/m and $n = 22$. When calculating the heat capacity and electrical resistivity of copper, the corresponding temperature dependencies were used, as above, by taking $\rho_{cu}(273\ \text{K}) = 1.55 \times 10^{-8}$ $\Omega \times$ m and RRR = 100. In Equation (8.22), the heat release in the tape $G = EJ$ was described by the formula

$$G(T) = \begin{cases} \eta J_{c0} E \dfrac{T_{cB} - T}{T_{cB} - T_0} \left(\dfrac{E}{E_c} \right)^{1/n} + \dfrac{E^2}{\rho_k}, & T < T_{cB} \\[3mm] \dfrac{E^2}{\rho_k}, & T \geq T_{cB} \end{cases}$$

where

$$\rho_k(T, B) = \frac{(a_s + a_{ag} + a_{cu})\rho_{ag}(T, B)\rho_{cu}(T, B)}{a_{cu}\rho_{ag}(T, B) + a_{ag}\rho_{cu}(T, B)}$$

To calculate the heat flux into the coolant, the Formula (8.24) was used at $h = 10$ W/(m$^2 \times$ K) in which the temperature of the coolant was set to $T_0 = 15$ K or $T_0 = 40$ K.

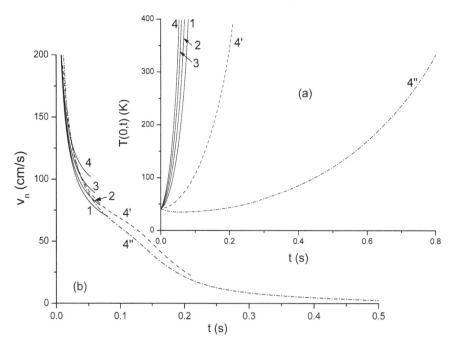

FIGURE 8.31 Temperature change in time of YBa$_2$Cu$_3$O$_7$ tape (a) and velocity of the thermal instability (b) as a result of the supercritical local temperature perturbation ($l_1 = 10^{-2}$ m, $T_1 = 40$ K) at $T_0 = 15$ K: 1 - $I = 160$ A, 2 - $I = 170$ A, 3 - $I = 180$ A, 4, 4', 4" - $I = 190$ A; (1–4) - $a_{cu} = 45$ μm, 4' - $a_{cu} = 100$ μm, 4" - $a_{cu} = 200$ μm.

As follows from the formulated condition for the occurrence of burnout, the avalanche self-heating mode of the Y123 tape can develop without the formation of a thermo-electric autowave (primarily due to the high density of the transport current). It is clearly shown in Figures 8.31 and 8.32. In this case, the ongoing modes including the propagating of the thermal instability are of a substantially nonstationary character as a result of which a rapid burnout of the tape may occur. On the other hand, an increase in the thickness of the copper coating or an increase in the temperature of the coolant will lead to an improvement in the thermal stabilization conditions of the Y123 tape since this reduces the overall density of the transport current. As a result, the intensity of the tape temperature rise decreases making it possible to take measures to prevent its burnout.

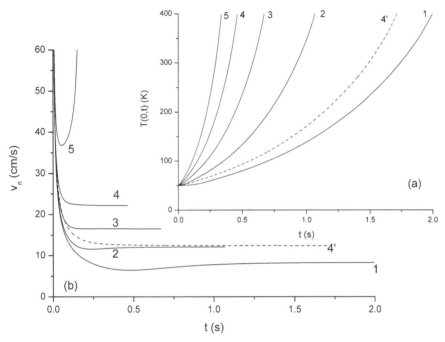

FIGURE 8.32 Temperature change in time of Y123 tape (a) and thermal instability velocity (b) as a result of the action of the supercritical extended temperature perturbation ($l_1 = 0.1$ m, $T_1 = 50$ K) at $T_0 = 40$ K: 1 - $I = 30$ A, 2 - $I = 40$ A, 3 - $I = 50$ A, 4, 4' - $I = 60$ A, 5 - $I = 70$ A; (1–5) - $a_{cu} = 45$ μm, 4' - $a_{cu} = 100$ μm.

Thus, the irreversible propagation of thermal instability along technical superconductors is characterized by the existence of special conditions of their thermal stabilization, if one takes into account the nonlinear temperature dependencies of the superconductor properties and the stabilizing matrix or coating and refrigerant. There may be no full recovery and full stabilization currents in intensively cooled current-carrying elements, which determine the range of currents when the superconducting state is stable for arbitrary thermal perturbations. This feature of the thermal stabilization conditions is a consequence of the nonlinear temperature dependence of the resistivity of stabilizing matrix, when the increase in the Joule heat release may exceed the heat flux into the coolant at temperatures above the critical temperature of superconductor. As a result, the autowave propagation velocities of the thermal instabilities in the entire range of the current variation do not take negative values.

The development of the thermal instability in technical superconductors can either lead or not lead to the formation of a thermo-electric autowaves. In the first case, the violation of the thermal stability after the occurrence of autowave propagating along with technical superconductors at a constant velocity occurs not only at subcritical currents but also in the region of supercritical currents which does not exceed the quenching value. The destruction of superconductivity without the formation of a thermo-electric wave is typical for technical superconductors with high values of the transport current density. In general, to correctly calculate the kinetics of the thermal perturbation along the composites with real V-I characteristics, it is necessary to take into account the nonlinear variation of the properties of superconductor and stabilizing matrix with temperature.

The development of the thermal instability in technical superconductors has the character of a chain reaction. The condition of its occurrence is written. According to this condition, the higher the current density in the superconductor or faster the increases with the temperature of the matrix resistivity, the higher the burnout probability of composite even under the intensive cooling conditions by liquid refrigerant.

8.5   JOULE HEAT RELEASE IN SUPERCONDUCTING COMPOSITES BASED ON LOW-T_C SUPERCONDUCTORS INITIATED BY THE DIFFUSION OF THE TRANSPORT CURRENT

The important role of the thermal stabilization theory for understanding the superconductivity conservation principles of composite superconductors was shown above. However, its main provisions are formulated within the framework of the model, according to which the current distribution over the cross-section of the composite is uniform and the Joule heat release is determined by the instantaneous change in the composite temperature. It allows one not only to simplify the analysis methods used but also to obtain analytical criteria for the thermal stability of the superconducting state. At the same time, the distribution of the transport current is nonuniform over the cross-section of the composite during current charging. Besides, in response to any external thermal perturbation, in the composite necessarily occurs a variation in the electromagnetic field, which must be accompanied by a corresponding heat release. However, the proposed above thermal models do not take into account the diffusion of the magnetic flux in the superconductor, especially, at the stage of its stable states. Let us formulate more general models to avoid these limitations and study the Joule heat release in superconducting composites based on low-T_c superconductors initiated by the diffusion of the transport current.

Let us first investigate the characteristic laws of energy dissipation in a superconducting composite with a circular cross-section, which occurs when current is charged at a constant rate using the nonisothermal model of the current diffusion (6.1-6.6). The results of numerical experiments carried out for a niobium-titanium superconductor in a copper matrix cooled by liquid helium are discussed below. The averaged values of the initial parameters are equal to

$$r_0 = 5 \times 10^{-4} \text{m}, \quad h = 10 \text{ W/(m}^2 \times \text{K}), \quad T_0 = 4.2 \text{ K}, \quad \eta = 0.5, \quad T_{cB} = 6.5 \text{ K}, \quad J_\delta = 4 \times 10^7 \text{A/m}^2,$$

$$\rho_m = 2 \times 10^{-10} \Omega \times \text{m}, \quad \lambda_k = 100 \text{ W/(m} \times \text{K}), \quad C_k = 10^3 \text{ J/(m}^3 \times \text{K}), E_c = 10^{-4} \text{ V/m} \qquad (8.28)$$

Define the Joule heat release that will take place in the composite during current charging at both stable and unstable states determining the stability boundary on the basis of the finite perturbation method of the initial state. The corresponding dependencies of the heat release density per unit of its length $G_s = \int_0^{r_0} EJrdr$ on the charged current are presented in Figure 8.33. The solid curves correspond to stable states and the dashed curves describe unstable states. The calculations were carried out for two values of the current charging rate, which cause instability with different filling of the composite cross-section by the current. Their distribution over the cross-section just before the occurrence of instability is shown by the solid curves on the inset to Figure 8.33. Here, the densities of the quenching currents J_q preceding the occurrence of the instability recalculated for each mode under consideration assuming a uniform distribution of the charged current over the cross-section of the composite are also shown in Figure 8.33. The presented curves show that both before and after the development of the instability, the intensity of the heat release in the composite depends on the current rate. In particular, considerable heat losses occur at high charging rates.

The heat generated in this case differs by several orders of magnitude from the amount of the heat released during its slow charging. Therefore, they must accordingly modify the dependence of the Joule losses on the temperature, which underlies the description of the thermal modes within the framework of the theory of thermal stabilization.

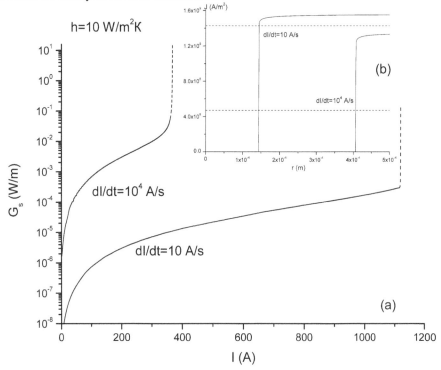

FIGURE 8.33 Dependence of the Joule heat release (a) in the composite on the current charged with different rate and current density distribution (b) just before the current instability:

(a): (———) – stable states, (- - - -) – unstable states;
(b): (———) – the current density distribution at the finite charging rate, (- - - -) – the corresponding equivalent current density distribution over the cross-section.

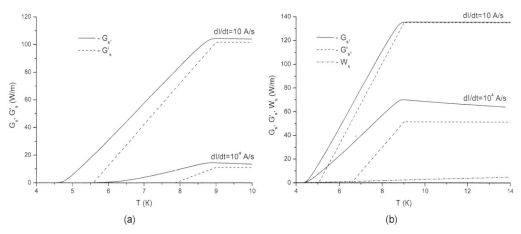

FIGURE 8.34 Joule heat releases G_s, G_s' and heat flux into the coolant W_s as a function of the temperature under stable and unstable current diffusion in the composite at different cooling conditions and current charging rates: (a) - $h = 10$ W/(m² × K), (b) - $h = 10^4$ W/(m² × K).

Figure 8.34 presents the temperature dependencies of the Joule heat release for the entire range of temperature variation (stable and unstable) calculated for different values of the heat transfer coefficient and current charging rates. The dash-dotted lines show the variation in the heat flux removed into the coolant. Its value is reduced to a unit length of the composite, which is $W_s = hr_0(T - T_0)$. The solid curves depict the temperature dependencies of the heat release in a composite superconductor reduced to a unit of its length G_s, which were calculated on the basis of the model (6.1-6.6). The dashed curves show the corresponding values of the heat release G'_s calculated by the formula adopted in the framework of the thermal stabilization theory. According to the results discussed above, the latter is described by the expression

$$G'_s = \frac{S}{2\pi} \frac{J_q^2 \rho_m}{1-\eta} \begin{cases} 1, & T > T_{cB} \\ (T - T_{cs})/(T_{cB} - T_{cs}), & T_{cs} \le T \le T_{cB} \\ 0, & T < T_{cs} = T_{cB} - (T_{cB} - T_0)J_q/(\eta J_{c0}) \end{cases} \qquad (8.29)$$

Here, J_q is the quenching current density recalculated under the assumption of its uniform distribution over the cross-section of the composite (Figure 8.34); J_{c0} is the critical current density of a superconductor with an ideal V-I characteristic ($J_\delta \to 0$) that appears under the assumption of zero exponent in the expression (1.4) for the real V-I characteristic. For the parameters specified above, it is equal to $J_{c0} = 4 \times 10^9$ A/m^2.

Comparing these results, it is easy to see that the Joule losses over the entire range of temperature variation (stable and unstable) increasingly deviates from the simplified dependence (8.29) as the current charging rate increases. This feature is based on the following regularities.

First, as noted above, the intense heat release takes place even in a stable state at high charging rates. Therefore, a significant decrease in the temperature range is observed, at which the Joule losses turn out to be negligible when dI/dt increases. At the same time, in the framework of the model (8.29), the Joule heat release in the temperature range from T_0 to T_{cs} is not taken into account. Second, the avalanche-like character of the instability development is accompanied by a sharp increase in the electric field induced in the composite. It decreases after reaching the maximum value. Therefore, when the Joule heat release generated in the composite during development of instability is correctly determined, then this will not only lead to higher values of the generated heat compared to those resulting from the Formula (8.29), but also to extremely high values $G_s(T)$ in the temperature range near the critical temperature of the superconductor. In particular, at $T = T_{cB}$ and $h = 10$ W/(m$^2 \times$ K), the difference between the values G_s and G'_s achieved at low current charging is equal to $G_s/G'_s = 1.04$, and it is equal to $G_s/G'_s = 1.29$ at fast current charging, that is the higher the current charging rate, the higher the difference.

Intense heat release can also be initiated by external thermal perturbations. Let us simulate the energy dissipation in a superconducting composite with a circular cross-section when the current is charged into a composite at a constant rate dI/dt and the external extended thermal pulse q located near the surface of the composite heats it. This approximation allows one to describe the operating modes of the current-carrying element of the superconducting magnetic system when the considerable mechanical stresses arising during current charging are accompanied by the corresponding heat release. Describe the distributions of the temperature T, electric field E and current density J in a cooled composite with a circular cross-section by the following system of equations

$$C_k(T)\frac{\partial T}{\partial t} = \frac{1}{r}\frac{\partial}{\partial r}\left(\lambda_k(T)r\frac{\partial T}{\partial r}\right) + EJ + q(r,t) \qquad (8.30)$$

$$\mu_0 \frac{\partial J}{\partial t} = \frac{1}{r}\frac{\partial}{\partial r}\left(r\frac{\partial E}{\partial r}\right) \qquad (8.31)$$

$$E = E_c \exp\left(\frac{J_s - J_c}{J_\delta}\right) = J_m \rho_m, \quad J = \eta J_s + (1-\eta)J_m \qquad (8.32)$$

with the following initial-boundary conditions

$$T(r,0) = T_0, \quad E(r,0) = 0, \quad \lambda_k \frac{\partial T}{\partial r} + h(T - T_0)\Big|_{r=r_0} = 0, \quad \frac{\partial E}{\partial r}\Big|_{r=r_0} = \begin{cases} \dfrac{\mu_0}{2\pi r_0} \dfrac{dI}{dt}, & 0 < t \le t_1 \\ 0, & t > t_1 \end{cases} \quad (8.33)$$

During numerical experiments, the initial parameters were set according to (8.28). The simplest operating modes are considered. Namely, the heat pulse of a rectangular shape acts in a thin surface layer of the composite $\Delta r = r_0 - r_q = 0.1 r_0$ during time $t_1 < t < t_1 + \Delta t$, and dI/dt was set equal to zero for all $t > t_1$. In this case, the time variation of the external thermal perturbation is determined as follows

$$q(r,t) = \begin{cases} q_0 = \text{const}, & r_q \le r \le r_0, t_1 \le t \le t_1 + \Delta t \\ 0, & 0 \le r \le r_q, \quad t < t_1 + \Delta t \\ 0, & t > t_1 + \Delta t \end{cases} \quad (8.34)$$

The temperature dependencies of the Joule heat release $G_s(T)$ are shown in Figure 8.35, which are initiated by subcritical (which do not cause the transition of the superconducting composite to the normal state) and supercritical (causing the transition of the superconducting composite to the normal state) perturbations of various duration after charging currents that are much smaller than the quenching current. The current distribution along the radius at the time before the thermal perturbation ($t = t_1$) is shown on the inset to Figure 8.35a. The corresponding values of the Joule heat release G'_s described by the Formula (8.29) (Curves 1) are also shown in Figure 8.35. During their calculations, the corresponding value J recalculated for the whole cross-section of the composite was used instead of the value J_q. They are also shown on the inset to Figure 8.35a.

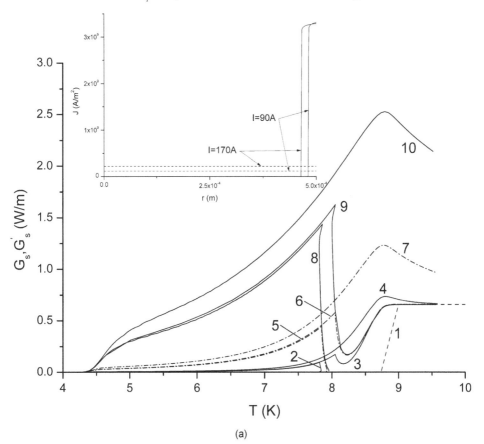

(a)

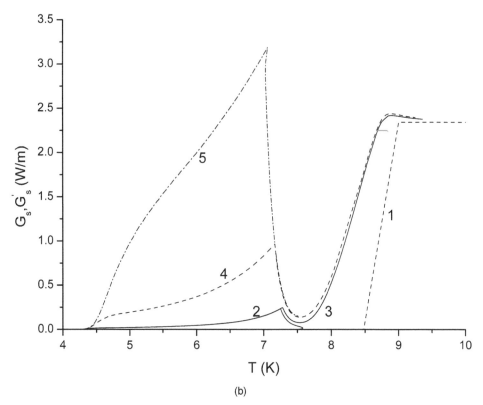

(b)

FIGURE 8.35   Temperature dependence of the Joule heat dissipation in the composite under the action of the external heat pulses of various duration and power after charging the currents I_0 = 90 A (a) and I_0 = 170 A (b). Here,

(a) Curve 1 - calculation by the Formula (8.29), Curves (2-10) - numerical calculation at various Δt and q_0: (2-4) - $\Delta t = 10^{-3}$ s, (5-7) - $\Delta t = 10^{-4}$ s, (8-10) - $\Delta t = 10^{-5}$ s; q_0: 2 - 0.19 × 10^8 W/m³, 3 - 0.2 × 10^8 W/m³, 4 - 0.3 × 10^8 W/m³, 5 - 1.9 × 10^8 W/m³, 6 - 2 × 10^8 W/m³, 7 - 3 × 10^8 W/m³, 8 - 19 × 10^8 W/m³, 9 - 20 × 10^8 W/m³, 10 - 30 × 10^8 W/m³;

(b) Curve 1 - calculation by the Formula (8.29), Curves (2-5) - numerical calculation at various Δt and q_0: 2, 3 - $\Delta t = 10^{-3}$ s, 4 - $\Delta t = 10^{-4}$ s, 5 - $\Delta t = 10^{-5}$ s; q_0: 2 - 0.13 × 10^8 W/m³, 3 - 0.14 × 10^8 W/m³, 4 - 1.4 × 10^8 W/m³, 5 - 14 × 10^8 W/m³.

Curves shown in Figure 8.35 demonstrate that the dissipative phenomena occurring in the composite already at stable states as a result of the external thermal perturbations may be accompanied by essential additional Joule heat releases, the intensity of which depends on the pulse power. Moreover, under the action of the perturbations with energies that are close to the critical, the heat losses increase significantly with decreasing pulse duration. This results not only in a noticeable deviation of the real values $G_s(T)$ from the simplified dependencies $G_s'(T)$ calculated by the Formula (8.29) but also leads to the occurrence in the dependence $G_s(T)$ of nonmonotonically changing sections. It reduces significantly the temperature range in which the effect of additional heat losses can be neglected under the action of the heat pulses. Therefore, the external thermal perturbations can markedly increase the background temperature of the composite before the instability occurrence. It should also be noted that the use of the numerical model (6.1-6.6) in determining the Joule heat release in the development of instability in the superconducting composite leads not only to the existence of a high extreme value in the dependence $G_s(T)$ but also to higher values of the heat release generated even if the temperature of the superconductor has exceeded its critical value. As was shown above, this is due to the avalanche-like character of the instability development accompanied by a sharp increase in the electric field developed in the composite, the attenuation of which occurs in a finite time after reaching the maximum value.

Thus, in superconducting composites in which the transport current diffusion is initiated by any thermal perturbations, there may be intense heat losses. They may occur in the stage of stable states. The heat releases generated in this case may differ by several orders of magnitude from the amount of heat release determined in the framework of the commonly used the Formula (8.29). They lead to a noticeable increase in the temperature of the composite. Therefore, the Expression (8.29) will lead to their significantly underestimated values over the entire range of the temperature variations of superconducting composite at high current charging rates. The consequence of such dissipative phenomena will be the corresponding decrease in the current-carrying capacity. This effect is discussed in the next section.

8.6 THERMAL DEGRADATION OF THE CURRENT-CARRYING CAPACITY OF SUPERCONDUCTING COMPOSITES UNDER THE ACTION OF THERMAL PERTURBATIONS

As already noted, the superconducting magnet is subject to numerous mechanical disturbances during charging current, which are accompanied by the corresponding heat dissipation. In this regard, let us consider the problem of the current state stability that is formed when a current is charged into a composite superconductor at a constant rate and the effect of external pulsed thermal perturbation. Correspondingly, determine the conditions when the boundary of stable states will be determined taking into account the thermal history of the formation of electrodynamic states, depending on the thermal perturbation character and the features of the charged current diffusion.

So, define the distributions of the temperature T, electric field E and current density J in a circular composite with a radius r_0 on the basis of the problem described by Equations (8.30)-(8.34) with the initial parameters (8.28) when the transport current increases at a constant rate dI/dt and the external source of the thermal perturbation $q(r, t)$ acts. In this case, the values of the quenching current and the critical energy of the heat release will be found in accordance with the method of the finite perturbation of initial state. Accordingly, for a given value of the charged current, two values of the heat source power shall be determined at which either a decrease or an irreversible increase in the electric field occurs after the termination of the perturbation. Then, an increase in the temperature of the composite is sustainable in the first case, and it leads to the transition of the composite to the normal state in the second case.

As an illustration, the calculation results of the currents of instability are shown in Figure 8.36. The simulation was made for two typical current charging modes and cooling conditions: in the case of a fast current charging into a nonintensively cooled composite (Figure 8.36a) and with a slow current charging into the intensively cooled composite (Figure 8.36b). Here, the solid curves describe the change in temperature of the composite surface at continuous current charging and the dashed and dashed-dotted curves describe the change in temperature when the charged current was fixed after reaching the specified value I_0. For the considered modes, the formation of the states describing the change in temperature of the composite surface in the case of an act of the disturbances of different duration at a charged current I_0, which is less than the corresponding quenching current, is shown in Figure 8.37. They show the temperature dynamics of the composite under the action of the thermal perturbations that are close to the critical.

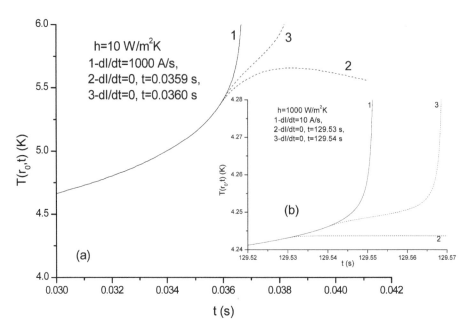

FIGURE 8.36 Determination of the current instability in a superconducting composite under different conditions of its cooling and charging current rate.

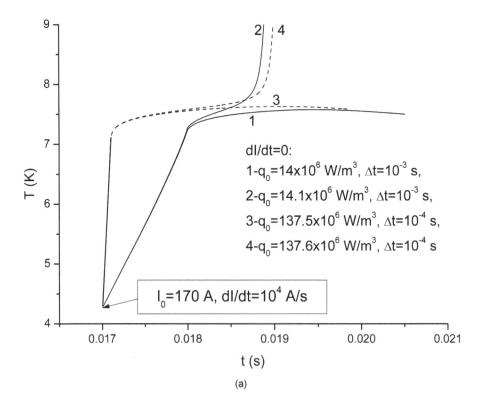

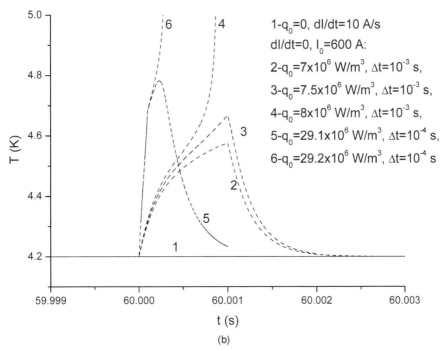

FIGURE 8.37 Determination of the thermal stability boundary of a superconducting composite under various conditions of its cooling, charging current rate and charged current under the thermal perturbations of different power: (a) - $I_0 = 170$ A, $dI/dt = 10^4$ A/s, $h = 10$ W/m²K, (b) - $I_0 = 600$ A, $h = 1000$ W/m²K; (1) - $dI/dt = 10$ A/s, (2-6) - $dI/dt = 0$.

Figure 8.37a demonstrates the formation of stable and unstable states of a nonintensively cooled composite under the action of subcritical (Curves 1 and 3) and supercritical perturbations (Curves 2 and 4) when the superconducting state either persists or collapses as a result of their action after a fast charged current of 170 A. The results of determining the critical value of an external thermal perturbation during a slow current charging into the intensely cooled composite are presented in Figure 8.37b. Here, the solid Curve 1 is calculated at a continuous current charging when an external thermal perturbation is absent. The temperature variation of the composite under the action of the subcritical (Curves 3 and 5) and supercritical perturbations (Curves 4 and 6) was determined after a current charging of 600 A.

The results in Figure 8.37 show that the subcritical external thermal perturbation may have a significant influence on the dynamics of the current modes during continuous current charging. As follows from Figure 8.37, it depends on the intensity of the cooling, rate and magnitude of the charged current and the duration of an external thermal perturbation. As discussed above, this is due to the fact that the subcritical thermal perturbation can initiate significant additional heat release. Consequently, a corresponding decrease in the stability boundary of the charged currents may be observed. The results of the corresponding numerical experiment are presented in Figure 8.38. They demonstrate the temperature fluctuation of the considered nonintensively cooled superconducting composite during continuous current charging and the action of subcritical perturbations. (According to the results presented in Figure 8.37, the calculated value of the critical value of q_0 is equal to 14.1 $\times 10^6$ W/m³ at $t_1 = 0.017$ s and $dI/dt = 10^4$ A/s). It is seen that the current instability begins to develop at lower values of the charged current. As a result, the quenching current decreases with increasing thermal perturbation energy. It should be underlined that the instability occurs at external thermal disturbances, which in themselves do not transfer the composite superconductor to the normal state. In other words, subcritical perturbations can lead to the degradation of the current-carrying capacity of superconducting composite.

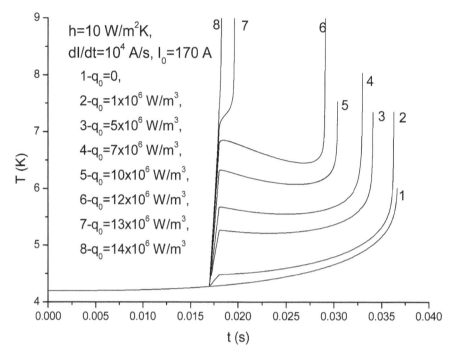

FIGURE 8.38 Instability of the current continuously charged into the nonintensively cooled composite under the action of the subcritical thermal perturbations at current I_0.

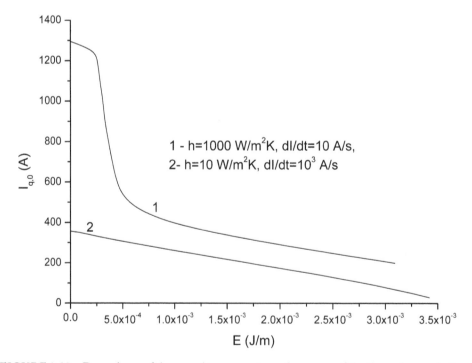

FIGURE 8.39 Dependence of the quenching currents on the energy of the thermal perturbations.

The dependence of the instability current on the energy of an external thermal perturbation is shown in Figure 8.39. The calculations were carried out at $\Delta t = 10^{-3}$ s in the framework of the following approximation to simplify the simulations. The critical value of the perturbation energy per unit length ($E = q_0 S$) was determined for a fixed current I_0. Therefore, the calculated dependence $I_{q,0}(E)$ corresponds to the upper limit of the thermal stability of the charged currents since the corresponding values of the currents of instability will be smaller than these values during continuous charging, as follows from Figures 8.37 and 8.38. Figure 8.39 shows that the thermal degradation effect can affect the current-carrying capacity of both intensely and nonintensively cooled composite superconductors. At the same time, the intensively cooled composites have the characteristic value of thermal perturbation energy. Once this value is exceeded, a significant improvement in heat transfer conditions will not be accompanied by a noticeable increase in instability currents even when the current charging rate decreases.

Thus, under the action of external thermal perturbations that do not transfer the superconductor into the normal state, the premature instability of the charged current is possible, as a result of which the instability currents become smaller when the external thermal perturbation energy increases.

8.7 CONCLUSIONS

The nonlinearity type of the V-I characteristic of superconductors significantly affects the dissipative phenomena occurring in composite superconductors. As a result, its account in the theory of the thermal stabilization of superconducting composites leads to the expansion of the conclusions that were discussed in Chapter 7.

First, in composite superconductors, the stable thermal modes are not limited by a priory set critical current of the superconductor since stable supercritical currents may exist. Moreover, the temperature of the resistive transition also has no physical meaning. In a composite with a real V-I characteristic, the current sharing occurs in such a way that the current in the superconducting core is always greater than the critical current even in supercritical current modes. Second, the analysis of the stability conditions of superconducting state with respect to external thermal disturbances leads not only to more realistic estimates of the critical energies of perturbations but also to their finite values in the range of supercritical currents. Thirdly, the smaller the *n*-value, the lower the velocities of the irreversible propagation of thermal instabilities along a composite superconductor. The thermal instability propagates stably along the composite in the form of a thermo-electric autowave in both subcritical and supercritical current modes.

The condition of the full thermal stabilization of technical superconductors, which takes into account the nonlinear nature of the increase in their V-I characteristics expands the range of currents, in which the operating modes are stable to external thermal disturbances at intensive cooling. Therefore, the known condition of Stekly ($\alpha = 1$), according to which the superconducting composite is completely stable to thermal perturbations with any energy, does not allow to correctly determine the conditions providing the full thermal stabilization of composites with real V-I characteristics.

An analysis of the irreversible propagation of thermal instability in the composite superconductors shows that it can be stabilized under the nontrivial conditions if the nonlinear temperature dependencies of the properties of superconductor and stabilizing matrix are taken into account. It is shown that the full stabilization currents, at which the superconductivity is stable against all thermal perturbations, may be absent in the intensely cooled current-carrying elements of superconducting magnetic systems. This feature of the thermal stabilization conditions is a result of the nonlinear temperature dependence of the stabilizing matrix resistivity when the rate of the Joule heat release increases with temperature faster than the rate of the heat transfer into a coolant at temperatures above the critical temperature of the superconductor. As a consequence, the full stabilization region is absent, and the autowave velocity of the thermal instability propagation turns out to be positive over the entire current range.

The development of the thermal instabilities in technical superconductors may or may not cause the formation of a thermo-electric autowave. In the first case, a thermal instability occurs in the autowave form, which propagates at constant velocity both at subcritical and supercritical currents below the quenching current. The destruction of the superconductivity without the formation of a thermo-electric autowaves is typical for technical superconductors with a high density of the transport current.

The quenching currents of superconductors with real V-I characteristic cooled by liquid helium or hydrogen are the result of the violation of the nucleate boiling regime. As a result, the quenching currents are many times higher than a priori set critical current when cooling technical superconductors with liquid helium or hydrogen.

The thermal modes of composites can approach the adiabatic ones even with intensive cooling. This regularity determines the peculiarities of the conditions under which technical superconductors may be burned when they are cooled by liquid coolant. The probability of their burnout increases with the increase of the critical current density of the superconductor. In this case, the development of thermal instability in technical superconductors has the form of an explosive-type chain reaction. The condition for its occurrence is written, from which it follows that the higher the current density in the superconductor or the higher the rate of rising of the matrix resistivity with temperature, the higher the probability of burnout even with intensive cooling. These results should be taken into account when analyzing the conditions of the thermal stability of current-carrying elements in superconducting magnetic systems based on both LTS and HTS.

References

Abrikosov, A.A. 1957. On the magnetic properties of second kind superconductors. Soviet Physics – JETP. 5: 1174 -1182.

Akachi, T., T. Ogasawara and K. Yasukochi. 1981. Magnetic instability in high field superconductors. Japanese J. Appl. Phys. 10: 1559–1571.

Amemiya, N., S. Murasawa, N. Banno and K. Miyamoto. 1998. Numerical modelings of superconducting wires for AC loss calculations. Physica C: Superconductivity. 310: 16–29.

Anashkin, O.P., V.E. Keilin and V.V. Lyikov. 1979. Stability of compound superconductors under localized heat pulse. Cryogenics. 19: 77–80.

Anderson, P.W. 1962. Theory of flux creep in hard superconductors. Phys. Rev. Letts. 9: 309–311.

Anderson, P.W. and Y.B. Kim. 1964. Hard superconductivity: Theory of the motion of Abrikosov flux lines. Rev. Mod. Phys. 36: 39–43.

Andrianov, V.V., V.P. Baev, S.S. Ivanov, R.G. Mints and A.L. Rakhmanov. 1983. Current-carrying capacity of composite superconductors. IEEE Trans. on Mag. 19: 240–243.

Awaji, S., K. Watanabe, N. Kobayashi, et al. 1996. High field properties of irreversibility field and pinning force for $YBa_2Cu_3O_7$ film. IEEE Trans Appl Supercond. 32: 2776–2779.

Bardeen, J., L.N. Cooper and J.R. Schrieffer. 1957. Theory of superconductivity. Phys. Rev. 108: 1175–1204.

Bean, C.P. 1962a. Magnetization of hard superconductors. Phys. Rev. Lett. 8: 250–253.

Bean, C.P. 1964b. Magnetization of high-field superconductors. Rev. Mod. Phys. 36: 31–39.

Bednorz, J.G. and K.A. Müller. 1986. Possible high-T_c superconductivity in the Ba-La-Cu-O system. Zeitschrift für Physik. Condenced Matter. B64: 189–193.

Blatter, G., M.V. Feigel'man, V.B. Geshkenbein, A.I. Larkin and V.M. Vinokur. 1994. Vortices in high-temperature superconductors. Reviews of Modern Physics, 66: 1125–1388.

Bottura, L. 1999a. A practical fit for the critical surface of Nb-Ti. IEEE Trans. on Mag. 10: 1054–1057.

Bottura L. 2002b. Critical surface for BSCCO-2212 superconductor. Note-CRYO/02/027, CryoSoft library, CERN.

Brandt, E.H. 1996. Superconductors of finite thickness in a perpendicular magnetic field: Strips and slabs. Phys. Rev. B54: 4246–4264.

Brechna, H. 1973. Superconducting Magnet Systems. Springer-Verlag. Berlin-New York.

Brentari, E.G. and R. Smith. 1965. Nucleate and film pool boiling design correlations for O_2, N_2, H_2 and He. Adv. Cryo. Engn. 10: 325–341.

Buckel, W. and R. Kleiner. 2004. Superconductivity: Fundamentals and Applications, 2nd Edition. Wiley-VCH Verlag GmbH & Co. KGaA.

Carr W. J., Jr. 1983. AC loss and macroscopic theory of superconductors. New York: Gordon & Breach.

Chabanenko, V.V., A.I. D'yachenko, K. Szymczak, et al. 1996. Reversible mechanism of magnetothermal instabilities in melt-textured YBaCuO. Physica C: Superconductivity and its Applications. 273: 127–134.

Chabanenko, V.V., V.F. Rusakov, A.I. D'yachenko, et al. 2002. Role of the field dependence of the heat capacity for the flux jump process in HTSC materials. Physica C: Superconductivity and its Applications. 369: 227–231.

Chen, K., Y.C. Chen, S.W. Lu, et al. 1991. Magnetic and thermal instabilities in a textured $YBa_2Cu_3O_x$ super-conductor with high transport J_c. Physica C: Superconductivity and its Applications. 173: 227–231.

Chen, W.Y. and J.R. Purcell. 1978. Numerical study of normal zone evolution and stability of composite superconductors. J. Appl. Phys., 49: 3546–3553.

Cherry, W.H. and J.I. Gittelman. 1960. Thermal and electrodynamic aspects of the superconductive transition process. Solid-State Electronics. 1: 287–305.

Däumling, M. 1998. AC power loss for superconducting strips of arbitrary thickness in the critical state carrying a transport current. Supercond. Sci. Technol. 11: 590–593.

De Haas, W.J. and J.J. Voogd. 1929. Disturbance of the superconductivity of the compound Bi_5Tl_3 and of the alloys Sn-Bi and Sn-Cd by magnetic field. Commun. Phys. Lab. Univ. Lieden. 199d. 31–40.

Del Gastillo, G. and L.O. Oswald. 1969. Magnetic and thermal instabilities observed in commercial Nb_3Sn superconductors. Proc. 1968 summer study on superconducting devices and accelerators. 601–611.

Ding, S.Y., H. Luo, Y.H. Zhang, et al. 2000. Voltage relaxation and its influence on critical current measurements. J. of Superconductivity. 13: 453–458.

Dresner L. 1979a. Analytic solution for the propagating velocity in superconducting composites. IEEE Trans. on Mag. 15: 328–330.

Dresner, L. 1993b. Stability and protection of Ag/BSCCO magnets operated in the 20-40 K range. Cryogenics. 33: 900–909.

Dresner, L. 2002c. Stability of Superconductors. Kluwer Academic Publishers. New York.

Duchateau, J.J. and B. Turk. 1975. Dynamic stability and critical currents in superconducting multifilamentary composites. J. Appl. Phys. 46: 4989–4995.

Elrod, S.A., J.W. Lue, J.R. Miller and L. Dresner. 1981. Metastable superconductive composites: dependence of stability on copper-to-superconductor ratio. IEEE Transactions on Magnetics. 17: 1083–1086.

Eskildsen, M.R., M. Kugler, S. Tanaka, J. Jun, S.M. Kazakov, J. Karpinski and Ø. Fischer. 2002. Vortex imaging in the π band of magnesium diboride. Physical Review Letters. 89: 187003.

Essmann. U. and H. Träuble. 1966. The direct observation of individual flux lines in type II superconductors. Phys. Lett. 24A: 526–527.

Evetts, J.E., A.M. Campbell and D. Dew-Hughes. 1964. Flux instabilities in hard superconductors. Philosophical Magazine. 10: 339–343.

Feigel'men, M.V., V.B. Geshkenbein and V.M. Vinokur. 1991. Flux creep and current relaxation in high-T_c superconductors. Phys. Rev. B43: 6263–6265.

Fisher, D.S., M.P.A. Fisher and D.A. Huse. 1991. Thermal fluctuations, quenched disorder, phase transitions, and transport in type-II superconductors. Phys. Rev. B43: 130–159.

Fisher, L.M., P.E. Goa, M. Baziljevich, et al. 2001. Hydrodynamic instability of the flux-antiflux interface in type-II superconductors. Phys. Rev. Lett. 87: 247005.

Fisher, M.P.A. 1989. Vortex-glass superconductivity: A possible new phase in bulk high-T_c oxides. Phys. Rev. Lett. 61: 1415–1418.

Funaki, K., K. Yamafuji, M. Takeo, et al. 1986. Influence of thermal coupling between neighboring turns on normal-zone propagation in a single-layer superconducting coil. Proc. of MT-9. Zürich. Switzerland. 601–604.

Gerber, A., J.N. Tarnawski, Z. Li, et al. 1993. Magnetic instabilities in high-temperature superconductors under rapidly varying magnetic fields. Phys. Rev. B47: 6047–6053.

Gilchrist, J. and C.J. van der Beek. 1994. Nonlinear diffusion in hard and soft superconductors. Physica C: Superconductivity and its Applications. 231: 147–156.

Ginzburg, V.L. and L.D. Landau. 1950. On the theory of superconductivity. Soviet Physics JETP. 20: 1064–1082 (in Russian).

Grilli, F., S. Stavrev, B. Dutoit and S. Spreafico. 2003. Numerical modeling of a HTS cable. IEEE Transactions on Applied Superconductivity. 13: 1886–1889.

Gurevich, A. 1995a. Nonlinear flux diffusion in superconductors. Int. J. of Modern Phys. B9: 1045–1065.

Gurevich, A. and E.H. Brandt. 1997b. AC response of thin superconductors in the flux-creep regime. Phys. Rev. B55: 569–577.

Gurevich, A.V., R.G. Mints and A.L. Rakhmanov. 1997c. The physics of composite superconductors. Beggel House, NY.

Hancox, R. 1965. Stabilization against flux jumping in sintered Nb_3Sn. Phys. Lett. 16: 208–209.

Herrmann, P.F, C. Albrecht, J. Bock, et al. 1993. European Project for the development of high-T_c current leads. IEEE Trans. Appl. Supercon. 3: 876–880.

Hess, H., R.B. Robinson and J.V. Waszczak. 1991. STM spectroscopy of vortex cores and the flux lattice. Physica B. 169: 422–431.

Hong, Z. and T.A. Coombs. 2010. Numerical Modelling of AC Loss in Coated Conductors by Finite Element Software Using H Formulation. J. of Superconductivity and Novel Magnetism. 23: 1551–1562.

Hug, H.J., A. Moser, I. Parashikov, B. Stiefel, O. Fritz, H.-J. Guentherodt and H. Thomas. 1994. Observation and manipulation of vortices in a $YBa_2Cu_3O_7$ thin film with a low temperature magnetic force microscope. Physica C: Superconductivity and its Applications. 235–240: 2695–2696.

Inoue, M., T. Kiss, D. Mitsui, et al. 2007. Current Transport Properties of 200 A-200 m-Class IBAD YBCO Coated Conductor Over Wide Range of Magnetic Field and Temperature. IEEE Trans. Appl. Supercond. 17: 3207–3210.

Ishibashi, K., et al. 1979. Thermal stability of SC high current density magnets pulse. Cryogenics. 19: 633–638.

Ishiyama, A., M. Yanai, T. Morisaki, et al. 2005. Normal transition and propagation characteristics of YBCO tape. IEEE Trans. Appl. Supercon. 15: 1659–1662.

Junod, A., K.O. Wang, T. Tsukamoto, et al. 1994. Specific heat up to 14 tesla and magnetization of a $Bi_2Sr_2CaCu_2O_8$ single crystal. Physica C: Superconductivity and its Applications. 229: 209–230.

Kantrowitz, A.R. and Z.J.J. Stekly. 1965. A new principle for the construction of stabilized superconducting coils. Appl. Phys. Letters. 6: 56–57.

Keilin V.E., E.Yu. Klimenko, M.G. Kremlev, et al. 1967. Stability criteria for current in combine (normal+superconducting) conductors. Les Champs magnetiques intenses. Paris. CNRS. p. 231–236.

Keilin, V.E. and V.R. Romanovsky. 1982. The dimensionless analysis of the stability of composite superconductors with respect to thermal disturbances. Cryogenics. 22: 313–317.

Khene, S. and B. Barbara. 1999. Flux jump in $YBa_2Cu_3O_7$ single crystals at low temperature and fields up to 11 T. Solid State Commun. 109: 727–731.

Kim, Y.B., C.F. Hempstead and A.R. Strnad. 1963. Flux creep in hard superconductors. Phys. Rev. 131: 2486–2495.

Kiss, T., M. Inoue, T. Kuga, et al. 2003. Critical current properties in HTS tapes. Physica C: Superconductivity and its Applications. 392–396: 1053–1062.

Klimenko, E.Yu. and N.N. Martovetsky. 1992. Stability of the superconducting wires. Modern state of the theory. IEEE Trans. on Mag., 28: 842–845.

Kolmogorov, A.N., I.G. Petrovsky and N.S. Piskunov. 1937. Investigation of the diffusion equation connected with an increase in the amount of a substance, and its application to a single biological problem. Bull. Moscow University. Ser. Mathematics and Mechanics. 1: 1–26 (In Russian).

Kremlev, M.G. 1974. Damping of flux jumps by flux-flow resistance. Cryogenics. 14: 132–134.

Labusch, R. and T.B. Doyle. 1997. Macroscopic equations for the description of the quasi-static magnetic behaviour of a type II superconductor of arbitrary shape. Physica C: Superconductivity and its Applications. 290: 143–147.

Lahtinen, M., J. Paasi and L. Kettunen. 1996. Diffusion model for the computation of current density distributions in composite superconductors. Cryogenics. 36: 951–956.

Legrand, L., I. Rosenman, Ch. Simon, et al. 1993. Magnetothermal instabilities in $YBa_2Cu_3O_7$. Physica C: Superconductivity and its Applications. 211: 239–249.

Legrand, L., I. Rosenman, R.G. Mints, et al. 1996. Self-organized criticality effect on stability: magneto-thermal oscillations in a granular YBCO superconductor. Europhys. Lett. 34: 287–292.

London, F. and H. London. 1935. The electromagnetic equations of the supraconductor. Proc. Roy. Soc. A149: 71–88.

Lubell, M.S. 1983. Empirical scaling formulas for critical current and critical cield for commercial Nb-Ti. IEEE Trans. on Mag. 19(3): 754–757.

Lvovsky, Yu.M. 1984. Velocity of normal zone propagation in a superconductor with temperature-dependent temperature properties and heat transfer. Cryogenics. 24: 261–273.

Maddock, B.J., G.B. James and W.T. Norris. 1969. Superconducting composites: heat transfer and steady state stabilization. Cryogenics. 9: 261–273.

Majoros M., A.M. Campbell, B.A. Glowacki, et al. 2004. Numerical modeling of heating and current-sharing effects on I-V curves of $YBa_2Cu_3O_7$ and MgB_2 conductors. Physica C: Superconductivity and its Applications. 401: 40–145.

Martinelli, A.P. and S.L. Wipf. 1972. Investigation of cryogenic stability and reliability of operation of Nb_3Sn coils in helium gas environment. Proc. Appl. Supercon. Conf. USA 331–340.

Meissner, W. and R. 1933. Ochsenfeld Ein neuer effekt bei eintritt der supraleitfahigkeit. Die Naturwissenschaften. 33: 787–788

Milner, A. 2001. High-field flux jumps in BSCCO at very low temperature. Physica B. 294–295: 388–392.

Mints, R.G. and A.L. Rakhmanov. 1975a. Flux jump and critical state stability in superconductors. J. Phys. D: Appl. Phys. 8: 1769–1782.

Mints, R.G. and A.L. Rakhmanov. 1982b. Current-voltage characteristics and superconducting state stability in composite. J. Phys. D: Appl. Phys. 15: 2297–2306.

Muller, K.H. and C. Andrikidis. 1994. Flux jumps in melt-textured Y-Ba-Cu-O. Phys. Rev. B.49: 1294–1307.

Murray, L.R. 2009. Nuclear Energy, Sixth ed. Elsevier Inc, UK.

Narlikar, A.V. 2014. Superconductors. Oxford University Press. UK.

Nattermann, T. 1990. Scaling approach to pinning: Charge density waves and giant flux creep in superconductors. Phys. Rev. Lett. 64: 2454–2457.

Nibbio N., S. Stavrev, and B. Dutoit. 2001. Finite element method simulation of AC loss in HTS tapes with B-dependent E-J power law. IEEE Transactions on Applied Superconductivity. 11: 2631–2634.

Nick, W., H. Krath and J. Ries. 1979. Cryogenic stability of composite conductors taking into account transient heat transfer. IEEE Trans. on Mag. 15: 359–362.

Norris, W.T. 1970. Calculation of hysteresis losses in hard superconductors carrying ac: isolated conductors and edges of thin sheets. J. Phys., D: Appl. Phys. 3: 489–507.

Onnes, H.K. 1911a. Further experiments with liquid helium. C. On the change of electric resistance of pure metals at very low temperatures, etc. IV. The resistance of pure mercury at helium temperatures. Leiden Comm., 120b: 13–18.

Onnes, H.K. 1911b. Further experiments with liquid helium. D. On the change of electric resistance of pure metals at very low temperatures, etc. V. The disappearance of the resistance of mercury. Leiden Comm. 122b: 13–15.

Onnes, H.K. 1911c. Further experiments with liquid helium. G. On the electrical resistance of pure metals, etc. VI. On the sudden change in the rate at which the resistance of mercury disappears. Comm. Phys. Lab. Univ. Leiden. 124c: 21–25.

Onnes, K.H. 1913d. Further experiments with liquid helium: The sudden disappearance of the ordinary resistance of tin and the supraconductive state of lead. Commun. Phys. Lab. Univ. Lieden. 133d: 51 P.

Paasi, J. and M. Lahtinen. 1998. AC losses in high-temperature superconductors: revisiting the fundamentals of the loss modeling. Physica C: Superconductivity and its Applications. 310: 57–61.

Paasi J., J. Lehtonen, T. Kalliohaka, et al. 2000. Stability and quench of a HTS magnet with a hot spot. Supercond. Sci. Technol. 13: 949–954.

Polak, M., I. Hlasnik and L. Krempasky. 1973. Voltage-current characteristics of Nb-Ti and Nb_3Sn superconductors in flux creep region. Cryogenics. 13: 702–711.

Poole, C.P., H.A. Farach, R.J. Creswick and R. Prozorov. 1995. Superconductivity. Academic Press, New York.

Prigozhin, L. 1997. Analysis of critical-state problems in type-II superconductivity. IEEE Trans. Appl. Supercond. 7: 3866–3873.

Qin, M.J. and X.X. Yao. 1996. AC susceptibility of high-temperature superconductors. Phys. Rev. B54: 7536–7544.

Rhyner, J. 1993. Magnetic properties and AC-losses of superconductors with power law current—voltage characteristics. Physica C: Superconductivity and its Applications. 212: 292–300.

Rjabinin, J.N. and L.W. Schubnikow. 1935. Magnetic properties and critical currents of superconducting alloys. Nature. 135: 581–582.

Romanovsky V.R. 1985a. Regularity of thermal stability conditions of composite superconductors postulated by the theory of minimum propagating zone. J. Phys. D: Appl. Phys. 18: 121–127.

Romanovskii, V.R. 1985b. Influence of volume fraction of superconductor on the stability of superconducting composites with respect to thermal disturbances of finite extent. Cryogenics. 25: 327–333.

Romanovskii, V.R. 1987c. On mathematical modelling of transition processes of the composite superconductors into the normal state. IEEE Transactions on Magnetics. 32: 1569–1571.

Romanovskii, V.R. 1997d. Nonlinear dynamics of the critical state in hard superconductors and composites based on them. Tech. Phys. 42: 1011–1015.

Romanovskii, V.R. 1997e. Solution of conjugated heat-conduction problems with unknown phase boundaries. High Temperature. 35: 501–503.

Romanovskii, V.R. 1998f. Stability of the critical state of a hard superconductor under coolant-temperature variations. Doklady Physics. 43: 80–83.

Romanovskii, V.R. and K. Watanabe. 2005a. Basic formation peculiarities of the stable and unstable states of high-T_c composite superconductors at applied fully penetrated currents. Physica C: Superconductivity and its Applications. 425: 1–13.

Romanovskii, V.R. and K. Watanabe. 2006b. Size dependence of the thermo-electrodynamics states of composite high-T_c superconductors and its effect on the current instability conditions. Physica C: Superconductivity and its Applications. 450: 88–95.

Romanovskii, V.R. and K. Watanabe. 2009c. Thermal stability characteristic of high temperature superconducting composites. pp. 293–399. In: H. Tovar and J. Fortier [eds]. Superconducting Magnets and Superconductivity: Research, Technology and Applications. Nova Science Publishers, NY, USA.

Romanovskii, V.R., K. Watanabe, S. Awaji and G. Nishijima. 2006. Current-carrying capacity dependence of composite $Bi_2Sr_2CaCu_2O_8$ superconductors on the liquid coolant conditions. Supercond Sci and Technol 19: 703–710.

Rusinov, A.I. 1994. On propagation of plane fronts of normal zone within superconducting winding of magnet. IEEE Trans. Mag. 30: 2681–2684.

Schmidt, C. 1978. The induction of a propagating normal zone (quench) in a superconductor by local release. Cryogenics. 18: 605–610.

Schnack, H.G. and R. Griessen. 1992. Comment on "Exact solution for flux creep with logarithmic $U(j)$ dependence: Self-organized critical state in high-T_c superconductors". Phys. Rev. Lett. 68: 2706–2707.

Seeber, B. 1998. Handbook of Applied Superconductivity. Institute of Physics Publishing, Bristol-Philadelphia.

Seto, T., S. Murase, S. Shimamoto, et al. 2001. Thermal stability of Ag/Bi-2212 tape at cryocooled condition. Teion Kogaku. 36: 60–67.

Shimamoto, S. 1974. Experiments on flux jumps in superconducting tapes. Cryogenics. 14: 568–573.

Silsbee, F.B. 1916. A note on electrical conduction in metals at low temperatures. Journal of the Washington Academy of Science. 6: 597–602.

Sizoo, G.J., W.J. de Haas and K.H. Onnes. 1926. Measurements on the magnetic disturbance of the superconductivity with tin. Commun. Phys. Lab. Univ. Lieden. 180c: 29–53.

Smith, P.F. 1963. Protection of superconducting coils. Rev. Sci. Instrum. 34: 368–373.

Stavrev, S., F. Grilli, B. Dutoit, et al. 2001. Comparison of numerical methods for modeling of superconductors. IEEE Transactions on Magnetics. 38: 849–852.

Stavrev, S., B. Dutoit and P. Lombard. 2003. Numerical modeling and AC losses of multifilamentary Bi-2223/Ag conductors with various geometry and filament arrangement. Physica C: Superconductivity and its Applications. 384: 19–31.

Stekly, Z.J.J. and J.L. Zar. 1965. Stable superconducting coils. IEEE Trans. Nucl. Sci. NS-12: 367–372.

Summers, L.T., M.W. Guinan, J.R. Miller, et al. 1991. A model for the prediction of Nb_3Sn critical current as a function of field, temperature, strain and radiation damage. IEEE Trans. on Mag. 27: 2041–2044.

Tholence, J.L., H. Noel, J.C. Levet, et al. 1988. Magnetization jumps and critical currents in $HoBa_2Cu_3O_7$ single crystals up to 18 T. Physica C: Superconductivity and its Applications. 153–155: 1479–1480.

Turk B. 1980. About the propagating velocity in superconducting composites. Cryogenics. 20: 146–150.

Tuyn, W.K. and H.K. Onnes. 1926. The disturbance of supra-conductivity by magnetic field and currents. The hypothesis of Silsbee. Commun. Phys. Lab. Univ. Lieden. 174a: 3–39.

Uher, C. 1990. Thermal conductivity of high-T_c superconductors. J. of superconductivity and Novel Magnetism. 3: 337–350.

van der Beek, C.J., G.J. Nieuwenhuys and P.H. Kes. 1991. Physica C: Superconductivity and its Applications. Numerical calculations on flux-creep in high temperature superconductors. 185-189, Part 1: 2241–2242.

van der Laan, D.C., H.J.N. van Eck, B. ten Haken, et al. 2001. Temperature and magnetic field dependence of the critical current of $Bi_2Sr_2Ca_3Cu_2O_x$. IEEE Trans Appl Supercond. 11: 3345–3348.

Vinokur, V.M., M.V. Feigel'man and V.B. Geshkenbein. 1991. Exact solution for flux creep with logarithmic $U(J)$ dependence: self-organized critical state in high-T_c superconductors. Phys. Rev. Lett. 67: 915–918.

Vinot, E., G. Meunier and P. Tixador. 2000. Different formulations to model superconductors. IEEE Trans. Magnetics. 36: 1226–1229

Wakuda, T., T. Nakano, M. Iwakuma and K. Funaki. 1997. E-J characteristics and a.c. losses in a superconducting Bi(2223) hollow cylinder. Cryogenics. 37: 381–388.

Watanabe, K., V.R. Romanovskii, Ken-ichiro Takahashi, G. Nishijima and S. Awaji. 2004. Current-carrying capacity properties in low resistivity state for Ag-sheathed $Bi_2Sr_2CaCu_2O_8$ tape. Supercond Sci and Technol 17: S533–S537.

Wesche, R. 1995a. Temperature dependence of critical currents in superconducting Bi-2212/Ag wires. Physica C: Superconductivity and its Applications. 246: 186–194.

Wesche, R. 1998b. High-Temperature Superconductors: Materials, Properties, and Applications. Kluwer Academic Publisgers. Boston-Dordrecht-London.

Wesche, R. 2011c. HTS Conductors for Fusion. Thermal Stability and Quench. HTS Workshop for Fusion Conductor. Karlsruhe.

Wetzko, M., M. Zahn and H. Reiss. 1995. Current sharing and stability in a Bi-2223/Ag high temperature superconductor. Cryogenics. 35: 375–386.

Wilson, M.N., C.R. Walters, J.D. Lewin and P.F. Smith. 1970. Experimental and theoretical studies of filamentary superconducting composites, part 1, basic ideas and theory. J. Phys. D. 3: 1517–1585.

Wilson, M.N. and Y. Iwasa. 1978. Stability of superconductors against localized disturbances of limited magnitude. Cryogenics. 18: 17–25.

Wilson, M.N. 1983. Superconducting magnets. Clarendon Press, Oxford.

Wipf, S.L. and M.S. Lubell. 1965. Flux jumps in Nb-25%Zr under nearly adiabatic conditions. Phys. Lett. 16: 103–105.

Wipf, S.L. 1967. Magnetic instabilities in type-II superconductors. Phys. Rev. 161: 404– 416.

Wipf, S.L. 1978. Stability and degradation of superconducting current-carrying devices. Technical Report LA-7275. Los Alamos Sci. Lab., Los Alamos.

Wolgast, R.C., H.P. Hernandez, P.R. Aron, H.C. Hitchcock and K.A. Solomon. 1963. Superconducting critical currents in wire samples and some experimental coils. Advan. Cryog. Eng. 8: 601–605.

Yamafuji K. and Y. Mawatari. 1992. Electromagnetic properties of high T_c superconductors: relaxation of magnetization. Cryogenics. 32: 569–577.

Yeshurun, Y., A.P. Malozemoff and A. Shaulov. 1996. Magnetic relaxation in high-temperature superconductors. Reviews of Modern Physics. 68: 911–950.

Zebouni, N.H., A. Vencataram, G. Rao, et al. 1964. Magnetothermal effects in type-II superconductors. Phys. Rev. Lett. 13: 606–609.

Zeldov, E., N.M. Amer, G. Koren, A. Gupta, R.J. Gambino and M.W. McElfresh. 1989. Optical and electrical enhancement of flux creep in $YBa_2Cu_3O_{7-d}$ epitaxial films. Phys. Rev. Lett. 62: 3093–3096.

Zeldov, E., M.N. Amer, G. Koren and A. Gupta. 1990. Flux creep in $Bi_2Sr_2CaCu_2O_8$ epitaxial films. Appl. Phys. Lett. 56: 1700–1702.

Zhou, Y. and X. Yang. 2006. Numerical simulations of thermomagnetic instability in high-T_c superconductors: Dependence on sweep rate and ambient temperature. Phys. Rev. B74: 054507.

Index

A

Abrikosov 2, 20
AC 19, 60, 244, 246, 248, 250
Adiabatic stability 91, 94, 96, 99, 101, 124, 158, 270
Alloy 2, 7
Anderson-Kim model 5-7, 24
Anisotropic 3, 6, 146, 148, 160-161, 168, 173, 176, 213, 244, 293, 295, 300, 305, 326
Autowave 265, 268, 280-283, 286-288, 291, 293, 295-296, 299, 303, 311-314, 323, 328, 341, 343-347, 350-352, 354-355, 360, 362-364, 369, 379-380

B

Bean model 5, 7, 21-22, 24, 44, 50, 60, 72, 86
Bifurcation 275, 326
$Bi_2Sr_2CaCu_2O_{8+x}$ (Bi2212) 2, 8, 193, 198-199, 201, 203, 206, 225-226, 229, 251, 347-348, 350-356, 359, 368,
$Bi_2Sr_2Ca_2Cu_3O_{10+x}$ (Bi2223) 2
Boiling crisis 199, 206, 210, 212, 254, 262
Boiling 2, 199, 206, 208-210, 212-213, 217, 220, 222, 253-254, 262-263, 347, 350-352, 354, 358, 364, 367, 380
    film 2, 199, 206, 210, 212-213, 217, 220, 222, 263, 347, 351-352, 354, 364, 367
    nucleate 199, 206, 208-210, 212-213, 217, 220, 222, 253, 262-263, 347, 350-352, 354, 358, 380
Boltzmann 4, 26
Boundary condition 10, 18, 22, 25, 27, 29, 34, 37, 40-42, 52, 62, 79, 83, 90, 93, 102, 110, 147, 155, 157, 169-170, 176, 183, 228, 259-260, 266, 281, 296-297, 305, 308, 314-315, 330, 332, 359, 373
Bronze 146, 316-319
Bulk 1-3, 11, 123, 125-127, 132-133, 140-141, 143, 295
Burnout 13, 350, 354-355, 358, 361, 364-367, 369-370, 380

C

Ceramic 2
Chain reaction 11, 89, 173, 370, 380

Characteristic 2-7, 10-12, 14-16, 20-22, 24-53, 55, 58-60, 62, 65-66, 68-70, 77-78, 83, 85-86, 88-89, 91, 93-96, 99-101, 110-111, 113, 124-125, 128-129, 133, 144-147, 150, 152, 154-156, 158-162, 165-166, 168-170, 172-174, 176-177, 180-183, 186-193, 195-197, 199-206, 209, 213-214, 217, 219, 223-227, 229, 231-234, 237-238, 240, 243-244, 246, 250, 253-255, 257-258, 260-263, 265, 267, 269, 274-277, 279-280, 286-288, 291, 295-297, 299, 302-303, 307, 309, 317, 321-322, 324-327, 329-342, 345-359, 362-368, 370, 372, 379-380
Coated 15, 146, 212, 250, 358
Coolant 14, 16-17, 79, 81, 86-87, 93-94, 99-100, 104, 109-110, 113, 124-125, 128, 136, 145, 150, 154, 156, 158, 160, 164, 174-175, 183-184, 188, 193, 195, 197, 199, 201, 203-206, 210-213, 217, 220, 222, 225, 227, 229, 233, 237, 244, 247, 249, 253, 257, 259-262, 264-267, 270, 273, 275, 296-297, 304-306, 308, 314-315, 319, 321, 323, 327, 329, 337, 342, 347-352, 358-359, 361-362, 364, 366-369, 371-372, 379-380
Copper 1-2, 6, 13, 146, 149, 155, 171, 177, 212-213, 215, 217-220, 222, 261, 275, 287, 316-319, 324, 368-370
Creep 4-6, 8, 14, 21, 25-26, 28, 31-34, 36-37, 39, 48, 50, 58-59, 112, 129, 175, 192, 197, 200, 253
    strong 8, 32, 34, 58-59, 69, 96, 192, 197, 212, 253, 320, 342
    weak 3, 32, 36, 48, 50, 58, 85, 106, 152, 192, 229, 255-256, 293, 304, 320, 322
Critical 1-18, 20-26, 34, 37, 39, 42, 44, 50-52, 58-60, 64-70, 72, 77, 79-81, 83, 85-102, 109-112, 125-126, 128-129, 132-134, 137-138, 141, 144-147, 149-150, 154-156, 158-161, 168, 173-174, 176, 181, 186-188, 190, 192-193, 195-199, 201, 205-207, 210, 212-213, 217, 219, 222-223, 225, 227, 231-233, 237, 240, 242-244, 246-247, 250-256, 258-259, 261-262, 264, 268-279, 283-284, 286, 288, 293, 303, 307, 309-311, 315-323, 325-329, 331, 333, 335-343, 345-352, 354-355, 357-359, 364-365, 368-369, 372, 374-375, 377, 379-380
    state model 5-7, 14, 20, 24, 26, 39, 42, 50, 59, 64-65, 70, 93, 101, 128, 159, 173